Prolonged seasonal drought affects most of the tropics, including vast areas presently or recently dominated by 'dry forests'. These have received scant attention although humans have used and changed them more than wet forests. This volume reviews the available information, often making contrasts with wetter forests. The world's dry forest heterogeneity of structure and function is shown regionally. In the neotropics, biogeographic patterns differ from those of wet forests, as does the spectrum of plant life forms in terms of structure, physiology, phenology and reproduction. Biomass distribution, nutrient cycling, belowground dynamics and nitrogen gas emission are reviewed. Exploitation schemes are surveyed, and examples are given of non-timber product economies. This volume aims to stimulate research leading to more conservative and productive management and will be of interest to those working in conservation, land management, forestry and wildlife-related disciplines.

SEASONALLY DRY TROPICAL FORESTS

SEASONALLY DRY TROPICAL FORESTS

Edited by

STEPHEN H. BULLOCK
Departamento de Ecología,
Centro de Investigación Científica y de Educación Superior de Ensenada,
Ensenada, Baja California, México

HAROLD A. MOONEY
Department of Biological Sciences,
Stanford University, Stanford, California, U.S.A.

and

ERNESTO MEDINA
Centro de Ecología y Ciencias Ambientales,
Instituto Venezolano de la Investigaciones Científicas,
Caracas, Venezuela

CAMBRIDGE UNIVERSITY PRESS
Cambridge, New York, Melbourne, Madrid, Cape Town, Singapore, São Paulo, Delhi

Cambridge University Press
The Edinburgh Building, Cambridge CB2 8RU, UK

Published in the United States of America by Cambridge University Press, New York

www.cambridge.org
Information on this title: www.cambridge.org/9780521112840

First published 1995
This digitally printed version 2009

A catalogue record for this publication is available from the British Library

Library of Congress Cataloguing in Publication data

Seasonally dry tropical forests/edited by Stephen H. Bullock. Harold A. Mooney, and
Ernesto Medina.
p. cm.
Includes indexes.
1. Forest ecology–Tropics. I. Bullock, Stephen H. II. Mooney, Harold A. III. Medina,
Ernesto.
QU936.S39 1995
581.5′2642′0913–dc20 95-13640 CIP

ISBN 978-0-521-43514-7 hardback
ISBN 978-0-521-11284-0 paperback

This book is dedicated to the memory of a marvelous colleague and friend, Alwyn H. Gentry, who inspired and challenged us with his ideas and energy and enthusiasm. His unmatched experience in all varieties of tropical forest contributed to an extraordinarily extensive impact on conservation while greatly enriching basic science.

Contents

Contributors

Luc Abbadie
Laboratoire d'Ecologie, École Normale Supérieure, 46 rue d'Ulm, 75230 Paris, France

Kansri Boonpragob
Department of Biology, Faculty of Sciences, Ramkhamhaeng University, Bangkok 10240, Thailand

Stephen H. Bullock
Departamento de Ecología, Centro de Investigación Científica y de Educación Superior de Ensenada, Apartado Postal 2732, 22800 Ensenada, Baja California, México

Robert Bye
Instituto de Biología, Universidad Nacional Autónoma de México, Apartado Postal 70-233, 04510 México, D.F., México

Gerardo Ceballos
Centro de Ecología, Universidad Nacional Autónoma de México, Apartado Postal 70-275, 04510 México, D.F., México

Elvira Cuevas
Centro de Ecología y Ciencias Ambientales, Instituto Venezolano de Investigaciones Científicas, Apartado 21827, Caracas 1020-A, Venezuela

David Dilcher
Florida Museum of Natural History, University of Florida, Gainesville, FL 32611, USA

Rodolfo Dirzo
Centro de Ecología, Universidad Nacional Autónoma de México, Apartado Postal 70-275, 04510 México, D.F., México

César A. Domínguez
Centro de Ecología, Universidad Nacional Autónoma de México, Apartado Postal 70-275, 04510 México, D.F., México

The late Alwyn H. Gentry
Formerly of Missouri Botanical Garden, St Louis, MO 63166, USA

Alan Graham
Department of Biological Sciences, Kent State University, Kent, OH 44242, USA

N. Michele Holbrook
Department of Biological Sciences, Stanford University, Stanford, CA 94305 USA

Victor J. Jaramillo
Centro de Ecología, Universidad Nacional Autónoma de México, Apartado Postal 70-275, 04510 México, D.F., México

Michel Lepage
Laboratoire d'Ecologie, École Normale Supérieure, 46 rue d'Ulm, 75230 Paris, France

Ariel E. Lugo
USDA Forest Service, International Institute of Tropical Forestry, Call Box 25000, Río Piedras, Puerto Rico 00928, USA

J. Manuel Maass
Centro de Ecología, Universidad Nacional Autónoma de México, Apartado Postal 70-275, 04510 México, D.F., México

Angelina Martínez-Yrízar
Centro de Ecología, Universidad Nacional Autónoma de México, Apartado Postal 1354, 83000 Hermosillo, Sonora, México

Pamela A. Matson
Ames Research Center, MS 239-20, National Aeronautics and Space Administration, Moffett Field, CA 94035 USA

Ernesto Medina
Centro de Ecología y Ciencias Ambientales, Instituto Venezolano de Investigaciones Científicas, Apartado 21827, Caracas 1020-A, Venezuela

Jean-Claude Menaut
Laboratoire d'Ecologie, École Normale Supérieure, 46 rue d'Ulm, 75230 Paris, France

Harold A. Mooney
Department of Biological Sciences, Stanford University, Stanford, CA 94305, USA

Peter G. Murphy
Department of Botany and Plant Pathology, Michigan State University, East Lansing, MI 48824, USA

Philip W. Rundel
Laboratory of Biomedical and Environmental Science, University of California, Los Angeles, CA 90024, USA

Everardo V. S. B. Sampaio
Departamento de Energía Nuclear, Universidade Federal de Pernambuco, Recife, Pernambuco 50730, Brazil

Robert L. Sanford, Jr
Department of Biological Sciences, University of Denver, Denver, CO 80208, USA

Peter M. Vitousek
Department of Biological Sciences, Stanford University, Stanford, CA 94305, USA

Julie L. Whitbeck
Department of Ecology, Evolution and Organismal Biology, Tulane University, New Orleans, LA 70118, USA

Acknowledgements

Most of the chapters in this volume were presented at a symposium held at the Estación de Biología Chamela, Jalisco, México. We are grateful to our hosts of the Instituto de Biología, Universidad Nacional Autónoma de México, for their gracious hospitality. Financial support was kindly provided by the Fundación Ecológica de Cuixmala, A.C., as well as by parent institutions of the participants and U.N.A.M. This book has also benefitted from the comments of Nora Martijena, Carlos Saravia-Toledo and Jane Bulleid as well as the many reviewers acknowledged in each chapter.

1
Introduction

HAROLD A. MOONEY, STEPHEN H. BULLOCK
& ERNESTO MEDINA

Most ecosystems of the tropical and subtropical latitudes are seasonally stressed by drought (Schimper, 1898; Köppen, 1931; Murphy & Lugo, 1986). Research on population and ecosystems dynamics, and conservation efforts, however, rarely address these ecosystems, but rather concentrate on what is usually understood as tropical wet forest or rain forest. There has been enormous scientific and public attention directed toward documenting the effects of destruction of wet forests on soil fertility, biotic diversity, and global biogeochemistry. These concerns are certainly justified as the rates of forest and species loss accelerate. In contrast, relatively little attention has been given to forests subject to prolonged dry seasons (Ridpath & Corbett, 1985), and to their changing status. Degradation and conversion of 'dry forest' is far more advanced than that of wet forest: only a small fraction remains intact (Murphy & Lugo, Chapter 2; Sampaio, Chapter 3; Menaut, Lepage & Abbadie, Chapter 4; Rundel & Boonpragob, Chapter 5; Gentry, Chapter 7), and the area explicitly conserved is hardly perceptible. This is unfortunate because the forests with prolonged annual drought occupy more area than wet forests, have been of greater use to humans, and are still poorly known over most of their distribution.

The extent of forest in the drier tropics, and even its character, are difficult subjects for debate and research. Particularly in Africa, India and Asia, the relations between savannas, woodlands and dry forests (of various leaf habits) are notoriously complex (Furley, Proctor & Ratter, 1992). Savannas and their degradation are certainly priority subjects of tropical and global ecology, but as with wet forests, they are well studied compared with dry forests. The present volume reviews the available information on tropical forests under climates with highly seasonal rainfall, often making contrasts with wet and moist forests, as a stimulus to furthering our understanding and hence management of these systems. Several regional overviews are provided. Largely in accord with the available literature, the coverage by discipline and by site is uneven; thus any synthesis is preliminary. The principal regional bias is to the neotropics, or portions thereof, except for systems ecology.

What are tropical dry forests? In the simplest terms they are forests occurring in the tropical regions where there are several months of severe, even absolute, drought. Drought stress has been indexed mostly by precipitation and temperature (representing evaporation), for the purposes of correlating physiological and phenological patterns with climate and of mapping the distribution of dry forests. The fundamental variable of soil moisture has been sparsely documented, and this substantially hinders intersite comparisons, because soil physical characteristics as well as topography take on great importance in drier forests as determinants of spatial heterogeneity in water availability. Drought normals also are affected by other factors, such as rainfall intensity, cloudiness, continentality, latitude and elevation. The response in terms of leaf habit is predominantly drought-deciduousness, but there are extensive dry evergreen forests, and a noticeable increase of evergreen and succulent plants in very dry deciduous forests.

Reviews of the distribution and structure of dry forests in northern Latin America, Africa and Thailand make clear that the only unifying characteristic of the dry forest climate is the strong seasonality of the rainfall distribution (Murphy & Lugo, Chapter 2; Sampaio, Chapter 3; Menaut *et al.*, Chapter 4; Rundel & Boonpragob, Chapter 5). Differences in rainfall amount, and the average duration of the rainy season, may account for the large differences in canopy height, total biomass, and productivity found among forests.

Seasonally dry forests and savannas occur under the same climatic conditions. While the former are essentially tree-dominated systems, the savannas never have a continuous tree canopy and are characterized by the presence of a xeromorphic, fire-tolerant grass layer. Separation of seasonal forests and savannas under little-disturbed environmental conditions is possible on the basis of the fertility status of the soil. Deciduous dry forests are found on soils of significantly higher fertility than savannas (Ratter *et al.*, 1973; Furley, Ratter & Gifford, 1988). Sarmiento (1992) developed a simple model based on the number of months with water deficit and soil fertility which allows the separation of evergreen forests, deciduous forests, and seasonal savanna systems in South America. However, in most places the balance between forest and savanna is regulated by human intervention and the fire regime (Menaut *et al.*, Chapter 4; Rundel & Boonpragob, Chapter 5). Agriculture and grazing have greatly decreased the extent of neotropical deciduous forests, converting them to exotic grasslands of one or a few exotic species. Deforestation and frequent burning lead to nutrient impoverishment and soil loss, among other biological, physical and geochemical changes (Maass, Chapter 17).

Neotropical dry forests are generally less species-rich than moister forests in terms of plants and vertebrates (Gentry, Chapter 7; Ceballos, Chapter 8), but the most diverse dry forests are not the wettest ones. Centers of diversity and endemism are strikingly not equatorial but nearer to the tropics, in western México and southeast Bolivia. The geography of species diversity in arthropod taxa remains poorly known. Diversity of plant life-forms appears to be greater in dry than in wet forests (Medina, Chapter 9). Diversity of life forms is both structural (wood specific gravity, plant habit) and physiological (photosynthetic types, water relations, growth seasonality). Adaptations for reproduction are not peculiar in terms of sexuality or compatibility (Bullock, Chapter 11), but the spectrum of mechanisms of gene and seed dispersal is distinct: conspicuous flowers and wind-dispersed seeds are relatively more frequent in trees of drier forests. Among animals, physiological adaptations have not been widely studied, but evidence from mammals suggests no distinctive features of dry forest species.

Dry forest elements have been recorded for the Mid to Late Eocene in Panamá, probably growing in edaphically controlled habitats within an otherwise mesic environment (Graham & Dilcher, Chapter 6). These elements appear to have increased in diversity throughout the Tertiary. In the southern United States tropical dry forest developed in the Mid Eocene but disappeared later in response to increasingly cooler climates and other factors. Unfortunately, the critical expanse of México lacks study.

Leaf water relations and gas exchange characteristics do not seem to differ markedly from those of plants of wetter regions (Holbrook, Whitbeck & Mooney, Chapter 10). Interactions between water availability and structural and physiological characteristics such as rooting depth, stem water storage, hydraulic architecture, and sensitivity to water stress lead to a wide variety of phenological behaviors. The relationship of reproductive to vegetative phenology and to water availability depends on life form, with trees and other high water-storage plants showing the most diversity (Bullock, Chapter 11). However, as the duration of soil water deficit increases flowering periods are shorter and more synchronous within and among species. Leaf damage by insects is higher in deciduous than in evergreen species (Dirzo & Domínguez, Chapter 12), but folivory is intense only at the beginning of the rainy season. Levels of folivory can be severe in some cases, and can strongly affect reproduction, but the majority of trees lose less than 20% of leaf area.

The responses of the forest as a system to seasonal droughts are analysed in this volume in relation to biomass distribution, nutrient cycling, dynamics of belowground processes (both fine root and microbiological dynamics) and the emission of nitrogen-containing gases (Martínez-Yrízar, Chapter 13; Jaramillo & Sanford, Chapter 14; Cuevas, Chapter 15; Matson & Vitousek, Chapter 16). Dry forests have consistently lower biomass than wetter forests, because their production of organic matter is limited by the length of the growing season and their stature is smaller. These forests in general allocate larger fractions of photosynthates to development of underground biomass than wetter forests. On the basis of nutrient use-efficiency estimates these forests appear not to be especially nutrient limited, and as in wetter forests, phosphorus use-efficiency is the highest among the nutrients studied. All biological activity in the underground system is strongly limited by water availability and interactions with carbon and nutrient availability. Synchronization between activity of decomposers and fine root production, at the beginning of the rainy season, seems essential for an efficient balance between nutrient uptake by the plants and nutrient immobilization in microbial biomass. Nitrification and N_2O fluxes are related to soil moisture, but are high only at the beginning of the wet season. Conversion of forest to maize or pasture apparently does not affect gas fluxes much, except where accompanied by fertilization.

The array of oscillating phenomena is not only impressive but complex. Activation of the soil microbiota leads to rapid cascades of decomposition of above- and belowground litter produced during the previous growing season, and consequently, fluxes of gases derived from soil respiration. There is some evidence that decomposition rates are grossly unequal between species (Singh, 1969; Malaisse *et al.*, 1975; Martínez-Yrízar, 1982), so that system behavior may depend on species composition. A simplistic view of water supply driving plant development must be adapted to the notoriously diverse patterns of plant development present in dry forests, particularly if we expect to understand or predict the effects of minor and major climatic variations.

Progress has been made towards general models (Holbrook *et al.*, Chapter 10), but wide testing is needed. Drivers of some important developmental events, such as secondary growth, internal nutrient and energy fluxes, and meristem differentiation are not well known. The phenology of other organisms not subject to rapid dehydration is poorly known. Activity of insects generally is greater in the rainy season, but some groups are less seasonal than others, e.g. bark, seed and pollen feeders vs folivores.

Dry forest is also characterized by the occurrence of phenomena determined by sudden increases and by slow reductions in resource availability. These are driven by pulses in water availability, determined by the principal seasons, as well as by anomalous rains in the dry seasons and frequent drought spells in the rainy season. Some phenomena might be entrained by photoperiod (Peacock & McMillan, 1965; Stubblebine, Langenheim & Lincoln, 1978), but cues of moisture and temperature may be most common.

Relative to wet forests, efforts at modeling ecosystem interactions of dry forests are scarce (Bandhu *et al.*, 1973), because many important processes and their geographic variations are not well understood. For example, there is no general consensus on the importance of different nutrient cycling mechanisms, or even basic information on aboveground respiration. Also, crucial animal groups have not been studied in all major regions. Moreover, despite much work on the dynamic relationship between herbaceous and woody vegetation, there is no scheme for the internal dynamics of dry forest comparable to 'gap' theories in wetter forests.

The possibility of modeling water-limited tropical deciduous forest by analogy with temperature-limited northern deciduous forests has not received much attention. In fact, inherent flaws in the analogy are not difficult to find. In the tropics, water stress varies at very local and regional scales; among other consequences, there are typically several tree life-forms at any site, and the frequency of different leaf habits varies greatly among sites. Growing periods in the tropics may vary greatly between years due to flushing in response to anomalous rain in the dry season, or variations in the rate of drying out. This also applies to decomposers in the soil. Shade is less important with lower precipitation because forest stature is reduced. Fully hydrated long-lived organisms probably experience large losses in carbohydrate and fat reserves during prolonged dry periods, due to high temperatures. Also, the activity of many animal species – phytophagous, pollinating, and predacious – is not in phase with precipitation. These and other problems are recognizable, although poorly studied; but to direct new research efforts discussions of modeling will be useful. A general need is for more attention to interannual variations, rather than 'typical' yearly cycles, because they probably dominate population and system properties.

Tropical dry forests have been exploited in several ways. One is the extraction of a variety of plant and animal products, for local use or international commerce (Saravia-Toledo, 1985; Sampaio, Chapter 3; Bye, Chapter 18), sometimes transforming the structure of the forest or leading to local extinction of useful species. Local inhabitants use a variety of forest species for purposes of consumption, medicine or fuel, and some make a living commercializing such resources. Medicinal products are important where traditional knowledge of the flora is profound (Arenas, 1987; Schmeda-Hirschmann, 1993; Bye, Chapter 18). Control of land use by external economies, e.g. export markets for meat, timber or charcoal, has rapidly transformed large regions (Schofield & Bucher, 1986). The effective agents of such change are not only technologies of steel: patterns of herbivory and seed dispersal may also be crucial (Saravia-Toledo & Del Castillo, 1988). These forests also have been annihilated to be replaced by herbaceous, succulent or tree crops, or pastures (Sprague, Hanna & Chappell, 1978; Maass, Chapter 17). The progress of integrated plans for a rational and sustainable use of dry forest (e.g. Eiten & Goodland, 1979; Bucher & Schofield, 1981; Hardesty, 1988) awaits review.

We may conclude that within the tropics there are resource gradients affecting, among other things, moisture availability. Temporal limitations on development by moisture availability are widespread, and the resulting dry forest ecosystems are intrinsically complex and varied. These ecosystems are not so outstanding for their species diversity as for their organization, and response to stress and disturbance. Alas, this evolutionary milieu is fast disappearing from the Earth. If we are interested in preserving the world's biodiversity, for direct exploitation or insurance against the unknowns of global change, this exceptionally endangered portion of the tropics deserves its own research priority, more conservative management, and adequate preservation.

References

Arenas, P. (1987). Medicine and magic among the Maka Indians of the Paraguayan Chaco. *Journal of Ethnopharmacology* **21**: 279–95.
Bandhu, D., Gist, C. S., Ogawa, H., Strand, M. A., Bullock, J., Malaisse, F. & Rust, B. (1973). Seasonal model of the tropical forests. In *Modeling Forest Ecosystems*, ed. L. Kern, pp. 285–95. Oak Ridge National Laboratory, Oak Ridge.
Bucher, E. H. & Schofield, C. J. (1981). Economic assault on Chagas disease. *New Scientist* **29**: 320–4.
Eiten, G. & Goodland, R. (1979). Ecology and management of semi-arid ecosystems in Brazil. In *Management of Semi-Arid Ecosystems*, ed. B. H. Walker, pp. 277–300. Elsevier, Amsterdam.
Furley, P. A., Proctor, J. & Ratter, J. A. (1992). *Nature and Dynamics of Forest–Savanna Boundaries*. Chapman & Hall, London.
Furley, P. A., Ratter, J. A. & Gifford, D. R. (1988). Observations on the vegetation of eastern Mato Grosso. III. Woody vegetation and soil of the Morro de Fumaca, Torixoreu, Brazil. *Proceedings of the Royal Society of London, Series **B** **203***: 191–208.
Hardesty, L. H. (1988). Multiple-use management in the Brazilian caatinga. *Journal of Forestry* **86**: 35–7.
Köppen, W. P. (1931). *Grundriss der Klimakunde*. Walter de Gruyter & Co., Berlin.
Malaisse, F., Freson, R., Gaffinet, G. & Malaisse-Mousset, M. (1975). Litterfall and litter breakdown in Miombo. In *Tropical Ecological Systems*, ed. F. B. Golley and E. Medina, pp. 137–52. Springer-Verlag, New York.
Martínez-Yrízar, A. (1982). Tasa de descomposición de materia orgánica foliar de especies arbóreas de selvas en clima estacional. Subsecretaría Forestal y de la Fauna, México.
Murphy, P. G. & Lugo, A. E. (1986). Ecology of tropical dry forest. *Annual Review of Ecology and Systematics* **17**: 67–88.
Peacock, J. T. & McMillan, C. (1965). Ecotypic differentiation in *Prosopis* (mesquite). *Ecology* **46**: 35–51.
Ratter, J. A., Richards, P. W., Argent, G. & Gifford, D. R. (1973) Observations on the vegetation of northeastern Mato Grosso. I. The woody vegetation types of the Xavantina–Cachimbo expedition area. *Philosophical Transactions of the Royal Society of London, Series B* **266**: 449–92.
Ridpath, M. G. & Corbett, L. K. (1985). *Ecology of the Wet-Dry Tropics*. Proceedings of the Ecological Society of Australia 13. Darwin Institute of Technology, Casuarina.
Saravia-Toledo, C. J. (1985). La tierra pública en el desarrollo futuro de las zonas áridas: estado actual y perspectivas. In *IV Reunión de Intercambio Tecnológico en Zonas Aridas* **I**: 115–40. Centro Argentino de Ingenieros Agrónomos, Buenos Aires.
Saravia-Toledo, C. J. & Del Castillo, E. M. (1988). Micro y macro tecnologías. Su impacto en el bosque chaqueño en los últimos cuatro siglos. In *Actas del VI Congreso Forestal Argentino* **III**: 853–5. Editorial Talleres Gráficos el Liberal, Santiago del Estero.
Sarmiento, G. (1992) A conceptual model relating environmental factors and vegetation formations in the lowlands of tropical South America. In *Nature and Dynamics of Forest–Savanna Boundaries*, ed. P. A. Furley, J. Proctor and J. A. Ratter, pp. 583–601. Chapman & Hall, London.
Schimper, A. F. W. (1898). *Pflanzen-Geographie auf Physiologische Grundlagen*. G. Fischer, Jena.

Schmeda-Hirschmann, G. (1993). Magic and medicinal plants of the Ayoreos of the chaco boreal (Paraguay). *Journal of Ethnopharmacology* **39**: 105–11.
Schofield, C. J. & Bucher, E. H. (1986). Industrial contributions to desertification in South America. *Trends in Ecology and Evolution* **1**: 78–80.
Singh, K. P. (1969). Studies in decomposition of leaf litter of important trees of tropical deciduous forests at Varanasi. *Tropical Ecology* **10**: 292–311.
Sprague, M. A., Hanna, W. J. & Chappell, W. E. (1978). Alternatives to a monoculture of henequen in Yucatán: the agriculture, climate, soil and weed control. *Interciencia* **3**: 285–90.
Stubblebine, W., Langenheim, J. L. & Lincoln, D. (1978). Vegetative response to photoperiod in the tropical leguminous tree *Hymenea courbaril* L. *Biotropica* **10**: 18–29.

2

Dry forests of Central America and the Caribbean

PETER G. MURPHY & ARIEL E. LUGO

Introduction

Holdridge (1947, 1967) developed a bioclimatic classification system by which the world's terrestrial biota may be categorized into approximately 120 life zones, each distinguished by climatic parameters that coincide with particular vegetational characteristics. Approximately 68 life zones are in the tropics and subtropics, of which 30 are dominated by forest of various types. Lugo, Schmidt & Brown (1981) estimated that 28 tropical and subtropical forested life zones are represented in Central America and the Caribbean, and 13 are found on the islands of the Caribbean. Despite this diversity, approximately half of the vegetation of Central America and the Caribbean is within the dry forest life zone (*sensu* Holdridge, 1967).

Dry forests are not infrequently referred to as deciduous forests, but the degree of deciduousness varies greatly (see below). Not all dry forests are conspicuously deciduous, and not all deciduous forests are dry forest. By Holdridge's criteria, tropical and subtropical dry forests are found in frost-free areas where mean annual biotemperature (a special calculation that reduces the effects of extreme temperatures) is above 17 °C, annual rainfall ranges from 250 to 2000 mm, and the ratio of potential evapotranspiration to precipitation is greater than one, to a maximum value of two. By these criteria, 49% (8.2×10^5 km^2) of the vegetation of Central America and the Caribbean is considered dry forest (Brown & Lugo, 1980). Africa has the most dry forest (16.5×10^6 km^2; 73% of the continent's vegetation); worldwide, about 42% of all intratropical vegetation is dry forest. Global patterns in dry forest distribution and overall ecological characteristics relative to wetter tropical and subtropical forest ecosystems were reviewed by Murphy & Lugo (1986a).

In the general literature, many different names have been applied to

Holdridge's dry forest and adjoining life zones. The names tend to emphasize different features: overall water limitation (e.g. dry or subhumid forest); seasonality (e.g. seasonal wet or seasonal drought forest); foliage longevity (e.g. evergreen, semievergreen, semideciduous or deciduous forest); vegetation structure (e.g. forest, woodland or thicket); substrate (e.g. limestone or alluvial forest); or some combination of these (e.g. dry limestone forest). Some of the equivalencies in nomenclature among forest classification systems have been reviewed previously (Rzedowski, 1978; Hartshorn, 1988). Hartshorn (1988) grouped related life zones into major vegetation types. For example, the designation lowland subhumid forest encompasses tropical dry forest, subtropical dry forest, and subtropical thorn forest life zones. Included in the 'low mountains' subhumid forest category are subtropical thorn woodland, subtropical dry forest, and subtropical lower montane dry forest.

Beard (1955) classified tropical American vegetation types based on their structural characteristics and primary environmental determinants. His system includes associations (floristic groupings), formations (physiognomic groupings, e.g. 'deciduous seasonal forest'), and formation series (habitat groupings, e.g. 'montane formations'). Of particular relevance to this review are his seasonal formation series and his dry evergreen formation series. The former consists of five formations, ranging from evergreen seasonal forest on the moister sites to cactus scrub on the driest; the intermediate formations constitute semi-evergreen seasonal forest, deciduous seasonal forest, and thorn woodland. Among the formations of this series, the number of tree stories, the horizontal continuity of each story, and other aspects of forest structure are presumed to decrease in a regular fashion as annual moisture supply decreases and seasonality of rainfall increases. The five formations of Beard's dry evergreen formation series include, from most to least developed, dry rain forest, dry evergreen forest, dry evergreen woodland, dry evergreen thicket, and evergreen bushland. In this formation series, edaphic factors (e.g. porous white sands), as well as annual rainfall, are responsible for the differences in forest structure among formations.

Beard's approach, particularly as reflected in his diagrammatic profiles of formation series, brings emphasis to the fact that all of these vegetation units occur as segments of continua along various environmental gradients, rather than as distinct, easily defined entities. This becomes apparent in attempting to condense the literature on the broad array of tropical and subtropical dry forest ecosystems (see Rundel & Boonpragob, Chapter 5). For these reasons, we use the term dry forest in a distinctly general manner, defined primarily by Holdridge's (1967) climatic criteria. In the Holdridge model the transition from tropics to subtropics occurs at *c.* 12 to 13° N latitude (Nicaragua), but in this review we make no attempt to distinguish forest types based on this criterion. Rather, we consider tropical and subtropical dry forests as a unit.

Origin of Central America and the Caribbean islands

There have been land bridge or island stepping-stone opportunities for the dispersal of plants and animals between South America and North America since before the Cretaceous (Coney, 1982; Brown & Gibson, 1983). A land bridge connecting México with South America appears to have been severed in the Cretaceous. The Greater Antilles originated in the Cretaceous as a magmatic arc of islands formed between southern México and Colombia, and eventually moved eastward on the Caribbean Plate. The southern portion of the arc became the Caribbean mountains of Venezuela and the islands of Aruba, Bonaire and Curaçao. Small fragments may have given rise to the original Lesser Antilles although the present islands are thought to have originated more recently, over the past 3 million years. The present position of the Greater Antilles had been achieved by the Eocene. Subsequently, the backbone of Central America was created by a new series of volcanic islands. Most likely, a continuous land bridge did not exist until the islands fused in the Pliocene, and the Isthmus of Panamá and northwestern Colombia emerged from the sea. It is regarded as unlikely that the Greater Antilles were ever connected with North America and, in fact, Jamaica was completely submerged as recently as the Miocene.

Thus, it appears that only during the past 4 million years has there been a continuous land bridge between North and South America, although opportunities for dispersal across islands have existed for over 63 million years. Ceballos (Chapter 8) comments on the zoogeographic consequences. As the islands of the Caribbean have been isolated from any mainland since the Cretaceous, it can be assumed that their flora and fauna were derived by overseas dispersal and differentiation on the islands.

The overall flora of the Caribbean islands is composed of four basic elements: cosmopolitan (mainly pan-tropical and pan-Caribbean); West Indian (Greater Antilles, Lesser Antilles, or both); endemic (on the larger islands); and continental (Asprey & Robbins, 1953; see also Gentry, Chapter 7). The continental group can in turn be subdivided into South, Central, and North American. In the Greater Antilles, most species in the continental group have their origin in Central America (Asprey & Robbins, 1953) although Seifriz (1943) reported that in Cuba the floristic connections are strongest with South America. At the other end of the island chain, Trinidad and Tobago have their closest floristic affinities with South America (Venezuela).

Physical environment

The prevailing air mass movement throughout the region under consideration is from northeast or east to southwest or west. Consequently, areas to the north or east (windward) of mountain systems tend to be wet whereas those to the south or west (leeward) are much drier and more seasonal in precipitation because of the orographic rain shadow. The effect is an important one because there are major mountain systems in every country of the region and on all except the smallest islands. México's Citlaltépetl, at 5700 m, is the highest North American mountain south of Alaska. In the Caribbean, mountain peaks reach elevations of 3177 m, 2256 m, 2006 m and 1338 m on Hispaniola, Jamaica, Cuba and Puerto Rico, respectively.

Annual rainfall on the windward slopes of northeastern Puerto Rico ranges from 1780 mm in the lowlands to over 4000 mm at 1000 m elevation. In contrast, the southwestern coast of the island, only 100–200 km distant, receives less than 1000 mm. Similar patterns apply to the Central American mainland, where the Pacific side is drier than the Caribbean. The Pacific coast of Panamá, supporting semideciduous forest, receives about 1780 mm of annual rainfall whereas the evergreen forest of the Caribbean coast receives 3300 mm. On the Caribbean side, minimum monthly rainfall is normally $\geq$38 mm while the Pacific coast receives $<$13 mm during the cooler months of February and March. In the Caribbean, the number of months having $<$100 mm of rainfall appears to be a criterion in delimiting scrub from forest vegetation. Areas receiving $<$100 mm for 11 months, for example, may support evergreen bushland and cactus shrub whereas ares with only 5–6 such months usually support better developed dry evergreen thicket and deciduous seasonal forest (Table 2.1; see Medina, Chapter 9).

In much of this region, the rainfall pattern is bimodal, with a short dry period (1–2 months) during summer and a long dry period (2–6 months) during winter (Richards, 1952). At the margin of the tropics there may be only one dry season but it can be as long as 8 months. Climate diagrams for a sample transect across Central America provide examples of the

Table 2.1. *Caribbean vegetation in relation to annual rainfall and number of dry months*

Rainfall	Dry evergreen thicket and deciduous seasonal forest	Evergreen bushland and cactus scrub
Annual	800–1300 mm	700 mm
Monthly		
>100 mm	6–7 months	1 month
25–100 mm	5–6 months	11 months
<25 mm	2–3 months	4 months

Source: Loveless & Asprey (1957).

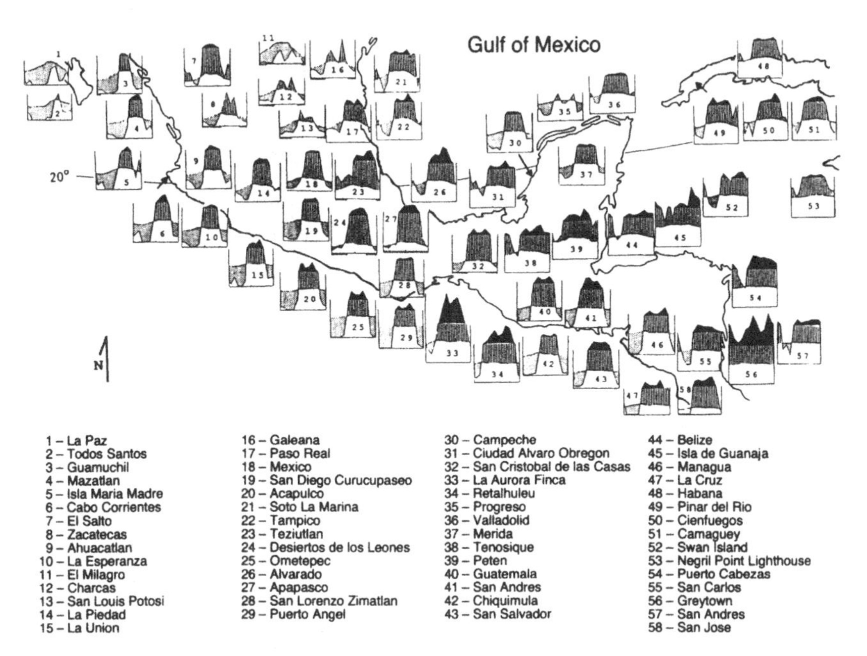

Figure 2.1. Climate diagrams for a representative portion of México and Central America. Modified from Walter *et al.* (1975).

various seasonal patterns in the region (Figure 2.1). On the dry Pacific side there are typically about six dry months each year. In the Caribbean, a representative diagram for the dry forest life zone of Puerto Rico (Figure 2.2) shows that in total, approximately six months of the year are dry, with the driest months being from December to March and May to July.

In Central America, Pacific storms may contribute large amounts of rainfall over very short periods of time, particularly during the months of

August–October. Hurricanes are an important climatic influence, from the eastern lowlands of Costa Rica to the Yucatán Peninsula, throughout the West Indies, and west of the Istmo de Tehuantepec in México. In Central America, hurricanes largely affect the eastern wet forests but the dry forests of Yucatán and the Caribbean islands are affected as well. Over the 92 year period 1885–1977, an estimated 761 tropical cyclones of tropical storm or greater intensity passed through the Caribbean region (Neumann *et al.*, 1978). The island of Puerto Rico has been struck by

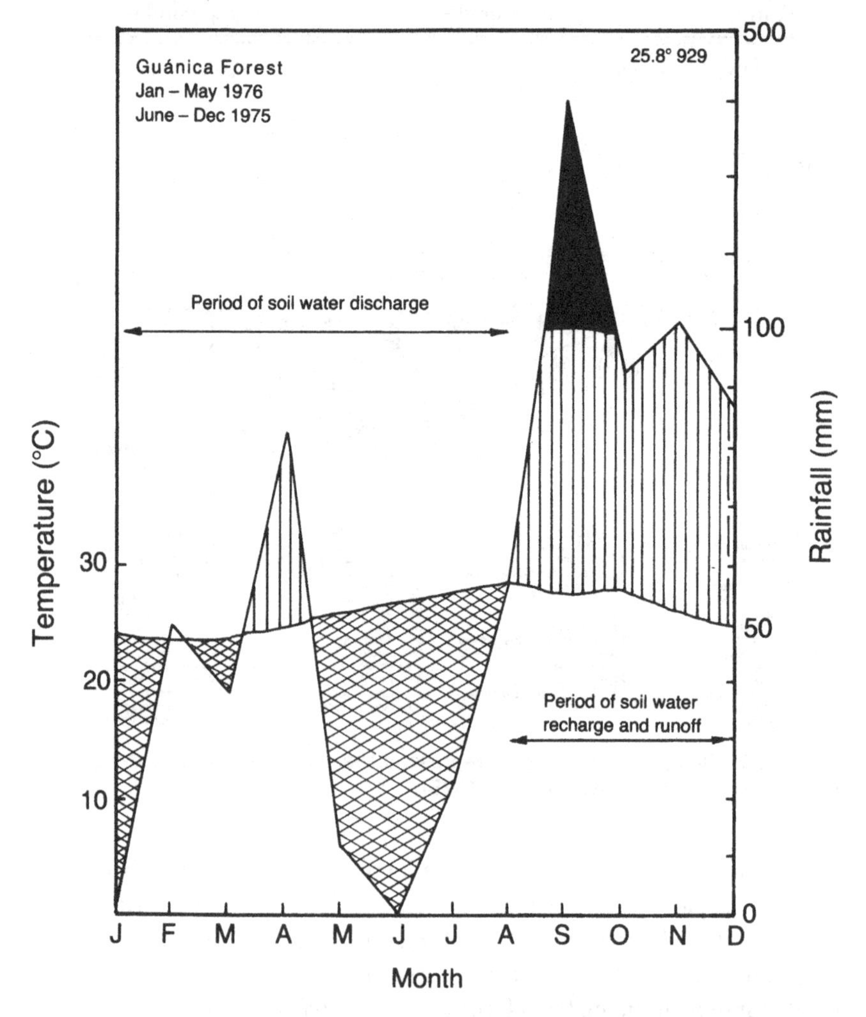

Figure 2.2. Climate diagram for Guánica Forest, southwestern Puerto Rico. [vertical hatching], moist periods; [solid black], wettest period with surplus water, [cross-hatching], periods of moisture deficit. Modified from Lugo *et al.* (1978) and Walter & Lieth (1967).

about 70 hurricanes since the early 1700s. Most of these seem to have affected mainly the wet or moist forest of the north coast (Weaver, 1989).

One of the least understood aspects of climatic effects on vegetation is the year-to-year variability in rainfall for which the drier areas of the tropics are well known. Whereas the coefficient of variation for annual rainfall in the temperate forest zone is typically around 15%, in the moderately dry tropics it may be 30% or more (Ruthenberg, 1980). Over a period of 70 years, annual rainfall in southwestern Puerto Rico varied from 500 to 1500 mm (Murphy & Lugo, 1990). Because of the absence of long-term ecological monitoring, the biological consequences of this variability are not well known. In the Yucatán Peninsula, annual increase in tree basal area was found to be directly related to amount of rainfall during the previous two years. Whigham *et al.* (1990) found that over the range of 1000–2500 mm, a doubling of the previous two years of rainfall increased annual basal area increment approximately five-fold. Although the effects of unusually dry years are not well known, much less understood, it is likely that such years are important in limiting the extent of system development with respect to structure and taxonomic composition.

As elevation increases, temperature (and potential evapotranspiration) decreases and rainfall usually increases. The lapse rate is on the order of 6 °C per 1000 m. Although it is difficult to assign regional elevational limits to specific vegetation types, Hartshorn (1988) found it useful to consider the vegetation of Mesoamerica in three elevational zones: lowlands (0–500 m), low mountains (500–2000 m), and high mountains (>2000 m). In Central America, tropical dry forest is limited to the lowlands and low mountains zones below 2000 m. Temperature varies relatively little throughout the year and thus seasonal differences in the region are related primarily to rainfall patterns.

Lowland coastal vegetation near the shoreline, especially on small islands with low relief, may be subject to stressful salt spray effects, although this has not been studied in the context of dry forest ecology. In Puerto Rico, vegetation of the Guánica Forest ranges from sparse, relatively species-poor scrub near the coast at elevations below 40 m, where salt spray is very likely a factor, to well-developed semideciduous forest at elevations of 125 m and more (Lugo *et al.*, 1978; Murphy & Lugo, 1990). However, lowland sites are generally the most heavily disturbed, so it can be difficult to determine the purely elevational effects.

Because dry forests are primarily climatically determined, they occur on a wide variety of soil types, including limestone, volcanic and alluvial substrates. Nonetheless, as in all forest situations, soil and other substrate factors are important in modifying vegetation structure, composition and function (see also Sampaio, Chapter 3; Rundel & Boonpragob, Chapter 5). Shallow, relatively infertile soils tend to have sclerophyllous, evergreen vegetation whereas deeper, more fertile soils tend to have deciduous vegetation. On some of the islands of the Caribbean, as much as 25% or more of the ground may be exposed limestone rock. Highly xerophytic plants, such as cacti, tend to dominate the thinnest and rockiest soils in southwestern Puerto Rico. In some instances the appearance of the ground can be deceiving in that the subsurface limestone may be a soft mollisol readily penetrated by roots.

In Puerto Rico the soil is relatively nutrient-rich but it is unclear how much of the phosphorus is chemically bound (Lugo & Murphy, 1986). Organic matter content may reach or exceed 30% in the upper soil (Lugo & Murphy, 1986; Whigham *et al.*, 1990). Hartshorn (1988) points out that edaphic factors may be the cause of single-species or monodominant dry forest stands that occasionally arise in Central America, e.g. on pumice and montmorillonite clay. (But see Martijena, 1993.)

Fire as a factor is of greatest importance in human-disturbed areas (see Sampaio, Chapter 3; Menaut, Abbadie & Lepage, Chapter 4; Rundel & Boonpragob, Chapter 5; Maass, Chapter 17), especially in the vicinity of anthropogenic rangeland. There is little evidence to suggest natural burning of intact forest in our region. Other common disturbances, in places where the forest is not totally eliminated for agriculture or other purposes, include grazing and wood gathering for fuel, fenceposts, and other similar purposes (see Sampaio, Chapter 3; Bye, Chapter 18).

Dry forest distribution and overall structure

The primary areas of dry forest and other environments of México and Central America are mapped in Figure 2.3, based upon a vegetation map by Rzedowski (1978) for México and a compilation of country maps for Central America. A summary of the spatial representation of tropical and subtropical forest life zones in Central America and the Caribbean is given in Table 2.2. Detailed maps of Holdridge life zones are available for several countries: Colombia (Espinal & Montenegro, 1963), Costa Rica (Tosi, 1969), Dominican Republic (Tasaico, 1967), El Salvador (Holdridge, 1977), Guatemala (Holdridge *et al.*, 1978), Haiti (OAS, 1972), Honduras (Holdridge, 1962a), Nicaragua (Holdridge, 1962b), Panamá (Tosi, 1971), Puerto Rico (Ewel & Whitmore, 1973), and Venezuela (Ewel, Madriz & Tosi, 1968). México has not been mapped using the Holdridge system,

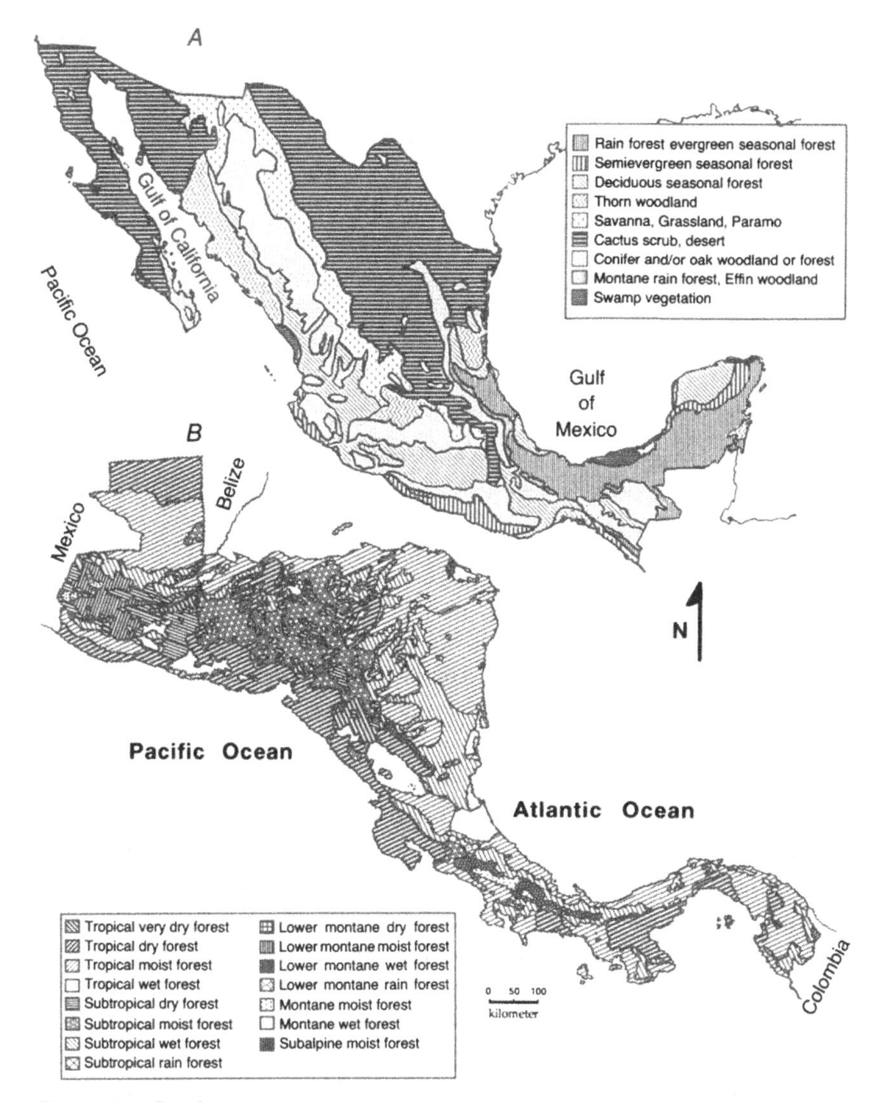

Figure 2.3. Dry forest and other environments of México and Central America. *A*. Vegetation of México (adapted from Rzedowski, 1978). *B*. Life zones of Central America. Compiled from maps in the files of the US Forest Service, International Institute for Tropical Forestry, Rio Pierdras, Puerto Rico.

Table 2.2. *Distribution of life zones in Central America and the Caribbean islands*

	Tropical forest (%)			Subtropical forest (%)		
Country/Island	Wet	Moist	Dry	Wet	Moist	Dry
Belize	15	–	–	35	50	–
Costa Rica	66	27	7	–	–	–
El Salvador	–	–	–	–	28	71
Guatemala	–	–	–	35	30	35
Honduras	–	–	–	25	35	40
México	–	–	–	5	25	70
Nicaragua	15	15	–	–	60	10
Panamá	52	48	–	–	–	–
Bahamas	–	–	–	–	50	50
Cuba	–	–	–	20	80	–
Dominican Republic	–	–	–	22	55	23
Puerto Rico	–	–	–	26	60	14
Trinidad	10	80	10	–	–	–

Source: Brown & Lugo (1980).

although Rzedowski (1978) provides a comprehensive account of México's many vegetation types and the associated environmental conditions.

As described previously, the prevailing winds from the northeast (cool season) and east (warm season) cause the preponderance of dry forest in México and Central America to be on the Pacific coast in the orographic rain shadow. Much of the Yucatán Peninsula is relatively dry in spite of its windward position, largely because of its overall low elevation (Trewartha, 1961); the dominance of limestone and soils derived from marine sediments are also important in the Yucatán (Hartshorn, 1988) and generally around the Caribbean Sea. On the islands, the occurrence of dry forests primarily on the southern (lee) sides can be exemplified by Puerto Rico (Figure 2.4). The small patches of dry forests in an unusual windward position in Puerto Rico can be attributed to low topographic relief. Similarly, many of the smaller islands of low relief (e.g. US Virgin Islands) were predominantly dry forest prior to clearing.

The vegetation of most of the Greater Antilles has been fairly well described, and there are many similarities among the islands in terms of composition and structure of dry forests. The vegetation and its floristic origins in Jamaica have been described in some detail (Asprey & Robbins, 1953; Loveless & Asprey, 1957; Kapos, 1986). Seifriz (1943) provided a comprehensive description of the vegetation of Cuba, and Gleason & Cook (1927) contributed a relatively early but astute account for Puerto

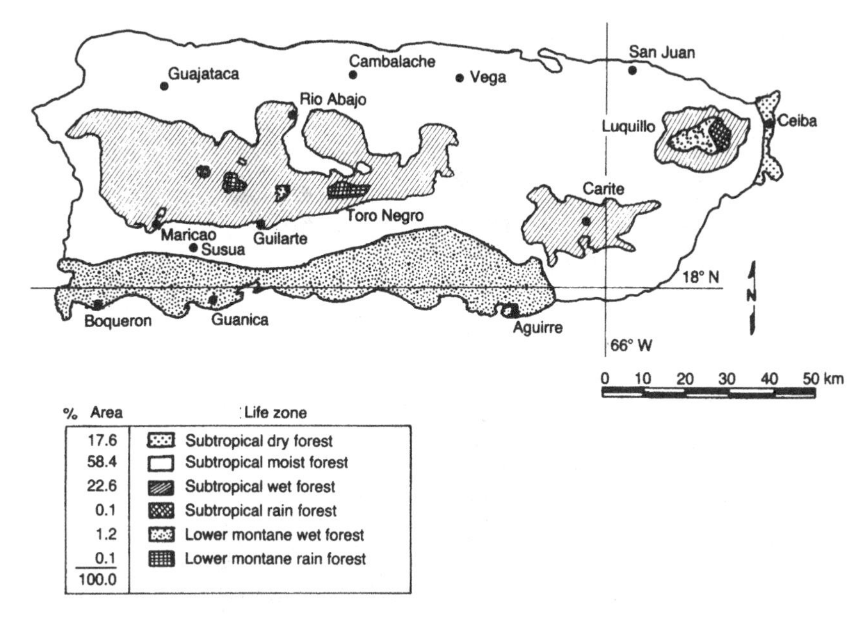

Figure 2.4. Life zones of Puerto Rico. Modified from Ewel & Whitmore (1973).

Rico and the Virgin Islands. Unfortunately, the islands have been extensively disturbed, so that tracts of native vegetation are true rarities.

Generally, dry forests are smaller in structure and simpler in composition than wetter forests of a given region, but there is enormous geographic variation in most features due to differences in climate, soil, biogeographic considerations, and disturbance history. Among representative dry forest sites in Jamaica, Puerto Rico, Costa Rica and México, forest canopy height ranges from about 2 m in the drier, more exposed and disturbed forests to 40 m in the more mature forests on relatively favorable sites (Table 2.3). Tree species richness is difficult to compare because of the different sample sizes but also varies geographically (Table 2.3; see Gentry, Chapter 7). The dispersion of trees within a population was studied by Hubbell (1979) in the province of Guanacaste in Costa Rica; the species showed either clumped or random dispersion patterns, with rare species more often clumped than common species.

The Costa Rican sites in Table 2.3 are among the best documented in Central America. The variation in structural traits for the four sites, all of which receive approximately 1400 mm of rainfall annually, is largely due to differences in soils and flooding periodicity. The great variation in overall structure is apparent from the profiles of these stands (Figure 2.5).

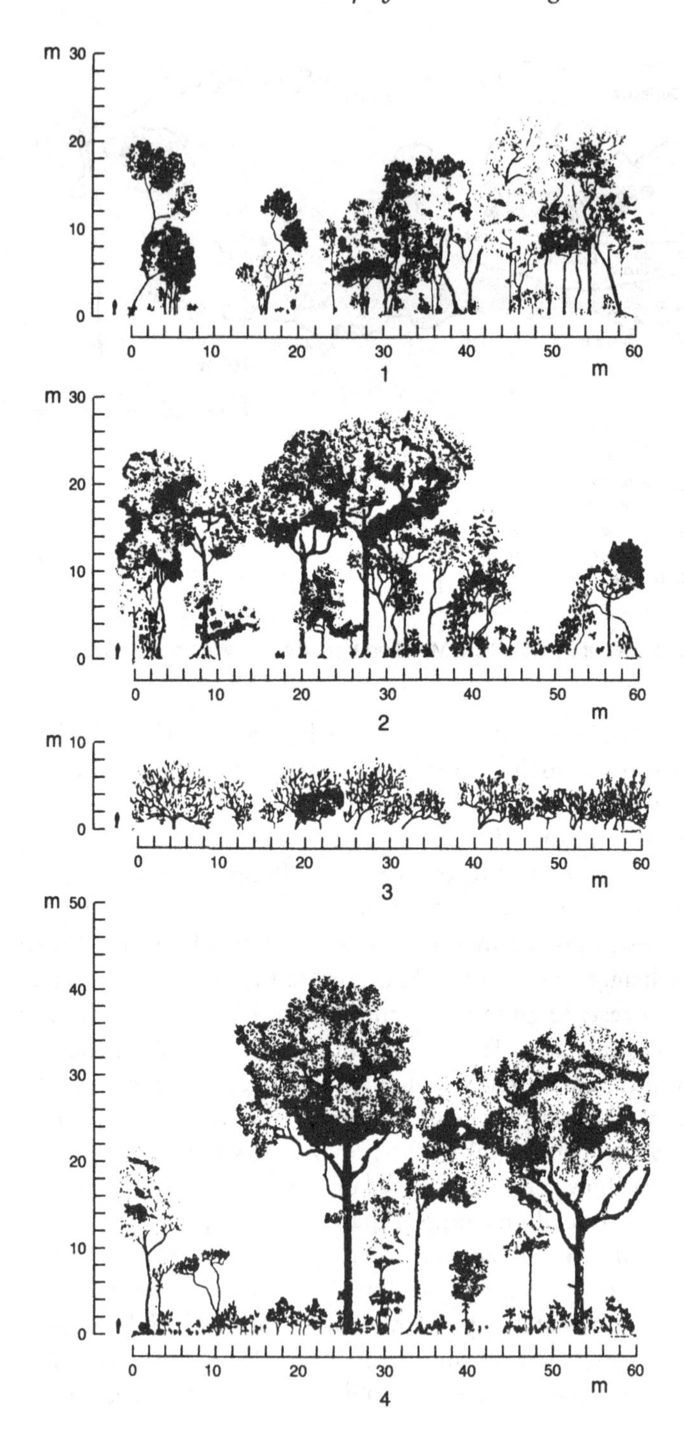
m 30
20
10
0
0 10 20 30 40 50 60
m
1
m 30
20
10
0
0 10 20 30 40 50 60
m
2
m 10
0
0 10 20 30 40 50 60
m
3
m 50
40
30
20
10
0
0 10 20 30 40 50 60
m
4

Table 2.3. *Structural characteristics of dry and seasonal vegetation in Central America and the Caribbean islands*

Vegetation	Annual rainfall (mm)	Number of tree species	Sample area (m^2)	Height: Canopy (m)	Height: Emergent (m)
Jamaica[a]					
Evergreen bushland	694	55	500	4	8
Dry evergreen thicket	1118	58	500	5–11	16
Semievergreen forest	1200	32	500	8–15	20
Puerto Rico[b]					
Semideciduous dry forest	860	33	1500	2–6	9
Costa Rica[c]					
Semideciduous dry forest	1400	15	1000	8–18	22
Semideciduous dry forest	1400	19	1000	10–33	33
Semideciduous dry forest	1400	2	1000	3–6	7
Semideciduous dry forest	1400	9	1000	12–40	44
México					
Semideciduous dry forest[d]	857	55	4800	10–20	20
Semideciduous dry forest[e]	714	80	1000	10	20
Deciduous dry forest[e]	714	83	1000	6	–

Sources: [a]Kapos (1986); [b]Murphy & Lugo (1986b); [c]Holdridge *et al.* (1971); [d]Whigham *et al.* (1990); [e]Lott, Bullock & Solís (1987).

Figure 2.5. Profiles representing four tropical dry forest sites studied by Holdridge *et al.* (1971) in northwestern Costa Rica (all units are meters). All sites receive approximately 1400 mm of annual rainfall but differ in other respects, as described by Holdridge *et al.* (1971).
Site 1. Bagaces Río Potrero. Well-developed forest on flat upland with residual soils of low fertility and moderate drainage. Possibly some disturbance from grazing.
Site 2. Bagaces Río Cabuyo. Well-developed forest on flat terrain with alluvial soils. Moderately high water table during wet season.
Site 3. Bagaces Bajuras Cortes. Low open stand of largely one species in a basin area of black clay soils with seasonal flooding and severe droughts.
Site 4. Taboga Río Lajas. Well-developed 'cathedral' forest on level terrain with moderately high water table, rarely flooded.

On the Caribbean islands, the 4000 ha Guánica Forest in Puerto Rico is one of the best remaining tracts of dry forest and has been the site for our own studies. Structurally and compositionally, it is similar to Jamaica's dry evergreen thicket, as described by Loveless & Asprey (1957), even though it is partially deciduous during the dry season. The low mean canopy height (4.3 m) and presence of many multiple-stem trees reflect the past cutting and grazing disturbance to which the forest had been subjected prior to the 1930s. Approximately half of the forest's relatively small biomass of 98 Mg ha^{-1} is below ground (Murphy & Lugo, 1986b). Gentry (Chapter 7) considers the dry forests of the Antilles to be generally denser and smaller in stature than those of the mainland. The effects of disturbance frequently complicate structural comparisons among sites. In our studies and literature reviews of dry forest ecology, we have found that undisturbed examples of this ecosystem type are extremely rare. Bush *et al.* (1992) and Gómez-Pompa, Flores & Sosa (1987) showed evidence of human alteration of Central American vegetation as far back as 11,000 years ago. In fact, most of the vegetation of their region appears to be a product of human activity. Hartshorn (1988) indicated that there are only three significant sites of protected lowland subhumid dry forest in Mesoamerica: Parque Nacional Santa Rosa and Refugio Nacional de Vida Silvestre Palo Verde in Costa Rica, and the Estación de Biología Chamela in México.

In Costa Rica, wet forest appears to be roughly twice as rich in tree species as dry forest. The wet forest at La Selva on the Caribbean side has approximately 98 tree species per hectare (Hartshorn & Hammel, 1982) whereas the dry forest at Palo Verde on the Pacific side has about 52 (Hartshorn & Poveda, 1983). Among moist/wet forest sites, tree species diversity is directly related to annual precipitation, but this does not hold true for dry forests (Gentry, Chapter 7). Overlap of tree species between wet and dry forests seems to be minimal. Out of 298 tree species sampled in Costa Rica, Frankie, Baker & Opler (1974) found only 11 that occurred in both wet and dry life zones. At the generic and family levels, Gentry (Chapter 7) concludes that dry forest has an impoverished subset of the flora of moist/wet forests.

In the Caribbean as well, tree species richness is lower in dry than in wet life zones. In Puerto Rico, Guánica Forest (860 mm of annual rainfall) has 169 tree species whereas Susua State Forest (1413 mm) has 218 and Maricao State Forest (2550 mm) has 265. Of the 33 tree species found in a 1500 m^2 sample in Guánica Forest, about two-thirds are restricted primarily to coastal and foothill areas receiving less than 2000 mm

(Murphy & Lugo, 1986b). The remaining third extends to moister foothill areas. None of these species is found in areas receiving more than 4000 mm annual rainfall. Plant diversity and floristic patterns in neotropical dry forest are covered more comprehensively by Gentry (Chapter 7).

Table 2.4 compares the global ranges in values of tree species richness, forest canopy height, basal area, total biomass, root biomass proportion, and net primary productivity for dry forest and wet forest. The effects of water limitation in reducing forest structure are apparent. Not surprisingly, because of annual drought stress, dry forests tend to have a higher proportion of belowground biomass but we consider the data on root biomass to be sketchy, and in some cases unreliable (Murphy & Lugo, 1986b; see Martínez-Yrízar, Chapter 13; Cuevas, Chapter 15).

It is also clear from Table 2.4 that dry and wet forests overlap in all properties; indeed, there are wet forests in the American tropics that superficially resemble dry forests in aboveground structure. An example is the Bana Forest of the San Carlos de Río Negro area of southern Venezuela (Medina & Cuevas, 1989). This undisturbed forest receives >3000 mm of annual rainfall, but is composed of trees similar in height and diameter at breast height (dbh) to those of Guánica. The reason is the extremely coarse, well-drained and infertile sandy substrate. Another case is the cloud forest in the Luquillo Experimental Forest of northeastern Puerto Rico, which is also very small in stature and relatively poor in tree species. In this case the causes may relate to excessive soil moisture (from rainfall up to 5000 mm y^{-1}), low transpiration, and possibly wind damage (Weaver, Byer & Bruck, 1973). Thus, factors other than low annual rainfall and rainfall seasonality can generate some elements of forest structure (and probably function) similar to those of the dry forest life zone.

Based on counts of bird species per 1000 individuals, the Guánica Forest has a 50% higher richness (31 species) than the wet forest (20) of Puerto Rico (Kepler & Kepler, 1970). It also has more than three times the number of individuals. Kepler & Kepler (1970) suggested that this may reflect the absence of competition from lizards and frogs which are far more abundant in the wet forest. They also point to the greater relative ability of xeric-habitat land birds to disperse among the many basically xeric island stepping-stones. The importance of dry forest to migratory birds is reviewed by Ceballos (Chapter 8).

In an analysis of six major habitats in South America, Mares (1992) found that the 'upland semideciduous forest' category contains 44% as

Table 2.4. *Comparison of dry and wet forests with respect to selected ecosystem properties*

Ecosystem property	Global range in values	
	Dry	Wet
Tree species on 1–3 ha	33–90	50–200
Canopy height (m)	10–40	20–84
Basal area of trees (m^2 ha^{-1})	17–40	20–75
Total biomass (Mg ha^{-1})	98–320	269–1186
Root biomass		
% of total live biomass	8–50	5–33
Primary production		
annual aboveground net (Mg ha^{-1})	6–16	10–22

Source: Murphy & Lugo (1986a).

many mammalian species as found in the 'lowland Amazon rain forest', even though the moist/wet forest encompasses about seven times as much land area. The 'drylands' category (including savanna, grassland, scrubland, thorn forest and desert) contains almost 20% more mammal species than lowland moist/wet forest, at least partially because of its vast area on the continent. The drylands also have the largest proportion of endemics (Mares, 1992; Ceballos, Chapter 8). These patterns emphasize the importance of including dry areas in the formulation of strategies for conservation (Mares, 1992; Ceballos, Chapter 8).

Forest function and response to seasonal rainfall

Seasonality in rainfall distribution largely controls the temporal patterns of growth, productivity, organic matter turnover, reproduction, and other functional traits of moisture-limited forest systems (Medina, Olivares & Marín, 1985). For example, in Puerto Rico, where the climate of Guánica Forest is characterized by a major dry season during the cooler months (December–March) and a minor dry season in the warmer part of the year (June–July), the pattern of tree growth corresponds closely to the rainfall seasonality, with growth during the wet periods and no growth or even shrinkage during the dry seasons (Murphy *et al.*, 1995). Reduced growth during dry seasons accounts for the relatively low net primary productivity in dry forest (6–16 Mg ha^{-1} y^{-1}), relative to wet forest (10–22 Mg ha^{-1} y^{-1}) (Murphy & Lugo, 1986b; Martínez-Yrízar, Chapter 13). On a growing-season-day basis, tree diameter increment in southwestern Puerto Rico dry forest trees is similar to that of wet forest trees in the northeast (Murphy & Lugo, 1986b). Even below ground, the effects of seasonal rainfall can be detected in the growth patterns of trees. Kummerow *et al.* (1990) showed that in a dry forest at Chamela, México, fine roots are shed during the dry season and restored rapidly at the beginning of the wet season.

In Guánica Forest, pulses in leaf litter production correlate closely with the occurrence of the two annual dry seasons (Murphy *et al.*, 1995). In this forest, an estimated 33% of the trees are described by Little & Wadsworth (1964) as deciduous or semideciduous with the remainder, including the local dominant *Gymnanthes lucida* (Euphorbiaceae), being evergreen (at least during years of normal rainfall). In the dry forests of Guanacaste, an estimated 60–75% of all tree species are deciduous (Frankie *et al.*, 1974; Hubbell, 1979), and the percentage is still higher in the hill forest at Chamela (Bullock & Solís, 1990; see Medina, Chapter 9). Over a 30-month period in Guánica Forest, canopy leaf area index ranged from a maximum of 4.3 during the wet seasons to a minimum of 2.1 during the dry seasons (Murphy & Lugo, 1986b). Year-to-year variation has not been well documented but is thought to be considerable, based on reports of local residents. Our observations during three years suggest the variation is high at both the community and population levels. Similarly, extraordinary rains in the dry season can cause massive production of new leaves (Bullock & Solís, 1990). Defoliations and episodes of severe herbivory also have been reported in these forests (Janzen, 1972; Dirzo & Domínguez, Chapter 12).

Reich & Borchert (1984) found that in the dry forest of Guanacaste, the timing of leaf fall and bud break in 12 tree species was determined mainly by changes in tree water status. Although water status was correlated with environmental water availability, it was also related to the functional state of the tree. Trees with growing shoots incurred water deficits during the wet season whereas leafless trees were able to rehydrate during periods of drought. Changes in stomatal behavior related to leaf age, and differences among species, were suggested as having important implications for differences in patterns of foliage retention among tree species (Reich & Borchert, 1988; see Holbrook, Whitbeck & Mooney, Chapter 10). Reich & Borchert (1984) also concluded that tree development was not controlled by seasonal variation in photoperiod or temperature.

Loveless & Asprey (1957) pointed out that in Jamaica the deciduous seasonal forest commonly occurs on the deeper, moister soils whereas dry evergreen (sclerophyllous) thicket, smaller in stature than the deciduous forest, is found on the shallower, drier soils where moisture stress can occur year-round (see also Table 2.1). Sclerophylly can be a response to oligotrophic soil conditions and may have little effect on drought tolerance (Medina, García & Cuevas, 1990). On the Pacific slopes of the Sierra Madre Occidental in Sonora, México, Goldberg (1982) found that evergreen oak woodland grows on extremely acid, infertile soils while deciduous forest is found on the more favorable, fertile substrates. Goldberg (1982) suggested that the evergreen oak could grow on the fertile soils but could not compete with deciduous species because of lower rates of photosynthesis. For other reasons, the evergreen habit also dominates the most favorable sites. For example, in the dry forest climate zone of Costa Rica, trees on sites with accessible groundwater may be evergreen and relatively tall, and the species in these special situations may be representative of wetter life zones (Hartshorn, 1988). Generally, the designation 'deciduous' in describing dry forests is fairly uninformative and even misleading unless quantification is provided. With the exception of the evergreen highly xerophytic vegetation on some of the harshest sites, all dry forest is deciduous to some extent. But the degree of deciduousness varies greatly among sites, years and species.

In lowland Guanacaste, there are two peak periods in flowering activity (Frankie *et al.*, 1974): primarily during the long dry season, but also at the onset of the rainy season. Borchert's (1983) studies of five tree species suggested that anthesis is determined largely by the periodicity of tree water status (not merely environmental water status), and therefore is unlikely to be the result of selection for optimum tree–pollinator interaction. In Guánica Forest, with two dry seasons, the frequency and timing of fruiting contrast with those in Guanacaste and at Chamela. Fruits were produced at least twice per year by half the species, and more or less continuously by 25% of the species; most fruits were borne at the end of the spring and fall wet seasons, rather than during the driest months (Murphy & Lugo, 1986a). At the mainland sites, a single fruiting period in the long dry season is typical (Frankie *et al.*, 1974; Bullock & Solís, 1990).

Use, disturbance and recovery

Bush *et al.* (1992) have shown evidence for human modification of vegetation in Panamá as early as 11,000 years ago. Early hunters may have had a role in reducing the megafauna of Central America as much as 9000 years ago (Janzen, 1988c). Agricultural alteration of the landscape began about 5000 years ago. Subsequent impacts of humans on the forests of Central America and the Caribbean have increased most markedly since the arrival of Europeans at the end of the 15th century, to the present when entire ecosystems, including many component plant and animal populations, are threatened with local or global extinction. According to Cruz & Fairbairn (1980), the Caribbean islands account for the highest rates of animal extinction in historical times. This is not surprising in view of the virtual elimination of most of the original lowland vegetation throughout the Caribbean, coupled with the problems of island isolation.

In reviewing the issue of overall deforestation in the region, Lugo *et al.* (1981) reported the average rate of deforestation for seven countries to be 0.5% y^{-1} (in Colombia, Costa Rica, Guatemala, Nicaragua, Puerto Rico, St Lucia and Venezuela: see Maass, Chapter 17). They also showed an inverse logarithmic relationship between population density and proportion of land forested. The islands of the Caribbean have higher human population densities and correspondingly lower percentages of forested areas. Not surprisingly, a similar relationship was found between human-related energy consumption and proportion of land forested. The availability of energy allows the more rapid exploitation and conversion of the landscape. Ultimately, the only portions of the landscape to survive intact are those too rugged in topography or too wet or too dry for agriculture and related activities.

The potential for sustained and profitable management of dry forests for forest products is questionable. However, Weaver & Francis (1988) found that in spite of growth limitations due to low rainfall, quality lumber could be produced on St Croix, possibly on a sustained yield basis. In the Dominican Republic, Jennings (1979) and Maxfield (1985) investigated the fuelwood production potential of dry forest. Maxfield estimated that one forest area contained the energy equivalent of 54–102 barrels of oil per hectare. Knudson, Chaney & Reynoso (1988) indicated that the dry forests of the Dominican Republic '. . . appear to respond quickly and positively to management'. They recommended that any managed dry forest (e.g. for fuelwood or charcoal production) be protected from livestock to avoid soil compaction and be subject to weed control and thinning as necessary to optimize tree growth. They also discussed the need to assure the survival of the dry forest as an ecosystem. Similar and other issues relating to the resource values of dry forest are under increasing discussion (Eiten & Goodland, 1979; Saravia-Toledo, 1985; Murphy & Lugo, 1986a, 1990; Hardesty, 1988; Murphy *et al.*, 1995; Sampaio, Chapter 3; Maass, Chapter 17; Bye, Chapter 18).

Some of the patterns of forest destruction appear to occur as boom-and-bust cycles, correlated with socioeconomic conditions. In Puerto Rico, for example, some abandoned agricultural areas on the south coast are currently being invaded by successional dry forest. Similarly, in Costa Rica, some of the worn-out pastureland presents the potential for dry forest re-establishment if the pastures are abandoned. Because of the virtual absence of truly original dry forest and associated wetter ecosystems in the vicinity, no-one can say with what success, through what ecological routes, or with what character, the dry forest will return to such areas, especially in view of the presence of competitive, exotic species.

Why are these problems particularly severe in the dry forest life zone? In the American tropics, most of the human population centers are in climatic areas where the ratio of potential evapotranspiration to precipitation has a value fairly close to one, as are the dry forests. Tosi & Voertman (1964) showed that in Central America, about 79% of the human population lived in the dry and moist forest life zones (ratio of potential evapotranspiration to precipitation, PET:P = 0.5–2.0). For Guatemala, Honduras, Nicaragua and Costa Rica taken together, the population density in wet, moist and dry life zones averaged 4.6, 14.4, and 30.5 people km^{-2}, respectively. The consequence is that dry forest ecosystems have been subject to massive disturbance. Most if not all of the surviving forest has at least been affected by the harvesting of trees for small lumber, fenceposts, firewood and charcoal, as well as by grazing in the understory.

Although the ecological effects of these uses on the remaining dry forests are not always apparent, many are conspicuous. For example, Guánica Forest is composed of 12,000–14,000 woody stems ha^{-1} (dbh $\geq$ 1 cm). Of these, $\geq$57% are stump or root sprouts, partly a consequence of past cutting. Studies in experimentally cut sites have shown that sprouts rapidly reestablish the original species complement (Murphy *et al.*, 1995). After 13 years, the system recovered about 42% of its original biomass. However, the structure of the vegetation appears to be limited by the high density of small stems (27,000 stems $\geq$1 cm dbh ha^{-1}: Dunevitz, 1985).

In the same study, we found no evidence of long-lived soil seed pools, suggesting that long-term disturbance, leading to the death of root systems, might greatly limit the rate of forest recovery and locally jeopardize some plant species. The roles of seed dispersal and dormancy in forest recovery have been the subject of interesting speculations. In

Central America, the extinctions of more than 15 genera of large mammalian herbivores approximately 10,000 years ago (Janzen & Martin, 1982) must have diminished the dispersal potential of many plant taxa (but see Howe, 1985), thus limiting their capacity to recolonize today's abandoned pasture land. Gómez-Pompa, Vázquez-Yanes & Guevara (1972), on the other hand, discussed the possible tendency for dry forest tree species with seed dormancy and tolerance of dry, open conditions, to invade deforested wet forest areas in México.

It seems fitting to end this review on an upbeat note by mentioning the large-scale dry forest restoration studies sited in Guanacaste on the Pacific side of Costa Rica. The challenge of restoration is a challenge to the real depth of our ecological understanding (Bradshaw, 1987). Janzen (1987, 1988a, b) has experimented with various techniques for re-establishing dry forest in old pastures, some of which are dominated by an aggressive African grass, *Hyparrhenia rufa*. The roles of various reforestation techniques are under investigation: tree seedling planting, controlled burning (to control grass competition), nuclear trees (as attractants to seed-dispersing animals), grazing animals (for dispersing seeds), among others. The ultimate goal of this effort is the re-establishment of 700 km^2 of dry forest and associated habitats known as Parque Nacional Guanacaste. The development of techniques that may be applied in other regions will be an equally significant achievement.

The forest rehabilitation work in Guanacaste builds on the pioneering work of Leopold in temperate zones (Jordan, Gilpin & Aber, 1987). Two other approaches to dry forest rehabilitation, as applied in Puerto Rico, include the use of plantations (Wadsworth, 1990; see also Rundel & Boonpragob, Chapter 5), and allowing natural recovery through protection and removal of human disturbance. The rehabilitation of forest conditions on degraded lands is a fundamental aspect of tropical forest conservation because it is an active effort to reverse the negative effects of human activity (Lugo, 1988). Perhaps the most important aspect of the widely publicized Guanacaste project is that it has captured the world's attention. The plight of the disappearing tropical dry forest, although still in the shadow of moist/wet forest issues, is now far more visible.

Summary

About half of the Central American and Caribbean land area is characterized by a tropical or subtropical dry forest climate. Dry forests occur most commonly on low islands or on the lee side of mountainous islands, on coastal areas of low relief, and on the Pacific (lee) side of the Central American land mass at elevations below 2000 m. In such areas, annual rainfall levels typically are below 2000 mm and vary substantially from year to year. Major and minor annual dry seasons totalling six months are characteristic. Dry forests occur on a wide variety of soil types.

Dry forests vary from deciduous to evergreen or semievergreen, and vary considerably in structure and composition. They range from 2 m tall woodlands in the drier, more exposed areas to 40 m forests on more favorable sites. Island examples tend to include the smallest and densest forests. Total biomass increases with annual precipitation and ranges from <98 to 320 Mg ha^{-1}, with a relatively large proportion below ground. Annual aboveground net primary productivity ranges from <6 to 16 Mg ha^{-1}. Based on sample sizes of 1–3 ha, these forests contain from 30 to 90 tree species, roughly half that of wet forests. The species richness of some animal groups, such as birds and mammals, may in some cases exceed that of wet forest.

Tree growth and primary productivity, leaf-out, litterfall, reproductive phenology, and other aspects of forest function closely track the seasonal availability of water.

In Central America, human influences in the form of hunting and modification of the vegetation cover have been factors for as long as 11,000 years. Agricultural alteration of the landscape began about 5000 years ago. Rates of animal extinction are particularly high on islands of the region; on many islands, and in some continental areas, few or no examples of natural lowland ecosystems remain. The tendency for Central American human populations to concentrate in drier climates has hastened the rate of dry forest degradation.

The potential for the sustained and profitable extraction of products from dry forest is not known but is likely to be low on a per unit area basis. The need to develop rehabilitation and management strategies for dry forest ecosystems, and to protect the relatively few remaining tracts of undisturbed examples, is of the highest priority.

References

Asprey, G. F. & Robbins, R. G. (1953). The vegetation of Jamaica. *Ecological Monographs* **23**: 359–412.

Beard, J. S. (1955). The classification of tropical American vegetation-types. *Ecology* **36**: 89–100.

Borchert, R. (1983). Phenology and control of flowering in tropical trees. *Biotropica* **15**: 81–89.

Bradshaw, A. D. (1987). Restoration: an acid test for ecology. In *Restoration Ecology: a synthetic approach to ecological research*, ed. W. R. Jordan, M. E. Gilpin and J. D. Aber, pp. 23–9. Cambridge University Press, Cambridge.
Brown, J. H. & Gibson, A. C. (1983). *Biogeography*. C. V. Mosby Co., St Louis.
Brown, S. & Lugo, A. E. (1980). Preliminary estimate of the storage of organic carbon in tropical ecosystems. In *Carbon Dioxide Effects Research and Assessment Program: the role of tropical forests on the world carbon cycles*, ed. S. Brown, A. E. Lugo and B. Liegel, pp. 65–117. Center for Wetlands, University of Florida, Gainesville.
Bullock, S. H. & Solís Magallanes, J. A. (1990). Phenology of canopy trees of a tropical deciduous forest in México. *Biotropica* **22**: 22–35.
Bush, M. B., Piperno, D. R., Colinvaux, P. A., de Oliveira, P. E., Krissek, L. A., Miller, M. C. & Rowe, W. E. (1992). A 14,300-yr paleoecological profile of a lowland tropical lake in Panamá. *Ecological Monographs* **62**: 251–75.
Coney, P. J. (1982). Plate tectonic constraints on biogeographic connections between North and South America. *Annals of the Missouri Botanical Garden* **69**: 432–43.
Cruz, A. & Fairbairn, P. (1980). Conservation of natural resources in the Caribbean: the avifauna of Jamaica. In *Forty-fifth North American Wildlife Conference*, pp. 438–44. American Wildlife Institute, Washington.
Dunevitz, V. L. (1985). Regrowth of clearcut subtropical dry forest: mechanisms of recovery and quantification of resilience. M.S. thesis, Michigan State University, East Lansing.
Espinal, L. S. & Montenegro, E. M. (1963). *Formaciones Vegetales de Colombia*. Instituto Geográfica Agustín Codazzi, Bogotá.
Eiten, G. & Goodland, R. (1979). Ecology and management of semi-arid ecosystems in Brazil. In *Management of Semi-Arid Ecosystems*, ed. B. H. Walker, pp. 277–300. Elsevier, Amsterdam.
Ewel, J. J., Madriz, A. & Tosi, J. A. (1968). Zonas de vida de Venezuela. República de Venezuela, Ministerio de Agricultura y Cría, Dirección de Investigación, Caracas.
Ewel, J. J. & Whitmore, J. L. (1973). The ecological life zones of Puerto Rico and the U.S. Virgin Islands. *Forest Service Research Paper* ITF-18, Institute for Tropical Forestry, Río Piedras.
Frankie, G. W., Baker, H. G. & Opler, P. A. (1974). Comparative phenological studies of trees in tropical wet and dry forests in the lowlands of Costa Rica. *Journal of Ecology* **62**: 881–919.
Gleason, H. A. & Cook, M. T. (1927). Plant ecology of Porto Rico. *Science Survey of Porto Rico and the Virgin Islands* **7**: 1–173.
Goldberg, D. E. (1982). The distribution of evergreen and deciduous trees relative to soil type: an example from the Sierra Madre, México, and a general model. *Ecology* **63**: 942–51.
Gómez-Pompa, A., Flores, J. S. & Sosa, V. (1987). The 'Pet kot': a man-made tropical forest of the Maya. *Interciencia* **12**: 10–15.
Gómez-Pompa, A., Vázquez-Yanes, C. & Guevara, S. (1972). The tropical rain forest: a nonrenewable resource. *Science* **177**: 762–5.
Hardesty, L. H. (1988). Multiple-use management in the Brazilian caatinga. *Journal of Forestry* **86**(8): 35–7.
Hartshorn, G. S. (1988). Tropical and subtropical vegetation of Meso-America. In *North American Terrestrial Vegetation*, ed. M. G. Barbour and W. D. Billings, pp. 365–90. Cambridge University Press, Cambridge.
Hartshorn, G. S. & Hammel, B. (1982). *Trees of La Selva*. Organization for Tropical Studies, San José.
Hartshorn, G. S. & Poveda, L. J. (1983). Plants: checklist of trees. In *Costa Rican Natural History*, ed. D. H. Janzen, pp. 158–83. University of Chicago Press, Chicago.

Holdridge, L. R. (1947). Determination of world plant formations from simple climatic data. *Science* **105**: 367–8.
Holdridge, L. R. (1962a). *Mapa ecológico de Honduras*. Informe Oficial Misión 105 á Honduras, Organization of American States, Washington.
Holdridge, L. R. (1962b). *Mapa ecológico de Nicaragua*. US Agency for International Development, Managua.
Holdridge, L. R. (1967). *Life Zone Ecology*. Tropical Science Center, San José.
Holdridge, L. R. (1977). *Mapa ecológico de El Salvador*. Ministerio de Agricultura y Ganadería, DGRN and UNDP/FAO, San Salvador.
Holdridge, L. R., Grenke, W. C., Hatheway, W. H., Liang, T. & Tosi, J. A., Jr (1971). *Forest Environments in Tropical Life Zones: a pilot study*. Pergamon Press, Oxford.
Holdridge, L. R. and others (1978). *Mapa ecológico de Guatemala*. INCOFOR, Guatemala.
Howe, H. F. (1985). Gomphothere fruits: a critique. *American Naturalist* **125**: 853–65.
Hubbell, S. P. (1979). Tree dispersion, abundance, and diversity in a tropical dry forest. *Science* **203**: 1299–1309.
Janzen, D. H. (1972). Patterns of herbivory in a tropical deciduous forest. *Biotropica* **13**: 271–82.
Janzen, D. H. (1987). How to grow a tropical national park: basic philosophy for Guanacaste National Park, northwestern Costa Rica. *Experientia* **43**: 1037–38.
Janzen, D. H. (1988a). Biocultural restoration of a tropical forest. *Bioscience* **38**: 156–61.
Janzen, D. H. (1988b). Management of habitat fragments in a tropical dry forest: growth. *Annals of the Missouri Botanical Garden* **75**: 105–16.
Janzen, D. H. (1988c). Tropical dry forests, the most endangered major tropical ecosystem. In *Biodiversity*, ed. E. O. Wilson, pp. 130–7. National Academy Press, Washington.
Janzen, D. H. & Martin, P. S. (1982). Neotropical anachronisms: the fruits the gomphotheres ate. *Science* **215**: 19–27.
Jennings, P. (1979). Dry forests of the Dominican Republic and their energy production capacity. In *Biological and Sociological Basis for a Rational Use of Forest Resources for Energy and Organics*, ed. S. G. Boyce, pp. 149–59. USDA Forest Service, Asheville.
Jordan, W. R., Gilpin, M. E. & Aber, J. D. (1987). Restoration ecology: ecological restoration as a technique for basic research. In *Restoration Ecology: a synthetic approach to ecological research*, ed. W. R. Jordan, M. E. Gilpin and J. D. Aber, pp. 3–21. Cambridge University Press, Cambridge.
Kapos, V. (1986). Dry limestone forests of Jamaica. In *Forests of Jamaica*, ed. D. A. Thompson, P. K. Bretting and M. Humphreys, pp. 49–58. Jamaican Society of Scientists and Technologists, Kingston.
Kepler, C. B. & Kepler, A. K. (1970). Comparison of bird species diversity and density in Luquillo and Guánica forests. In *A Tropical Rain Forest*, ed. H. T. Odum and R. F. Pigeon, pp. E183–E191. US Atomic Energy Commission, Oak Ridge.
Knudson, D. M., Chaney, W. R. & Reynoso, F. A. (1988). Fuelwood and charcoal research in the Dominican Republic. Purdue University, Department of Forestry and Natural Resources, Lafayette.
Kummerow, J., Castellanos, J., Maass, M. & Larigauderie, A. (1990). Production of fine roots and the seasonality of their growth in a Mexican deciduous dry forest. *Vegetatio* **90**: 73–80.
Little, E. L. & Wadsworth, F. H. (1964). *Common Trees of Puerto Rico and the Virgin Islands*. Agriculture Handbook 249. US Department of Agriculture, Washington.
Lott, E. J., Bullock, S. H. & Solís Magallanes, J. A. (1987). Floristic diversity and structure of upland and arroyo forests of coastal Jalisco. *Biotropica* **19**: 228–35.

Loveless, A. R. & Asprey, G. F. (1957). The dry evergreen formations of Jamaica. I. The limestone hills of the south coast. *Journal of Ecology* **45**: 799–822.

Lugo, A. E. (1988). The future of the forests: ecosystem rehabilitation in the tropics. *Environment* **30**(7): 16–20, 41–5.

Lugo, A. E., Gonzalez-Liboy, J. A., Cintrón, B. & Dugger, K. (1978). Structure, productivity, and transpiration of a subtropical dry forest in Puerto Rico. *Biotropica* **10**: 278–91.

Lugo, A. E. & Murphy, P. G. (1986). Nutrient dynamics of a Puerto Rican subtropical dry forest. *Journal of Tropical Ecology* **2**: 55–72.

Lugo, A. E., Schmidt, R. & Brown, S. (1981). Tropical forests of the Caribbean. *Ambio* **10**: 318–24.

Mares, M. A. (1992). Neotropical mammals and the myth of Amazonian biodiversity. *Science* **255**: 976–9.

Martijena, N. (1993). Establecimiento y sobrevivencia de plántulas de especies árboreas en un bosque tropical deciduo de baja diversidad, dominado por una sola especie. Tesis Doctoral, Centro de Ecología, Universidad Nacional Autónoma de México, México.

Maxfield, D. G., Jr. (1985). Biomass, moisture contents and energy equivalents of tree species in the subtropical dry forest of the Dominican Republic. M.S. thesis, Ohio State University, Columbus.

Medina, E. & Cuevas, E. (1989). Patterns of nutrient accumulation and release in Amazonian forests of the upper Rio Negro basin. In *Mineral Nutrients in Tropical Forest and Savanna Ecosystems*, ed. J. Proctor, pp. 217–40. Blackwell Scientific Publications, Oxford.

Medina, E., García, V. & Cuevas, E. (1990). Sclerophylly and oligotrophic environments: relationships between leaf structure, mineral nutrient content, and drought resistance in tropical rain forests of the upper Rio Negro. *Biotropica* **22**: 51–64.

Medina, E., Olivares, E. & Marín, D. (1985). Eco-physiological adaptations in the use of water and nutrients by woody plants of arid and semi-arid tropical regions. *Medio Ambiente* **7**(2): 91–102.

Murphy, P. G. & Lugo, A. E. (1986a). Ecology of tropical dry forest. *Annual Review of Ecology and Systematics* **17**: 67–88.

Murphy, P. G. & Lugo, A. E. (1986b). Structure and biomass of a subtropical dry forest in Puerto Rico. *Biotropica* **18**: 89–96.

Murphy, P. G. & Lugo, A. E. (1990). Dry forests of the tropics and subtropics: Guánica Forest in context. *Acta Científica (San Juan)* **4**(1–3): 15–24.

Murphy, P. G., Lugo, A. E., Murphy, A. J. & Nepstad, D. (1995). The dry forests of Puerto Rico's south coast. In *Tropical Forests: Management and Ecology*, ed. A. E. Lugo and C. Lowe. Springer-Verlag, New York (in press).

Neumann, C. J., Cry, G. W., Caso, E. L., & Jarvinen, B. R. (1978). *Tropical Cyclones of the North Atlantic Ocean, 1871–1977*. National Climatic Center, National Oceanic and Atmospheric Administration, Asheville.

Organization of American States (1972). Haiti, Mission d'Assistance Technique Integrée. OAS, Washington.

Reich, P. B. & Borchert, R. (1984). Water stress and tree phenology in a tropical dry forest in the lowlands of Costa Rica. *Journal of Ecology* **72**: 61–74.

Reich, P. B. & Borchert, R. (1988). Changes with leaf age in stomatal function and water status of several tropical tree species. *Biotropica* **20**: 60–9.

Richards, P. W. (1952). *The Tropical Rain Forest: an ecological study*. Cambridge University Press, Cambridge.

Ruthenberg, H. (1980). *Farming Systems in the Tropics*, 3rd edition. Clarendon Press, Oxford.

Rzedowski, J. (1978). *Vegetación de México*. Editorial LIMUSA, México.

Saravia-Toledo, C. J. (1985). La tierra pública en el desarrollo futuro de las zonas áridas: estado actual y perspectivas. In *IV Reunión de Intercambio Tecnológico en Zonas Aridas y Semiáridas, I*, pp. 115–40. Centro Argentino de Ingenieros Agrónomos, Buenos Aires.
Seifriz, W. (1943). The plant life of Cuba. *Ecological Monographs* **13**: 375–426.
Tasaico, H. (1967). Reconocimiento y evaluación de los recursos naturales de la República Dominicana. Unión Panamericana, Washington.
Tosi, J. A., Jr (1969). *Mapa ecológico de Costa Rica*. Tropical Science Center, San José.
Tosi, J. A., Jr (1971). Zonas de vida de Panamá: una base ecológica para investigaciones silvícolas e inventariación forestal en la República de Panamá. FAO and UN Development Program, FO-SF/PAN 6, Informe técnico 2.
Tosi, J. A., Jr & Voertman, R. F. (1964). Some environmental factors in the economic development of the tropics. *Economic Geography* **40**: 189–205.
Trewartha, G. T. (1961). *The Earth's Problem Climates*. University of Wisconsin Press, Madison.
Wadsworth, F. H. (1990). Plantaciones forestales en el bosque estatal de Guánica. *Acta Científica (San Juan)* **4**: 61–8.
Walter, H., Harnickell, E. & Mueller-Dombois, D. (1975). *Climate-diagram maps of the individual continents and the ecological climatic regions of the earth* (supplement to the vegetation monographs). Springer-Verlag, New York.
Walter, H. & Lieth, H. (1967). *Klimadiagramm-Weltatlas*. G. Fischer Verlag, Jena.
Weaver, P. L. (1989). Forest changes after hurricanes in Puerto Rico's Luquillo Mountains. *Interciencia* **14**: 181–92.
Weaver, P. L., Byer, M. D. & Bruck, D. L. (1973). Transpiration rates in the Luquillo Mountains of Puerto Rico. *Biotropica* **5**: 123–33.
Weaver, P. L. & Francis, J. K. (1988). Growth of teak, mahogany, and Spanish cedar on St. Croix, U.S. Virgin Islands. *Turrialba* **38**: 308–17.
Whigham, D. F., Zugasty Towle, P., Cabrera Cano, E., O'Neill, J. & Ley, E. (1990). The effect of variation in precipitation on growth and litter production in a tropical dry forest in the Yucatán of México. *Tropical Ecology* **31**: 23–34.

3

Overview of the Brazilian caatinga

EVERARDO V.S.B. SAMPAIO

Introduction

The dry forest and scrub vegetation in Brazil, generally called 'caatinga', covers an estimated area of $6–9 \times 10^5$ km^2 in the northeastern region. It is conditioned by the prevailing semiarid climate, with high potential evapotranspiration throughout the year (1500–2000 mm y^{-1}) and low rainfall (300–1000 mm y^{-1}), which is usually concentrated in 3–5 months and is very erratic (Reddy, 1983). Drought years are common and severe droughts lasting 3–5 years have occurred every 3–4 decades.

The area has been inhabited for more than 10,000 years, mainly in the river valleys and humid mountains, but according to early colonial sources, population density was generally low. Cattle raising spread in the 18th century and still is the main economic activity. From that period on, population pressure has increased in more favorable areas, where subsistence agriculture is practised in fenced plots. Until the middle of this century, cattle roamed freely on the non-agricultural land, independently of land ownership, but most properties are now fenced.

Land productivity is low and since resources are limited and birth rates have been high, the area has been a center of continuous migration to more favorable places in the same region, mainly the coastal area, or to other regions in the country. Migration increases during catastrophic drought periods. Nonetheless, population has steadily increased in the area and most of it has remained at a bare subsistence level. Social and economic parameters are the worst in the country, from lowest per capita income to highest illiteracy.

Government efforts to foster economic development in the region have centered on the coastal area, except for large irrigation projects. Although these projects have been highly successful, land suitable for irrigation comprises only 2–3% of the whole semiarid region. Therefore, the pattern of land use will probably continue to be similar to the present one, perhaps with less subsistence agriculture and more extensive livestock production.

Physical environment

Climate

The semiarid region is influenced by three major systems in the atmospheric circulation: the Continental Equatorial, the Intertropical Convergence and the Atlantic Polar Front. Fronts gradually lose humidity as they penetrate into the region and the residual rainfall depends on the strength of each front in a particular period. This strength is connected to the global atmospheric circulation, notably to the phenomena called 'El Niño-Southern Oscillation' ('ENSO': Molion, 1989). Thus, average annual rainfall decreases from the borders of the region to its interior (Figure 3.1*A*). The limits of the semiarid region are usually set at 1000 mm annual rainfall.

The effect of the fronts is modified by orographic characteristics. Elevations are mostly 400–500 m and reach about 1000 m. Although small, this elevation difference is enough to create more humid microclimates on the mountain sides facing the front and drier ones on the opposite side. Several of these humid mountain sides, called 'brejos', dot the area. According to the fronts and orographic features the Northeastern region has three rainfall systems (Figure 3.1*B*).

Over most of the Northeastern region, average annual rainfall is between 500 and 750 mm (37%) but a small proportion receives less than 250 mm (<250 mm = 0.3%, 250–500 mm = 11% and 750–1000 mm = 20%). However, annual averages are a poor climatic parameter in an area where local rainfall may vary from close to zero to as much as ten times the long-term average, and deviation from the normal rainfall may be higher than 55% (Figure 3.1C). The main characteristic of rainfall in the region is its erratic nature. It varies greatly in amount from place to place and from year to year, and in the beginning and duration of the rainy season (Reddy, 1983). As a general figure, 20% of the annual rainfall occurs on a single day and 60% in a single month. The contribution to the annual precipitation of the three months with the most rainfall is shown in Figure 3.1*D*.

Average temperatures are 23–27 °C, with less than 5 °C monthly variations and 5–10 °C daily variations, and average relative humidities of about 50%. Insolation periods are about 2800 h y^{-1}. Therefore,

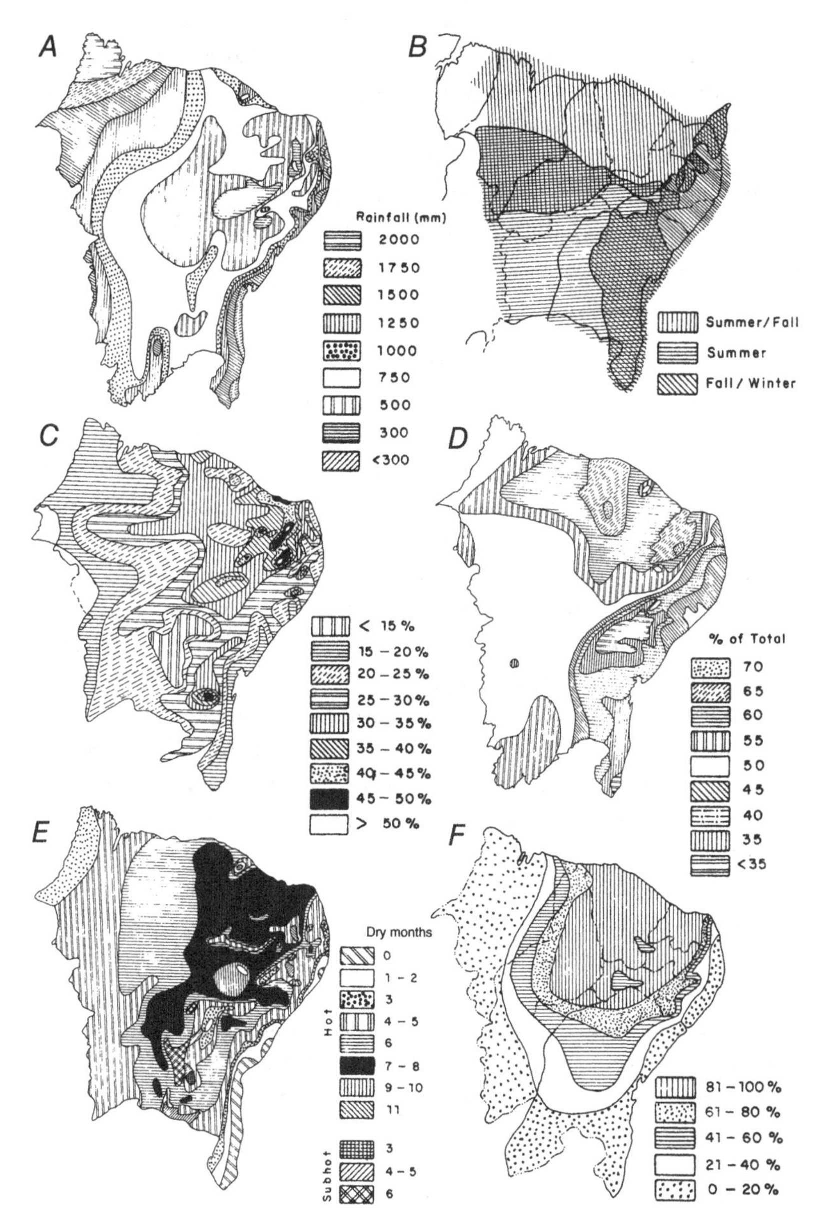

Figure 3.1. Characteristics of precipitation in northeastern Brazil. *A*. Average annual rainfall (from IBGE, 1985). *B*. Rainfall patterns (from Andrade, 1977). *C*. Annual average rainfall deviation in relation to the normal (from IBGE, 1977). *D*. Maximum contribution to the annual rainfall of three consecutive months (from IBGE, 1985). *E*. Climatic differentiation (from IBGE, 1977). *F*. Drought probabilities (from SUDENE, 1985).

evapotranspiration is high, generally above 2000 mm y^{-1} (Brazil, SUDENE, 1985). The combination of these features results in low available humidity for the plants during 7–11 months (Figure 3.1*E*).

Drought periods are a notorious characteristic of the region (Figure 3.1*F*). They are usually defined as those years in which the annual rainfall is less than 30% of the average. However, it is common that in years with 50–70% of the average, rainfall is so erratic that it does not provide moisture for crop growth.

Geology and soils

The main geologic units in the Northeastern region are the Proterozoic crystalline basement and the Paleozoic and Mesozoic sedimentary basins (Figure 3.2*A*). Most of the semiarid area is located on the crystalline basement which was successively raised and eroded until the Tertiary, forming a large flattened surface of elevations between 300 and 500 m. This surface is divided by higher elevation mountains and plateaus (900–1000 m), respectively representing residual formations of the Proterozoic surface, and Paleozoic and Mesozoic sedimentary layers (Andrade, 1977). The mountain range occupies the center of the dry area dividing the drainage in a radial pattern, to the north, east and south.

This formation resulted in a complex mosaic of soil types with extremely different characteristics (Figure 3.3). Soils on the crystalline basement tend to be shallow, clayey and rocky, usually classified as lithosols, regosols and non-calcic brown soils. Those on sedimentary material tend to be deep and sandy, usually classified as latosol, podzolic and quartz sand soils. However, exceptions are so frequent that it is difficult to establish patterns. There are two additional problems in using the available soil information for vegetation studies. (1) Soil surveys group large areas under the same classification, but there are important variations in soil type within short distances. (2) Some soil classes include soil profiles with different depths, and soil depth is the most important factor in soil water availability (Sampaio, Alves & Carneiro, 1987).

The variation of soils can be illustrated by the characteristics of 10 soils studied in an area of 5 × 5 km (Table 3.1).

Agriculture and plant products

The semiarid region includes a great variety of agricultural production patterns, different levels of development and different degrees of integra-

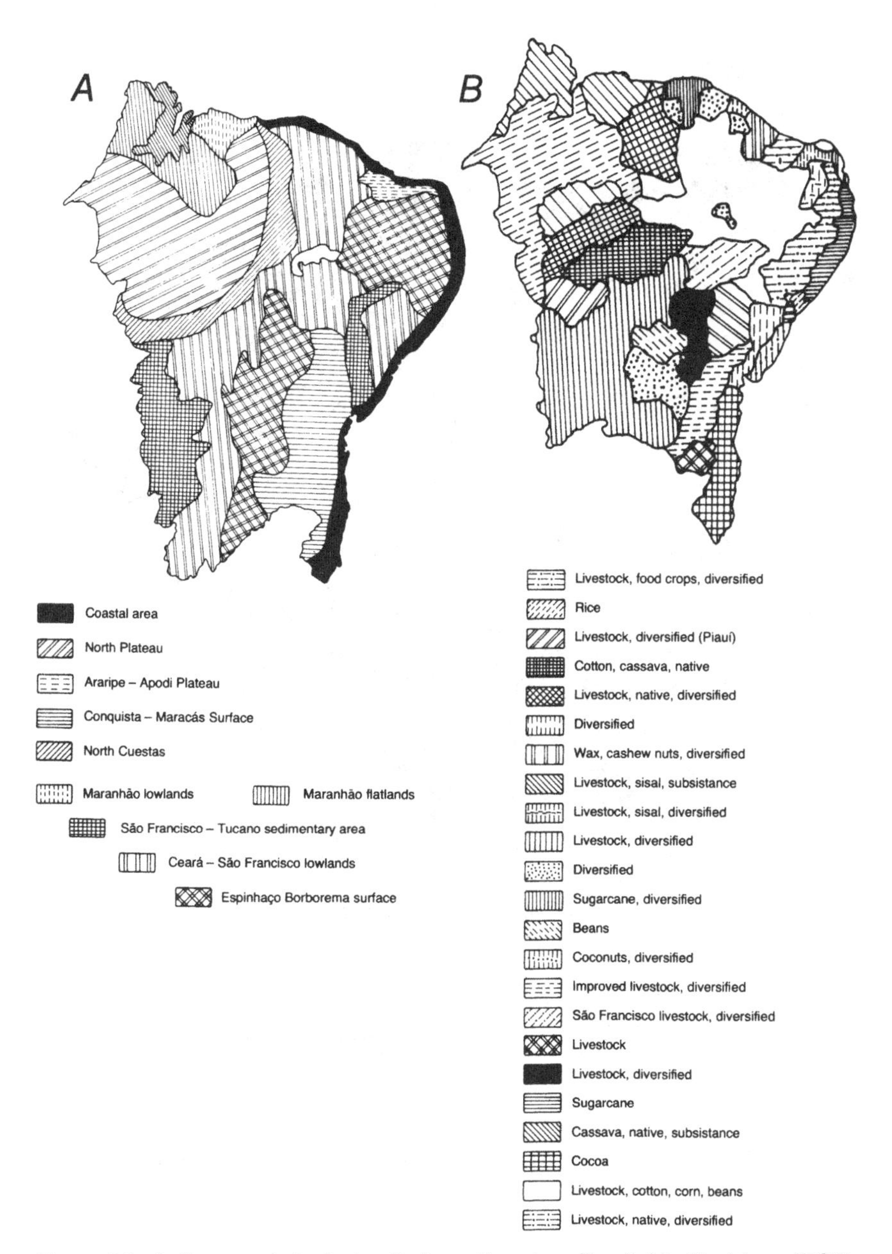

Figure 3.2. *A*. Geomorphological units in northeastern Brazil. Modified from IBGE (1985). *B*. Production systems. Modified from Sampaio, Sampaio & Bastos (1987).

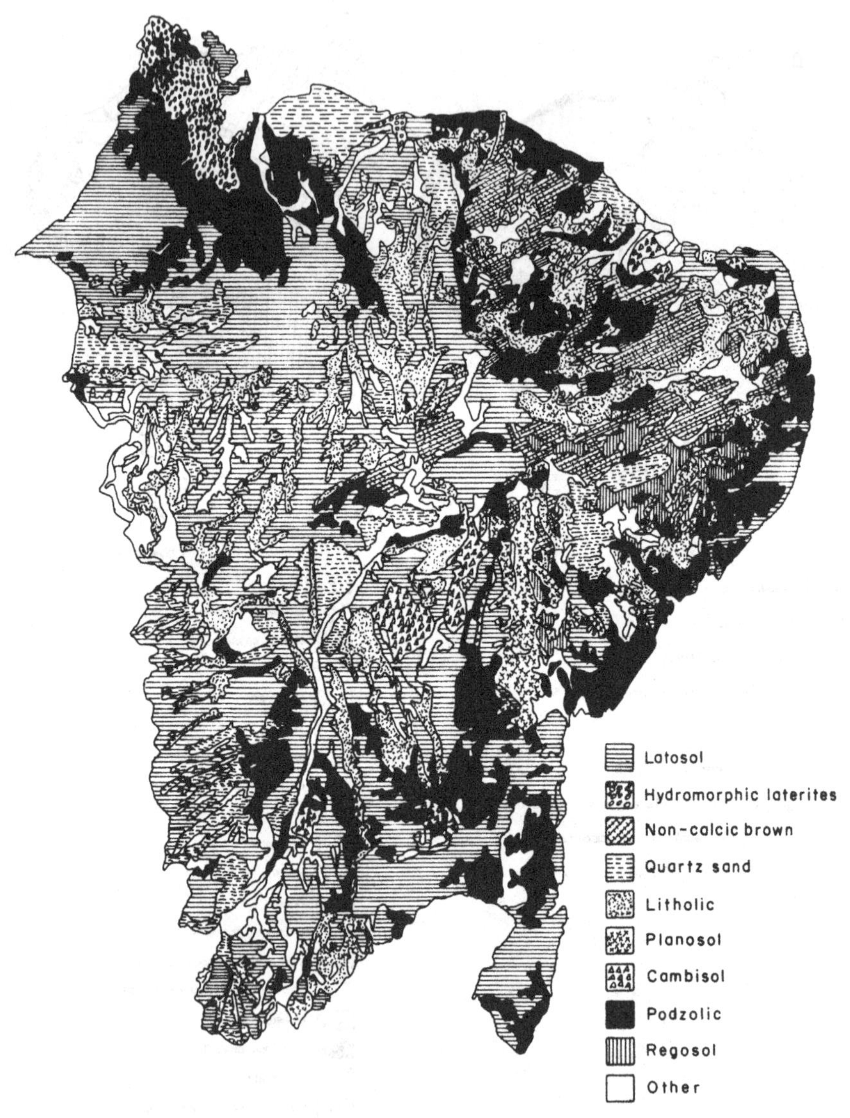

Figure 3.3. Main soil types in northeastern Brazil. Modified from IBGE (1985).

tion of agriculture into the urban economies. Agriculture is usually based on three types of land holdings (Bastos, 1980). (1) On traditional latifundia, production relies on sharecroppers; mixed cropping is commonly practised, and extensive areas may be rotated through bush fallow while some plots are used continuously. (2) Commercial farms employ

Table 3.1. *Soil diversity in a 5 × 5 km area in the semiarid region of Pernambuco*

Soil type	Parent material	Land use	Total C (%)	Total N (%)	Total P ($\mu g\ g^{-1}$)	Mineralized Pib[a] ($\mu g\ g^{-1}$)	Mineralized N[b] ($\mu g\ g^{-1}$)
Alluvial	Sediment	Cropped	0.75	0.08	220	6.2	37
Alluvial	Sediment	Cropped	1.01	0.09	534	45.6	13
Alluvial	Sediment	Cropped	1.31	0.10	323	9.7	34
Non-calcic	Gneiss	Uncropped	0.92	0.12	234	1.4	43
Non-calcic	Micaschist	Uncropped	0.50	0.04	175	0.7	10
Planosol	Gneiss	Uncropped	0.93	0.13	173	1.9	32
Podzolic	Megmatite	Uncropped	0.48	0.05	105	1.9	27
Regosol	Gneiss	Uncropped	0.68	0.06	187	3.3	39
Regosol	Granite	Uncropped	0.28	0.04	107	0.8	16
Vertisol	Gneiss	Uncropped	0.64	0.06	72	1.0	20

[a] Inorganic P extracted with bicarbonate.
[b] Laboratory incubation for 20 weeks.
Source: Alves (1989).

landless hired labor and have a high degree of modernization. Monoculture is typical and there may be rotation of crops and cropping areas. In some cases soil fertility is maintained with fertilizer inputs. (3) Small farms rely on family labor and operate near the subsistence level. When yields are low the family may rotate cropped areas if the landholding is large enough to permit bush fallow; otherwise, they may abandon the land.

The number of major products is small (beef cattle, goats, cotton, corn, beans, cassava and firewood), but many others with a small share in the regional production are important at a local level. Based on socioeconomic parameters and relative value of the main agriculture products, the semiarid region has been divided into 11 different zones (Figure 3.2*B*; Sampaio, Sampaio & Bastos, 1987). Zone differences reflect changes in climate, soils and socioeconomic structure. The zones have different combinations of major products and include some products that may be atypical under semiarid conditions (Table 3.2). The latter are usually produced under special conditions, like irrigation, in valleys, river and dam margins, floodplains or brejos. These areas of special conditions are small relative to the total area of a given zone but can have a large influence on agricultural production.

Average productivities for most common crops are low compared with

Table 3.2. *Main products of the different agricultural systems in the semiarid region of NE Brazil*

System	Products
1	Carnauba, cashew, beans, cassava, corn, coconuts, sugarcane
2	Cattle, coffee, castorbeans, beans, cassava, corn, annual cotton, sugarcane, banana, tomato, rice
3	Cattle, perennial cotton, corn, beans
4	Cattle, perennial cotton, corn, rice, beans
5	Cattle, carnauba, annual cotton, beans, cashew, corn, banana, sweet potato, rice
6	Cattle, sisal, cassava, annual cotton, corn, beans
7	Cattle, annual cotton, corn, beans, sugarcane, rice, cassava
8	Cattle, onions, grapes, perennial cotton, corn, beans
9	Cattle, beans, licury palm, castorbeans, cassava
10	Beans, corn
11	Cattle, cassava, sugarcane, annual cotton, beans, tobacco

Source: Y. Sampaio *et al.* (1987).

those attained in other regions of the country (Table 3.3). These low yields are most commonly due to water deficiency during some period of the growing cycle, but nutrient deficiencies are also important as shown by increased productivities with fertilizer additions (Table 3.4). However, fertilizer use in the semiarid region is almost non-existent, due to the general uncertainty of moisture availability and the associated economic risk.

Agriculture has influenced native vegetation in four main ways. (1) Most of the land with more favorable conditions has been cleared to establish crop plants. (2) A small part of the land with less favorable conditions has also been cleared, mostly in recent decades, to plant introduced forage grasses. (3) An even smaller portion has been managed to increase forage production by selectively reducing the density of shrubs. (4) The largest portion of the area is used as native pasture, with a stocking rate higher than advisable.

The region has a population of more than 20 million cattle, most of them in semiarid areas. They have low productivity indices: birth, death and slaughter rates around 50, 10 and 8%; 600 l y^{-1} of milk per lactating cow; first birth at 4–5 years and birth intervals of 20–24 months; 16 kg ha^{-1} of meat production, and slaughter age of 4–5 years. The main problem is forage availability during the dry season. Forage conservation practices, like silage and hay, are almost non-existent but the animals

Table 3.3. *Soils usually used for cultivation, and regional averages and averages of highest experimental productivities of main agricultural products in the semiarid region of NE Brazil*

Products	Productivities ($kg\ ha^{-1}$)		Usual soil types
	Average	Experimental	
Perennial cotton	120	1000	Non-calcic, Vertisol
Annual cotton	300	2000	Non-calcic, Vertisol, Alluvial
Beans	350	2500	Non-calcic, Cambisol
Castorbeans	500	1500	Non-calcic, Cambisol
Sisal	700	1500	Non-calcic, Regosol
Corn	600	4000	Non-calcic
Cassava	10,000	30,000	Podzolic, Regosol
Humid-land crops	–	–	Alluvial, irrigated
Beef	15	100	Litholic, Planosol, Regosol

Source: Sampaio *et al.* (1990b).

may feed on crop residues. They may also receive fodder produced on agricultural sites, like elephant grass (*Pennisetum purpureum* Schumach.), or planted Cactaceae (*Opuntia* species). Movement of herds to more humid places during the dry season also occurs but the extent of this practice is not known. In general, the animals gain weight during the rainy season and lose most of it during the following dry season.

Animals kept only on native pasture need 10–15 ha per head and have an average annual weight gain of about 10 $kg\ ha^{-1}$. They feed on the herbaceous layer during the rainy season but also on shrub leaves and even on dried leaves on the ground when forage becomes scarce. Available forage production has been estimated at 1000–3000 $kg\ ha^{-1}\ y^{-1}$ (Lima, 1984; Araújo, 1986; Moura, 1987; Schacht *et al.*, 1989). Replacing the native vegetation with introduced grasses, mainly buffel (*Cenchrus ciliaris* L.), increases productivity to around 100 $kg\ ha^{-1}$ but establishing the pasture may not be profitable (Freire *et al.*, 1982). Recently, it has been recommended to manage native pastures by (1) selectively cutting bushes and trees to reduce their density, allowing more light to penetrate to the herbaceous layer; (2) lowering the height of shrubs to make leaf production more accessible to the animals; and (3) seeding introduced grasses between bushes and in open spaces (Araújo, 1986). These three procedures have increased production to 60, 35 and 100 $kg\ ha^{-1}$, respectively, when done independently. Schacht *et al.*

Table 3.4. *Response to fertilizer of corn and beans, sole cropped and intercropped. Averages and standard deviations for 59 experiments*

Crop	Unfertilized yield (kg ha^{-1})	Fertilized yield (kg ha^{-1})	LER[a]
Sole crop			
Corn	810 ± 402	3229 ± 1691	–
Beans	569 ± 265	1201 ± 561	–
Intercrop			
Corn	482 ± 328	2441 ± 1575	0.70
Beans	447 ± 248	647 ± 319	0.62
Total	929 ± 435	3088 ± 1609	1.32

[a] Land Equivalent Ratio.
Source: Rao & Morgado (1984).

(1989) showed that understory herbaceous production does not increase with clearing more than 50% of the woody vegetation and recommended this treatment because it provides herbaceous production in the rainy season and high yields of leaf litter during the dry season. Recovery after clearing was fast: canopy cover after one, two and three years was 30, 78 and 96% (Schacht *et al.*, 1989).

Crop residue production is highly variable, depending on the crop and on growth conditions. Good production figures are 2–7 Mg ha^{-1}, with corn tending to the upper limit and beans and cassava to the lower. Cotton fields may support 2 head ha^{-1} for 3–4 months. Data on the overall contribution of residues to livestock production in the semiarid region are not available. Residues may be responsible for the abnormally high stocking rates of most farms, 0.5–1 head ha^{-1}, compared with those obtained on native pasture alone.

Another contribution is by *Opuntia* species, which are planted as a reserve for drought periods. Production is about 150 Mg ha^{-1} of fresh fodder (15 Mg ha^{-1} dry mass) every two years. A large part of the dairy production also depends on *Opuntia* species, each cow receiving 30–60 kg day^{-1}. These cacti cover about 300,000 ha in the region.

Besides agriculture the population has influenced native vegetation by (1) clearing to sell firewood or to produce charcoal; (2) selectively cutting large shrubs and trees for fenceposts and construction; and (3) harvesting plant parts to sell or for domestic use.

Data on firewood and charcoal production shows a more or less stable situation over the last decades for the whole region (Y. Sampaio *et al.*,

1987). Cooking and small bakeries are still the major consumers but some large consumers (metal and cement industries) have been installed lately. Use of gas ovens has increased but is still too expensive for a large part of the population. Areas around industries have been overcut and replanting is required by law but not enforced. Replanting is very expensive and usually not successful; a better alternative could be rotation management for regrowth of native vegetation. Information on this subject is very scarce. Firewood productivity in the region was estimated to be 40–80 m^3 ha^{-1} (Brazil, SUDENE, 1985) and actual measurements in 184 sites yielded averages of 85 and 111 stereo ha^{-1} (bulk-loaded m^3, with air spaces) in open and dense caatinga, respectively (Ururahy & Oliveira, 1986). The effect of repeated cutting has not been measured; native vegetation may be slowly deteriorating or changing in composition but there is no clear evidence. A rough estimate of clearing for firewood and charcoal is 0.5–1.0% of the whole caatinga area each year. If the vegetation took 20 years to regrow, 10–20% of the region would have to be used to sustain this exploitation.

Some of the caatinga plants (*Schinopsis brasiliensis* Engl., *Astronium urundeuva* Engl.) produce wood with excellent physical characteristics (Paula & Alves, 1980) but their production is very low. Also, they grow slowly and have been eliminated from the most favorable sites by agriculture. Commercial exploitation is very small but large trees are continuously cut for local use. This selective elimination may affect the composition of the vegetation, or reproduction, but the effects have not been documented.

Besides the types of plant use described above, eight others have been listed: human consumption, popular medicine, production of fixed oil, wax, latex, fiber and essential oils, and general chemical extraction. The following text is based on the review by Y. Sampaio *et al.* (1987).

Human consumption is mainly of fruits and their use is usually so low that they are not even sold at the local markets. The only exception is *Spondias tuberosa* Arr. Cam. which is included in official production statistics (Y. Sampaio *et al.*, 1987). In spite of its importance (Pires, 1990), planting is limited to a few plants around houses. On the other hand, native plants are usually spared when the vegetation is cleared. All other fruits have little edible material (e.g. *Bumelia sartorum* Mart., *Zizyphus joazeiro* Mart.) or a taste that is not much appreciated (*Cereus jamacaru* DC.).

Several native plants are used in popular medicine. Those used most frequently are grown in house gardens, so their exploitation has little effect on the vegetation. Occasionally, some plants are advertised as wonder cures and are harvested intensively until the abnormal interest declines. The impact of these fads on the plants has not been documented.

The main oil-producing plants in the semiarid region are *Licania rigida* Benth., *Syagrus coronata* (Mart.) Becc. (licury palm) and *Cnidoscolus* species. The first two are commercially important. *L. rigida* grows along the borders of small rivers, but its population has declined because of clearance of land for agriculture when prices are low. The plants are not cultivated because they do not begin production for 8–10 years. *S. coronata* occurs with relatively high density in a few special places with sandy soils and average humidity. Seed oil of *Cnidoscolus* species has high quality and the plants are abundant in the caatinga. However, they are not commercially exploited because the fruits are dehiscent and the plants have stinging spines.

Wax is extracted from the leaves of *Copernicia prunifera* (Miller) H. E. Moore and *S. coronata*. *C. prunifera* grows in heavy soil in seasonally flooded valleys, with a density of about 500 plants ha^{-1}, covering $1.8–2.5 \times 10^5$ ha (SUDENE, 1972). Most of this is native vegetation but some planting has been done. Latex is not exploited commercially now; a small production was obtained in the past from *Manihot* species. *Neoglaziovia variegata* (Arr. Cam.) Mez. leaves are cut to extract long tough fibers but the exploitation has little importance; imported *Agave sisalana* Perr. substituted *N. variegata* as a fiber-producing plant 40–50 years ago.

Several chemicals could be extracted from native plants but presently only a few are commercially exploited. There is a growing interest in this potential but current usage has virtually no effect on the vegetation.

Vegetation structure and function

Vegetation types

The caatinga includes several local vegetation types, referred to by specific names (Duque, 1980). Water deficit during a large part of the year is the factor that conditions the vegetation and, because the intensity of this deficit varies greatly, the physiognomy and the flora also vary. Superimposed on the large-scale variation due to climatic and orographic patterns there are small-scale variations due to topography and soil type, making the caatinga a mosaic of vegetation types. Nonetheless, some broad classifications have been made, usually based on physiognomy and/or flora, but frequently without an accompanying map. The two most important works are those produced by the group in charge of a general survey of Brasil (RADAMBRASIL, 1983) and by Andrade-Lima (1981).

RADAMBRASIL (1983) divided the caatinga into three types: dense tree steppe, open tree steppe and park steppe. The first type occurs on higher, more humid places, and the second type on the pediplains, where the soils are shallow. The third type occurs in some river valleys and is subject to higher human impact. Andrade-Lima (1981) divided the caatinga into six units with 12 subunits; some of them can be located but others cannot. The first unit occurs in the southern portion, with annual rainfall between 850 and 1000 mm, and is a dense vegetation, 25–30 m high with at least three strata. The second unit, with four subunits, is the typical caatinga forest on crystalline rocks, with a less dense tree stratum 7–15 m high, many thorn-bearing species, and a frequently sparse ground cover. The third unit covers part of the sedimentary low places with deep sandy soils; it is only 5–7 m high, and the disposition of its slender branches and small leaves allows light to pass through easily. The fourth unit, with four subunits, occupies the largest area, on shallow soils derived from crystalline rocks; it is a shrubby vegetation of varying density. It is difficult to determine if this is a natural or a human-induced type. The fifth unit occurs in scattered places with low rainfall (<400 mm y^{-1}) and shallow sandy soils, derived from metamorphic rocks, and is a low vegetation. The sixth unit is a palm forest restricted to heavy alluvial soils on the fringe of the main rivers in the north.

From the above, it can be stated that most of the region is covered by a mixture of shrubs and trees, the former being dominant. Tree-dominated places are a small proportion of the whole area, including the places presently used for agriculture. Some authors have claimed that most of the area was formerly covered with trees and degenerated because of repeated fires and cutting for firewood since prehistoric times. No evidence exists for or against such claims but certainly conditions are now not appropriate to support a tree community over a large area.

Flora

A complete inventory of plant species in the region has not been made although several herbaria have been collecting material during the last century. As usual for harsh environments, the flora has been considered to have low diversity. The number of woody plant species in single localized studies normally varies from 21 to 195 (Table 3.5). A recent

Table 3.5. *Total number of woody species at several localities in the semiarid region of NE Brazil, and the range among sites at the same locality of species number, plant density and maximum plant height. Note the different sampling areas, methods and criteria of plant inclusion*

Locality	Number of sites	Area[a] (m^2)	Criteria[b]	Number of species		Density (plants ha^{-1})	Height (m)	Source
				Total	per site			
Ceará								
Aiuaba	12	6000	all	161	10–47	2220–28,020	8–17	1
Barbalha	1	50,000	k 3 cm	195	195	975	–	2
Quixadá	1	50,000	k 3 cm	43	43	923	–	3
Tauá	1	50,000	k 3 cm	28	28	534	–	4
Rio Grande do Norte								
Açu	2	pq	b 5 cm	21	21	742–966	10	5
Reg. Salineira	8	4000	s 5 cm	43	6–17	560–1380	10	6
Piranhas, Açu	2	90,000	k 3 cm	?	21–28	671–772	–	7
Paraíba								
Cariris Velhos	10	10,000	s 5 cm	32	7–32	670–3190	–	8
Pernambuco								
S.M.B. Vista	1	4290	b 5 cm	26	26	459	–	9
Jatobá	1	30,000	k 3 cm	38	38	836	–	10
Custódia	3	pq	s 3 cm	59	22–27	3098–5385	7–15	11
Floresta	4	10,000	s 3 cm	55	22–28	1076–2172	10–19	12
Petrolina	1	pq	>shrubs	31	31	8589	–	13
Fazenda Nova	1	3000	h >20 cm	44	44	25,233	9	14
Parnamirim	7	3500	all	52	18–31	619–4356	5–12	15

[a] Total sample area, or sampling method (pq = point quarter).
[b] Height of stem diameter measurement and minimum diameter for inclusion. Levels: s, soil; k, 50 cm; b, breast height; h, plant height.
Sources: 1. UFCE (1982); 2. Tavares *et al.* (1974b); 3. Tavares *et al.* (1969); 4. Tavares *et al.* (1974a); 5. Ferreira (1988); 6. Figueiredo (1987); 7. Tavares *et al.* (1975); 8. Gomes (1979); 9. Drumond *et al.* (1982); 10. SUDENE (1979); 11. Araújo (1990); 12. Rodal (1992); 13. Albuquerque *et al.* (1982); 14. Lyra (1982); 15. Santos (1987).

review of 38 published studies found a total of 339 woody species, belonging to 161 genera and 48 families (Rodal, Sampaio & Pereira, 1988). The families or subfamilies with most species were Caesalpinoideae, Mimosoideae, Euphorbiaceae, Papilionoideae and Cactaceae (45, 43, 32, 30 and 14 species, respectively) and the genera were *Cassia*, *Mimosa* and *Pithecellobium* (14, 10 and 9 species, respectively). No species was found at all sites studied. About half of the species (164) were cited in only one study and 94 species in 2–4 studies. The species with broadest distribution were *Astronium urundeuva* (35 studies), *Bursera leptophloeos* Mart. (35) and *Anadenanthera macrocarpa* (Benth.) Brenan (34). Of course, the number of herbaceous species must be much higher than the number of woody species but few studies have dealt with them. Santos (1987) found 52 woody and 136 herbaceous species at one site, and Silva (1985) found that of 257 species in her study area, the majority were herbaceous.

The large difference between the number of woody species in the region and in localized places is certainly due to the size of the region and the limited geographic distribution of some species. Rodal, Sampaio & Pereira (1990) have compared floras on the basis of locality, soil (Santos, 1987) and geomorphological types (UFCE, 1982). There was greater similarity between the different substrates at a given locality than between matched substrates between localities. Some species are present only in the northern part of the caatinga (*Auxemma oncocalix* Taub.) or in the southern part (*Schinopsis brasiliensis*), irrespective of soil type. Other species occur in the caatinga but are more common in the border vegetation types, mainly dry forest and cerrado. These vegetation types also occur within the semiarid region, mainly in the higher places (brejos); dry forest usually occurs in brejos derived from crystalline material and cerrado in those derived from sedimentary material (Chagas & Figueiredo, 1987; Figueiredo & Barbosa, 1988; Figueiredo & Diógenes, 1988). Their vegetation may be completely different from that of the surrounding caatinga (Lyra, 1983), but their proximity may be responsible for a relatively higher number of species in the caatinga, as found in Barbalha and Aiuaba (Table 3.5).

The most important attempt to relate plant species to caatinga types was made by Andrade-Lima (1981) (Table 3.6). His vegetation units were characterized by the following genera (and most important species): (1) *Tabebuia* (*avellanedae* Lor.), *Aspidosperma* (*pyrifolium* Mart.), *Astronium* (*urundeuva*) and *Cavanillesia* (*arborea* K. Schum.); (2) *Astronium*, *Aspidosperma*, *Cereus*, *Mimosa*, *Schinopsis* (*brasiliensis*), *Caesalpinia*

Table 3.6. *Caatinga subdivisions, according to xerothermic index, soil origin, physiognomy and characteristic genera*

Type	Index	Soil origin	Physiognomy	Characteristic genera
1	100–150	Limestone	25–30 m	*Tabebuia, Aspidosperma, Astronium, Cavanillesia*
2.1	150–200	Crystalline	7–15 m	*Astronium, Schinopsis, Caesalpinia*
2.2	150–200	Crystalline	7–15 m	*Caesalpinia, Spondias, Bursera, Aspidosperma*
2.3	150–200	Crystalline	5–7 m	*Mimosa, Syagrus, Spondias, Cereus*
2.4	150–200	Crystalline	5–7 m	*Cnidoscolus, Bursera, Caesalpinia*
3	150–200	Sandstone	5–7 m	*Pilosocereus, Poeppigia, Dalbergia, Piptadenia*
4.1	200–300	Crystalline	Shrubs	*Caesalpinia, Aspidosperma, Jatropha*
4.2	200–300	Crystalline	Open scrub	*Caesalpinia, Aspidosperma*
4.3	200–300	Crystalline	Open scrub	*Mimosa, Caesalpinia, Aristida*
4.4	200–300	Crystalline	Open scrub	*Aspidosperma, Pilosocereus*
5	150–200	Metamorphic	Open scrub	*Calliandra, Pilosocereus*
6	100–200	Alluvial	Palms	*Copernicia, Geoffrea, Licania*

Source: Andrade-Lima (1981).

(*pyramidalis* Tul.), *Spondias* (*tuberosa*), *Bursera* (*leptophloeos*), *Syagrus* (*coronata*) and *Cnidoscolus* (*phyllacanthus* [Muell. Arg.] Pax. et Hoffm.); (3) *Pilosocereus* (cf. *piauhiensis* [Gurke] Byl. et Rowl.), *Poeppigia* (*procera* Presl.), *Dalbergia* (*cearensis* Ducke) and *Piptadenia* (*obliqua* [Pers.] Macbr.); (4) *Caesalpinia, Aspidosperma, Pilosocereus, Mimosa, Aristida* and *Jatropha* (*pohliana* Muell. Arg.); (5) *Pilosocereus* (*gounellei* [Weber] Byl. et Rowl.) and *Calliandra* (*depauperata* Benth.); and (6) *Copernicia* (*prunifera*), *Licania* (*rigida*) and *Geoffroea* (*spinosa* Jacq.).

Several papers refer to certain species as being associated with specific soil characteristics, usually texture. Species of *Jatropha* and *Cnidoscolus*, for instance, are said to occur only on heavy-textured vertisol. However, these statements are usually based on personal experience and lack quantitative data to support them. Other authors have claimed that the species may be present on different soil types and what changes is their relative density. Fotius & Sá (1988), determining the presence of species on latosol and vertisol sites, stated that *Piptadenia obliqua* was characteristic of the latosol and *Aristida elliptica* (Ness.) Kunth., *Combretum leprosum* Mart. and *Aspidosperma pyrifolium* of the vertisol.

Rodal (1983) determined the flora of 275 plots in three different situations in the same locality: (1) hill tops and hillsides; (2) low crystalline sites; and (3) low sedimentary sites. Considering the species present in most of the plots at one site, she found that *Cnidoscolus phyllacanthus* was only at site 2 and *Cassia granulata* Rizzini, *Hohenbergia catingae* Ule and *Turnera mycrophylla* only at site 3; *Caesalpinia pyramidalis* and *Cordia leucocephala* Moric. were mostly at sites 1 and 2 and *Piptadenia obliqua* mostly at site 3.

Santos, Ribeiro & Sampaio (1992) determined the similarities of seven soil types (see Table 3.1), based on presence and density of woody and herbaceous species. Three groups were formed (Figure 3.4), and two more isolated types were also noted. The non-calcic brown soil (NC) seemed to have a vegetation similar to that of all other soils except the very sandy regosols. Similarity in the density of perennial species, mainly *Neoglaziovia variegata* and *Croton* species (marmeleiro), was notable among NC, planosol, podzol and, to a lesser extent, deep regosol. Similarity in the density of herbaceous species, mainly *Selaginella convoluta* Spring and *Aristida* species, comprised NC, vertisol and litholic non-calcic brown. The presence of woody species comprised a third group, of NC, vertisol, planosol and, to a lesser extent, shallow regosol. The most isolated soils types were the regosols.

Vegetation structure

Several studies have measured woody plant height and density (Table 3.5). Fewer have measured basal and crown projection area. These parameters depend on growing conditions but also on management of the stand, mainly cutting. Few records of vegetation management are kept, even on research stations, so it is difficult to select study sites. It is likely that selection has been biased toward the tallest, most dense communities, based on the assumption that they represent the best preserved sites. Maximum plant height varied from 7 to 19 m (Table 3.5), but usually was attained by only a few emergents. Average height data also have been published but the comparison of different works is difficult because of different sampling criteria. In one study involving 10 sites, average height was correlated with rainfall, soil depth and soil permeability (Sampaio, Andrade-Lima & Gomes, 1981). The tallest species are usually *Astronium urundeuva* and *Anadenanthera macrocarpa*.

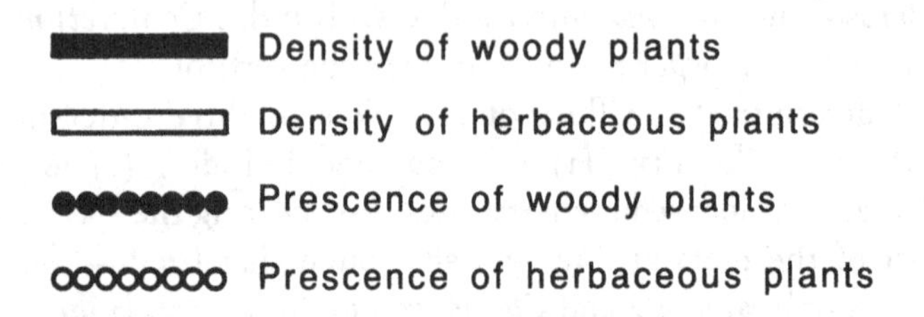

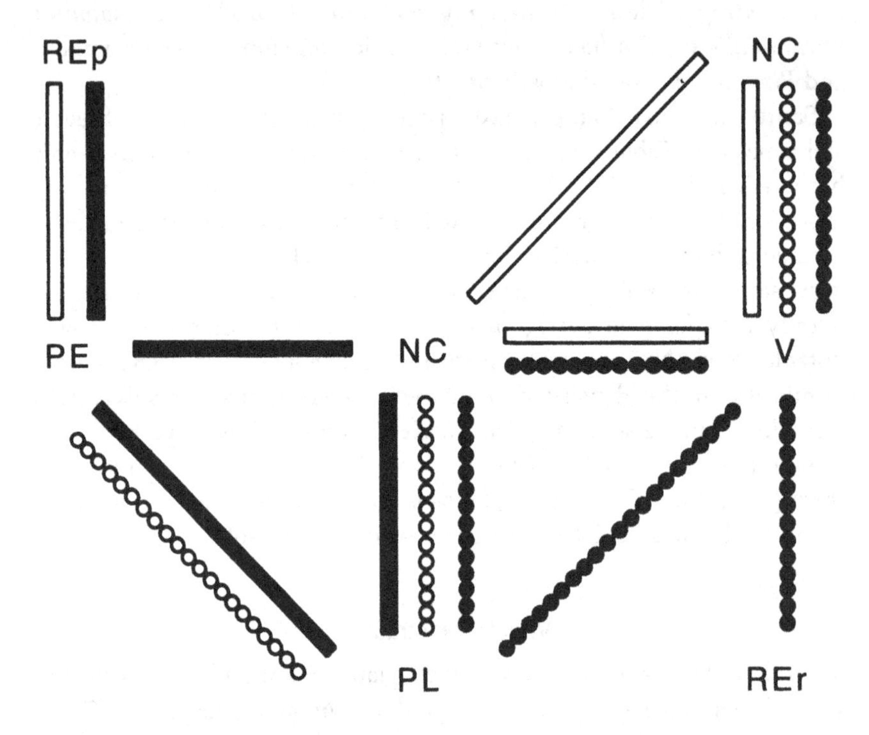

REp Deep eutrophic regosol
PE Red–Yellow eutropic podzol
V Vertisol
NC Non-calcic brown
PL Solidic planosol
NC1 Litholic non-calcic brown
REr Shallow eutrophic regosol

Figure 3.4. Similarities of plant communities growing on seven different soil types within an area of 5 × 5 km in Parnamirim, PE. From Santos *et al.* (1992).

Estimates of plant density depend strongly on the criteria for which plants are included. The number of plants per hectare with stem diameter at soil level ≥3 cm was 1000–5000, but increasing the limit to 5 cm decreased the densities to 500–3000 (Table 3.5). Using the diameter limit of 3 cm but measured 50 cm above soil level, the densities fell in the range of 500–1000. This lower number is partly due to the decrease in stem diameter with height but a more important factor may be the branching of shrubs a few centimeters above soil level. The density of all woody plants may reach 20,000–30,000 plants ha^{-1}. Human action and less favorable environmental conditions usually result in lower density of larger plants but may have little effect on, or even increase, total density (Lima *et al.*, 1988b).

The relative density of the single most abundant species in one place is usually high, frequently above 50%. The woody species most frequently listed among the ones with highest densities are *Caesalpinia pyramidalis, Aspidosperma pyrifolium, Croton* species and *Mimosa* species (jurema), representing the shrubs, and *Neoglaziovia variegata* and *Bromelia laciniosa* Mart. ex Schult. representing the low stratum. It is commonly believed that the high density of *Croton* and *Mimosa* indicate previous site perturbation. Data from one locality do not support this idea, because the density of *Croton* was as high on a preserved site as on sites subject to different degrees of human action (Lima *et al.*, 1988a; Sampaio, Rodal & Lima, 1988). However, *Mimosa* (jurema preta) sprouted much more vigorously after cutting and its mortality after burning was much less than that of all other species (Sampaio, Salcedo & Kauffman, 1990a). This could explain the presence of almost pure jurema preta stands on some previously disturbed sites.

Certainly, some places have low densities of woody plants, being almost open fields with a continuous herbaceous layer. Some of these fields may result from intentional management for forage production, but some may be natural, for example on sites with very shallow soils or partially covered with rocks. Santos (1987), comparing the vegetation on seven different soils, found a density of herbaceous plants about ten times higher (1134 plants m^{-2}) on a shallow regosol than on other soils. The vertisol and litholic non-calcic brown sites had several bare patches with hard surface crusts and low densities of both woody and herbaceous plants. The planosol site also had a low density of herbaceous plants, perhaps due to shading by the tall closed canopy.

Total basal area also depends on the criteria for which plants are included, but less so than average height or density. Few measurements have been made: in one place total basal area varied from 8.7 to 10.8 m^2 ha^{-1} (Ferreira, 1988) but in other places it reached 30–40 m^2 ha^{-1} (Sampaio *et al.*, 1988; Araújo, 1990; Rodal, 1992). This is the best parameter to indicate human action (Lima *et al.*, 1988b). The woody species with highest densities are usually the ones with highest relative basal area, because of their number and because they usually have average diameter equal to or above the average diameter of the community. However, they seldom have the plants with largest diameter. These are the tallest ones, plus *Bursera leptophloeos* and *Spondias tuberosa*, which are not so tall but have large crowns. The largest diameter measurements have been in the range of 33–79 cm (Ferreira, 1988; Araújo, 1990; Rodal, 1992).

Crown measurements were found in only one study (Albuquerque, Soares & Araújo, 1982). *S. tuberosa* (105 m^2), *Pseudobombax simplicifolium* A. Robyns. (69 m^2) and *B. leptophloeos* (52 m^2) were the species with largest average crown projection areas, but except for the latter they were based on only a few plants. Total crown area of trees was 14,000 m^2 ha^{-1} and that of shrubs 16,000 m^2 ha^{-1}, indicating a large degree of crown interpenetration.

Total aboveground biomass was determined in only one place, 84 Mg ha^{-1} (Sampaio *et al.*, 1990a).

Considering its broad distribution, high density and basal area *C. pyramidalis* (caatingueira) could be considered the most typical plant of the caatinga.

Vegetation function

Studies on phenology, physiology and ecology of caatinga plants are very few, mainly because they require more field time and few research groups are located within the semiarid region. Some mechanism of dealing with water deficit during a large part of the year must occur in all the plants. The most conspicuous adaptations are the shedding of leaves during the dry season, the death of herbaceous plants and Crassulacean acid metabolism (CAM).

Shedding of leaves is common to all shrubs and trees except Cactaceae. However, some species maintain their leaves longer than others. *Zizyphus joazeiro* and *Maytenus rigida* Mart. may even maintain green leaves throughout the year. *Z. joazeiro* usually occurs in low places with deep soil and develops a deep root system; the adaptations of *M. rigida* have not been studied. The period with leaves must be related to water availability but since these periods have been determined for only one place and one year (Barbosa *et al.*, 1989), even a simple correlation with rainfall cannot be made. In the study of Barbosa *et al.* (1989) four species maintained their leaves throughout the year: the two above and *Schinopsis brasiliensis* and *Bumelia sartorum*. These species started forming new leaves during the dry season. Another six species remained leafless for 2–3 months at the end of the dry season (*Cassia excelsa* Schrad., *Caesalpinia pyramidalis, Aspidosperma pyrifolium, Anadenanthera macrocarpa, Astronium urundeuva* and *Spondias tuberosa*). This study was conducted in a transition zone to dry forest and in a year with a higher than average rainfall, which may explain the short leafless periods.

Barbosa *et al.* (1989) also followed flowering and fruit set in these species. *C. pyramidalis* and *A. pyrifolium* flowered twice during the year, at the beginning of the rainy and the dry seasons. *S. tuberosa, A. urundeuva* and *Z. joazeiro* started flowering by the end of the dry season and fruiting during the rainy season. The other species flowered during the rainy season and produced fruits in the dry season. *B. sartorum* had the shortest interval between the start of flowering and the fruit maturation, four months. For the other species this interval varied from six to eight months and was spaced such that, during the whole year, there were always flowers and fruits present of one or more species.

Machado (1990), working in the same site with 21 species, found that *Melochia tomentosa* L. flowered during the whole year, 13 other species flowered during the rainy season and six flowered during the dry season (*Lonchocarpus aff. campestris* Benth., *Ruellia asperula* Lindau, *R. aff. paniculata* L., *Serjania comata* Radlk., *S. tuberosa* and *Z. joazeiro*). Flowering of *S. tuberosa* during the dry season has been confirmed by other authors (Monte, Silva & Silva, 1990).

Although there is no information on the dynamics of caatinga plant populations, some data are available on reproductive mechanisms, germination and seedling establishment.

Although species growing under unfavorable conditions tend to be autogamous, six of the 10 species studied by Machado (1990) were self-incompatible; among the other species *Ruellia aff. paniculata* and *Angelonia pubescens* Benth. were colonizing plants. Regarding phenology, Machado (1990) determined that eight of 13 species had steady-state flowering and five had cornucopia flowering. Flowers were most commonly visited by bees (84% of 21 plant species), butterflies (29%), birds (26%) and wasps (16%).

Germination rates of most species studied were usually high (>60%), with the exception of intermediate rates (20–30%) in *Mimosa caesalpinifolia* Benth., *M. malacocentra* Mart. and *Piptadenia obliqua*, and low rates (<10%) in *Cassia excelsa* and *Schinopsis brasiliensis* (Alves & Prazeres, 1980; Prazeres, 1982, 1986; Prazeres, Barbosa & Silva, 1984; Prazeres *et al.*, 1987; Prado & Barbosa, 1990). A high germination rate has also been reported for *S. brasiliensis* (Silva, Silva & Silva, 1988). Germination rates in the field, maintenance of viability with time and other influences of environmental conditions have not been determined.

Barbosa and co-workers (Barbosa, 1980, 1987) have followed the germination and seedling establishment of *A. macrocarpa, B. sartorum, M. rigida* and *Z. joazeiro*. In all of these, seeds germinated at the beginning of the rainy season and the seedlings died progressively during the subsequent dry period. However, the species differed greatly in maximum seedling density and also in the number of surviving seedlings. *B. sartorum* had an initial high density under the canopy (140 seedlings m^{-2}), *A. macrocarpa* an intermediate density (25 seedlings m^{-2}) and *M. rigida* and *Z. joazeiro* had low initial densities (0.3 seedlings m^{-2}). By the end of the following dry period the number of surviving seedlings had decreased to about 2 m^{-2} for the first two species and 0.1 for the others.

Since leaves of most species are present only during the rainy season they have not developed any special mechanism of moisture retention and have high transpiration rates when water is available (Grisi, 1976). However, some species, like *Croton* species (Nepomuceno, 1987), have leaves with dense hairs and abundant small stomata. Several species form tubers with high water content (*Spondias tuberosa, Carica corumbaensis*) or swollen trunks (*Cavanillesia arborea* and *Chorizia glaziovii* [O. Kuntze] Em. Santos). The development of *S. tuberosa* tubers has been studied by Alves (1986). Their function is not very clear but may be related to flowering at the end of the dry season. Some species, including most Cactaceae, have thorns and others have urticating leaves but the effect on species survival is not known. Several species have chemicals in their leaves that inhibit seed germination of other plants (*A. macrocarpa, A. pyrifolium, C. pyramidalis, M. rigida, S. brasiliensis, S. tuberosa, Z. joazeiro, Croton sonderianus* M. Arg. and *Caesalpinia ferea* Mart.: Tavares, 1982).

Animals

Animals in the semiarid area have been less studied than plants. Lists of species present in the area have been published for mammals (Mares *et al.*, 1981; Streilein, 1982a), birds (Sick, 1965; Coelho, 1987) and reptiles (Vanzolini, Ramos-Costa & Vitt, 1980). It has been concluded that only a few species are endemic to the caatinga, which shares most of its fauna with the neighboring cerrado and humid forest areas (Sick, 1965; Mares, Willig & Lacher, 1985). In fact, most of the species seem to have evolved under more mesic conditions than those of caatinga and inhabit special sites within the caatinga which provide more favorable living conditions. Apparently, they have not developed special physiological mechanisms of adaptation to xeric conditions as have the fauna of other semiarid areas in the world (Streilein, 1982b). In general, little has been written about their reproductive biology, behavior, population ecology and relationships with flora and environment. Although it has been repeatedly stated that their number is declining and some species are threatened, due to hunting and habitat modification, this idea has not been well documented.

More information has been published about mammals than other classes, particularly by Streilein (1982a–e). A total of 86 species were listed in the area, although several were related to human-created habitats, mainly buildings and abandoned agricultural fields, and only 46 occurred in the caatinga (Mares *et al.*, 1981). A large proportion of these mammals are Chiroptera, both in relation to the total (50 species) and to those in rock outcrops and open caatinga (36 species). The number of species belonging to other orders were 4 Marsupialia, 2 Primates, 3 Edentata, 1 Lagomorpha, 18 Rodentia, 7 Carnivora and 1 Artiodactyla. Mares *et al.* (1985) state that the number of species of small non-volant mammals found in any abundance in the caatinga is only five, thus the caatinga support one of the most depauperate faunas in the tropics. Only one species was recognized as being endemic of the caatinga, *Kerodon rupestris* (Wied, 1820), a small rodent usually inhabiting rock outcrops and *Wiedomys pyrrhorhinos* (Wied, 1821) was the only small mammal restricted to open caatinga habitats.

Most of the small mammals bred throughout the year, like *K. ruprestis*, *Galea spixii* (Wagler, 1831), *Monodelphis domestica* (Wagner, 1842), *Bolomys lasiurus* (Lund, 1841), *Calomys callosus* (Rengger, 1830) and *Thrichomys apereoides* (Lund, 1841) but *Didelphis albiventris* (Lund, 1841) exhibited synchronous, monestrous reproduction, with births occurring by the end of the dry season (Streilein, 1982c). The population levels of these species at Exu, Pernambuco, were very low (Streilein, 1982c), although a previous mark–release experiment estimated the population of *B. lasiurus* as 187 animals ha^{-1}. Their home range sizes were relatively small, less than 0.2 ha. Streilein concluded that these short-lived species have difficulty adapting to severe conditions which occur at widespread, irregular intervals.

Vanzolini *et al.* (1980) listed 47 species of reptiles in the caatinga, including 25 snakes, 18 lizards, three chelonia and one amphisbenia. The snakes belonged to five of the eight families occurring in South America: Colubridae (18 species), Leptotyphlopidae, Boidae, Viperidae (2 species each) and Elapidae (1 species). The lizards belonged to five families: Gekkonidae (7 species), Teiidae (5 species), Iguanidae (4 species), Scincidae and Anguidae (1 species each). The only endemic reptile in the caatinga is the iguanid *Platynotus semitaeniatus* which inhabits rock outcrops, usually flat surfaces covering relatively large areas, locally called 'lajeiros' (Vanzolini *et al.*, 1980).

Birds seem to be the least studied of all classes. Only a preliminary, incomplete list of birds was found (Coelho, 1987); other lists refer mainly to the cerrado (Sick, 1965). Sick (1965) states that cerrado and caatinga have fewer than 200 species, about 11% endemic in the two regions and only two endemic to the caatinga, *Augastes scutatus* and *A. lumachellus*. However, these two hummingbird species seem to occur mainly in the hilly areas of Bahia and Minas Gerais. An abundant species, *Zenaida auriculata*, migrates through the caatingas in large flocks (Azevedo, Antas & Nascimento, 1987). It feeds on native plant seeds, especially those of *Croton* species (Azevedo & Antas, 1988a). This dove reproduces in large open areas, usually at the beginning of the rainy season (Azevedo & Antas, 1988b). The density of nests in these areas has been found to range from 0.5 to 1.9 nests m^{-2}, usually with two eggs each. A great number of the doves and eggs are consumed by men and other predators.

Summary

The caatinga, including tropical dry forest, related scrub vegetation and derived ecosystems, covers an area of 6–9 × 10^5 km^2 in northeastern Brazil. This semiarid region has very erratic rainfall averaging 300–1000 mm y^{-1} and concentrated in 3–5 months, and has high potential evapotranspiration (1500–2000 mm y^{-1}). Most soils are shallow, clayey and rocky, and plant-available moisture is low for 7–11 months.

Agriculture has occupied the favorable land with crops of cotton, corn, beans and cassava. Most land is used as native pasture for beef cattle and goats, and for firewood/charcoal extraction. Available annual forage is 1000–3000 kg ha^{-1}. Animals kept only on native pasture need 10–15 ha per head and gain about 10 kg ha^{-1} y^{-1}. Clearing for firewood and charcoal affects 0.5–1.0% of the caatinga each year and regrowth takes 15–20 years. Useful native plants include medicinal species and a few commercial species, e.g. *Spondias tuberosa* (fruits), *Licania rigida* (oil) and *Copernicia prunifera* (wax).

The native flora is not well known but includes at least 339 species of trees and shrubs; families with the most species are Leguminosae, Euphorbiaceae and Cactaceae. Adaptations to water deficit are mainly deciduousness, CAM metabolism and perennation as seeds. Maximum plant height is 7–19 m and the number of plants with stem diameter at soil level ≥ 3 cm is 1000–5000 ha^{-1}. The relative density of the most abundant species in any place is frequently above 50%. Total basal area reaches 30–40 m^2 ha^{-1} and is the best indicator of human activity.

The fauna includes 86 mammal species (with 50 species of Chiroptera and 18 Rodentia) and 47 reptiles. A list of cerrado and caatinga birds has fewer than 200 species. The caatinga shares most of its vertebrate fauna with neighboring cerrado and humid forest areas; few species are endemic, and few show physiological adaptations to xeric conditions.

References

Albuquerque, S. G., Soares, J. G. G. & Araújo F°, J. A. (1982). Densidade de espécies arbóreas e arbustivas em vegetação de caatinga. *Pesquisa em Andamento* **16**. EMBRAPA-CPATSA, Petrolina.

Alves, G. D. (1989). Mineralização do carbono e do nitrogênio em 20 solos do Estado de Pernambuco e absorção de nitrogênio pelo sorgo. Dissertação de Mestrado Universidade Federal Rural de Pernambuco, Recife.

Alves, J. L. H. (1986). Ontogênese do tuber (cunca) do 'umbu' – *Spondias tuberosa* Arr. Cam. In *Resumos XXXVII Congresso Nacional de Botânica*, p. 2. Sociedade Botânica do Brasil, Ouro Preto.

Alves, J. L. H. & Prazeres, S. M. (1980). Estudo da morfologia e fisiologia da germinação da semente de plantas ocorrentes em região de caatinga – *Triplaris pachau* Mart. (Poligonaceae). *Brasil Florestal* **10**: 85–94.

Andrade, G. O. (1977). *Alguns Aspectos do Quadro Natural do Nordeste*. MINTER-SUDENE, Recife.

Andrade-Lima, D. (1981). The caatinga dominium. *Revista Brasileira de Botânica* **4**: 149–63.

Araújo F°, J. A. (1986). Manipulação da vegetação lenhosa da caatinga com fins pastoris. In *Anais Simpósio sobre a Caatinga e sua Exploração Racional*, pp. 327–43. EMBRAPA-DDT, Brasilia.

Araújo, E. L. (1990). Composição florística e estrutura da vegetação em três áreas de caatinga de Pernambuco. Dissertação de Mestrado, Universidade Federal Rural de Pernambuco, Recife.

Azevedo, S. M. Jr., & Antas, P. T. Z. (1988a). Novas informações sobre a alimentação da *Zenaida auriculata* no Nordeste do Brasil. In *Anais IV ENAV*, pp. 59–64. Universidade Federal Rural de Pernambuco, Recife.

Azevedo, S. M. Jr., & Antas, P. T. Z. (1988b). Observações sobre a reprodução da *Zenaida auriculata* no Nordeste do Brasil. In *Anais IV ENAV*, pp. 65- 72. Universidade Federal Rural de Pernambuco, Recife.

Azevedo, S. M. Jr., Antas, P. T. Z. & Nascimento, J. L. X. (1987). Censo da *Zenaida auriculata noronha* fora da época de reprodução no Nordeste. *Caderno Omega Universidade Federal Rural de Pernambuco, Sér. Biol.* **2**: 157–68.

Barbosa, D. C. A. (1980). Estudos ecofisiológicos em *Anadenanthera macrocarpa* (Benth.) Brenan – aspectos da germinação e crescimento. Tese de Doutoramento, Universidade de São Paulo, São Paulo.

Barbosa, D. C. A. (1987). Ocorrência de plantas jovens de *Bumelia sartorum* Mart. (quixabeira) sob a copa da planta adulta, durante as estações chuvosa e seca em região de caatinga (Alagoinha-PE). In *Resumos XXXVIII Congresso Nacional de Botânica*, p. 418. Sociedade Botânica do Brasil, São Paulo.

Barbosa, D. C. A., Alves, J. L. H., Prazeres, S. M. & Paiva, A. M. A. (1989). Dados fenológicos de 10 espécies arbóreas de uma área de caatinga (Alagoinha-PE). In *Anais XL Congresso Nacional de Botânica*, pp. 109–17. Sociedade Botânica do Brasil, Cuiabá.

Bastos, E. G. (1980). Farming in the Brazilian Sertão: social organization and economic behavior. Ph.D. thesis, Cornell University, Ithaca.

Chagas, F. G. & Figueiredo, M. A. (1987). O ambiente e a vegetação da chapada do Araripe. In *Resumos XI Reunião Nordestina de Botânica*, p. 28. Sociedade Botânica do Brasil, Fortaleza.

Coelho, A. G. M. (1987). *Aves da Reserva Biológica de Serra Negra (Floresta – PE). Lista Preliminar*. Universidade Federal de Pernambuco, Recife.

Drumond, M. A., Lima, P. C. F., Souza, S. M. & Lima, J. L. S. (1982). Sociabilidade das espécies florestais da caatinga em Santa Maria da Boa Vista-PE. *Boletim de Pesquisa Florestal* **4**: 47–59.

Duque, J. G. (1980). *O Nordeste e as Lavouras Xerófilas*. CNPq-ESAM – Fundação Guimarães Duque, Brasilia.

Ferreira, R. (1988). Análise estrutural da vegetação da Estação Florestal de Experimentação de Açu-RN, como subsídio básico para o manejo florestal. Dissertação de Mestrado, Universidade Federal de Viçosa, Viçosa.

Figueiredo, M. A. (1987). A microrregião salineira Norte-riograndense no domínio das caatingas. *Coleção Mossoroense* **353**. ESAM, Mossoró.

Figueiredo, M. A. & Barbosa, M. A. (1988). A cobertura vegetal do sul do Ceará – 6°00′ a 7°50′. In *Resumos XII Reunião Nordestina de Botânica*, p. 19. Sociedade Botânica do Brasil, João Pessoa.

Figueiredo, M. A. & Diógenes, M. B. (1988). Tipos de vegetação da faixa 4°30′ a 6°00′ no estado do Ceará. In *Resumos XII Reunião Nordestina de Botânica*, p. 20. Sociedade Botânica do Brasil, João Pessoa.

Fotius, G. A. & Sá, I. B. (1988). *Prospecção Botânica em Área de Exploração Petrolífera no município de Pendências, RN*. Série Documentos 47. EMBRAPA-CPATSA, Petrolina.

Freire, L. C., Albuquerque, S. G., Soares, J. G. G., Salviano, L. M. C., Oliveira, M. C. & Guimaraes F°, C. (1982). Alguns aspectos econômicos sobre a implantação e utilização de capim buffel em área de caatinga. *Circular Técnica* **9**. EMBRAPA-CPATSA, Petrolina.

Gomes, M. A. F. (1979). Padrões de caatinga nos Cariris Velhos, Paraiba. Dissertação de Mestrado, Universidade Federal Rural de Pernambuco, Recife.

Grisi, B. M. (1976). Ecofisiologia da caatinga: comportamento hídrico de *Caesalpinia pyramidalis* Tul. e *Schinopsis brasiliensis* Engl. *Ciência e Cultura* **28**: 417–25.
IBGE (1977). *Geografia do Brasil. Região Nordeste.* Fundação Instituto Brasileiro de Geografia e Estatística, Rio de Janeiro.
IBGE (1985). *Atlas nacional do Brasil. Região Nordeste.* Fundação Instituto Brasileiro de Geografia e Estatística, Rio de Janeiro.
Lima, G. F. C. (1984). Determinação da fitomassa aérea disponível ao acesso animal em caatinga pastejada – região de Ouricuri, PE. Dissertação de Mestrado, Universidade Federal Rural de Pernambuco, Recife.
Lima, M. J. A., Sampaio, E. V. S. B., Andrade, S. L. S. & Araújo, E. L. (1988a). Caracterização fitossociológica em 3 áreas antropizadas na microregião ecológica de Soledade – PB. In *Resumos XXXIX Congresso Nacional de Botânica*, p. 417. Sociedade Botânica do Brasil, Belém.
Lima, M. J. A., Sampaio, E. V. S. B., Sales, M. F., Rodal, M. J., Araújo, E. L. & Andrade, S. L. S. (1988b). Seleção de indicadores para determinação do nível de antropização em vegetação de caatinga. In *Resumos XXXIX Congresso Nacional de Botânica*, p. 385. Sociedade Botânica do Brasil, Belém.
Lyra, A. L. R. T. (1982). A condição de 'brejo' e o efeito do relevo na vegetação de duas áreas no município de Brejo da Madre de Deus–PE. Dissertação de Mestrado, Universidade Federal Rural de Pernambuco, Recife.
Lyra, A. L. R. T. (1983). Efeito do relevo na vegetação de duas áreas do município do Brejo da Madre de Deus (PE). III Diversidade florística. In *Anais XXXIV Congresso Nacional de Botânica*, pp. 287–96. Sociedade Botânica do Brasil, Porto Alegre.
Machado, I. C. S. (1990). Biologia floral de espécies de caatinga no município de Alagoinha (PE). Tese de Doutoramento, Universidade de Campinas, Campinas.
Mares, M. A., Willig, M. R. & Lacher, T. E., Jr (1985). The Brazilian caatinga in South American zoogeography: tropical mammals in a dry region. *Journal of Biogeography* **12**: 57–69.
Mares, M. A., Willig, M. R., Streilein, K. E. & Lacher, T. E., Jr (1981). The mammals of Northeastern Brazil: a preliminary assessment. *Annals of the Carnegie Museum* **50**: 81–137.
Molion, L. C. B. (1989). ENOS e o clima no Brasil. *Ciência Hoje* **10**(58): 23–9.
Monte, H. M., Silva, A. Q. & Silva, H. (1990). Fenologia de plantas de umbu na região do Curimataú Paraibano. In *Resumos XIV Reunião Nordestina de Botânica*, p. 54. Sociedade Botânica do Brasil, Recife.
Moura, J. W. S. (1987). Disponibilidade e qualidade de pastos nativos e de capim Buffel (*Cenchrus ciliaris* L.) diferido no semi-árido de Pernambuco. Dissertação de Mestrado, Universidade Federal Rural de Pernambuco, Recife.
Nepomuceno, V. A. G. (1987). Características anatomo-ecológicas de seis espécies do gênero *Croton* (Euphorbiaceae). In *Resumos XI Reunião Nordestina de Botânica*, p. 51. Sociedade Botânica do Brasil, Fortaleza.
Paula, J. E. & Alves, J. L. H. (1980). Estudos das estruturas anatômica e algumas propriedades físicas da madeira de 14 espécies ocorrentes em áreas de caatinga. *Brasil Florestal* **43**: 47–58.
Pires, M. G. M. (1990). Estudo taxonômico e área de ocorrência de *Spondias tuberosa* Arr. Cam. (umbuzeiro) no estado de Pernambuco – Brasil. Dissertação de Mestrado, Universidade Federal Rural de Pernambuco, Recife.
Prado, M. C. G. & Barbosa, D. C. A. (1990). Germinação e crescimento de *Schinopsis brasiliensis* Engl. (baraúna). In *Resumos XLI Congresso Nacional de Botânica*, p. 431. Sociedade Botânica do Brasil, Fortaleza.
Prazeres, S. M. (1982). Morfologia e germinação de sementes e unidades de dispersão das caatingas. Dissertação de Mestrado, Universidade Federal Rural de Pernambuco, Recife.

Prazeres, S. M. (1986). Avaliação da qualidade fisiológica das sementes de 10 espécies das caatingas. In *Resumos XXXVII Congresso Nacional de Botânica*, p. 189. Sociedade Botânica do Brasil, Ouro Preto.

Prazeres, S. M., Barbosa, D. C. A., Alves, G. D. & Paiva, A. M. A. (1987). Aspecto da produção e fisiologia da maturação do fruto e semente de *Aspidosperma pyrifolium* Mart. 'pereiro do Sertão'. In *Resumos XXXVIII Congresso Nacional de Botânica*, p. 54. Sociedade Botânica do Brasil, São Paulo.

Prazeres, S. M., Barbosa, S. A. & Silva, M. G. V. (1984). Viabilidade e vigor de sementes de Bromeliaceae das caatingas do Nordeste – *Bromelia laciniosa* Mart. 'macambira'. In *Resumos XXXV Congresso Nacional de Botânica*, p. 103. Sociedade Botânica do Brasil, Manaus.

RADAMBRASIL (1983). Vegetação. In *Folhas SC. 24/25. Aracaju/Recife. Geologia, geomorfologia, pedologia, vegetação, uso potencial da terra. Levantamento de Recursos Vegetais 30*, pp. 573–643. RADAMBRASIL, Rio de Janeiro.

Rao, M. R. & Morgado, L. B. (1984). A review of maize-beans and maize-cowpea intercrop systems in the semiarid northeast Brazil. *Pesquisa Agropecuaria Brasileira* **19**: 179–92.

Reddy, S. J. (1983). Climatic classification: the semiarid tropics and its environment – a review. *Pesquisa Agropecuaria Brasileira* **18**: 823–47.

Rodal, M. J. N. (1983). Fitoecologia de uma área do Médio Vale do Moxotó, Pernambuco. Dissertação de Mestrado, Universidade Federal Rural de Pernambuco, Recife.

Rodal, M. J. N. (1992). Fitosociologia da vegetação arbustivo-arbórea em quatro áreas de caatinga em Pernambuco. Tese de Doutoramento, Universidade de Campinas, Campinas.

Rodal, M. J. N., Sampaio, E. V. S. B. & Pereira, R. C. (1988). Revisão dos levantamentos florísticos no domínio das caatingas. In *Resumos XII Reunião Nordestina de Botânica*, p. 27. Sociedade Botânica do Brasil, João Pessoa.

Rodal, M. J. N., Sampaio, E. V. S. B. & Pereira, R. C. (1990). Similaridade florística entre localidades de caatinga. In *Resumos XLI Congresso Nacional de Botânica*, p. 373. Sociedade Botânica do Brasil, Fortaleza.

Sampaio, E. V. S. B., Alves, G. D. & Carneiro, C. J. G. (1987). Disponibilidade de água na caatinga. In *Resumos XXXVIII Congresso Nacional de Botânica*, p. 413. Sociedade Botânica do Brasil, São Paulo.

Sampaio, E. V. S. B., Andrade-Lima, D. & Gomes, M. A. F. (1981). O gradiente vegetacional das caatingas e áreas anexas. *Revista Brasileira de Botanica* **4**: 27–30.

Sampaio, E. V. S. B., Rodal, M. J. & Lima, M. J. A. (1988). Caracterização da vegetação em área preservada nos Cariris Velhos, Paraíba. In *Resumos XXXIX Congresso Nacional de Botânica*, p. 418. Sociedade Botânica do Brasil, Belém.

Sampaio, E. V. S. B., Salcedo, I. H. & Kauffman, J. B. (1990a). Efeito da intensidade de fogo na rebrota de plantas de caatinga em Serra Talhada, PE. In *Resumos XIV Reunião Nordestina de Botânica*, p. 55. Sociedade Botânica do Brasil, Recife.

Sampaio, E. V. S. B., Salcedo, I. H. & Tiessen, H. (1990b). Agriculture in Brazil's semiarid Northeast: a nutrient budget under shifting cultivation. In *Proceedings of the International Conference on Soil Quality in Semiarid Agriculture, Vol. II*, pp. 323–31. Saskatchewan Institute of Pedology, Saskatoon.

Sampaio, Y., Sampaio, E. V. S. B. & Bastos, E. (1987). *Parâmetros para a Determinação de Prioridades de Pesquisas Agropecuárias no Nordeste Semi-Árido*. Departamento de Economia, PIMES/UFPE, Recife.

Santos, M. F. A. V. (1987). Características dos solos e da vegetação em sete áreas de Parnamirim, Pernambuco. Dissertação de Mestrado, Universidade Federal Rural de Pernambuco, Recife.

Santos, M. F. A. V., Ribeiro, M. R. & Sampaio, E. V. S. B. (1992). Semelhanças vegetacionais em sete solos de caatinga. *Pesquisa Agropecuaria Brasileira* **27**: 305–14.

Schacht, W. H., Mesquita, R. C. M, Malechek, J. C. & Kirmse, R. D. (1989). Response of caatinga vegetation to decreasing levels of canopy cover. *Pesquisa Agropecuaria Brasileira* **24**: 1421–6.
Sick, H. (1965). A fauna dos cerrados. *Arquivos de Zoologia* **12**: 71–93.
Silva, A. Q., Silva, H. & Silva, M. A. (1988). Aspectos morfológicos da germinação da baraúna (*Schinopsis brasiliensis* Engl.) e aroeira (*Astronium urundeuva*). In *Resumos XII Reunião Nordestina de Botânica*, p. 5. Sociedade Botânica do Brasil, João Pessoa.
Silva, G. C. (1985). Flora e vegetação das depressões inundáveis da Região de Ouricuri–PE. Dissertação de Mestrado, Universidade Federal Rural de Pernambuco, Recife.
Streilein, K. E. (1982a). Ecology of small mammals in the semiarid Brazilian caatinga. I. Climate and faunal composition. *Annals of the Carnegie Museum* **51**: 79–107.
Streilein, K. E. (1982b). Ecology of small mammals in the semiarid Brazilian caatinga. II. Water relations. *Annals of the Carnegie Museum* **51**: 109–26.
Streilein, K. E. (1982c). Ecology of small mammals in the semiarid Brazilian caatinga. III. Reproductive biology and population ecology. *Annals of the Carnegie Museum* **51**: 251–69.
Streilein, K. E. (1982d). Ecology of small mammals in the semiarid Brazilian caatinga. IV. Habitat selection. *Annals of the Carnegie Museum* **51**: 331–43.
Streilein, K. E. (1982e). Ecology of small mammals in the semiarid Brazilian caatinga. V. Agonistic behavior and overview. *Annals of the Carnegie Museum* **51**: 345–69.
SUDENE (1972). *Estudo de Mercado de Produtos Agropecuários do nordeste – Carnaúba*. (mimeograph) SUDENE-DAA, Recife.
SUDENE (1979). *Projeto para o Desenvolvimento Integrado da Bacia Hidrogeológica do Jatobá: levantamento dos recursos da vegetação*. SUDENE, Recife.
SUDENE (1985). *Recursos Naturais do Nordeste. Investigação e potencial*. SUDENE, Recife.
SUDENE (1985). *Recursos Naturais do Nordeste; investigação e potencial (sumário das atividades)*. SUDENE, Recife.
Tavares, M. C. R. (1982). Ocorrência de inibidores de germinação em espécies de caatinga. Dissertação de Mestrado, Universidade Federal Rural de Pernambuco, Recife.
Tavares, S., Paiva, F. A. F., Tavares, E. J. S. & Carvalho, G. H. (1975). Inventário florestal na Paraíba e no Rio Grande do Norte. I Estudo preliminar das matas remanescentes do Vale do Piranhas. *Recursos Naturais* **3**. SUDENE, Recife.
Tavares, S., Paiva, F. A. F., Tavares, E. J. S. & Lima, J. L. S. (1969). Inventário florestal do Ceará. II Estudo preliminar das matas remanescentes do município de Quixadá. *Boletim de Recursos Naturais SUDENE* **7**: 93–111.
Tavares, S., Paiva, F. A. F., Tavares, E. J. S. & Lima, J. L. S. (1974a). Inventário florestal do Ceará. II Estudo preliminar das matas remanescentes do município de Tauá. *Boletim de Recursos Naturais SUDENE* **12**: 5–19.
Tavares, S., Paiva, F. A. F., Tavares, E. J. S. & Lima, J. L. S. (1974b). Inventário florestal do Ceará. II Estudo preliminar das matas remanescentes do município de Barbalha. *Boletim de Recursos Naturais SUDENE* **13**: 20–46.
Universidade Federal do Ceará (1982). Estudo de comunidades de caatinga na Estação Ecológia de Aiuaba; relatório técnico. Convênio SUBIN 049/79-UFCE/UFRN/UFPE, Fortaleza.
Ururahy, J. C. & Oliveira, L. C., Jr (1986). Estimativa do volume de fitomassa parcial das formaçoes arbóreas da caatinga. In *Anais Simpósio sobre a Caatinga e sua Exploração Racional*, pp. 243–69. EMBRAPA-DDT, Brasília.
Vanzolini, P. E., Ramos-Costa, A. M. M. & Vitt, L. J. (1980). *Répteis das Caatingas*. Academia Brasileira de Ciências, Rio de Janeiro.

4

Savannas, woodlands and dry forests in Africa

JEAN-CLAUDE MENAUT, MICHEL LEPAGE
& LUC ABBADIE

Introduction

In Africa, the term dry forest covers vegetation types dominated by a more or less continuous tree cover (70%), experiencing pronounced drought during more than three months per year, and occurring within the savanna biome. They may be called (open) woodlands or (dense) dry forests according to tree density and understory structure (Menaut 1983). The Yangambi classification establishes the following (Boughey, 1957a, b; Monod, 1963; Aubréville, 1965).

- A woodland has an upper stratum of deciduous trees of small or medium size, with their crowns more or less touching above a sparse woody understory. Tree density is high enough to affect the herbaceous stratum which differs floristically from the adjacent savanna. The ground layer consists of grasses, herbs and suffrutescent plants in sufficient density to allow for annual burnings. The canopy of a woodland tends to be dominated by one or very few species.
- A dry forest, strictly speaking, is defined as a closed stand with several woody strata. The grass layer, when present, is weak and discontinuous, only allowing for episodic and sparse fires. In most cases, the trees of the upper stratum are deciduous whereas the understory is composed of evergreen and/or deciduous shrubs which differ from the canopy floristically. The canopy is multispecific and often devoid of woodland dominants. In both dry forest and woodland, the tree species which make up most of the canopy are present but never dominant in the surrounding savanna.

Some authors have considered dry forests to be a tropophilous extension of the rain forest, with adaptations to xeric conditions in characteristics of stems but not of leaves (Schnell, 1976–7). This view is supported by the presence of a large number of woody genera in common between these types (cf. Gentry, Chapter 7). However, woodlands and dry forests may still include a few representatives of the very old dry flora of Africa (e.g. *Encephalartos*, Cycadaceae). Savannas, woodlands and dry forests are plant formations which might even have antedated the rain forests themselves (Aubréville, 1949). Dry forest is bound to certain climatic conditions (under which a number of other vegetation types occur) and primarily to the absence of fire.

Distribution and main types

According to Griffith (1961), African woodlands and dry forests should cover *c.* 13×10^6 km^2, an area divided into two distinct regions: in the Northern Hemisphere, there is an elongate area of about 5.25×10^6 km^2, situated between 6° and 13° N; in the Southern Hemisphere, there is a more massive area, between 5° and 20° S, totalling 7.75×10^6 km^2. These values are certainly greatly overestimated, and at best correspond to the area in which woodlands and dry forests might occur (White, 1983). More probably, their area is restricted to about 3.63×10^6 km^2 (12% of the total land area of Africa), according to the map by Aubréville *et al.* (1958) and Keay (1959a) (Figure 4.1). In the Northern Hemisphere (Sudanian region), these types have tended to disappear through the impact of climate and human activities, and now occur as relatively small patches within the savanna area. In the Southern Hemisphere (Zambezian region), they still occur as much larger blocks in spite of their increasing utilization by humans. Northern woodlands and dry forests, although growing in more humid conditions than the southern ones, experience a higher atmospheric saturation deficit. The deficit is attributable to higher ambient temperatures caused by lower elevations (*c.* 200 m vs 1000 m) and also to the effects of the dry winds blowing from the Sahara during the dry season. Northern woodlands and dry forests are therefore more open, shorter and less species-rich than the southern ones.

Both woodland and dry forest are extensive in continental climatic areas, which are arid during the dry season. They are never in contact with the rain forests, always being isolated by a belt of humid to mesic savannas. These savannas, with a depauperate tree stratum, might result from intense burning regimes sustained by high grass biomass and climatic conditions. Some botanists think that the humid savanna belt results from repeated clearing of a former rain forest zone, and is maintained by fire. Others consider that the rain forest receded during a dry episode (Aubréville, 1961; Bellier *et al.*, 1968; Van der Hammen, 1983), and that fire now prevents forest regeneration (Menaut, 1983). In any case, human density and activity have always been much higher in the mesic and arid savannas than in the humid zones. Figure 4.2 illustrates the changes in vegetation along an aridity gradient, emphasizing the separation of core woodland and dry forest from the rain forest.

The traditional view was that 'dense' dry forests once occupied, under a wetter climate, most of the present savanna regions in Africa. Broadly speaking, the core area of woodlands and dry forests is located in mesic climatic conditions: annual rainfall of 800–1100 mm y^{-1}, number of dry

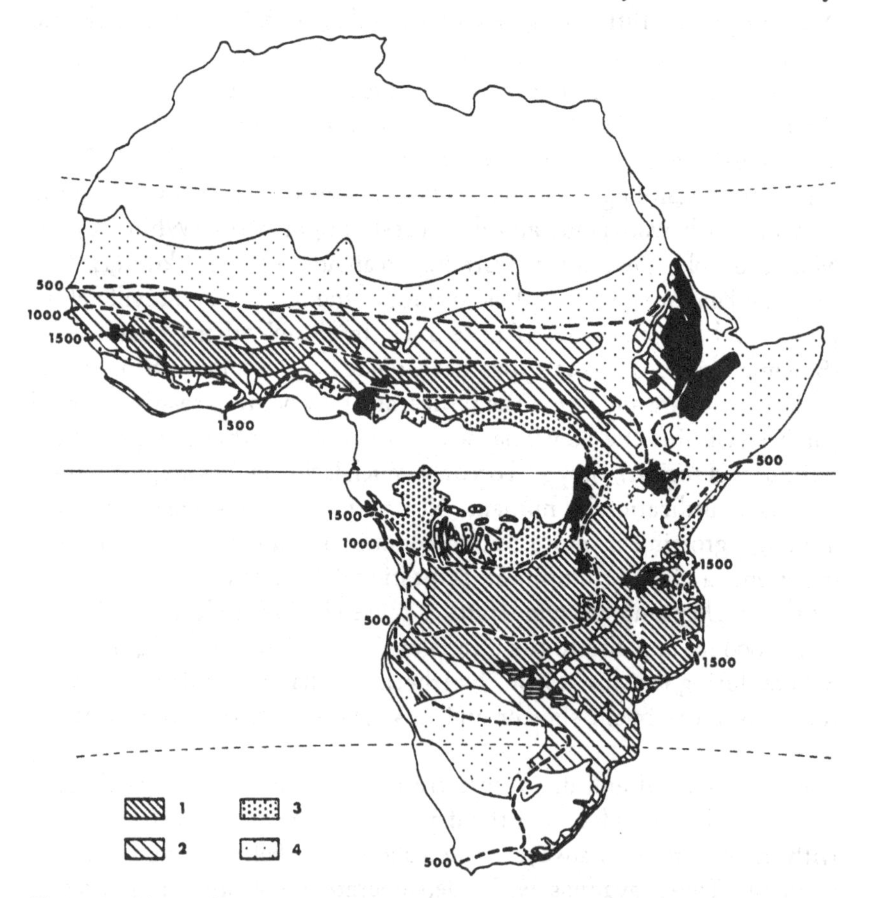

Figure 4.1. Vegetation types in the seasonally dry tropics of Africa. 1, densely wooded savannas, woodlands and dry forests; 2, tree/shrub savannas; 3, humid savannas; 4, arid savannas (modified from Aubréville *et al.*, 1958). Broken lines: mean annual isohyets.

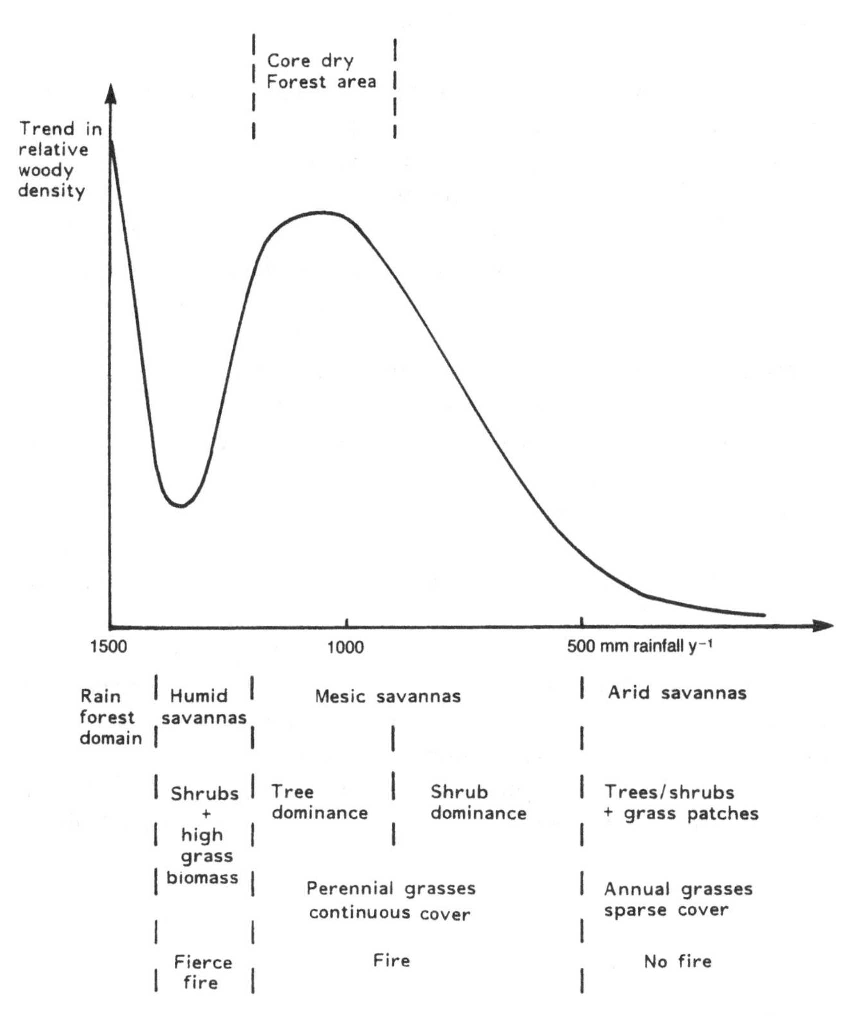

Figure 4.2. Idealized scheme of vegetation types and woody plant density along a gradient of rainfall.

months per year between 3 and 6, ratio of precipitation:potential evaporation between 0.4 and 0.8. However, depending on soil conditions, some woodlands may extend into the arid zone (rainfall *c.* 500 mm y^{-1}; number of dry months per year >7: Walker, 1981).

At a broad level, there is no clear evidence for an edaphic origin of woodlands or dry forests, and Schmitz (1962) has shown that dry forest occurs on any soil type. One can state only that most of them are associated with old lateritic planation surfaces. They also extend onto redistributed laterite and exposed bedrock, and onto detritic substrates (Cole, 1986). As seen in Figure 4.3, in mesic conditions they often occur either on shallow lateritic horizons or on dismantled hard-pan or bedrock. At a local level with a given rainfall, plant biomass (i.e. tree density) as opposed to production (i.e. grass domination), tends to be higher in conditions of lower nutrient availability (Bell, 1982), but the idea that woody formations occur on nutrient-poor soils (Bell, 1982) may be an artifact. Once established, trees incorporate nutrients in their biomass and in conditions of relatively low productivity may impoverish the soil (Menaut, Abbadie & Vitousek, 1993), contributing to the maintenance of a vegetation type which originated in other conditions (Germain, 1965).

Interactions with climate and humans are important. Woodlands resist fire because of limited fuel loads (sparse grass biomass and shrub understory). They do not seem to regenerate easily after tree felling and cultivation when fire is maintained, however, and subsequently degrade into savanna-like formations. Dry forests are highly sensitive to clearing which results in the loss of atmospheric and soil humidity, and to burning with the loss of the dominant non-fire-resistant species. Through regressive succession, dry forests turn into increasingly open types and subsequently to savannas (Freson, Goffinet & Malaisse, 1974). Fire damage varies in relation to intensity and the species concerned, but also with the time of burning: 'late' fires, occurring during the vegetation's development, always have a strong depressive effect; 'early' fires, taking place at the end of the vegetative period, generally have a lesser impact (Gillon, 1983).

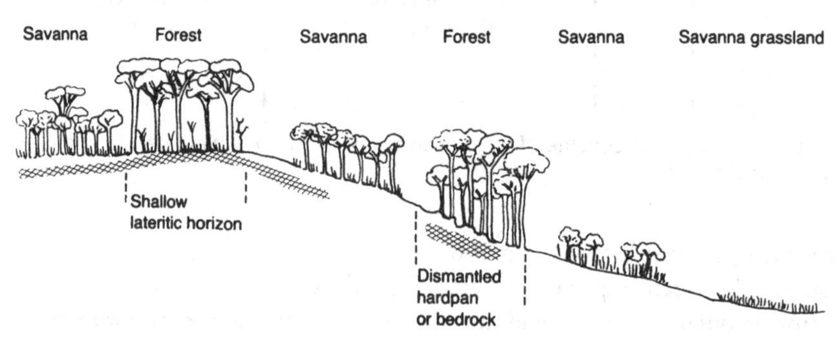

Figure 4.3. Relationships of vegetation with soil characteristics at the local level for mesic climatic conditions.

Woodlands

The structure of woodlands consists of small to medium-sized trees (8–20 m), with a canopy cover of 70–90%, light foliage of compound leaves, sparse woody undergrowth and a more or less continuous cover of sun-loving grasses. The woody flora consists mostly of Leguminosae and is relatively species-poor. All species are deciduous in the annual drought period. Although the drought lasts for several months, defoliation does not last more than a few weeks, because tree root systems are able to explore the soil horizons underneath the hardpan. Grass and tree leaf litter provide fuel for fires which destroy most seedlings.

In the Sudanian region, the core type is the *Isoberlinia* woodland which mostly occurs on shallow plateau soils (Menaut, 1983; Figure 4.4*A*). The trees are 9–14 m tall, and there are few or no understory species. The main species are *Isoberlinia doka* and *I. tomentosa*. The *Isoberlinia* type is replaced by *Monotes kerstingi* and *Uapaca somon* on eroded slopes (with *Isoberlinia dalzielii* on soils rich in quartzite), and by *Terminalia macroptera* on poorly drained clay soils or heavily farmed and burnt areas. Other woodlands dominated by savanna trees (e.g. *Daniellia oliveri* and *Lophira alata*) exist in the area. Although fire-tolerant, these woodlands cannot resist regular intensive burning: tree density and canopy cover diminish, and they rapidly become similar to the adjacent tree savannas. *Anogeissus leiocarpus* groves, of more or less large extent, develop on abandoned village sites or farmlands (Jones, 1963; Sobey, 1978). Keay (1949 in White, 1965) suggested that 'the climax vegetation of the Sudan zone may, before the advent of human influence, have consisted of a dense understory of deciduous shrubs and lower stratum trees forming a rather open canopy.'

Northward, in the Sahelian region, there is an arid woodland type characterized by *Acacia seyal* on clayey soils in small depressions, by *A. nilotica* on floodplains, and by *A. raddiana* on valley sandy soils. The type consists of small trees (<10 m tall) with a light microphyllous canopy cover and no woody understory. Such *Acacia* woodlands thrive in most valley floors of East Africa. Short woodlands of *Boswellia* are sometimes found on rocky sites.

In the Zambezian region, the core type is the *Brachystegia* woodland, also called 'miombo' (Figure 4.4*B*). This type occupies most of the plateau of southern Tanzania, western Mozambique, northern Zimbabwe, southern Zaïre and eastern Angola. The elevation range is 900–1500 m, with rainfall of 600–1300 mm y^{-1} and a dry season extending 6 to 8 months. This type occurs mostly on lateritic soil, but may also be found on stony escarpments and on sandy to clayey soils. It is characterized by a relatively dense woody understory. Its physiognomy does not change much in spite of the local variations in species composition imposed by edaphic conditions. The main species are *Julbernardia paniculata* and *Isoberlinia angolensis* on all sites, *Brachystegia spiciformis* on deep soils (but absent from hydromorphic soils), *B. longifolia* on deep sandy soils, *B. utilis* on sandy-clayey soils, *B. floribunda* on coarsely textured soils, and *B. boehmi* on heavy clay soils (Malaisse, 1979). Compared with the *Isoberlinia* woodlands of the Sudanian region, the miombo exhibits a much higher species richness and diversity, and a more complex structure and phenology, probably as a consequence of the mixed influence of more varied rainfall, lower temperatures, and complex soil patterning.

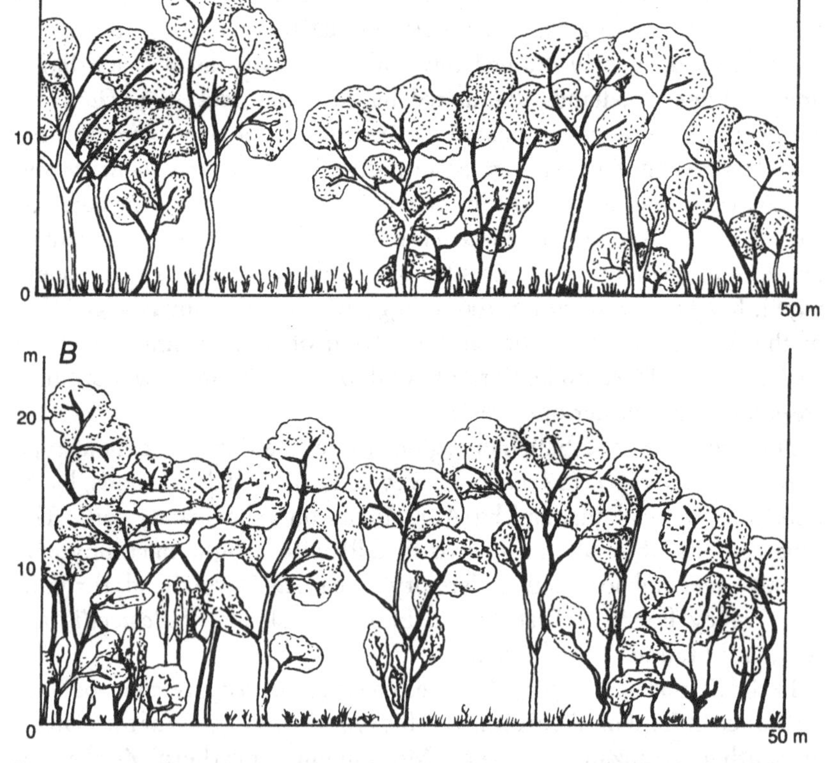

Figure 4.4. Profiles of core types of woodland from the northern (Sudanian) and southern (Zambezian) regions. *A.* northern *Isoberlinia* woodland (adapted from Devineau, 1985). *B.* southern *Brachystegia* woodland (adapted from Malaisse, 1974).

There are two other main variants of woodlands in the southern region. First, the *Baikiaea* woodlands are similar to miombo, but grow on dystrophic soils of the deep Kalahari sands. They receive only 500–750 mm annual rainfall, and have a fairly dense understory. The main species are *Guibourtia coleosperma, Pterocarpus angolensis, Baikiaea plurijuga, Pterocarpus stevensoni, Burkea africana* and *Copaifera coleosperma.* Second, the *Colophospermum mopane* woodlands occur further south, and at lower elevation and rainfall. They are associated with eutrophic soils on deep black clays. They are markedly xerophytic, and the understory is sparse.

Dry forests

The dense forest types occur in the humid parts of the Sudano-Zambezian regions. They show no general relationship to soil characteristics, occurring either on plinthic horizons or on sandy humic loams. The flora is species-rich, and Leguminosae are less dominant. The woody vegetation often has three strata: a closed canopy of semideciduous species up to 25 m tall, a loose understory 10–15 m tall, and a dense layer of shrubs and scramblers below 5 m. Climbers may occur in all three strata. The ground is covered with litter and sparsely distributed shade-loving grass species. Fire, when not deliberately set, is accidental or restricted to creeping ground fires.

In the northern region, the core type is the *Afraegle* forest (Figure 4.5*A*), which occurs on lateritic plateau soils (Devineau, 1985). Leaf phenology is linked to the annual drought period. There is a diverse understory. The main species are *Diospyros mespiliformis, Malacantha alnifolia, Lecaniodiscus cupanioides, Canthium multiflorum, Afraegle paniculata, Khaya senegalensis, Drypetes floribunda, Mimusops kumel, Monodora tenuiflora, Kigelia africana, Oncoba spinosa, Tamarindus indica* and *Manilkara multinervis.* Some particular types occur on well-watered ground in semiarid areas. They may be composed of a mixture of semideciduous humid forest species (e.g. *Antiaris africana, Morus mesozygia*) and of dry savanna trees such as *Adansonia digitata* (baobab) and *Parkia biglobosa.* It is impossible to know if the humid flora is relictual or has been able to develop in favorable conditions (Schnell, 1976–7).

Some evergreen types occur as scattered remnants on rocky sites without fire. They have humid forest affinities, and may contain a few rain forest species, although the dominant tree species (*Gilletiodendron glandulosum* or *Guibourtia copallifera*) belong to the dry flora and are well adapted to the constraints of the Sudanian climate (Jaeger, 1956). The understory is easily penetrable; the herb cover is sparse, almost without grasses.

In the southern region, the core type is the *Entandophragma* forest (Figure 4.5*B*). It occurs on soils with more clay than those under open woodlands (*c.* 50 vs 25%). The dominant story is deciduous, the understory is either deciduous or evergreen. The phenology, marked by the dry season, is simpler than in the adjacent woodlands in spite of a greater species richness. The main species are *Entandophragma delevoyi*,

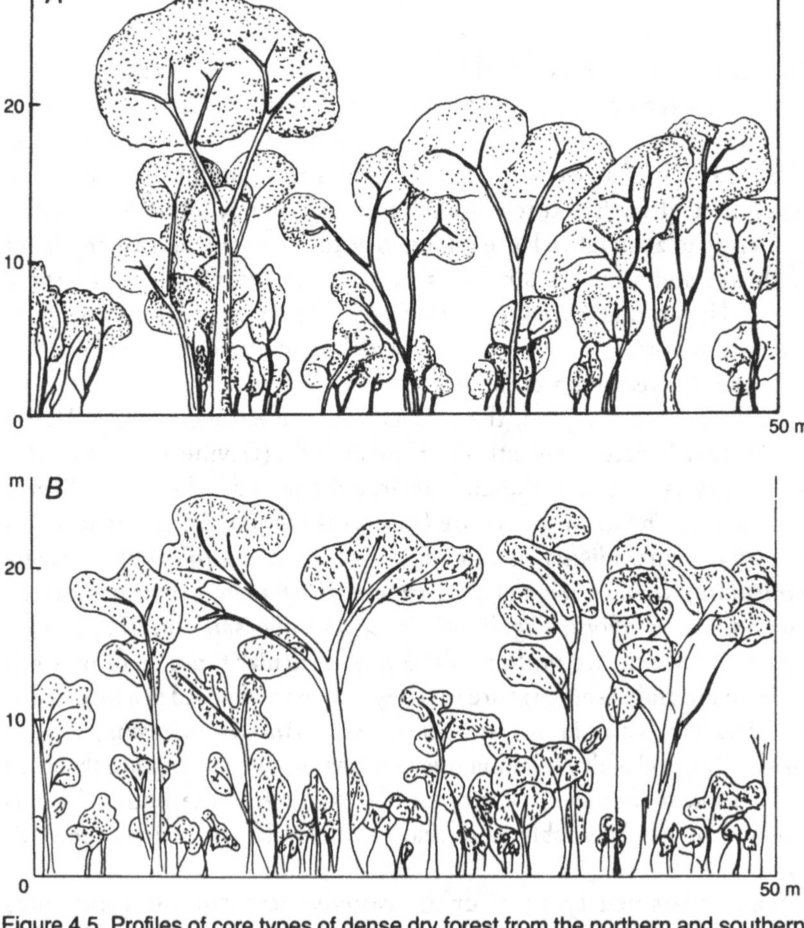

Figure 4.5. Profiles of core types of dense dry forest from the northern and southern regions. *A*. northern *Afraegle* forest (adapted from Devineau, 1985). *B*. southern *Entandophragma* forest (from Malaisse, 1984).

Brachystegia spiciformis, *B. taxifolia*, *Parinari holstii*, *Aida micrantha*, *Diospyros hoyleana* and *Dichapetalum bangii*. Floristically, the dense dry forests are much less variable than the woodlands of the region. Their composition remains identical whatever the soil type (Schmitz, 1962).

Three dry evergreen forest types, often restricted to relict patches, may be distinguished (Fanshawe, 1961). The *Parinari–Syzigium* forests occur either on granular sands, sandy loams or sandy clays; they probably covered much larger areas before being transformed by humans into woodlands and savannas. Dense, low *Cryptosepalum* forests occur on almost sterile sands; the canopy is often tied together with lianas. *Marquesia* forests occur on deep loamy sands; they have clear dynamic relationships with the surrounding deciduous woodlands. The understory of all three types is characterized by a high density of shrubs and scramblers, and sparse humid forest grasses.

Climatic conditions and soil water dynamics

Fairly extensive studies of climate and soil moisture have been conducted in the Zambezian region. Malaisse (1979) noted that the *Brachystegia* woodlands in Zaïre occur under conditions where the rainy season extends from November to March and the dry season from May to September. The annual rainfall averages 1300 mm, but with considerable interannual variations (700–1550 mm). Annual average temperature is 20 °C; yearly mean insolation is 2800 h with a daily mean of 8 h (Freson, 1971, in Malaisse, 1979). According to Griffith (1961), the Zambezian woodlands occur under rainfall of 625–1270 mm y^{-1}, with a dry season of 6–8 months, and annual mean temperatures of 16–27 °C (with occasional cold temperatures in the dry season). Partly due to the elevation, these features differ somewhat in character and in parameter values from the worldwide criteria proposed by Holdridge (1967) for dry tropical and subtropical forests: frost-free areas where mean annual temperature is >17 °C; mean annual rainfall 250–1000 mm; and a ratio of potential evapotranspiration to precipitation greater than one.

Ernst (1971) and Ernst & Walker (1973) found that the limit of woodland corresponds to the tolerance of the main tree species (*Brachystegia*) to heat (48 °C for leaves and buds) and cold (−2 °C for stems). When temperature regularly drops below −2 °C, miombo woodlands are replaced by more open woodlands and savannas (Wild, 1968; Werger & Coetzee, 1978). Fire following frost seems highly damaging to miombo species.

On the other hand, the vegetative cover also influences the local climate. Freson *et al.* (1974) distinguish the mesoclimate from the macroclimate prevailing along a gradient from the evergreen forest to the savanna (Table 4.1). The values obtained demonstrate the effect of the density of the canopy on temperature, total solar radiation, atmospheric humidity and throughfall precipitation. Mean annual temperature increases from dense evergreen forest to woodland and savanna, and air moisture decreases. A comparison of soil water reserves indicated that the substitution of dry evergreen forest by woodland produces a clear decrease of moisture in the upper 2 m of the soil (64 mm or 5% of the mean annual rainfall: Malaisse & Kapinga, 1987). Potential evaporation (Penman's formula) in the miombo varies between 100 and 200 mm per month with a total of 1600 mm y^{-1}.

Soil moisture distribution and flux are also affected by termite activity (see below). Abandoned termitaria, which often cover large areas, have a drier cone in the center of the mound, display a large evaporation surface and hence create more xeric conditions (Schmitz, 1971). On another hand, termite mounds often occur in seasonally waterlogged savannas and enable trees to establish. Under favorable circumstances, trees may progressively dry out the soil, perpetuate themselves over a weak grass cover and create some type of woodland (Tinley, 1982). Overall, termite activity in woodlands and dry forests seems beneficial to the soil water status. Below ground, termite chambers and galleries increase porosity and water circulation. Above ground, surface workings by termites appear to be a major factor in loosening up the soil surface, thus reducing erosion and favoring water infiltration (Trapnell *et al.*, 1976). These effects are particularly important on plots burned late in the dry season, where the soil is exposed to the full force of torrential rains, and is liable to compaction and development of sheet erosion.

Soil moisture regime and vertical distribution appear to be major determinants of vegetation structure in the savanna biome (Frost *et al.* 1986, and references therein). Following Walter's hypotheses (1971), Walker *et al.* (1981) modeled the savanna tree/grass dynamics as a function of soil water availability for both plant types, and showed how trees may dominate and even eliminate grasses. In particular, this occurs when water percolates rapidly to the deeper horizons, in stony soils or under a lateritic hardpan: (1) on deep soils where water is available only in the deeper horizons for much of the year, (2) on rocky escarpments, and (3) over lateritic hardpans where water either is trapped underneath

Table 4.1. *Climatic parameters of savannas, woodlands and dry forests in southeastern Zaïre*

Climatic parameter	Savanna	Woodland	Dry forest
Mean annual temperature (°C)	22.1	20.6	19.2
Mean daily range of temperature (°C)	20.8	16.5	10.4
Solar radiation at 1.3 m height (%)	100	26.8	2.3
Throughfall percentage of total rainfall	100	78.8	57.7
Mean annual relative humidity (%)	64.0	71.8	81.7

Source: Freson *et al.* (1974).

and only accessible to tree roots or stagnates in the shallow upper horizon and is preferentially absorbed by established trees.

Structure and function

Floristic composition

From a comparison between the Sudanian and Zambezian woodlands and dry forests, White (1965) recorded 426 tree species from West Africa and Zambia, belonging to 155 genera, of which 79 (51%) are common to both regions. Fourteen West African species are absent from Zambia, whereas 62 of the Zambezian genera do not occur in West Africa. As stated earlier, forests from the southern region are richer in species than their counterparts from the northern region. In the Zambezian region, Malaisse (1979) recorded 480 species for the miombo woodland, whereas Devineau (1985) found fewer than 100 species in the *Isoberlinia* woodland of the Sudanian region.

Table 4.2 shows that southeastern Zaïre woodland, as compared to savanna and dry forest, appears to be the richest vegetation type, both in grass species (112) and in tree species (72) (for intercontinental comparisons, see Gentry, Chapter 7). In West Africa, woodlands are much poorer than dry forests, sometimes even poorer than savannas themselves.

Habitat heterogeneity significantly increases tree diversity, primarily in dry forests but also in woodlands and savannas. For example, large termitaria have an important effect on the community structure, in terms of tree distribution, density and diversity. This has often been found in West Africa (Rose-Innes, 1977), East Africa (Lind & Morrison, 1974) and southern Africa (Malaisse, 1978). In southern Shaba (Zaïre), these

Table 4.2. *Vegetation characteristics of savannas, woodlands and dry forests in southeastern Zaïre*

Vegetation characteristics	Savanna	Woodland	Dry forest
Grass species	68	112	11
Tree species	37	72	67
including termitaria	74	112	95
Tree density (ha^{-1})	340	570	1465
including termitaria	385	662	1525
Height (m)	1–5	14–18	18–22
Canopy cover (%)	2–20	70–90	100
Basal area (m^2 ha^{-1})	1–5	15–25	30–40
Leaf area index	?	0.8–3.2	2.6–5.0
Tree biomass (Mg ha^{-1})	10	150	320
Grass biomass (Mg ha^{-1})	4.2	2.2	trace

Sources: Freson *et al.* (1974); Malaisse (1978).

mounds produce islets of xerophilous and eutrophic plants within the miombo woodland (Malaisse & Anastassiou-Socquet, 1977). Xerophily is related to the high clay content of mounds, and to their low water infiltration rate (runoff is enhanced by their conical shape). Eutrophy is related to the higher pH and total exchangeable bases of mound soil. Termitaria increase tree species richness by 40–100%, but tree density by less than 20% (or, only a few per cent in dry forests). However, shrub density may be considerably higher on termite mounds: some plant species, usually absent from the surrounding woodland, often constitute impenetrable thickets of erect and scandent shrubs dominated by one or a few emergent trees (Wild, 1952).

Phenology

The main phenological traits are influenced by rainfall and temperatures. Due to the elevation, there are five phenological seasons in the Zambezian plateaus (Malaisse, 1974): May–July (cold dry season), August–September (hot dry season), October–November (early rainy season), December–February (rainy season) and March–April (late rainy season). At all times of year there are some plants in flower, and the different strata display their own rhythm. The maximum leaf fall for trees and shrubs is in the hot dry season. For the herbaceous layer, there are two peaks: the main one in the middle of the rainy season (hemicryptophytes), and the second in September (geophytes). Fruiting takes place mainly in the hot dry season. Under hotter climates in the Sudanian region, the deciduous period is longer and more uniform for all species. It extends from November to March when there is 85–90% defoliation (Devineau, 1982).

In savannas, fire occurs with the dry season and, due to its intensity (high fuel load of grasses), renders phenology even more marked by defoliating all trees which have not yet shed their leaves. In woodlands, relatively weak fires burn the short grass cover every year; they have a much lower and direct impact on tree defoliation which mostly occurs only late in the dry season (Werger & Coetzee, 1978; Sanford, 1982). When dry forests are not accidentally (in exceptionally dry years) or deliberately burned, fire may creep in and consume the thin leaf litter layer on the ground, without noticeable effect on the vegetation.

Biomass and primary production

The main characteristics of the savanna, woodland and dry forest vegetation in southeastern Zaïre are shown in Table 4.2. Depending on the parameter considered, a woodland is closer to a savanna or to a forest. Some parameters, such as biomass and leaf area index, increase gradually from savanna to woodland to dry forest. Some others, such as canopy height and cover, show a marked increase from savanna to woodland, and a lesser increase from woodland to dry forest. Tree density increases slightly from savanna to woodland, but almost triples from woodland to dry forest: woodlands are open formations, made of a low number of large trees; dry forests are closed formations, made of a high number of more slender trees. Grass biomass shows the opposite trend: abundant in the savanna, it becomes negligible in the dry forest.

Aboveground grass and tree biomass of dry forests and woodlands, as well as litter production are compared in Table 4.3 for the northern and southern African regions. Tree biomass increases and grass biomass decreases in dry forests as compared to woodlands. For both vegetation types, the northern ones appear to have a lower aboveground tree biomass and litter layer.

The large differences in production components among savannas, woodlands, dry forests and rain forests are indicated in Table 4.4. There are considerable variations among years in these components; for example, fruit production in the miombo ranged from 0.16 Mg ha^{-1} in 1968 to 2 Mg ha^{-1} in 1972 (Malaisse, 1979). Also, in the Zambezian region there is a high correlation between annual rainfall and grass production (Freson, 1973).

Table 4.3. *Biomass characteristics and decomposition rates of northern and southern African dry forests and woodlands. All values in Mg ha^{-1} y^{-1}*

	Dry forest		Woodlands		
	southern	northern	southern	northern	northern (*Acacia*)
Aboveground herb biomass[a]	trace	trace	3.0	5.1	0.9–2.4
Aboveground tree biomass	320	140	150	100	?
Litter					
leaves	3.9–4.6	6.9	1.4–3.4	1.4	0.9–1.1
fruits	0.7–1.5	5.0	0.1–0.6	0.07	0.3–0.5
wood	2.9–3.5	1.7	0.7–1.0	0.2	0.2–0.7
total	7.5–9.6	13.6	2.2–5.0	1.7	1.4–2.3
Leaf litter on soil [a]	7.9	3.7	4.4	2.3	?
Decomposition rate of litter	0.5	1.8	0.5	0.6	?

[a] Maximum.
Sources: Devineau (1982, 1985); Fournier, Hoffman & Devineau (1982); Freson (1973); Malaisse (1978).

Table 4.4. *Biomass and productivity of savannas, dry forests and woodlands, and rain forests. All values in Mg ha^{-1} y^{-1}*

Components	Savannas	Dry forests and woodlands	Rain forests
Herb biomass[a]			
Aboveground	4–10	trace–5	trace
Belowground	9–25	?	trace
Tree biomass			
Aboveground	6–60	100–320	250–500
Belowground	3–30	?	50
Herb productivity			
Aboveground	5–15	?	trace
Belowground	7–20	?	trace
Litterfall			
Leaves	0.6–2.5	1–7	6–9
Fruits	?	0.1–1.5	0.5–1.2
Wood	?	0.2–3.5	1.1–2.6
Tree productivity			
Aboveground	0.5–6.3	?	13–20
Belowground	0.05–0.5	?	0.5–2.5

[a] Maximum.

Abandoned termite mounds have a significant impact on plant production (Menaut *et al.*, 1985). They reduce primary productivity when water availability is low (effect of higher clay content), and stimulate primary productivity when water availability is high (effect of higher nutrient content).

Secondary production and soil fauna

In comparison with the savanna, the epigeous fauna under the sparser herbaceous layer in the miombo is characterized by a higher mean population density (10.5 vs 6.9 individuals m^{-2}) and a higher biomass (0.95 vs 0.42 g fresh mass m^{-2}: Freson *et al.*, 1974). This seems to be related to the detrimental effects of fire in savanna.

The main characteristics of the soil fauna along the gradient savanna–woodland–dry forest in southeastern Zaïre are shown in Table 4.5. The pedofauna exhibits a decrease in average abundance and/or biomass from dry forest to woodland and savanna for most animal groups, with the exception of Oligochaeta (excluded from the dry forest by rapid drying of the soil) and some termite groups, like the epigeous humivorous *Cubitermes* (favored in savanna by a higher soil quality). The macroarthropods (other than termites) show a marked decrease in the woodland compared to either savanna or dry forest. Deforestation induces a progressive substitution of the large-mound epigeous colonies of fungus-growing termites (*Macrotermes falciger*) by humivorous termites (*Cubitermes*), as well as an increase of the humivorous hypogeous termites (*Anoplotermes, Crenetermes, Ophiotermes*). The large mounds scattered in the savanna are then left by their original builders and replaced by other species. In the dry forest, hypogeous termites increase their number and biomass.

The humivorous epigeous mounds apparently are very sensitive to the fire regime of the ecosystem. Trapnell *et al.* (1976) found in Zambia that the density of *Cubitermes* varied from 160 mounds per 0.4 ha under complete protection to 120 under early burning to 80 under late burning. Late burning has a negative effect on the topsoil because low vegetative cover allows a heavy impact by rain drops, resulting in soil compaction, which in turn strongly decreases termite activity.

In Africa, the abundance of large termitaria is often a main characteristic of dry forests and woodlands. In the miombo ecosystem, *Macrotermes falciger* mounds are about 8 m tall and 14–15 m wide. Malaisse (1978) counted 2.7–4.9 termitaria per hectare, with an approximate volume of

Table 4.5. *Biomass of some soil fauna of savannas, woodlands and dry forests in southeastern Zaïre and in Ivory Coast. All values in g fresh weight m*$^{-2}$

Soil fauna	Savanna	Woodland	Dry forest
Southeastern Zaïre			
Oligochaeta	30.3	10.8	9.9
Macroarthropods	11.3	3.5	13.3
Hypogeous termites	4.9	6.4	10.3
Ivory Coast			
Cubitermes (small mounds)	2.7	1.3	0.1
Macrotermes (large mounds)	trace	4.8	9.5

Sources: Lepage (1984); Goffinet (1973).

2200 m^3 ha^{-1} above the ground, and representing *c.* 6% of the land surface. Table 4.6 summarizes the importance of large termitaria in southeastern Zaïre, including the density of trees growing on the mounds.

Plant matter recycling

Decomposition rates can be estimated as the ratio between annual litterfall and maximum litter accumulation (Table 4.3). These estimates show no remarkable difference between woodlands in the northern and southern regions, or between these and the southern dry forests. In contrast, dry forest in the northern region appears to have a much higher decomposition rate. However, such values have to be considered with the greatest care due to the scarcity of reliable quantitative information on these ecosystems, their spatial heterogeneity and their interannual variability (Malaisse, 1979).

Apart from fire, the factors in litter disappearance may be grouped into two main categories (Freson *et al.*, 1974): soil microorganisms and lignivorous termites. These differ in capacity as well as in spatial and temporal distribution. Microorganisms show great differences in activity according to the chemical composition of tree litter. Biological activity is markedly attenuated during the dry season, and this effect is greater where the tree stratum is sparser: the ratio of rainy to dry season activity is 1.5 in dry evergreen forest, 3 in woodland, and 4 in savanna (Goffinet, 1973). The lignivorous termites, especially the fungus-growing group (*Macrotermes falciger, Odontotermes* species, *Pseudoacanthotermes spiniger, Microtermes* species), are the most important factor of litter

Table 4.6. *Characteristics of large termitaria* (Macrotermes) *of savannas, woodlands and dry forests in southern Zaïre. All values* ha^{-1}

Characteristic	Savanna	Woodland	Dry forest
Density	2.5	3.3	4.7
Basal area (m^2)	400	566	899
Aboveground volume (m^3)	387	831	1133
Tree density on mounds	1456	2202	2165

Source: Malaisse (1985).

breakdown. Their activity roughly follows litter production, but with a maximal intensity during the months of seasonal transition.

Soil mineral and organic status

Trapnell *et al.* (1976), working in a *Brachystegia* woodland in Zambia, noted that the mineral status of the topsoil is inversely related to the biomass on a plot, which depends on the burning regime. Plots protected from fire have the greatest tree biomass, but this biomass also holds a very large proportion of the mineral nutrients of the ecosystem. The fallen litter is largely consumed by termites and the bases are apparently concentrated in the termitaria. The soil elsewhere has scant input and remains in a state of extreme mineral deficiency, especially in Ca and Mg as shown in Table 4.7. With early burning, some proportion of the bases are still held in the canopy and some are consumed by termites in tree litter, but bases from the litter and ground layer are returned to the soil as ash. In contrast, under continuous late burning the whole supply of bases, from the canopy and grass layers, returns to the soil while litter consumption by termites is largely prevented. Soil organic matter tends to decrease in protected plots, probably in response to tree development (Menaut *et al.*, 1993). Nitrogen and cation exchange capacity seem to be insensitive to burning, while the status of Ca and Mg, and P to a lower degree, is improved on burned plots.

Whatever the burning regime, the topsoil is very nutrient deficient, except in termite mounds (e.g. C and N are 3–4 times richer) as a result of litter gathering and cementing with organic materials. Although termitaria represent less than 10% of the land surface, they contain 14.5% of the C, 16.8% of the total N, as much as 74.8% of the exchangeable Ca, 47.0% of the exchangeable Mg and 61.6% of the exchangeable K

Table 4.7. *Effect of burning regime on the mineral and organic status of the topsoil (0–15 cm depth) of a Zambian woodland*

	Burning regime		
Soil characteristic	Early	Late	No burning
pH	6.2	6.0	5.7
Total carbon (%)	0.96	0.98	0.85
Total nitrogen (%)	0.06	0.06	0.06
Calcium (meq)	0.45	0.73	0.21
Magnesium (meq)	0.48	0.41	0.28
Potassium (meq)	0.11	0.09	0.09
Cation exchange capacity (meq)	5.5	5.4	5.3
Phosphorus (ppm)	5.7	5.8	4.4

Source: Trapnell *et al.* (1976).

(calculated from data in Malaisse, 1978; Table 4.8). Carbonate accumulation in the mound has been repeatedly observed (Watson, 1962; Aloni, 1978). Each mound can contain up to 500 kg of carbonates. The slightly acidic soils of the woodlands become neutral to basic in the termite mounds.

Dynamics

Fire

Spatial and temporal variability of fire induces a large heterogeneity in all three burned ecosystems. Sensitivity to fire increases from savanna to woodland to dry forest. Undoubtedly, through regressive succession, dry forests turn into increasingly open types (not necessarily woodlands), and subsequently to savannas (Freson *et al.*, 1974).

Woodland is hardly affected by fire when burning takes place early in the dry season. Repeated late burnings kill most seedlings and saplings, and even affect mature trees. In Zambia, annual mortality for trees was 0.38% in fire protected plots, 0.64% in early burned, and 1.58% in late burned plots (Trapnell, 1959). However, under the usual fire regime, miombo can be considered a fire-maintained type of vegetation. Exclusion of fire may or may not modify the structure of the woodland. In Sudanian *Isoberlinia* woodlands, Keay (1959b) observed only a slight thickening of the canopy and understory, and opening of the herbaceous cover. In Zambezian woodlands, Trapnell (1959) found that the canopy

Table 4.8. *Nutrient content of soils and large termitaria in dry forests of southeastern Zaïre*

	C	N	Ca	Mg	K
	(Mg ha^{-1})		(10^6 meq 100 g^{-1})		
'Control soil'[a]	47.2	4.3	13.1	20.5	7.4
Termitaria[b]	8.0	0.9	39.0	18.2	11.8
Total	55.2	5.2	52.1	38.7	19.2
Percentage in termitaria	14.5	16.8	74.8	47.0	61.6

[a] 0–120 cm depth, 91% of the area.
[b] 1133 m^3, 9% of the area.
Source: Malaisse (1978).

remained unaltered after 20 years of protection from fire. White (1968) showed that it took almost 30 years for the dominant woodland trees to die, their saplings to be outcompeted by evergreen thicket, and for the herbaceous stratum to disappear. With long-term fire exclusion, a dense dry forest establishes (Schmitz, 1950).

Apart from creeping ground fires which consume the tree leaf litter and are not very damaging, dense dry forest normally does not experience fire. However, fires may penetrate a forest from adjacent savanna in the late dry season when the shrubby understory is no longer green (Schmitz, 1962). With regular fires, dry evergreen forest vegetation soon changes into woodland (Freson *et al.*, 1974; Werger & Coetzee, 1978).

It must be emphasized that the so-called 'annual' fires which run over African savannas and woodlands, burn annually only *c.* 60% of the Sudanian region (Menaut *et al.*, 1991). Large areas remain unburnt for one to several years, enabling some tree seedlings to reach a strength sufficient to resist fire. But protection against fire is another type of human disturbance. 'Natural' fires would be less frequent, perhaps once every 5–10 years, rather than annually. Fire in dense vegetation attained after long-term protection can be very destructive. Absolute protection against fire is very difficult to maintain, however, and fire early in vegetation development may have the same effect as protection, as far as physiognomy and floristic composition are concerned. Early fires eliminate only the weaker stems and thus reduce competition for the remaining individuals, as in woodlands (Delvaux, 1958).

It is commonly believed that protecting savannas and woodlands from fire should lead to the development of dry forest. Indeed, savannas protected from fire turn into dense woody formations (10–12 m tall) with a sparse grassy ground cover. For some authors, this corresponds to a successional stage towards woodland and/or dry forest (Schmitz, 1950; White, 1968). In some experiments, even after 20 years of protection, saplings are still suppressed by dominants (Keay, 1959b). Experiments have not lasted long enough to produce definite evidence, and are too few or conducted in too particular conditions to provide general conclusions.

Regeneration capacity

Regeneration capacity in woodland is strongly related to the fire regime, as suggested above. After fire, vegetative regeneration from root suckers and coppice shoots is common in both Sudanian (Aubréville, 1950) and miombo woodlands (Strang, 1966). Very few seedlings survive as they cannot withstand fire and the rapid fluctuations between cool, moist conditions and hot, dry ones (Strang, 1966). Sapling growth, and hence resistance to fire, are correlated with available soil moisture (Jeffers & Boaler, 1966). Strang (1966) also showed that seedlings of *Pterocarpus angolensis* which escaped fire had developed an extensive root system to compete for water. Seedling shoots die back in the dry season (Boaler, 1966) for seven years before the sapling stage (Groome, 1955). Until then, reserves are stored in the root system (Groome, Less & Wigg, 1957). When not submitted to fire or to drier conditions, dry forests regenerate well.

Secondary succession after clear-cutting and cultivation has been studied in Zimbabwe by Strang (1974). There is an initial rapid increase in basal area of woody vegetation up to about 50 years. Tree density increases for 20 years and then decreases. The woody understory opens and the remaining dominants are fewer but larger. The herbaceous layer decreases for 20 years and then remains constant. The main effect of fire is, on a regional basis, to maintain a higher ratio of grass to woody vegetation than would otherwise exist, but does not prevent woodland regeneration. In Zambia, abandoned agricultural lands may be invaded by thickets under which *Uapaca* species may establish and eventually outcompete the understory shrubs which formed the thicket. Depending on soil water availability and fire occurrence, these *Uapaca* formations may turn into either *Brachystegia–Julbernardia* woodlands or into *Marquesia* dry evergreen forests (Lawton, 1978).

Role of termites

Termite activity is a striking feature of woodlands and dry forests, affecting soil characteristics (structure, texture, and the status of water, organic matter and minerals) and hence vegetation productivity, structure and dynamics. Spatial heterogeneity within an ecosystem may appear to be exacerbated by termites, but their effects are also pervasive.

From deep in the soil, termites collect clay particles which have leached through the coarser-textured levels. They combine these particles with organic matter and nutrients to build their nests, in which the plant litter collected all over the ecosystem is stored and processed. When abandoned, the mounds are colonized by a dense and diverse tree community which does not burn (due to the exclusion of the grass cover) and which may serve as sources for future colonization of favorable spots. Colonization of mounds by trees is more rapid if the plant formation is dense and humid (e.g. dry forest). Colonization is slower in woodlands, and still slower in savannas where old levelled mounds may remain as barren patches for a long time. The higher clay content renders mounds more humid than the surrounding soil in humid environmental conditions and more arid in dry conditions.

Colonized or not by trees, the mounds are progressively dismantled, redistributing the organic matter and nutrients which sooner or later are stored by the woody plants. Clays are also redistributed on the soil surface by erosion and in the galleries that termites permanently build to collect litter. The continuous process of lifting and spreading of clay particles at the ground surface has led to termite-formed soils, characteristic of the area covered by woodlands and dry forests (Sys, 1955; Menaut *et al.*, 1985). Exclusion of termites might well lead to impoverished, weathered soils, and prevent tree regeneration.

The interaction of termites and fire is another example of the role played by termites in spatial heterogeneity and nutrient protection. In unburned plots (or dry forests), termites are very abundant and consume most of the litter which they concentrate in a small number of large termitaria. Nutrients are slowly and irregularly released over the soil surface. Such systems are highly heterogeneous. In early burned plots (or woodlands), termites are less abundant and only partly consume plant litter which partly burns. A substantial amount of nutrients returns in ashes to the soil. These systems display a lower spatial heterogeneity. In late burned plots (or savannas), termites are even less numerous. Most of the litter is burned and a major part of the nutrients returns in ashes to the soil. A higher number of smaller termitaria retains a minor part of the nutrients which are more rapidly and regularly redistributed to the soil surface.

Climax vs dynamics

There is no general agreement on the pre-existence of 'climax' woodlands or dry forests in savannas. Changes in supposed climax vegetation patterns (Tinley, 1982; Cole, 1986) may occur with environmental changes in the long term (climatic or geomorphologic change) and medium term (modification of the soil water balance), but are seldom documented. In the short term (5–100 years) there may be distinct types of factors which concurrently influence vegetation dynamics (Walker, 1981): stochastic, successional and interactive.

Stochastic events might occur a few to several times per century. For example, major droughts would allow fire to penetrate deeply into dense dry forests, while also allowing hydromorphic grassland to be invaded by woody vegetation, albeit temporarily (Tinley, 1982). Insect eruptions could also be considered stochastic events. There are examples of large areas (thousands of square kilometers) of *Colophospermum mopane* and *Burkea africana* woodlands being totally defoliated by moth larvae (Walker, 1981).

Successional change results from fire exclusion or secondary succession after clear cutting and cultivation. However, in order to reach a 'climax' woodland or dry forest, 'it is necessary that all of the vegetation stages claimed to be part of the succession can occur on any one site', which is far from being reliably documented.

Interactive change assumes interactions and feedbacks among factors, exemplified here by the fire–rainfall–termite interactions. Figure 4.6 summarizes the hypothesis about the role of termitaria in the woodland ecosystem, through their effect on soil properties, litter breakdown, landscape heterogeneity and tree dynamics. Other factors, such as herbivory, might be of major importance (Walker, 1981; Werger, 1983, and references therein). Some woodlands are heavily browsed, and even destroyed: a cycle lasting for *c.* 200 years of animal increase/tree destruction and animal decrease/tree resurgence might prevent any attainable natural equilibrium (Caughley, 1976).

As stated by Walker (1981), succession in savannas 'is multidirectional, occurring over different time scales, and is confounded by equally important non-directional changes of other kinds'. The savanna floristic

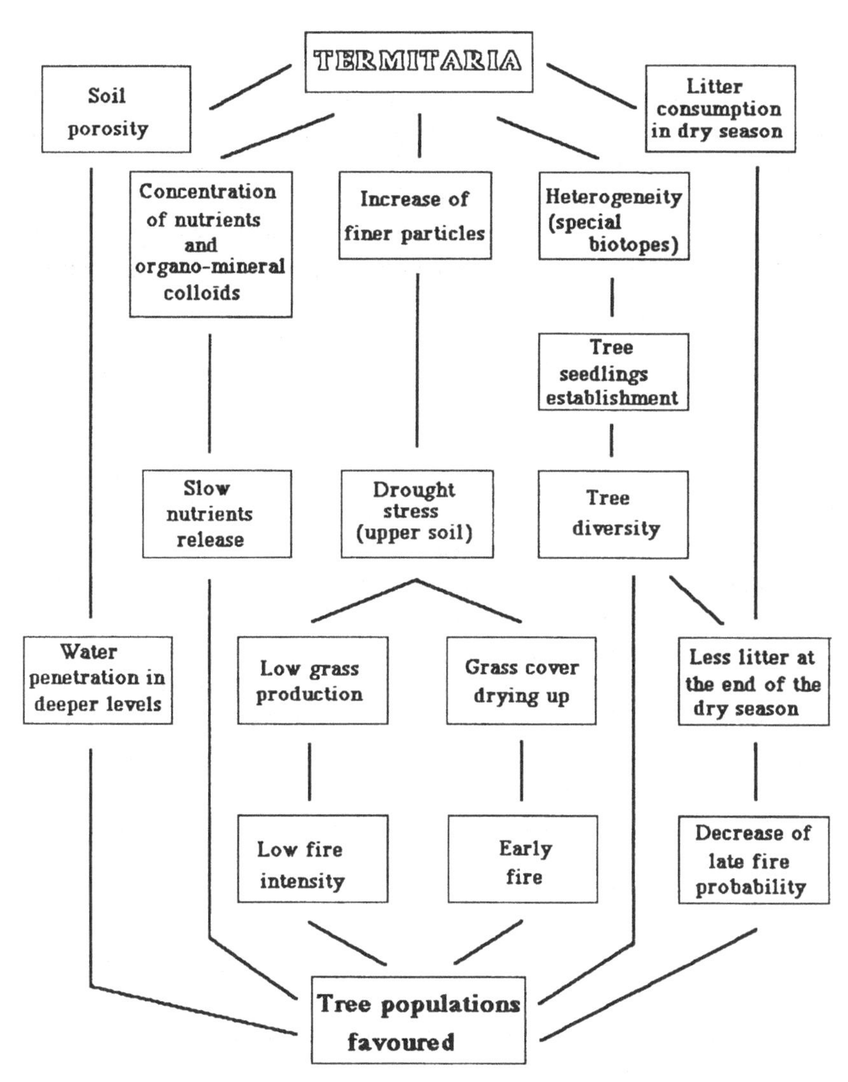

Figure 4.6. General hypothesis on the role of termitaria in woodland ecosystems.

pool, which includes the woodland and dry forest flora, is disseminated all over the biome. The grassy or woody component may survive and thrive here and there when conditions are favorable. Most probably, savannas, woodlands and dry forests have long occurred together, in proportions varying with time and region according to local conditions, climate and human activity.

Summary

African 'dry forests' vary from open woodlands made of a stratum of deciduous trees, dominating a grass layer, to closed species-rich forests displaying several woody strata excluding grasses. They belong to the savanna biome and still cover extensive areas in Africa, especially south of the Equator, and are separated from rain forests by a large savanna belt. Woodlands are often very extensive whereas dense dry forests often occur as smaller patches. Both are limited to mesic climatic conditions, although dense dry forests tend to develop in more humid sites than open woodlands, and woodlands can occur where the dry season exceeds six months. Woodlands often occur on lateritic soils, but both types may be found on a large variety of soils. At the local level, woodlands and forests tend to occur in conditions of lower nutrient availability than grassy formations.

Southern formations are markedly richer in species at both the local and regional scales, and display more complex structure and phenology. Northern woodlands are even poorer in species and are structurally simpler than adjacent savannas; their phenology is uniformly bound to the dry season. Dry forests and especially woodlands have low productivity, with most of the nutrients stored in the biomass. In the same climates, savannas have higher productivity, mostly due to the grass components.

Woodlands are fire-prone ecosystems. They regenerate mostly from root suckers and coppice shoots, and little from seedlings which suffer from fire and drought. Protection from fire leads to a slight thickening of the understory and loosening of the grass cover, but transformation into forest also requires favorable moisture over a very long period. Dense dry forests do not normally burn, but fierce fires can change them into woodlands and savannas. Clearing and burning by humans degrades both woodland and forest to savanna. Even before the advent of humans in Africa fires probably limited the development of woody communities.

The functioning and dynamics of woodlands and dry forests are strongly related to termite activity. Termites constantly move the finer clay particles from deeper horizons to the surface. Termite mounds are very abundant and when abandoned are colonized by a dense and diverse tree community. While the mounds disintegrate, redistributing clay particles, organic matter and nutrients to the surrounding surface, trees also invade the surroundings. The processes of lifting, accumulation and spreading lead to termite-formed soils, characteristic of woodland and dry forest areas.

Dry forests and woodlands cannot be considered climax formations. In the savanna biome, dynamic processes occur over different spatial and temporal scales in relation to the heterogeneity of environmental conditions. Such processes may be successional (temporary exclusion of fire), stochastic (severe drought, pest eruption) and interactive (termites, herbivores) and lead to a variety of vegetation structures varying with space and time.

Acknowledgements

Our deepest gratitude goes to S. Bullock and H. Mooney for their great help and patience in editing the manuscript.

References

Aloni, J. (1978). Le rôle des termites dans la mise en place des sols de plateau dans le Shaba méridional. *Géo-Eco-Trop* **1**: 81–93.

Aubréville, A. (1949). *Climats, Forêts et Désertification de l'Afrique Tropicale*. Larose, Paris.

Aubréville. A. (1950). *Flore Forestière Soudano-Guinéenne*. Editions géographiques, maritimes et coloniales, Paris.

Aubréville, A. (1961). Savanisation tropicale et glaciations quaternaires. *Adansonia* **2**: 16–84.

Aubréville, A. (1965). Principes d'une systématique des formations végétales tropicales. *Adansonia* **5**: 153–96.

Aubréville, A., Duvigneaud, P., Hoyle, A. C., Keay, R. W. J., Mendonça, F. A. & Pichi-Sermolli, R. (1958). *Vegetation Map of Africa*. Oxford University Press, Oxford.

Bell, R. H. V. (1982). The effect of soil nutrient availability on community structure in African ecosystems. In *Ecology of Tropical Savannas*, ed. B. J. Huntley and B. H. Walker, pp. 193–216. Springer-Verlag, Berlin.

Bellier, L., Gillon, D., Gillon, Y., Guillaumet, J. L. & Perraud, A. (1968). Recherches sur l'origine d'une savane incluse dans le bloc forestier du Bas-Cavally (Côte d'Ivoire) par l'étude des sols et de la biocénose. *Cahiers de l'ORSTOM, Série Biologie* **10**: 65–93.

Boaler, S. B. (1966). Ecology of a miombo site, Lupa North Forest Reserve, Tanzania. II. Plant communities and seasonal variation in the vegetation. *Journal of Ecology* **54**: 465–79.

Boughey, A. S. (1957a). The physiognomic delimitation of West African vegetation types. *Journal of the West African Science Association* **3**: 148–63.

Boughey, A. S. (1957b). The vegetation types of the Federation. *Proceedings and Transactions of the Rhodesian Science Association* **45**: 73–91.

Caughley, G. (1976). The elephant problem - an alternative hypothesis. *East African Wildlife Journal* **14**: 265–83.

Cole, M. (1986). *The Savannas: biogeography and geobotany*. Academic Press, London.

Delvaux, J. (1958). Effets mesurés des feux de brousse sur la forêt claire et les coupes blanc dans la région d'Elisabethville. *Bulletin d'Agriculture du Congo Belge* **44**: 683–714.
Devineau, J. L. (1982). Etude pondérale des litières d'arbres dans deux types de forêts tropophiles en Côte d'Ivoire. *Annales de l'Université d'Abidjan, Série E* **15**: 27–62.
Devineau, J. L. (1985). Structure et dynamique de quelques forêts tropophiles de l'ouest africain (Côte d'Ivoire). *Travaux des Chercheurs de Lamto* **5**, 295 pp.
Ernst, W. (1971). Zur Ökologie der Miombo-Wälder. *Flora* **160**: 317–331.
Ernst, W. & Walker, B. H. (1973). Studies on the hydrature of trees in the Miombo woodland in south central Africa. *Journal of Ecology* **61**: 667–73.
Fanshawe, D. B. (1961). Evergreen forest relics in Northern Rhodesia. *Kirkia* **1**: 20–4.
Fournier, A., Hoffman, O. & Devineau, J. L. (1982). Variations de la phytomasse herbacée le long d'une toposéquence en savane soudano-guinéenne. Ouango-Fitini (Côte d'Ivoire). *Bulletin de l'IFAN, Série A* **44**: 71–7.
Freson, R. (1973). Aperçu de la biomasse et de la productivité de la strate herbacée du miombo de la Luiswishi. *Annales de l'Université d'Abidjan, Série E* **6**: 265–77.
Freson, R., Goffinet, G. & Malaisse, F. (1974). Ecological effects of the regressive succession Muhulu–Miombo–savannah in Upper Shaba (Zaïre). In *Proceedings of the First International Congress of Ecology*, pp. 365–71. PUDOC, Wageningen.
Frost, P., Medina, E., Menaut, J. C., Solbrig, O., Swift, M. & Walker, B. H. (1986). Responses of savannas to stress and disturbance. *Biology International* **10** (special issue): 1–82.
Germain, R. (1965). Les biotopes alluvionnaires herbeux et les savanes intercalaires du Congo équatorial. *Académie Royale des Sciences d'Outre- mer, Classe des Sciences Naturelles et Médicales* **15**, 399 pp.
Gillon, D. (1983). The fire problem in tropical savannas. In *Tropical Savannas*, ed. F. Bourlière, pp. 617–41. Elsevier, Amsterdam.
Goffinet, G. (1973). Synécologie comparée des mileux édaphiques de quatre écosystèmes caractéristiques du Haut-Shaba (Zaïre). Dissertation, Université de Liège, Liège.
Griffith, A. L. (1961). Les forêts claires sèches d'Afrique au sud du Sahara. *Unasylva* **15**: 10–21.
Groome, J. S. (1955). Muninge (*Pterocarpus angolensis* D.C.) in the western province of Tanganika. *East African Agriculture Journal* **21**.
Groome, J. S., Less, H. M. N. & Wigg, L. T. (1957). A summary of information on *Pterocarpus angolensis*. *Forest Abstracts* **18**: 3–8 and 153–62.
Holdridge, L. R. (1967). *Life Zone Ecology*. Tropical Science Center, San José.
Jaeger, P. (1956). Contribution l'étude des forêts reliques du Soudan occidental. *Bulletin de l'IFAN, Série A* **18**: 993–1053.
Jeffers, J. N. R. & Boaler, S. B. (1966). Ecology of a miombo site, Lupa North Forest Reserve, Tanzania. I. Weather and plant growth. *Journal of Ecology* **54**: 447–63.
Jones, E. W. (1963). The Cece Forest Reserve, northern Nigeria. *Journal of Ecology* **51**: 461–6.
Keay, R. W. J. (1959a). *Vegetation Map of Africa: descriptive memoir*. Oxford University Press, Oxford.
Keay, R. W. J. (1959b). Derived savanna - derived from what? *Bulletin de l'IFAN, Série A* **21**: 28–39.
Lawton, R. M. (1978). A study of the dynamic ecology of Zambian vegetation. *Journal of Ecology* **66**: 175–98.
Lepage, M. (1984). Distribution, density and evolution of *Macrotermes bellicosus* nests (Isoptera, Macrotermitinae) in the North-East of Ivory Coast. *Journal of Animal Ecology* **53**: 107–17.
Lind, E. M. & Morrison, M. E. S. (1974). *East African Vegetation*. Longman, London.
Malaisse, F. (1974). Phenology of the Zambezian woodland area with emphasis on the

Miombo ecosystem. In *Phenology and Seasonality Modeling*, ed. H. Lieth, pp. 269–86. Springer-Verlag, Berlin.
Malaisse, F. (1978). High termitaria. In *Biogeography and Ecology of Southern Africa*, ed. M.J.A. Werger and A.C. van Bruggen, pp. 1279–1300. Dr. W. Junk, The Hague.
Malaisse, F. (1979). L'écosystème miombo. In *Ecosystèmes Forestiers Tropicaux*, pp. 641–59. UNESCO, Paris.
Malaisse, F. (1984). Structure d'une forêt dense sèche zambézienne des environs de Lubumbashi (Zaïre). *Bulletin de la Société Royale de Botanique de Belgique* **117**: 428–58.
Malaisse, F. (1985). Comparison of the woody structure in a regressive Zambezian succession with emphasis on high termitaria vegetation (Luiswishi, Shaba, Zaïre). *Bulletin de la Société Royale de Botanique de Belgique* **118**: 244–65.
Malaisse, F. & Anastassiou-Socquet, F. (1977). Phytogéographie des hautes termitières du Shaba méridional (Zaïre). *Bulletin de la Société Royale de Belgique* **110**: 85–95.
Malaisse, F. & Kapinga, I. (1987). The influence of deforestation on the hydric balance of soils in the Lubumbashi environment (Shaba, Zaïre). *Bulletin de la Société Royale de Belgique* **119**: 161–78.
Menaut, J. C. (1983). The vegetation of African savannas. In *Tropical Savannas*, ed. F. Bourlière, pp. 109–49. Elsevier, Amsterdam.
Menaut, J. C., Barbault, R., Lavelle, P. & Lepage, M. (1985). African savannas: biological systems of humification and mineralization. In *Ecology and Management of the World's Savannas*, ed. J. C. Tothill and J. J. Mott, pp. 14–33. Australian Academy of Science, Canberra.
Menaut, J. C., Abbadie, L., Lavenu, F., Loudjani, P. & Podaire, A. (1991). Biomass burning in West African savannas. In *Global Biomass Burning*, ed. J. S. Levine, pp. 133–42. Massachusetts Institute of Technology Press, Cambridge.
Menaut, J. C., Abbadie, L. & Vitousek, P. (1993). Nutrient and organic matter dynamics in tropical ecosystems. In *Fire in the Environment: the ecological, atmospheric and climatic importance of vegetation fires*, ed. P. J. Crutzen and J. G. Goldammer, pp. 215–31. John Wiley and Sons, Chichester.
Monod, T. (1963). Après Yangambi (1956): notes de phytogéographie africaine. *Bulletin de l'IFAN, Série A* **24**: 594–657.
Rose-Innes, R. (1977). *Manual of Ghana Grasses*. Ministry of Overseas Development, Surbiton, UK.
Sanford, W. W. (1982). The effects of seasonal burning: a review.In *Nigerian Savanna*, ed. W. W. Sanford, H. M. Yefusu and J. S. O. Ayeni, pp.160–88. New Bussa.
Schmitz, A. (1950). Principaux types de végétation forestière dans le Haut-Katanga. *Comptes Rendus des Travaux et Congrès Scientifiques d'Elisabethville*, 30 pp.
Schmitz, A. (1962). Les muhulus du Haut-Katanga méridional. *Bulletin du Jardin Botanique d'Etat de Bruxelles* **32**: 221–91.
Schmitz, A. (1971). La végétation de la plaine de Lubumbashi (Haut Katanga). *Publication INEAC, Série Scientifique* **113**: 1–338.
Schnell, R. (1976–7). *Introduction la Phytogéographie des Pays Tropicaux. La Flore et la Végétation de l'Afrique*. Vol. 3 and 4. Gauthiers-Villars, Paris.
Sobey, D. G. (1978). *Anogeissus* groves on abandoned village sites in the Mole Game National Park, Ghana. *Biotropica* **10**: 87–99.
Strang, R. M. (1966). The spread and establishment of *Brachystegia spiciformis* Benth. and *Julbernardia globiflora* (Benth.) Troupin in the Rhodesian highveld. *Commonwealth Forestry Review* **45**: 253–6.
Strang, R. M. (1974). Some man-made changes in successional trends on the Rhodesian highveld. *Journal of Applied Ecology* **11**: 249–63.
Sys, C. (1955). L'importance des termites sur la construction des latosols de la région d'Elisabethville. *Sols Africains* **3**: 393–5.

Tinley, K. L. (1982). The influence of soil moisture balance on ecosystem patterns in South Africa. In *Ecology of Tropical Savannas*, ed. B. J. Huntley and B. H. Walker, pp. 175–92. Springer-Verlag, Berlin.

Trapnell, C. G. (1959). Ecological results of woodland burning experiments in Northern Rhodesia. *Journal of Ecology* **47**: 129–68.

Trapnell, C. G., Friend, M. T., Chamberlain, G. T. & Birch, H. F. (1976). The effect of fire and termites on a Zambian woodland soil. *Journal of Ecology* **64**: 577–88.

Van der Hammen, T. (1983). The paleoecology and paleogeography of tropical savannas. In *Tropical Savannas*, ed. F. Bourlière, pp. 19–35. Elsevier, Amsterdam.

Walker, B. H. (1981). Is succession a viable concept in African savanna ecosystems. In *Forest Succession: concepts and application*, ed. D. West, H. Shugart and D. Botkin, pp. 431–47. Springer–Verlag, New York.

Walker, B. H., Judwig, D., Holling, C. S. & Peterman, R. M. (1981). Stability of semi-arid grazing systems. *Journal of Ecology* **69**: 473–98.

Walter, H. (1971). *Ecology of Tropical and Subtropical Vegetation*. Oliver and Boyd, Edinburgh.

Watson, J. P. (1962). The soil below a termite mound. *Journal of Soil Science* **13**: 46–51.

Werger, M. J. A. (1983). Tropical grasslands, savannas, woodlands: natural and man-made. In *Man's Impact on Vegetation*, ed. W. Holzner, M. J. A. Werger and I. Ikusima, pp. 107–37. W. Junk, The Hague.

Werger, M. J. A. & Coetzee, B. J. (1978). The Sudano-Zambezian region. In *Biogeography and Ecology of Southern Africa*, ed. M. J. A. Werger, pp. 301- 462. W. Junk, The Hague.

White, F. (1965). The savanna woodlands of the Zambezian and Sudanian domains. An ecological and phytogeographical comparison. *Webbia* **19**: 651–81.

White, F. (1968). Zambia. *Acta Phytogeographica Suecica* **54**: 208–15.

White, F. (1983). *The Vegetation of Africa: a descriptive memoir to accompany the UNESCO/AETFAT/UNSO vegetation map of Africa*. UNESCO, Paris.

Wild, H. (1952). The vegetation of southern Rhodesia termitaria. *Rhodesia Agricultural Journal* **49**: 280–92.

Wild, H. (1968). Phytogeography of South Central Africa. *Kirkia* **6**: 197–222.

5

Dry forest ecosystems of Thailand

PHILIP W. RUNDEL & KANSRI BOONPRAGOB

Introduction

Dry tropical forest vegetation in Thailand and adjacent parts of Southeast Asia exhibits structures and ecological processes very different from those characteristic of neotropical dry forests. Unlike the Pacific coast of Central America and México where dry forests are virtually all deciduous, evergreen forest types are widespread in the dry forest climatic regime of Southeast Asia. Evergreen dry forests in areas with 1200–1500 mm y^{-1} precipitation and deciduous forest in areas with up to 2300 mm y^{-1} are surprising occurrences in relation to climate–vegetation seen in dry neotropical forests. Many of these regional differences can be attributed to the poor nutrient status and low water-holding capacity of the shallow and infertile latosols and lithosols which predominate in Southeast Asia. Many other factors – historical, abiotic and genetic – are no doubt involved as well. In this review, we provide a broad biogeographic survey of the geography and dry forest communities of Thailand as an introduction to this region.

The literature on forest vegetation in Southeast Asia has been largely unavailable in Western libraries, thus limiting ecological interest in this important region. International concerns over global climate change and loss of biodiversity, however, have led to renewed interest in the structure and function of forest vegetation in Thailand (Round, 1988; Elliot, Maxwell & Beaver, 1989) as elsewhere. Also, future studies contrasting paleotropical and neotropical dry forests will undoubtedly lead to new perspectives on old problems, and to a much better understanding of the interactive nature of nutrient availability and seasonality in soil moisture which lead to broad patterns of forest dominance by evergreen or deciduous species.

Physical environment

Physiography

The Kingdom of Thailand (Figure 5.1*A*) is centrally located in Southeast Asia, extending from 5° 40′ to 20° 30′ N and covering 513,115 km^2 (*c.* 50% larger than Venezuela). Much of Thailand is delineated by the drainage of the Chao Phraya River. The other major drainage systems dividing Southeast Asia are the Irawaddy of Myanmar, and the Mekong River which drains Laos, Cambodia, Vietnam and the Khorat Plateau region of eastern Thailand. The physiography of Thailand has been discussed in detail by Pendleton (1962) and Thiramongkol (1984).

The western cordillera along the Burmese border are rugged mountains cut by steep canyons and narrow valleys; only a few peaks exceed 1800 m elevation. The Northern Highlands are made up of north–south oriented hills and ridges, alternating with elongate and flat valleys. Large river valleys with alluvial soils are found around Chiang Mai, Chaing Rai, Lampang, Phrae and Nan but younger landscapes are common, exhibiting narrow, steep-sided valleys. Steep limestone ridges are present in several areas, as are volcanic plateaus.

The Central Highlands have a complex physiography, consisting of hills and strongly incised plateaus and peneplains. Steep limestone ridges are present over much of this region and form a characteristic element of the landscape.

Northeastern Thailand is dominated by the Khorat Plateau which covers nearly one-third of the country (Figure 5.1*A*). In contrast to the geotopically complex Northern and Central Highlands, the Khorat Plateau is relatively simple geologically and topographically. The plateau slopes gently to the southeast from an elevation of about 150–200 m along its western and northern margins to about 60 m elevation to the southeast. The region is drained by a tributary of the Mekong River.

The southeast coast region of Thailand is an area of numerous small ranges of hills, generally having a NW–SE orientation, which are a continuation of the Cardamon Mountains of southwestern Cambodia. Numerous small ranges of NE–SW trending hills are present, and the sinuous coastline is fringed with rocky forested islands.

The peninsula region is formed from a series of distinct ranges of hills and mountains. Few areas exceed 1000 m elevation (Figure 5.1*A*). Coastal terraces and plains are narrow along the west coast, with

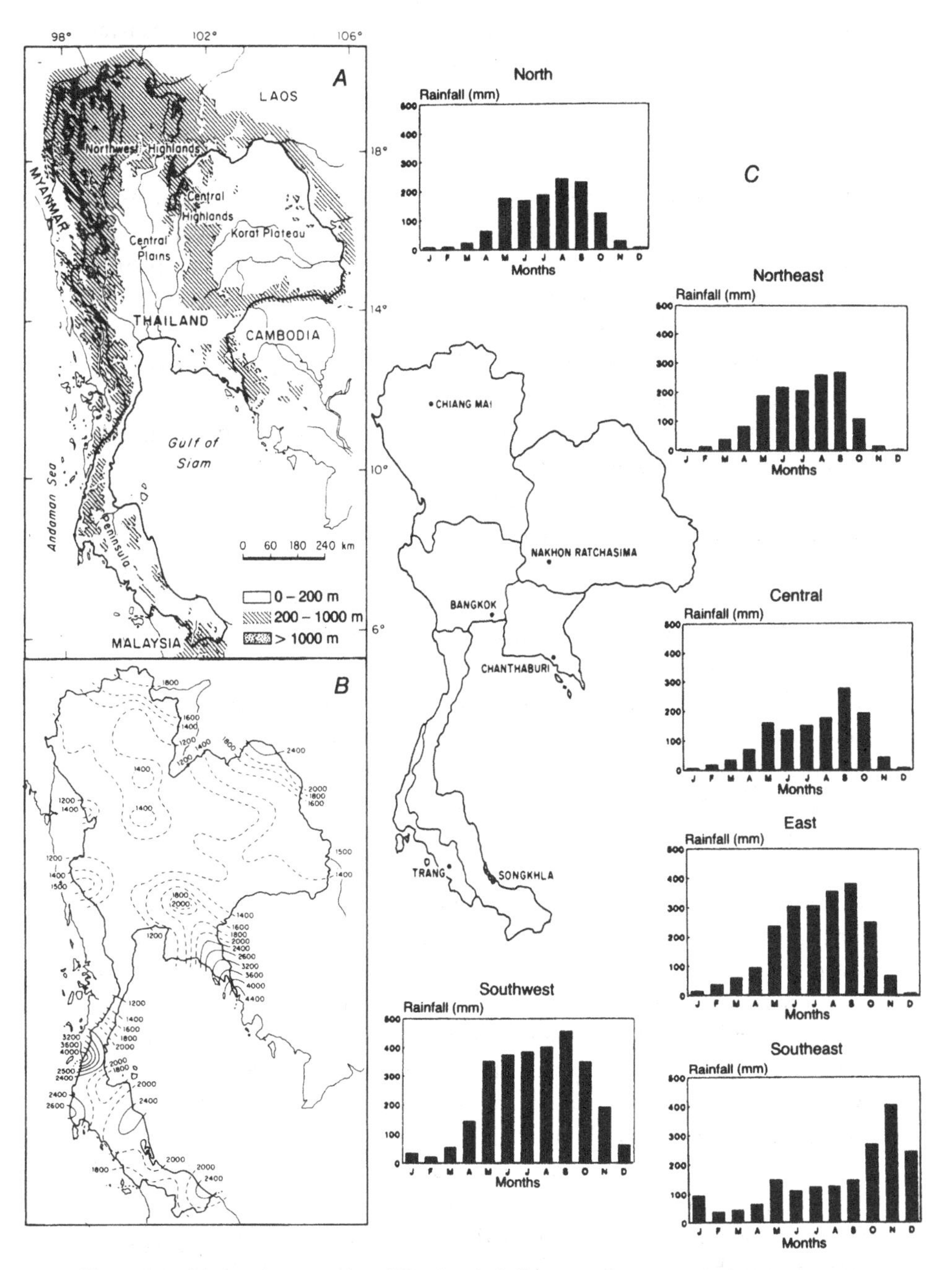

Figure 5.1. Physical geography of Thailand. *A*. Topography and major physiographic regions. *B*. Mean annual precipitation, 1951–75 (from Arbhabhirama *et al*., 1987). *C*. Seasonal distribution of rainfall (from Rundel, 1991).

mountains extending down to the sea in many places. Wide coastal terraces are characteristic of the east coast.

Climate

Seasonal monsoonal climates predominate over most of Southeast Asia, with wet conditions during the Northern Hemisphere summer and moderately dry conditions during the winter. This seasonality results from the dynamic movement of the Intertropical Convergence Zone (ITCZ), which separates the flow of air masses from the north and south. Low pressure belts associated with the ITCZ bring cloudiness and convective precipitation. From a January position of 5–10° S, the ITCZ moves steadily northward, passing peninsular Malaysia in April and May, and reaching its northernmost position at 20–25° N about August, before retreating again and passing over Malaysia in November and reaching its southern limit again in December or January.

The northerly position of the ITCZ in summer brings characteristic flows of the southwestern monsoon. From a high pressure center over the Indian Ocean flow is northeastward, bringing warm and humid air masses over much of Thailand, the northern coasts of Sumatra and Malaysia, and as far east as the Philippines. From a high pressure center over Australia, a flow of cool and dry air produces relatively arid conditions over much of southern Indonesia.

During the northern winter, high pressure over the Tibetan plateau influences Southeast Asia, while the ITCZ is toward its southern position. Winds from the north bring relatively cool and dry air over much of Thailand, northern Malaysia, and the northeastern coast of Sumatra. At the same time, northeasterly winds from the Pacific Ocean bring heavy precipitation to the east coast of Malaysia, peninsular Thailand, and the north coast of Sarawak. These air masses move below the tradewind inversion and thus do not penetrate to the southwestern sides of the major mountain systems. South of the Equator, generally northwestern winds bring heavy rains to the mountains in Java, Sulawesi, eastern Indonesia and the coast of Australia.

These seasonal patterns of air mass flow produce distinct rainfall regimes. A belt on both sides of the Equator (including most of Malaysia, Sumatra, Borneo and New Guinea) receives year-round precipitation. One or two short, relatively dry seasons associated with the passage of the ITCZ may occur in this zone, but rainfall is only reduced, not absent. Most of the mainland portions of Southeast Asia – Myanmar, Thailand,

Laos, Cambodia and Vietnam – have a seasonal monsoon climate (Gaussen, Legris & Blasco, 1967). Strongly seasonal climates with drought during the Northern Hemisphere summer are present in southern Indonesia and northern Australia. Transition areas between the perhumid and monsoon climate regions are present in peninsular Thailand and central portions of Indonesia.

Most of Thailand is classified under the Köppen system as a tropical monsoon climate (Aw), with high total rainfall, distinct wet and dry seasons, and the driest month having a mean precipitation of less than 6 mm (Figure 5.1*B*). However, topographic diversity has a strong influence on regional patterns of precipitation. Mean annual precipitation is about 1550 mm for Thailand overall. A rain forest climate (Af) prevails in the peninsular region, with annual precipitation commonly exceeding 2000 mm, and along the southeast coast where the mean may reach 4000 mm (Figure 5.1*B*).

Rainfall commonly peaks in August or September in northern Thailand and 1–2 months later in central and southern Thailand (Figure 5.1C). With the incursion of dry continental air from November to February, monthly precipitation drops to 10 mm or less in much of northern Thailand. Maximum temperatures in December are lower by 6–8 °C than in May and June in northern and central Thailand. As circulation continues to change, maximum temperatures peak at 35–39 °C in April, a month before declining as the monsoon brings more rain; cloudiness and other maritime influences keep temperatures lower in the south. By April, the entire country with the exception of the northeast is receiving at least 50 mm of precipitation monthly. During May, most parts of Thailand receive at least 100 mm of rainfall, while the western coast of the peninsular region and higher mountain areas have 200 mm. As the ITCZ reaches its northern limit in June and July, rainfall becomes more erratic. Distinctive dry periods are evident along the east coast of peninsular Thailand and in sheltered areas of the central plains. In August, higher precipitation levels return as the ITCZ moves southward again.

Soils

Soils in Thailand are mostly latosols and lithosols that are shallow and poor in available nutrients (Pendleton, 1962; Dudal & Moormann, 1964). The mountain areas of the west and north comprise extensive areas of undifferentiated latosols and lithosols, while latosols of greater depth extend through areas of more gentle topography in the peninsula

(Figure 5.2*A*). Sandy ferruginous latosols are extensive, covering much of the Khorat Plateau. These soils generally have a leached A horizon and clay accumulation (up to 40%) in the B horizon. Laterites are common, and indurated laterite horizons may occur at depth. These were used extensively in temple construction during the Khymer and early Thai empires (Pendleton, 1941, 1962). Water-holding capacity is poor and available nutrients are low. Iron concretions are common and iron compounds form insoluble complexes with phosphorus and other important elements (Pendleton & Montrakun, 1960).

Soils have been classified over 60% of Thailand, with nine soil orders represented (Arbhabhirama *et al.*, 1987). The remaining area is represented by unclassified soils of mountainous regions or complex slope structures. The dominant soil order is the ultisol group, formerly called red and yellow podzols, which covers two-thirds of the classified area. These reddish latosols are moderately acidic; cation exchange capacity is commonly low and phosphorus is bound in unavailable forms, particularly in soils with clayey upper horizons. Inceptisols are the second most important soil group, covering 10% of the country. Well-drained, non-volcanic inceptisols are widespread in Southeast Asia. Like the ultisols, these soils are generally reddish in color, moderately acidic, and

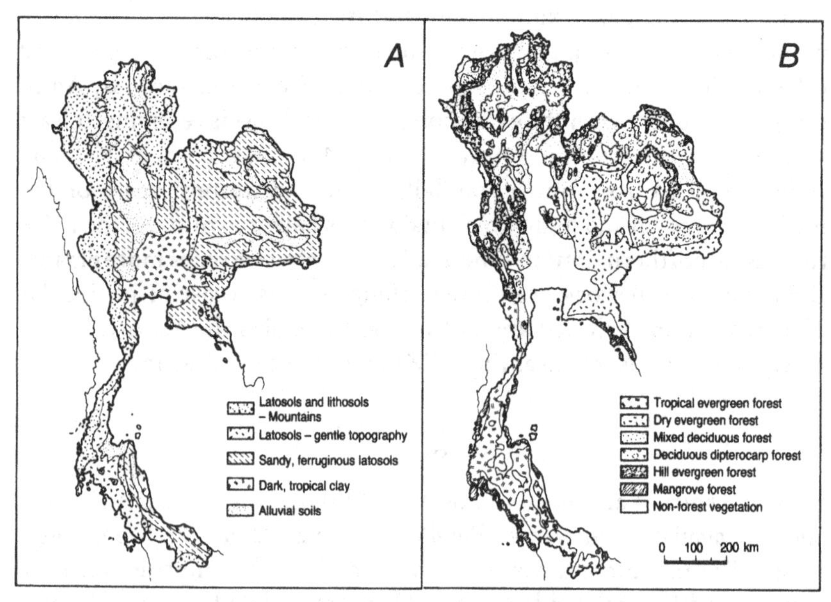

Figure 5.2. Geography of substrate and vegetation in Thailand. *A*. Major soil types (from Williams, 1967). *B*. Major forest types (from Ogino, 1976).

infertile. Moderately fertile soils cover only about 15% of the area of Thailand. Most widespread are alfisols (10%), largely in the Central Plains. These soils are commonly moderate to high in fertility, and now mostly support rice and other crops. Three other highly fertile soil types are also present: entisols (3%), largely in the form of fluvents along the major river valleys; and vertisols and mollisols (each about 1%). Spodosols, oxisols and histosols are also present but each cover only 0.1% or less of Thailand.

Soil structure appears to play a significant role in the delineation of many forest types in Thailand as discussed below (Figure 5.2*A*, *B*). The core areas of distribution for the dry dipterocarp forests and savannas across Southeast Asia are the sandy ferruginous latosols on plateaus and the shallow lithosols derived *in situ* over bedrock in areas where the laterite mantle has been removed (Cole, 1986). Soil catenas are a significant factor in explaining transitions in northeastern Thailand and Cambodia from open savannas and woodlands of deciduous dipterocarps on ridge tops through dry dipterocarp forest and/or dry evergreen forest on the slopes to evergreen gallery forests along major river valleys (Blasco, 1983). Such a mosaic of forest gradients is shown in Figure 5.3*A*. Many of these changes in forest type are remarkably abrupt, occurring over very short distances. Human impacts, however, have blurred many of the transitions, particularly with the combination of forest cutting and annual or semiannual ground fires.

Forest types

The classification of forest types in Thailand has been heavily influenced by European workers in India and Myanmar (Champion, 1936; Edwards, 1950; Champion & Seth, 1968; Table 5.1) and other parts of Southeast Asia. All of this region shares the monsoon climate which sharply delineates the vegetation, except in peninsular Thailand, from the wet tropical forests of Malaysia and Indonesia.

The level of detail utilized in recognizing major and minor forest types differs among authors (Aubert de la Rüe, 1958; Smitinand, 1977b). The most recent comprehensive vegetation map of Thailand was developed in 1982 by the Royal Forest Department, using Landsat data (Smitinand, 1989), and shows considerable detail. For this review we have distinguished six major forest types (Table 5.1, Figure 5.2*B*). Three of these – tropical wet forest, mangrove forest, and hill evergreen forest – are not dry forests and are mentioned only briefly here.

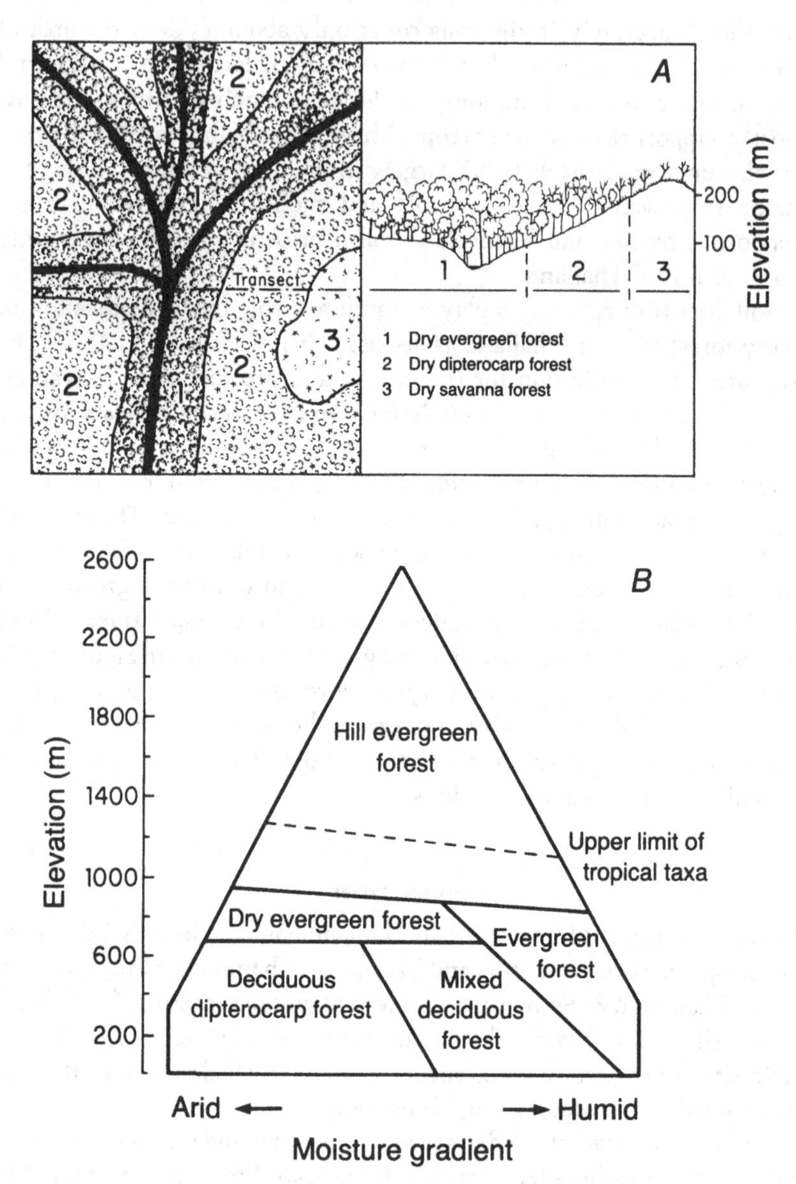

Figure 5.3. Distribution of major forest types in relation to elevation and moisture. *A*. Local patterns of vegetation change with topography typical of northeastern Thailand around the Khorat Plateau and in western Cambodia (adapted from Blasco, 1983). *B*. Schematic representation of the vegetation types in northern Thailand along gradients of elevation and moisture availability (adapted from Ogawa *et al*., 1961).

Table 5.1. *Major forest types recognized in Southeast Asia*

Thailand (this chapter)	Myanmar (Stamp, 1925)	Laos (Vidal, 1960)	India and Myanmar (Champion & Seth, 1968)
Evergreen			
Tropical wet	Evergreen dipterocarp	Dense humide	Tropical
Hill evergreen	–	Dense à Fagacées et Lauracées	Montane temperate
Dry evergreen	Pyinkado/semi-evergreen	Dense humide semi-décidue	Tropical semi-evergreen
Mangrove	–	–	Mangrove
Deciduous			
Mixed deciduous	Moist teak	Mixte décidue	Tropical moist deciduous
	Dry teak	–	–
	Dry deciduous without teak	–	–
Deciduous dipterocarp	Idaing	Claire à Dipterocarpacées	Tropical dry deciduous

Tropical wet (evergreen) forest is largely restricted to the peninsula of Thailand and represents the northern limit of rain forest conditions and the Indo-Malayan flora of the south. Structure and floristics of this type have been described in considerable detail for Thailand (Ogino, Sabhasri & Shidei, 1964; Ogawa *et al.*, 1965a, b; Kira *et al.*, 1967; Yoda, 1967; Yoda & Kira, 1967; Foxworth, 1979). While there is a gradual transition between perhumid and monsoonal climates in peninsular Thailand, Whitmore (1984) has established a line at the Kra Isthmus as the floristic boundary of the Indo-Malayan forest region. However, a small outlier of tropical wet forest is locally present at intermediate elevations in the mountains of Khao Yai National Park northeast of Bangkok (Smitinand, 1968; Suwannapinunt & Siripatanadilok, 1982). Ogino (1976) estimated that tropical wet forests covered approximately 9% of the area of Thailand.

Mangrove forests are evergreen communities growing in the zone of tidal influence along muddy seashores and stream estuaries. These forests cover nearly 3000 km^2 in Thailand and support a woody flora of 74 species (Arbhabhirama *et al.*, 1987).

Hill evergreen forest is another mesic type, often referred to as temperate evergreen forest. This forest type, largely restricted to elevations above 1000 m, is most widespread in northern Thailand (Ogawa, Yoda & Kira, 1961; Robbins & Smitinand, 1966; Smitinand, 1966; Kuchler & Sawyer, 1967; Sawyer & Chermsirivanthana, 1969; Santisuk, 1988), but also occurs on some mountains in the northeast (Smitinand, 1968). Temperate genera are often dominant in the canopy (*Quercus, Lithocarpus, Chrysophyllum, Cinnamomum* and *Magnolia*), while *Rhododendron* and *Vaccinium* are characteristic of the understory. Conifers are widespread, including *Pinus kesiya* and *P. merkusii, Dacridium elatum*, and *Podocarpus* species. The pines, however, are also common in dry dipterocarp forest (Werner, 1993), as described below. Few taxa of tropical origin extend above the lower margins of this community. Hill evergreen forest covered 9–10% of Thailand (Ogino, 1976).

Dry forests

The focus of this review is on dry evergreen forest, mixed deciduous forest, and dry or deciduous dipterocarp forest. These three types originally covered nearly two-thirds of the forest area of Thailand (Ogino, 1976; Figure 5.2*B*). Mosaics of several forest types occur in northern Thailand, according to local elevation and moisture gradients, as shown diagrammatically in Figure 5.3*B*. This diagram, however, is a generalized simplification of a complex vegetational pattern. Mixed deciduous forest and deciduous dipterocarp forest occur at similar elevations in northern Thailand, with the latter on the driest or steepest slopes. Slope exposure is often a major factor in transitions between these two types. Dry evergreen forest in northern Thailand occurs mostly at higher elevations as a transition between mixed deciduous and hill evergreen types, or as gallery forest along streams at lower elevations.

Mixed deciduous forest

Mixed deciduous forests are best developed in northwestern Thailand where they represent an extension from India and Myanmar of similar mixed deciduous forests dominated by teak (*Tectona grandis*). Mean annual rainfall in this type is commonly 1400–1800 mm, with 5–6 dry months. To the north of Chiang Mai, these forests were originally widespread, occurring on both slopes and flats at low to moderate elevation (Ogawa *et al.*, 1961; Kaosa-ard, 1981; Santisuk, 1988). However, it is impossible to determine accurately how far south mixed deciduous forests once occurred. Mixed deciduous forests have been heavily disturbed by the logging of teak trees. Outside the northwest highlands, mixed deciduous forest also has been replaced by secondary dipterocarp forest or agriculture (Santisuk, 1988). Extensive areas around the northern margin of the Central Plains, now covered by dry dipterocarp savannas, once supported teak forests (Ogawa *et al.*, 1961).

The canopy of mixed deciduous forest is closed and high, often at 30 m or above (Table 5.2). The general forest understory is relatively open, despite an understory canopy more than 7 m tall, a diverse assemblage of small trees and shrubs, and tall bamboos. Epiphytes and lianas are uncommon. Leaf fall normally begins in February, well after the onset of the dry season in early December, and continues at varying rates until the forest is leafless by the end of March. The leafless period extends for 3–4 months (Richards, 1952).

The name mixed deciduous forest derives from the diversity of tree species in these stands. *Tectona grandis* is usually the dominant (most important) species, while *Xylia kerrii* (Leguminosae) is often abundant, and is joined by codominant species of *Terminalia* (Combretaceae), *Lagerstroemia calyculata* (Lythraceae), *Dalbergia* (Leguminosae) and *Pterocarpus macrocarpus* (Leguminosae). Dipterocarpaceae are largely

Table 5.2. *Structural characteristics of mixed deciduous forest stands (northern and western Thailand) and deciduous dipterocarp forest. All data are for trees ≥ 10 cm dbh on 20 plots of 10 × 10 m in each stand (± SD)*

Forest type	Maximum height (m)	Basal area ($m^2\ ha^{-1}$)	Density (trees ha^{-1})	Elevation range (m)	Slope (%)	Species diversity (species per 0.2 ha)
Mixed deciduous[a] (33 stands)						
Tectona/Xylia	28.2 ± 3.8	42.4 ± 11.0	262 ± 67	150–640	21 ± 18	15.1 ± 6.9
Tectona/Xylia/Terminalia	25.2 ± 3.9	33.9 ± 8.4	396 ± 136	190–595	21 ± 18	20.4 ± 5.2
Lagerstroemia	24.4 ± 4.1	33.1 ± 14.3	360 ± 118	190–540	18 ± 16	23.2 ± 7.2
Deciduous dipterocarp (54 stands)						
Shorea siamensis	19.3 ± 6.4	20.3 ± 7.5	406 ± 168	–	–	18.4 ± 3.8
Shorea obtusa	17.5 ± 4.0	16.7 ± 5.6	417 ± 144	–	–	10.6 ± 3.9
Dipterocarpus obtusifolius/ Shorea obtusa	22.6 ± 7.3	23.5 ± 8.0	438 ± 100	–	–	19.1 ± 7.1
Dipterocarpus tuberculatus/ Shorea obtusa	22.7 ± 5.9	23.9 ± 8.1	470 ± 78.2	–	–	16.0 ± 0.8
Pine/dipterocarp	21.9 ± 4.6	24.4 ± 5.2	463 ± 96.7	–	–	10.4 ± 3.0

[a] *Tectona grandis* forests have *Xylia kerrii* and *Terminalia mucronata* as secondary dominants. The *Lagerstroemia calyculata* type may contain small numbers of *T. grandis*.
Source: Bunyavejchewin (1983a, b).

absent from these forests. Elliot *et al.* (1989) have recently described remarkable levels of species diversity in mixed deciduous forest on the slopes of Doi Suthep west of Chiang Mai.

The general composition of this forest type has been described from samples of 33 stands of 0.2 ha in northern and western Thailand (Sukwong, 1976, 1977; Bunyavejchewin, 1983b, 1985). These stands occurred on a variety of exposures and slopes from flat to steep, at elevations from 190–640 m. Basal area ranged from 13 to 63 m^2 ha^{-1} and the density of trees ≥10 cm dbh ranged from 17 to 59 trees per 0.1 ha. For trees ≥10 cm dbh, a total of 151 species was found; 30 of these were present in at least 20% of the stands (Figure 5.4*A*). Two main variants were distinguished on the basis of the most important species, a *T. grandis* type (with *X. kerrii*) and a *L. calyculata* type. The *T. grandis* type was divided into two subtypes, with or without *Terminalia mucronata* as a secondary dominant. Statistically, the three types were not significantly different in most stand characteristics (Table 5.2). The mean number of tree species ≥10 cm dbh ranged from 15.1 in the *Tectona/Xylia* subtype to 23.2 for the *L. calyculata* type. Individual stands had as few as six tree species and as many as 41 (Figure 5.5).

The upper canopy of the teak forests was 25–30 m tall, and individual trees reached 40 m on deep, well-drained soils. A middle canopy at 10–20 m filled openings in the upper canopy to form a continuous forest canopy. Two bamboo species dominated in the shrub layer but ground cover was rather open. The *L. calyculata* type, generally found below 400 m, was very similar to the *T. grandis* forests. Indeed, *T. grandis* was commonly present in these stands, as were the same associated canopy species; the characteristic feature of this type was the importance of *L. calyculata*.

Soil characteristics of this same set of mixed deciduous forest stands showed no significant differences in structure or chemistry separating the three types (Table 5.3). Soil pH averaged 5.5, but ranged widely from 4.4 to 7.4 in different stands. Cation exchange capacity was about 20 meq per 100 g of soil. Both organic matter and total nitrogen content were low. Root penetration averaged only about 0.6 m.

Several other studies of mixed deciduous forests are available. At Ping Kong north of Chiang Mai, *L. calyculata* was the dominant tree. The presence of *Dipterocarpus obtusifolius* and *Shorea obtusa* suggested that this was a relatively dry site, representing a transition to dry dipterocarp forest (Ogawa *et al.*, 1965a, b). Based on one 40 × 40 m plot, the density of trees ≥4.5 cm dbh was 713 ha^{-1} and the basal area was 35.4 m^2 ha^{-1}.

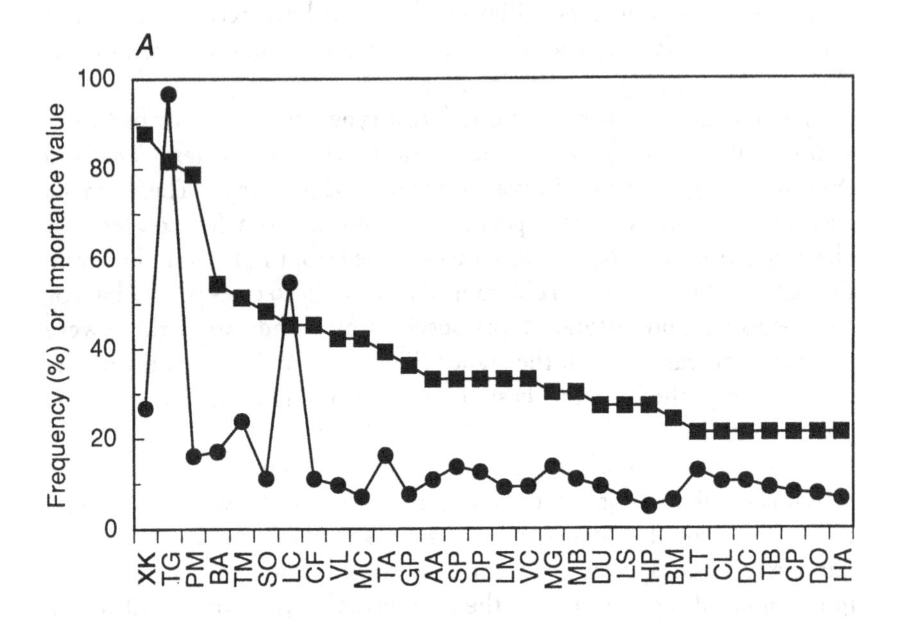

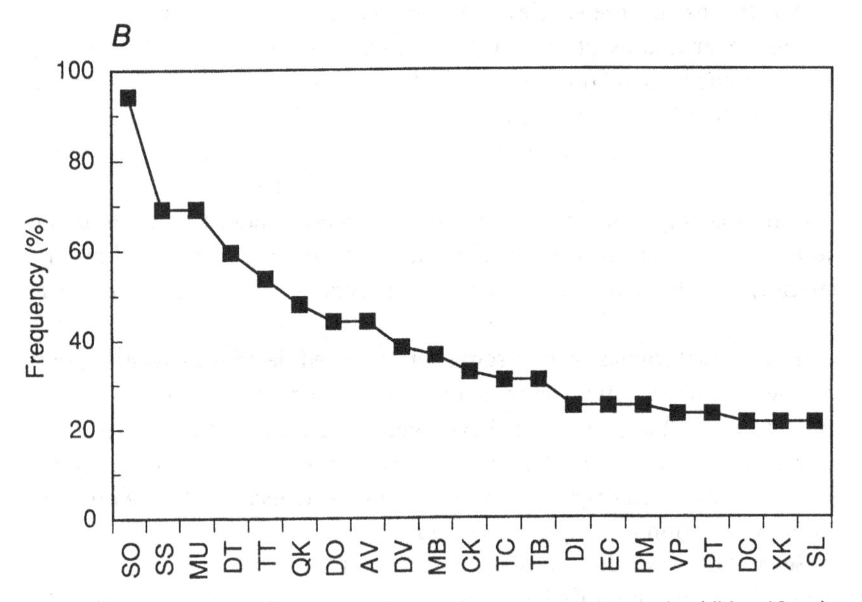

Figure 5.4. Frequency and importance values for common tree species (dbh ≥10 cm). *A*. Mixed deciduous forest (33 stands). *B*. Deciduous dipterocarp forest (52 stands). Only species with frequencies greater than 20% are included here. (Data from Sukwong, 1976, 1977; and Bunyavejchewin, 1983a.) Key to species: Anacardiaceae: LM *Lannea coramandelica*, MU *Melanorrhoea ustita*, SP *Spondias pinnata*; Annonaceae: CL *Cananga latifolium*; Apocynaceae: HA *Holarrhena antidysenterica*;

Thirty-one species of trees were present among the 114 individuals sampled in this small plot.

Forests on granite and limestone mountains near Chiang Mai have been described in considerable detail (Kuchler & Sawyer, 1967; Sawyer & Chermsirivanthana, 1969; Elliot *et al.*, 1989). On Doi Suthep, these forests are most common at 350–1000 m elevation on the lower south-eastern and southern slopes of the mountains. *T. grandis*, *Terminalia mucronata* and *Protium serratum* were the dominant species; *L. calyculata* was present but uncommon.

On limestone hills near Lampang in north central Thailand, mixed deciduous forests are well developed on deep soils of the foothills, while deciduous dry dipterocarp forests predominate along the rocky ridges (Sukwong & Kaitpraneet, 1975). Teak is the dominant species for most of the mixed deciduous stands, but the *L. calyculata* type is also present.

South of Lampang, mixed deciduous forests become less common and lower in stature, reflecting drier conditions. Teak drops out, although many other characteristic dominants are still present. These communities have been termed dry mixed deciduous forest (Smitinand, 1966), and represent a gradation into dry evergreen forest and deciduous dipterocarp forests, showing floristic elements of these latter types.

In Khao Yai National Park near Bangkok, mixed deciduous forest is present on northern slopes at 400–600 m (Smitinand, 1968, 1977a). The dominant tree species include *X. kerrii*, *L. calyculata*, *Pterocarpus macrocarpus*, *Afzelia xylocarpa* and *Terminalia bellirica* but teak is absent. Further south, near Chonburi in Khao Khieo Game Sanctuary, *X. kerrii*, *T. bellirica*, and *A. xylocarpa* remain as codominants in a forest which has been heavily disturbed by cutting and burning (Maxwell, 1980).

Figure 5.4 (*cont.*)
Bignoniaceae: HP *Heterophragma adenophyllum*; Bombacaceae: BA *Bombax anceps*; Burseraceae: CK *Canarium kerrii*, GP *Garuga pinnata*; Combretaceae: AA *Anogeissus acuminata*, TA *Terminalia alata*, TB *T. bellirica*, TC *T. chebula*, TM *T. mucronata*; Dilleniaceae: DV *Dillenia obovata*, DP *D. parviflora*; Dipterocarpaceae: DO *Dipterocarpus obtusifolius*, DT *D. tuberculatus*, SO *Shorea obtusa*, SS *S. siamensis*; Euphorbiaceae: AV *Aporusa villosa*; Fagaceae: QK *Quercus kerrii*, Guttiferae: CP *Cratoylum pruniflorum*; Leguminosae: MB *Milletia brandisiana*, DC *Dalbergia cana*, DU *D. cultrata*, DI *D. oliveri*, PT *Pterocarpus macrocarpus/parvifolia*, XK *Xylia kerrii*; Lythraceae: LC *Lagerstroemia calyculata*, LT *L. colletti*; Myrtaceae: EC *Eugenia cumini*, TT *Tristania burmanica*; Pinaceae: PM *Pinus merkusii*; Rubiaceae: MG *Mitragyna brunosis*, MC *Morinda coreia*; Sapindaceae: LS *Lepisanthes siamensis*, SL *Schleichera oleosa*; Tiliaceae: BM *Berrya mollis*, CF *Colona fragrocarpa*; Verbenaceae: TG *Tectona grandis*, VC *Vitex canescens*, VL *V. limonifolia*, VP *V. peduncularis*.

Dry evergreen forest

The name dry evergreen forest is misleading because a mixture of deciduous species may be present although evergreen trees predominate. Dry evergreen forest ranges from gallery forests along river channels, with virtually 100% evergreen canopies, to semievergreen closed forests on more xeric sites. Mean annual rainfall is commonly 1200–1500 mm, little different from that of the dry deciduous dipterocarp forests described below. Dry evergreen forests are widespread in the Central Highlands and in mesic areas around the south and southwestern slopes of the Khorat Plateau where the dry season is moderated by local moisture conditions and soils are less poor than on the Khorat Plateau itself (Ogawa *et al.*, 1961; Williams, 1967; Figure 5.2B). However, their distribution has been highly fragmented by human actions (Williams,

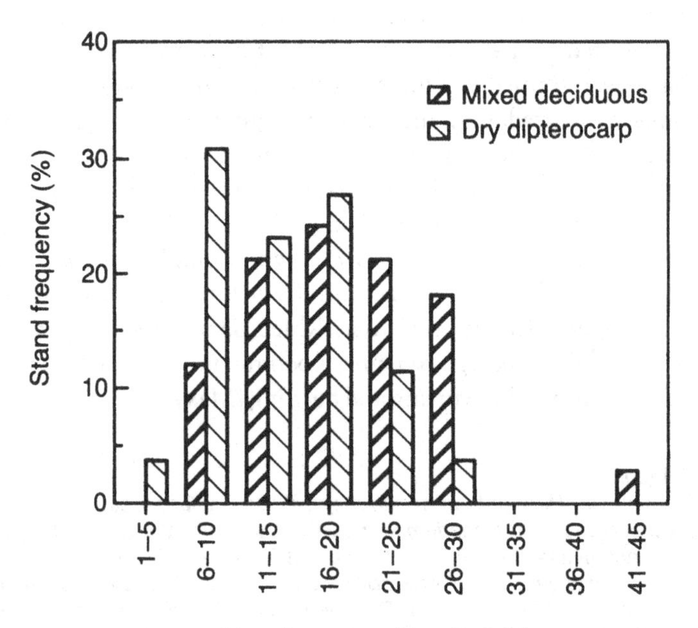

Figure 5.5. Diversity of tree species in mixed deciduous and dry deciduous dipterocarp forest of northern Thailand. Data are frequency of stands, each stand sampled with 20 plots of 10 × 10 m for trees ≥10 cm dbh (data from Sukwong, 1976, 1977).

Table 5.3. *Soil characteristics of mixed deciduous forest (34 stands) and dry evergreen forest (at Sakaerat) (± SD)*

	Mixed deciduous forest		Dry evergreen forest		
	Tectona/Xylia	*Tectona/Xylia/ Terminalia*	*Lagerstroemia*	*Hopea ferrea*	*Shorea henryana*
pH	5.88 ± 1.07	5.50 ± 0.54	5.33 ± 0.41	4.5 ± 0.8	3.7 ± 0.6
CEC (meq 100 g^{-1})	21.2 ± 17.3	19.3 ± 17.5	18.3 ± 5.8	7.0 ± 1.8	8.5 ± 2.0
Organic matter (%)	4.1 ± 1.1	3.5 ± 1.1	3.5 ± 0.9	3.2 ± 0.9	3.7 ± 1.6
Total N (%)	0.16 ± 0.04	0.14 ± 0.047	0.13 ± 0.04		
Available P (mg g^{-1})	11.5 ± 18.6	16.8 ± 15.8	9.1	5.3 ± 1.0	4.7 ± 2.0
Exchangeable cations (meq 100 g^{-1})					
K	0.52 ± 0.21	0.45 ± 0.13	0.45 ± 0.19	89.5 ± 14.0	108.3 ± 28.6
Ca	13.08 ± 12.05	7.65 ± 5.09	6.01 ± 3.47	111 ± 56.7	288 ± 179
Mg	4.66 ± 3.89	4.63 ± 3.12	4.47 ± 3.98	149 ± 55.4	210 ± 101
Na	0.54 ± 0.21	0.55 ± 0.14	0.52 ± 0.22	16.6 ± 5.8	17.2 ± 3.8
Root penetration (m)	0.61 ± 0.08	0.54 ± 0.14	0.69 ± 0.19		
Bulk density (g cm^{-3})	1.16 ± 0.15	1.17 ± 0.15	1.26 ± 0.16	1.16 ± 0.06	1.04 ± 0.08
Silt and clay (%)	52.8 ± 9.3	50.7 ± 10.0	47.9 ± 14.9	56.4 ± 6.2	54.6 ± 11.1

Source: Bunyavejchewin (1983b, 1987).

1967; Boulbet, 1982). This forest is also tall, with a closed canopy at 25–30 m. Floristically, it is derived from more mesic Indo-Malaysian elements to the south, although the composition is distinct in this monsoon climate. The most important family in the canopy is Dipterocarpaceae – with species of *Hopea*, *Shorea*, *Anisoptera* and *Dipterocarpus* – but the species present are largely distinct from those of wet forests to the south.

There are no broad geographical studies of the structure and floristic composition of dry evergreen forest in Thailand comparable to those for mixed deciduous and dry dipterocarp forests. One site has been described in detail, the Sakaerat Environmental Research Station, 300 km north-east of Bangkok (Sukapanpotharam, 1979; Bunyavejchewin, 1986a, b). The upper canopy trees at Sakaerat average 20–35 m in height, with a maximum dbh of 140 cm and individual crowns up to 18 m in diameter. A lower stratum at 5–17 m is also present, with some typical species as well as small individuals of canopy species. The understory, unlike that of the mixed deciduous forest, is rich in lianas, and horizontal visibility is often less than 20 m (Sukapanpotharam, 1979). The vertical gradient of irradiance as measured at Sakaerat suggested that three distinctive canopy layers are present (Yoda, Nishioka & Dhanmanonda, 1983). The dense upper canopy intercepts 90% of the irradiance in the dry season and 80% in the wet season. The intermediate level (7–23 m) is relatively open in contrast to the dense understory (Figure 5.6).

The dominant tree in the canopy at Sakaerat is *Hopea ferrea* (Dipterocarpaceae), with *Memeclyon ovatum* (Melastomataceae), *Hydnocarpus ilicifolius* (Flacourtiaceae) and *Walsura trichostemon* (Meliaceae) as subcanopy dominants. Dry evergreen forest dominated by *Shorea henryana* is also present at Sakaerat (Bunyavejchewin, 1986a, 1987).

The density of trees ≥20 cm dbh in *H. ferrea* forest was about 240 ha^{-1} (Table 5.4). Counting all woody species ≥1 cm dbh, the overall density was about 2500 individuals ha^{-1}. Floristic diversity was high, with 175 tree species in 120 genera and 55 families. A 1 ha plot had 56 species of trees ≥5 cm dbh (Sukapanpotharam, 1979). The Dipterocarpaceae formed 21.5% of these individuals, followed in importance by the Melastomataceae (20.0%), Meliaceae (18.4%), Flacourtiaceae (12.1%) and Rubiaceae (8.7%). The Leguminosae were surprisingly unimportant, with only two species and less than 1% of the individuals. Total aboveground biomass was found to be 394 Mg ha^{-1}, with *H. ferrea* accounting for about 80% of this total.

Structural characteristics of the undergrowth were described in detail

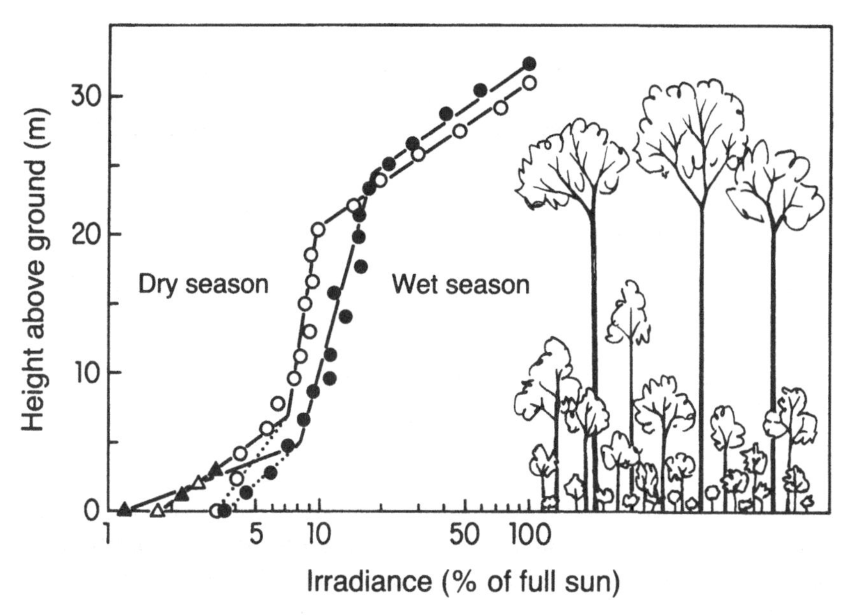

Figure 5.6. Vertical distribution of mean irradiance in dry evergreen forest for the dry and wet seasons at Sakaerat at a meteorological tower. Open symbols show values for the dry season, and closed symbols indicate the wet season. The triangles show values measured in sample plots away from the tower (adapted from Yoda *et al.*, 1983).

by Jantanee (1987). The density of saplings <4.5 cm dbh but taller than 1.3 m was more than 20,000 ha^{-1} with *H. ferrea* and *M. ovatum* as the most important components. These two species, joined by *W. trichostemon*, were generally dominant in gaps of all sizes as well as under the forest canopy, but sapling densities more than doubled in large gaps. Overall, 28 species of tree saplings and 11 species of vines were identified in this sapling pool. For individuals <1.3 m in height, densities averaged about 400,000 ha^{-1}, again with *H. ferrea* and *M. ovatum* as dominants. This group included 30 species of tree seedlings, 13 vines and seven herbs.

Soil characteristics of the dry evergreen forest at Sakaerat differ markedly from those of the mixed deciduous forests in northern and western Thailand. Sakaerat forests have more acidic soils, lower cation exchange capacities and lower available phosphorus than the mixed deciduous forest soils (Table 5.3).

Dry evergreen forests have been qualitatively described at Khao Yai National Park and Khao Khieo. At Khao Yai, this type occurs from 100 to 1000 m elevation. Up to seven species of Dipterocarpaceae are present, representing *Dipterocarpus*, *Vatica*, *Shorea* and *Hopea* (Smitinand,

Table 5.4. *Density of woody stems (>1 cm dbh) in dry evergreen and deciduous dipterocarp forest communities at Sakaerat*

dbh class (cm)	Number of stems ha^{-1}	
	Dry evergreen	Deciduous dipterocarp
<5	915	285
5–20	1350	687
>20	236	203
Total	2501	1175

Source: Sukapanpotharam (1979).

1968, 1977a; Suwannapinunt & Siripatanadilok, 1982). The only species with direct Malesian affinities is *H. ferrea* (Smitinand, 1969; Smitinand, Santisuk & Phengkhai, 1980). Other important canopy species were *L. calyculata* and *Afzelia xylocarpa*, both also present in mixed deciduous forest. Also common were *Tetrameles nudiflora* (Datiscaceae), *Lophopetalum wallichii* (Celastraceae), *Pterocymbium javanicum* (Sterculiaceae), *Parkia streptocarpa* (Leguminosae), *Erythrophloeum succirubrum* (Leguminosae), and *Carallia brachiata* (Rhizophoraceae). At Khao Khieo, dry evergreen forest occurs up to 600 m elevation with *Dipterocarpus alatus*, *Walsura angulata* (Meliaceae), *Radermachera pierrei* (Bignoniaceae), *Pterospermum diversifolium* (Sterculiaceae), *Irvingia malayana* (Simarubaceae), and *Ficus* species (Moraceae) (Maxwell, 1980).

Deciduous dipterocarp forest

Dry dipterocarp forest, or deciduous dipterocarp forest as it is more descriptively called, is a relatively low and open forest vegetation dominated by deciduous trees. This type ranges from nearly closed canopy to an open woodland structure, and merges with true savanna. In the Indian forest classification, this vegetation would be considered a tropical dry deciduous forest, but it is more open and xeric in structure than most such communities (Champion & Seth, 1968). The French term 'forêt claire à dipterocarpacées' (Vidal, 1960) is more specific to this type in Thailand.

Deciduous dipterocarp forest covers more area in Southeast Asia than any other forest type, extending from northeastern India and Myanmar

(Champion & Seth, 1968) through Thailand to the Mekong River region of Laos (Vidal, 1956–58, 1960, 1966), Cambodia (Rollet, 1953, 1972; Aubreville, 1957; Pfeffer, 1969) and Vietnam (Schmid, 1974). Its areas of occurrence are generally characterized by 5–7 dry months and about 1000–1500 mm total rainfall; evaporation may exceed precipitation for up to 9 months y^{-1}. Estimates of the relative area of deciduous dipterocarp forest in Thailand range from 33% (Ogino, 1976) to 45% (Neal, 1967) of total forest cover. Its range extends from drier slopes in the north and around the Central Highlands to the Khorat Plateau in the east (Figure 5.2*B*). More than half of the area of occurrence is mountainous or hilly, on relatively dry exposures or sites with shallow soil.

Transitions from mixed deciduous forest to deciduous dipterocarp forest in the mountains of northern Thailand are generally sharp, often following topographic patterns of exposure, slope angle and soil depth. These factors do not readily separate the forest types below 600 m. In eastern Thailand, sharp transitions from dry evergreen forest to deciduous dipterocarp forest occur on relatively flat topography (e.g. at Sakaerat), suggesting that edaphic factors play an important role. Human impacts on deciduous dipterocarp forest particularly fire, may be an important factor in producing such sharp boundaries. Forest clearing and subsequent increases in fire frequency can lead to secondary dominance by deciduous dipterocarp forests, savanna or savanna woodlands dominated by woody shrubs which are resistant to fire and grazing (Vidal, 1960; Sabhasri *et al.*, 1968; Boulbet, 1982; Smitinand 1989; Stott, 1990).

Despite the relatively large and thick leaves of many of the dominant species, virtually all of the species lose their leaves during the dry season. The six species of Dipterocarpaceae in this forest type are unique among members of this family in Southeast Asia in their deciduous habit. There are considerable species and site-specific differences in leaf fall, producing a notable lack of synchrony, unlike most temperate deciduous forests. Depending on the species, leaf fall may begin as early as December or as late as April (Figure 5.7). Most species are leafless from February to April (Sukwong, Dhamanitayakul & Pongumphai, 1975; Dhamanitayakul, 1984), but individuals growing in mesic microsites may retain leaves throughout the year (UNESCO, 1978). New leaves begin forming before the end of the dry season.

Fires are frequent events in these forests, at intervals of 1–3 years. The majority of these fires are human-caused, both accidental and deliberate. Fuel to feed such frequent fires is provided by the large biomass of leaves and the abundant herbaceous ground cover. Most fires occur between

December and early March. Benefits of these ground fires for the local population include new grass growth for grazing animals, production of edible new shoots and leaves prized in Thai cooking, concentration of nutrients and clearing for agriculture, and reduction of agricultural pests and diseases (Stott, 1986).

All of the dominant dipterocarp tree species in these forests exhibit adaptations to fire. These include morphological and life history characteristics such as thick rough bark which reduces damage to the cambium, roots insulated from high soil temperatures, and root crowns which resprout readily (Sukwong *et al.*, 1975; Stott, 1988; Stott, Goldammer & Werner, 1990). Patterns of reproductive phenology also may be important in seedling establishment following fire. Fruits of *Shorea* and *Dipterocarpus* are dispersed by wind after the peak season of fires and germinate at the beginning of the wet season (Sukwong *et al.*, 1975; Stott, 1988). Hard-seeded tree species such as *Sindora siamensis, Irvingia malayana, Quercus kerrii* and *Lithocarpus polystachyus* disperse fruits before the fire season. Fire may play a role in the germination of these seeds. However, fire frequency is also critical because tree seedlings are vulnerable for several years, e.g. as much as 7 years in slow-growing *Shorea talura* (Sukwong, 1982). Ground fires may scorch leaves to 1 m, or as much as 5 m above ground, depending on the fuel. Soils are

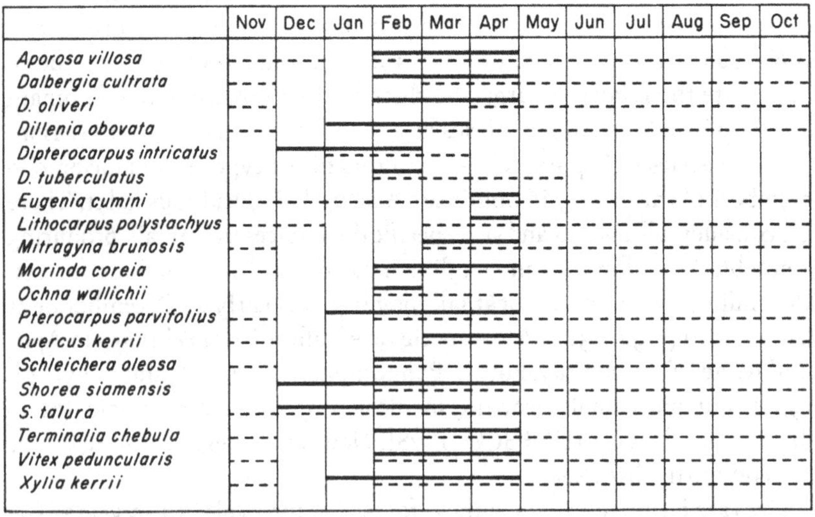

Nov Dec Jan Feb Mar Apr May Jun Jul Aug Sep Oct

Aporosa villosa
Dalbergia cultrata
D. oliveri
Dillenia obovata
Dipterocarpus intricatus
D. tuberculatus
Eugenia cumini
Lithocarpus polystachyus
Mitragyna brunosis
Morinda coreia
Ochna wallichii
Pterocarpus parvifolius
Quercus kerrii
Schleichera oleosa
Shorea siamensis
S. talura
Terminalia chebula
Vitex peduncularis
Xylia kerrii

Figure 5.7. Seasonal leaf phenology for 19 species of trees in the dry deciduous dipterocarp forest at the Sakaerat. Solid lines show the leafless period of leaf fall and dashed lines show months with initiation of new leaves. The dry season at this site normally extends from November to April (data from Dhamanitayakul, 1984).

well insulated: e.g. temperatures can exceed 700 °C at the surface but reach only 35–75 °C at 5 cm depth in sandy loam (Stott, 1986).

Like the tree species, the ground layer plants exhibit striking adaptations to fire. Perennial grasses resprout within a few weeks and restore full coverage within a few months (Stott, 1988). Dwarf palms and cycads have overlapping leaf bases which protect their meristems; damaged leaves are quickly replaced after a burn.

The structure and floristics of deciduous dipterocarp forests have been characterized on the basis of data from 52 plots of 0.2 ha in relatively undisturbed stands throughout Thailand (Sukwong, 1976, 1977; Bunyavejchewin, 1983a). Density of trees ≥10 cm dbh ranged from about 20 to 88 ha^{-1} in these stands, and basal area from 7 to 42 m^2 ha^{-1}. In comparison to the samples from mixed deciduous forest, these stands tended to have higher densities but lower stand basal areas. The median diversity was more than 20 tree species per plot (Figure 5.5), a value close to that for mixed deciduous forests. Individual sample stands had from 2 to 28 tree species present in this area. Deciduous dipterocarp forest at the Sakaerat station had a density of trees 1 cm dbh of 1175 ha^{-1} (Table 5.4), with a greater proportion of large stems than in dry evergreen forest. Aboveground biomass of another stand at Sakaerat was 184 Mg ha^{-1} with 5 Mg in leaves (Tongyai, 1980).

The typical tree flora can be represented by the number of species (22) present in at least 20% the stands sampled (Figure 5.4*B*). Four species of Dipterocarpaceae form the dominant component of deciduous dipterocarp forests. These are *Shorea obtusa, S. siamensis, Dipterocarpus obtusifolius* and *D. tuberculatus*. Other widespread and important tree species include two legumes that are also important in mixed deciduous forests (*X. kerrii* and *Pterocarpus macrocarpus*). Other important tree species include *Canarium subulatum* (Burseraceae), *Melanorrhoea ustita* (Anacardiaceae), *Quercus kerrii* (Fagaceae), *Careya sphaerica* (Myrtaceae), *Aporusa villosa* (Euphorbiaceae), *Dillenia obovata* (Dilleniaceae), and *Terminalia alata* (Combretaceae). Grasses such as *Arundinaria pusilla* and *Imperata cylindrica* dominate the open understory in these forests, and the low-growing *Cycas siamensis* (Cycadaceae) and *Phoenix acaulis* (Arecaceae) are widespread (Bunyavejchewin, 1983a).

Various authors have recognized a number of floristic associations within the deciduous dipterocarp forest type (Ogawa *et al.*, 1961; Khemnark *et al.*, 1972: Kutintara, 1975; Bunyavejchewin, 1983a). In the northern province of Chaing Mai, associations with *Pinus* are notable at

Table 5.5. *Areas of forested land in various regions of Thailand from 1961 to 1985*

Region	Area (km²)	Percentage of area forested					
		1961	1973	1976	1978	1982	1985
North	169 600	68.5	67.0	60.3	56.0	51.7	49.6
East	36 500	58.0	41.2	34.6	30.2	21.9	21.9
Northeast	168 900	42.0	30.0	24.6	18.5	15.3	14.4
Central Plain	67 400	52.9	35.6	32.4	30.3	24.5	25.6
South	70 700	41.9	26.1	28.5	24.9	23.2	21.9
Total	513 100	53.3	43.2	38.7	34.2	30.5	29.0

Source: Arbhabhirama *et al.* (1987).

450–1100 m elevation (most commonly above 750 m: Khemnark *et al.*, 1972; Kutintara, 1975; Bunyavejchewin, 1983a, Werner, 1993). Among subtypes based on relative importance of tree species (Bunyavejchewin, 1983a), species diversity appears to differ more than structural characteristics (Table 5.2).

Human impacts on forest cover

Historically, government policies for natural resource management in Thailand suggested that at least 50% of the total land area of the country remain under forest cover. Likewise, an FAO forestry mission to Thailand in 1949 recommended that 53% of the nation's area be preserved as forests. Of this total forest area, 43% would be in the north, 20% in the northeast, 13% in the central region and 17% in the south and east (Hafner & Apichatvullop, 1990). These goals were never seriously threatened until the 1950s when population growth and economic development put pressure on land use policies. While the relative percentages of forest cover by regions have been maintained, the total area of forest cover has dropped sharply.

In 1961, forested areas covered 53% of Thailand, with a range from 68% in the north to 42% in the south (Table 5.5). This cover has declined dramatically in recent years, however, with only 29% of the country forested in 1985. Between 1961 and 1985 while total forest cover declined 46%, reductions in the northeast and east were even greater, 66% and 62% respectively. There are multiple causes for the rapid reduction of forests: logging operations, clearance for agriculture in the highlands, population pressures and cultivation practices of the northern hill tribes all contribute.

Teak forests were extensively exploited in Myanmar by the British from the mid-19th century and logging spread to Thailand in the late 19th century (Dauphinot, 1905; Hofseus, 1907). Competition for teak leases in northern Thailand was intense among Burmese and Europeans, leading to heavy and indiscriminate cutting. Because logs were transported largely by water, teak and other mixed deciduous forest trees were cut along major rafting routes, without regard for regrowth. Problems with teak exploitation led to the establishment of the Royal Forest Department in 1896, and to transfer of ownership and control of all forests from feudal chiefs to the national government in 1899. At the same time, a number of acts were passed to regulate the teak industry. Nevertheless, heavy exploitation continued (Decamps, 1955; see also the lively account by Campbell, 1935). Commercial logging operations in northern Thailand have not been limited to teak. More than a dozen other mixed deciduous forest species are important lumber trees, including *X. kerrii*, *A. xylocarpa* and *L. calyculata*. Logging operations in wet evergreen forests in the peninsular region began in the 1920s (Letourneaux, 1952; Pendleton, 1962).

Population growth and expansion of agriculture have been serious problems in lowland forest areas. Encroachment and illegal settlement are widespread problems, even in protected reserves and national parks (Hafner & Apichatvullop, 1990; McNeely & Dobias, 1991). Government policies are now showing some success in developing economic incentives for villagers to promote forest protection and the maintenance of biological diversity in such areas. However, extensive human-caused fires still sweep through mixed deciduous forests at frequent intervals, and may dramatically affect the diversity and abundance of tree seedlings. In the north, particularly in hill evergreen forest, population growth among hill tribe villagers, the associated shifting cultivation, and influxes of refugees, have compounded the problems of deforestation and erosion. Appropriately, considerable attention has been given in recent years to managing forest resources in these highlands for sustainable yield (Kunstadter, Chapman & Sabhasri, 1978; Ives, Sabhasri & Voraurai, 1980). Forest destruction has been notably severe in the open dry dipterocarp forests of the Khorat Plateau in northeastern Thailand (Arbhabhirana *et al.*, 1987; Stott *et al.*, 1990).

A model to describe patterns of forest degradation for Thailand was developed by Blasco (1983). In this model, selective deforestation and burning of mixed deciduous or dry evergreen forest leads to open stands of deciduous dipterocarp forest dominated by *Shorea* and *Dipterocarpus*. If disturbance increases to annual burning, the deciduous forest becomes progressively more open and savanna-like, with increasing loss of biological diversity. The changes in plant community structure following human activities of land clearance, grazing and burning for deciduous dipterocarp forests have been described broadly for northeastern Thailand by Boulbet (1982) and for Laos and Cambodia by Vidal (1960), Rollet (1953, 1972) and Wharton (1966). Remote sensing techniques and aerial photography are being used with increasing effectiveness to monitor changes in human impacts on forest lands (Bruneau & Caboussel, 1973; National Research Council, 1991).

Deforestation has the potential to cause environmental problems of immediate importance to human economies. Tangtham & Sutthipibul (1988) carried out a statistical analysis of rainfall in northeastern Thailand in relation to diminishing forest cover from 1951 to 1984. Their data suggested a significant positive correlation between forest depletion and declines in both mean annual precipitation and the number of rainy days. When heavy rains occur, areas which have reduced forest cover may have severe problems of erosion (Zinke, 1989).

The current government policy of Thailand is to maintain forest cover over no less than 40% of the nation. This is 11% more than is covered today. Under government plans, 15% of the land area would be in conservation forests, including national parks, wildlife sanctuaries, headwater source areas, and other reserves. Commercial forest would cover 25% of the land area, made up of exploitable forest reserves for sustained yield and plantation areas (Arbhabrirama *et al.*, 1987). At the present time, national parks and wildlife sanctuaries cover approximately 8.8% of the land area (or 45,000 km^2: Round, 1988); not all of this is forested. Plantations of fast-growing tropical trees (especially *Eucalyptus camaldulensis* and *Casuarina junghuhniana*) are increasing in economic importance and cover 240 km^2, particularly in central and eastern Thailand (Arbhabhirama *et al.*, 1987). Also, improvements in teak plantation yield are continuing in northern Thailand. While there is considerable interest in reforestation of degraded areas with native tree species, progress has been slow. Less than 600 km^2 has been reforested in 40 years of work (Round, 1988), but the efforts are becoming increasingly effective.

Summary

Dry forest ecosystems comprise approximately two-thirds of the total forest area of Thailand, growing in a strongly seasonal climate that predominates over most of the country. Monsoonal conditions in Thailand result from the dynamic movements of the Intertropical Convergence Zone, which brings wet conditions in the Northern Hemisphere summer and dry conditions during the summer. Mean annual precipitation through the dry forest regions of Thailand is about 1550 mm, with tropical rain forest prevailing in areas which exceed 2000 mm. While substrate geology is varied, most soils are latosols and lithosols that are shallow and poor in nutrients. Three major categories of dry forest communities are present in Thailand. Mixed deciduous forests of northwestern Thailand represent an extension of similar deciduous forests extending into Burma and India. Biomass, stand basal area, and species diversity are all remarkably high at present. *Tectona grandis* (teak) was once the dominant species in these forests, but has been heavily logged in this century.

Dry evergreen forests of variable composition provide an unusual example of closed evergreen forest of large stature growing in areas of only 1200–1500 mm annual rainfall. Floristically, these forests are extensions of Malaysian rain forest, but with high levels of endemics among the dominant dipterocarps. Dry or deciduous dipterocarp forests are the third, and geographically most extensive, form of tropical dry forest in Thailand. These low and open forests extend across much of Southeast Asia. All of the species in this community lose their leaves during the dry season, although the length of the leafless period varies greatly between species. Fire is an important ecological factor in these ecosystems.

Deforestation has been a major problem over most of Thailand. While most of the country once had forest cover, little more than 20% has such cover today. Illegal logging, agricultural expansion, and encroachment of mountain settlements have all compounded environmental problems, even in protected reserves and national parks.

Acknowledgements

We thank the University of California Pacific Rim Research for supporting the research that lead to this review. We are appreciative of the help we have received from Drs T. Smitinand and S. Sabhasri.

References

Arbhabhirama, A., Phantumvanit, D., Elkington, J. & Ingkasuwan, P. (1987). *Thailand: natural resources profile*. Thailand Development Research Institute, Bangkok.

Aubert de la Rüe, E. (1958). Quelques aspects biogéographiques du Thailand et observations sur la vallée du Mae Ping. *Compte Rendu Sommaire des Séances de la Société de Biogeographie* **303**: 15–22.

Aubreville, A. (1957). An pays des eaux et des forêts. Impressions du Cambodge forestier. *Bois et Forêts des Tropiques* **52**: 49–56.

Blasco, F. (1983). The transition from open forest to savanna in continental Southeast Asia. In *Tropical Savannas*, ed. F. Bourlierè, pp. 167–82. Elsevier, Amsterdam.

Bruneau, M. & Caboussel, G. (1973). Le dynamique des paysages en zone tropical: essai de cartographie dans la region de Si Satchanalei (Thailande Septentrianale). *Travaux et Documents de Geographie Tropicale* **9**: 1–73.

Boulbet, J. (1982). Evolution des paysages végétaux en Thaïlande du nord-est. *Publications de l'Ecole Francaise d'Extréme-Orient* 136.

Bunyavejchewin, S. (1983a). Canopy structure of the dry dipterocarp forest of Thailand. *Thai Forest Bulletin* **14**: 1–93.

Bunyavejchewin, S. (1983b). Analysis of the tropical dry deciduous forest of Thailand. I. Characteristics of the dominance types. *Natural History Bulletin of the Siam Society* **31**: 109–22.

Bunyavejchewin, S. (1985). Analysis of the tropical dry deciduous forest of Thailand. II. Vegetation in relation to topographic and soil gradients. *Natural History Bulletin of the Siam Society* **33**: 3–20.

Bunyavejchewin, S. (1986a). Ecological studies of tropical semi-evergreen rain forest at Sakaerat, Nakhon Ratchasima, Northeast Thailand. I. Vegetation patterns. *Natural History Bulletin of the Siam Society* **34**: 35–57.

Bunyavejchewin, S. (1986b). Ecological studies of tropical semi-evergreen rain forest of Thailand. *Thai Forest Bulletin* **14**: 1–93.

Bunyavejchewin, S. (1987). Vegetation patterns in the tropical semi-evergreen forest at Sakaerat, Nakhon Ratchasima. *Thai Journal of Forestry* **6**: 36–50 (in Thai).

Campbell, R. (1935). *Teak-Wallah*. Oxford University Press, Oxford.

Champion, H. G. (1936). A preliminary survey of the forest types of India and Burma. *Indian Forest Records, New Series* **1**: 1–286.

Champion, H. G. & Seth, S. K. (1968). *The Forest Types of India: a revised survey*. Manager of Publications, New Delhi.

Cole, M. M. (1986). *The Savannas: biogeography and geobotany*. Academic Press, London.

Dauphinot, G. (1905). Les forêts de teck au Siam. *Bulletin Economique de l'Indochine* **29**: 625–36.

Decamps, A. (1955). Les exploitations de teck au Siam. *Bois et Forêts des Tropiques* **42**: 26–36.

Dhamanitayakul, P. (1984). The phenology of trees in dry dipterocarp forest and its application to timing for logging operations. *Thai Journal of Forestry* **3**: 151–62 (in Thai).

Dudal, R. & Moormann, E. R. (1964). Major soils of Southeast Asia: their characteristics, distribution and agricultural potential. *Journal of Tropical Geography* **18**: 54–80.

Edwards, M. V. (1950). Burma forest types according to Champion's classification. *Indian Forest Records, New Series* **7**: 135–73.

Elliot, S., Maxwell, J. F. & Beaver, O. P. (1989). A transect survey of monsoon forest in Doi Suthep-Pui National Park. *Natural History Bulletin of the Siam Society* **37**: 137–41.

Foxworth, F. W. (1979). Notes on a trip in peninsular Siam. *Natural History Bulletin of the Siam Society* **28**: 47–54.
Gaussen, H., Legris, P. & Blasco, F. (1967). Bioclimats du Sud-Est Asiatique. *Travaux de la Section Scientifique et Technique, Institut Francaise de Pondichery* **3**: 1–113.
Hafner, J. A. & Apichatvullop, Y. (1990). Farming the forest: managing people and trees in reserved forests in Thailand. *Geoforum* **21**: 331–46.
Hofseus, C. C. (1907). Das geakholz in Siam. *Tropenplanzen Beihefte* **8**: 378–91.
Ives, J. D., Sabhasri, S. & Voraurai, P. (1980). *Conservation and Development in Northern Thailand.* United Nations University, Tokyo.
Jantanee, P. (1987). *Structural Characteristics of Undergrowth of Dry Evergreen Forest, Sakaerat.* Department of Silviculture, Kasetsart University, Bangkok (in Thai).
Kaosa-ard, A. (1981). Teak (*Tectona grandis*): its natural distribution and related factors. *Natural History Bulletin of the Siam Society* **29**: 55–74.
Khemnark, C., Wacharakitti, S., Aksornkoae, S. & Kaewlaid, T. (1972). Forest production and soil fertility at Nikham Doi Chiangdao, Chiangmai Province. *Kasetsart University Forest Research Bulletin* No. 22.
Kira, T., Ogawa, H., Yoda, K. & Ogino, K. (1967). Comparative ecological studies on three main types of forest vegetation in Thailand. IV. Dry matter production, with special reference to the Khao Chong rain forest. *Nature and Life in Southeast Asia* **5**: 149–74.
Kuchler, A. W. & Sawyer, J. O. (1967). A study of the vegetation near Chiangmai, Thailand. *Transactions of the Kansas Academy of Science* **70**: 281–348.
Kunstadter, P., Chapman, E. C. & Sabhasri, S. (1978). *Farmers in the Forest.* University of Hawaii Press, Honolulu.
Kutintara, U. (1975). Structure of the dry dipterocarp forest. Ph.D. thesis, Colorado State University, Fort Collins.
Letourneaux, C. (1952). La situation forestière du Siam. *Bois et Forêts des Tropiques* **26**: 370–80.
Maxwell, J. F. (1980). Vegetation of Khao Khieo Game Sanctuary, Chonburi Province, Thailand. *Natural History Bulletin of the Siam Society* **28**: 9- 24.
McNeely, J. A. & Dobias, R. (1991). Economic incentives for conserving biological diversity in Thailand. *Ambio* **20**: 86–90.
National Research Council (1991). *Thailand from Space.* National Research Council of Thailand, Bangkok.
Neal, D. G. (1967). *Statistical Description of the Forests of Thailand.* Military Research and Development Center, Bangkok.
Ogawa, H., Yoda, K. & Kira, T. (1961). A preliminary survey on the vegetation of Thailand. *Nature and Life in Southeast Asia* **1**: 21–157.
Ogawa, H., Yoda, K., Kira, T., Ogino, K., Shidei, T., Ratanawangse, D. & Apasutaya, C. (1965a). Comparative ecological studies on three main types of forest vegetation in Thailand. I. Structure and floristic composition. *Nature and Life in Southeast Asia* **4**: 13–48.
Ogawa, H., Yoda, K., Ogino, K. & Kira, T. (1965b). Comparative ecological studies on three main types of forests in Thailand II. Plant biomass. *Nature and Life in Southeast Asia* **4**: 49–80.
Ogino, K. (1976). Human influences on the occurrence of deciduous forest vegetation in Thailand. *Memoirs of the College of Agriculture Kyoto University* **108**: 55–74.
Ogino, K., Sabhasri, S. & Shidei, T. (1964). The estimation of the standing crop of the forest in northeastern Thailand. *South East Asian Studies* **4**: 89–97.
Pendleton, R. L. (1941). Laterite and its structural uses in Thailand and Cambodia. *Geographical Review* **31**: 177–202.
Pendleton, R. L. (1962). *Thailand: aspects of landscape and life.* Deull, Sloan and Pearce, New York.
Pendleton, R. L. & Montrakun, S. (1960). The soils of Thailand. *Proceedings of the 9th Pacific Science Congress* **18**: 12–32.

Pfeffer, P. (1969). Considerations sur l'ecologie des forêts claires du Cambodge oriental. *Terre et Vie* **23**: 3–24.
Richards, P. W. (1952). *The Tropical Rain Forest.* Cambridge University Press, Cambridge.
Robbins, R. C. & Smitinand, T. (1966). A botanical ascent of Doi Inthanan. *Natural History Bulletin of the Siam Society* **21**: 205–27.
Rollet, B. (1953). Notes sur les forêts claires du sud de l'Indochine. *Bois et Forêts des Tropiques* **31**: 3–13.
Rollet, B. (1972). La végétation du Cambodge. *Bois et Forêts des Tropiques* **145**: 23–38 and **146**: 3–20.
Round, P. D. (1988). *Resident Forest Birds in Thailand: their status and conservation.* International Council for Bird Preservation, Monograph No. 2, Cambridge.
Rundel, P. W. (1991). Implications of global climate change on the ecophysiology of agricultural crop plants. In *Global Climate Change: effects on tropical forests, agricultural, urban and industrial ecosystems,* ed. P. Kongton, S. Bhumibhaman, H. Wood and K. Boonpragob, pp. 86–98. ITTO Technical Series No. 6. Bangkok.
Sabhasri, S., Boonnitee, A., Khemnark, C. & Aksornkoae, S. (1968). Structure and floristic composition of forest vegetation at Sakaerat, Pak Thong Chai, Nakhon Ratchasima. I. Variation of floristic composition along a transect through dry evergreen and dry dipterocarp forest. Advanced Research Projects Agency, Report No. 2. Bangkok.
Santisuk, T. (1988). *An Account of the Vegetation of Northern Thailand.* Franz Steiner Verlag, Weisbaden.
Sawyer, J. O. & Chermsirivanthana, C. (1969). A flora of Doi Suthep, Doi Pui, Chiang Mai, North Thailand. *Natural History Bulletin of the Siam Society* **23**: 99–132.
Schmid, M. (1974). Végétation du Vietnam. Le Massif Sud-Annamítique et les régions limitrophes. *Memoires ORSTOM* **74**: 1–243.
Smitinand, T. (1966). The vegetation of Doi Chiang Dao, a limestone massive in Chiang Mai, north Thailand. *Natural History Bulletin of the Siam Society* **21**: 93–128.
Smitinand, T. (1968). Vegetation of Khao Yai National Park. *Natural History Bulletin of the Siam Society* **22**: 289–297.
Smitinand, T. (1969). The distribution of the Dipterocarpaceae in Thailand. *Natural History Bulletin of the Siam Society* **23**: 67–75.
Smitinand, T. (1977a). *Plants of Khao Yai National Park.* Friends of Khao Yai National Park, Bangkok.
Smitinand, T. (1977b). Vegetation and ground cover of Thailand. Technical Paper No. 1, Department of Forest Biology, Kasetsart University, Bangkok.
Smitinand, T. (1989). Thailand. In *Floristic Inventory of Tropical Countries,* ed. D. G. Campbell and H. D. Hammond, pp. 63–82. New York Botanical Garden, New York.
Smitinand, T., Santisuk, T. & Phengkhai, C. (1980). *The Manual of the Dipterocarpaceae of Mainland Southeast Asia.* Royal Forest Department, Bangkok.
Stamp, L. D. (1925). *The Vegetation of Burma.* Thacker, Spink and Co., Calcutta.
Stott, P. (1986). The spatial pattern of dry season fires in the savanna forests of Thailand. *Journal of Biogeography* **13**: 345–58.
Stott, P. (1988). The forest as phoenix: towards a biogeography of fire in mainland Southeast Asia. *Geographical Journal* **154**: 337–50.
Stott, P. (1990). Stability and stress in the savanna forests of mainland Southeast Asia. *Journal of Biogeography* **17**: 373–83.
Stott, P., Goldammer, J. G. & Werner, W. L. (1990). The role of fire in the tropical lowland deciduous forests of Asia. In *Fire in the Tropical Biota: ecosystem processes and global challenges,* ed. J. G. Goldammer, pp. 32–44. Springer-Verlag, Berlin.

Sukapanpotharam, V. (1979). Scarab beetle communities in deciduous dipterocarp and dry evergreen forests in northeastern Thailand. *Natural History Bulletin of the Siam Society* **28**: 55–100.
Sukwong, S. (ed.) (1976). *Quantitative Studies of the Seasonal Tropical Forest vegetation in Thailand.* Annual Report No. 1, Faculty of Forestry, Kasetsart University, Bangkok.
Sukwong, S. (ed.) (1977). *Quantitative Studies of the Seasonal Tropical Forest vegetation in Thailand.* Annual Report No. 2, Faculty of Forestry, Kasetsart University, Bangkok.
Sukwong, S. (1982). Growth of dry dipterocarp forest tree species. *Thai Journal of Forestry* **1**: 1–13.
Sukwong, S., Dhamanitayakul, P. & Pongumphai, P. S. (1975). Phenology and seasonal growth of dry dipterocarp forest species. *Kasetsart Journal* **9**: 105–13.
Sukwong, S. & Kaitpraneet, W. (1975). Influence of environmental factors on species distribution in the mixed deciduous forest on a limestone hill. *Kasetsart Journal* **9**: 142–8.
Suwannapinunt, W. & Siripatanadilok, S. (1982). *Khao Yai Ecosystem Project. Vol. III. Soil and Vegetation.* Faculty of Forestry, Kasetsart University, Bangkok.
Tangtham, N. & Sutthipibul, V. (1988). Effects of diminishing forest area on rainfall amount and distribution in northeastern Thailand. *Thai Journal of Forestry* **7**: 141–56.
Thiramongkol, N. (1984). Reviews of geomorphology of Thailand. In *First Symposium on Geomorphology and Quaternary Geology of Thailand*, ed. N. Thiramongkol and V. Pisutha-Arnond, pp. 6–23. Department of Geology, Chulalongkorn University, Bangkok.
Tongyai, P. (ed.) (1980). *The Sakaerat Environmental Research Station.* Thailand MAB Committee, Bangkok.
United Nations Educational, Scientific and Cultural Organization (1978). The natural forest: plant biology, regeneration, and tree growth. In *Tropical Forest Ecosystems*, pp. 180–215. UNESCO, Paris.
Vidal, J. (1956–8). La végétation du Laos. *Travaux du Laboratoire Forestier de Toulouse* **1**: 1–120 and **2**: 1–415.
Vidal, J. (1960). Les forêts du Laos. *Bois et Forêts des Tropiques* **70**: 5–21.
Vidal, J. (1966). Types biologiques dans la végétation forestière du Laos. *Bulletin de la Société Botanique de France, Memoires* **1966**: 197–203.
Werner, W. (1993). *Pinus* in Thailand. Franz Steiner, Stuttgart.
Wharton, C. H. (1966). Man, fire and wild cattle in north Cambodia. *Tall Timbers Fire Ecology Conference* **5**: 23–65.
Whitmore, T. C. (1984). *Tropical Rain Forests of the Far East*, 2nd edition. Clarendon Press, Oxford.
Williams, L. (1967). *Forests of Southeast Asia, Puerto Rico, and Texas.* USDA Agricultural Research Service, Washington.
Yoda, K. (1967). Comparative ecological studies on three main types of forest vegetation in Thailand. III. Community respiration. *Nature and Life in Southeast Asia* **5**: 83–148.
Yoda, K. & Kira, T. (1967). Comparative ecological studies on three main types of forest vegetation in Thailand. V. Accumulation and turnover of soil organic matter with notes on the altitude soil sequence on Khao (mt.) Luang, peninsular Thailand. *Nature and Life in Southeast Asia* **5**: 83–110.
Yoda, K., Nishioka, M. & Dhanmanonda, P. (1983). Vertical and horizontal distribution of relative illuminance in the dry and wet seasons in a tropical dry-evergreen forest in Sakaerat, NE Thailand. *Japanese Journal of Ecology* **33**: 97–100.
Zinke, P. (1989). Forest influences on the floods of 2531 and flood hazard mitigation. In *Safeguarding the Future*, pp. 1–31. US Agency for International Development, Bangkok.

6

The Cenozoic record of tropical dry forest in northern Latin America and the southern United States

ALAN GRAHAM & DAVID DILCHER

Introduction

The history of a plant community is reconstructed from remains preserved in a fragmentary fossil record. The completeness of the reconstructed community is, in part, a function of whether the community grew under conditions favorable to the preservation of macro- and microfossils. Macrofossil assemblages (leaves, fruits, seeds, wood) generally record plants growing near the site of deposition, and these afford opportunities to determine the general vegetation type and the paleoclimate by comparisons with modern analogs and by the use of leaf physiognomy. Microfossils (pollen, spores, trichomes, cuticles, phytoliths, microscopic organisms) provide a record of the regional vegetation, and also include species often not represented by macrofossils, such as annual, suffrutescent and herbaceous plants. Each methodology has its own set of strengths, weaknesses, practitioners and advocates, but the most complete history of plant communities is produced when both macro- and microfossil floras are available.

Northern Latin America

In the case of the tropical dry forest in northern Latin America, the reconstruction of its history is made more challenging by the facts that (1) dry environments have fewer sites of deposition, and less water for the transport of remains to these sites, and (2) there are very few well-preserved Tertiary macrofossil floras of significant size or diversity known for northern Latin America. An exception is the Oligocene San Sebastian flora from Puerto Rico, but it has not been studied or revised since Hollick's (1928) original publication. Consequently, the history of the dry forest in northern Latin America presently must be read from microfossil assemblages that accumulated various distances from the forests.

Three other limitations are particularly relevant to the dry forest. At the **species** level many neotropical dry forests and moist/wet forests are distinct. Only 11 of 298 tree species sampled from dry forest in Costa Rica occur in both (Murphy & Lugo, Chapter 2). However, as noted by Gentry (Chapter 7), over two-thirds of the 350 most common dry forest woody **genera** are also widespread in moist and wet forests: 'Thus at the most superficial level the dry forest flora may be characterized as essentially a depauperate subset of the moist/wet forest flora.' Although some fossil pollen types can be recognized to species, identifications are made mostly to the level of genus. This means that, in many cases, it is not possible to determine whether an identified microfossil represents a dry or a moist/ wet species if both occur in a single genus (e.g. *Bunchosia, Byrsonima, Coccoloba*). In other instances, however, such as in *Casearia*, which has a relatively good microfossil record in northern Latin America, the generic identification is adequate to provide useful information.

Another limiting factor for pollen analysis of dry forest paleofloras derives from the 'prevalence in dry forests of conspicuously-flowered, specialist-pollinated woody species . . . ' (*c.* two-thirds to three-fourths of neotropical dry forest woody taxa: Gentry, Chapter 7). Although pollen of entomophilous plants is preserved in Tertiary deposits from the neotropics, its representation is limited and the assemblages are clearly biased toward wind-pollinated types.

A third limitation is the lack of geographic separation between moist and dry forests, the former insinuating extensively into the latter in hydrologically favorable sites. In such mosaics, good conditions for fossilization are likely to be surrounded by the moister forest elements, so that dry forest is under-represented.

These aspects of the dry forest fossil record define a level of expectation for finding and interpreting evidence of its presence in microfossil assemblages. An overemphasis on the relatively few pollen types recovered leads to interpretations that may not be justified by the meager data base, and that exceed the sensitivity of the palynological method; an over-tentative assessment would miss the valuable, albeit limited, information that clearly does exist. A realistic expectation is that a generalized model depicting the history of dry forest can be constructed and assessed through a combination of several available sources of information. The Cenozoic plant microfossil assemblages in northern Latin

America provide direct evidence for **elements** of dry forest during early stages of its development, and during times when it was restricted to edaphic, physiographic (slope), and exposure-controlled areas as climatic conditions favored more moist vegetation (i.e. in the Early Tertiary). The methodology can also detect changes in composition from sequences of floras, and changes in environmental factors favoring expansion and coalescence of these elements into a vegetation type qualifying as one of the many possible versions of dry forest (i.e. as occurred in the Pliocene). It is then possible to look at vegetational trends elsewhere to determine whether the timing of the change in the paleobotanical signal is consistent with global or regional events favoring more extensive development of dry forest. These three components (direct evidence of elements, trends favoring coalescence, and global context) provide the most accurate form of analysis possible within the existing data, and given the strengths and limitations of the methods available.

Early and Mid Tertiary: elements of the tropical dry forest

Information on the history of dry forest in northern Latin America comes principally from the localities shown in Figures 6.1 and 6.2, and the plant microfossil assemblages listed in Table 6.1. The fossil evidence left by dry forest in these deposits can be read against the generic analysis provided by Gentry (Chapter 7, Table 7.5, columns pertaining to sites in northern Latin America: Chamela, México; Sian Ka'an, México; Santa Rosa, Costa Rica). In the following discussion an asterisk indicates that pollen

		Panama	Mexico	Puerto Rico	Costa Rica	Guatemala	Jamaica	Haiti
		Gatun	Paraje solo			Guanstatoya San Jacinto		
Miocene	U					Borrios		Artibonite
	M	La Boca						
	L	Cucaracha Culebra			Uscari			
Oligocene	U		Simojovel					
	M			San Sebastian				
	L							
Eocene	U	Gatuncillo						
	M						Chapelton	
	L							
Paleocene								

Figure 6.1. Age of Tertiary microfossil floras in northern Latin America. Stippling indicates study of the assemblage is complete; no stippling indicates study is in progress.

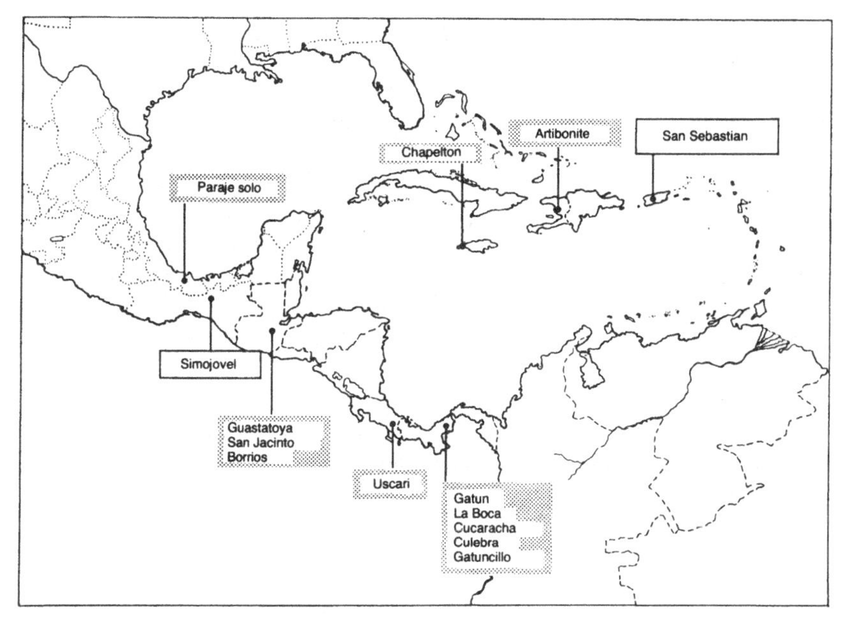

Figure 6.2. Distribution of Tertiary microfossil floras in northern Latin America. See legend of Figure 6.1 for explanation.

similar to that of the modern genus also occurs in the microfossil record (see Table 6.2 for notes on the identifications).

The Leguminosae is the dominant family in neotropical dry forests, and includes such genera as *Acacia**, *Bauhinia**, *Caesalpinia*, *Calliandra**, *Cassia*, *Centrosema*, *Crotalaria*, *Desmodium*, *Lonchocarpus*, *Mimosa**, *Phaseolus*, *Pithecellobium* and *Rhynchosia*. The second most prominent family, represented by a strong liana component, is the Bignoniaceae (*Arrabidaea** (as *Paragonia/Arrabidaea*), *Cydista*, *Tabebuia*). Others include the Sapindaceae (*Paullinia**, *Serjania**), Euphorbiaceae (*Acalypha**, *Euphorbia*, *Croton**, *Phyllanthus*, *Jatropha**), Flacourtiaceae (*Casearia**), Capparidaceae (*Capparis*), Apocynaceae (*Aspidosperma*, *Forsteronia*), Nyctaginaceae (*Guapira*, *Neea*, *Pisonia*), Polygonaceae (*Ruprechtia*, *Triplaris*, *Coccoloba**), and Malpighiaceae (*Bunchosia*, *Byrsonima**, *Malpighia**).

In Gentry's analysis, nine genera are listed that occur in all samples from the six phytogeographic regions surveyed (including the three neotropical localities noted above): *Casearia**, *Croton**, *Erythroxylum*, *Hippocratea*, *Randia**, *Serjania**, *Tabebuia*, *Trichilia* and *Zanthoxylum**. Fifteen more are found in dry forests in all but one region:

Table 6.1. *Data base for tracing the history of tropical dry forest in northern Latin America. Only relatively recent publications with reports of dry forest genera are included*

Assemblage/formation/locality	Age	Source
Guatemala	Quaternary	Leyden (1984)
Costa Rica	Quaternary	Horn (1985)
Panamá	Quaternary	Bartlett & Barghoorn (1973)
Gatun, Panamá	Mid Pliocene	Graham (1991a, b, c)
Paraje Solo, Veracruz, México	Mid Pliocene	Graham (1976a)
Uscari sequence, Costa Rica	Early Miocene	Graham (1987)
La Boca, Panamá	Early Miocene	Graham (1989a)
Cucaracha, Panamá	Early Miocene	Graham (1988b)
Culebra, Panamá	Early Miocene	Graham (1988a)
Simojovel group, Chiapas, México	Oligo-Miocene	Langenheim *et al.* (1967)
San Sebastian, Puerto Rico	Late Oligocene	Graham & Jarzen (1969)
Gatuncillo, Panamá	Late Eocene	Graham (1985)

*Acacia**, *Arrabidaea** (as *Paragonia/Arrabidaea*), *Bauhinia**, *Bursera**, *Capparis, Celtis**, *Coccoloba**, *Combretum** (as *Combretum/ Terminalia*), *Cordia**, *Eugenia** (as *Eugenia/Myrcia*), *Ficus**, *Macfadyena, Paullinia**, *Pithecellobium* and *Pterocarpus**. Five are found in several forest types, but are more prevalent in dry forest: *Bursera**, *Capparis, Erythroxylum, Randia**, and *Serjania**. The highly speciose genera *Ipomoea** and *Erythrina** also include species with representatives in dry forest.

These 51 genera provide a general characterization of the composition of dry forest in northern Latin America, and pollen similar to that of 25 (*c.* 50%) is present in Cenozoic deposits (Table 6.2; some adjacent northern South American occurrences are also included).

The only Paleogene microfossil flora from northern Latin America is the Mid (?) to Upper Eocene Gatuncillo Formation of central Panamá (Graham, 1985). Forty-seven palynomorphs have been identified, and these sort into three principal paleocommunities (following the classification and terminology of Holdridge, 1947; Holdridge *et al.*, 1971): tropical moist forest, tropical wet forest, and premontane wet forest. Other vegetation types, trending toward higher elevation (montane) or drier habitats, are poorly represented. On the basis of the similarity of the Gatuncillo flora with the modern lowland vegetation along the Atlantic coast of Panamá, and from other lines of evidence, the paleophysiography was insular and low-lying, and the paleoclimate was tropical moist to wet (comparable to that now present in the Atlantic lowlands of southern Central America). The climatic reconstruction is consistent with independent paleotemperature estimates based on $^{16}O:^{18}O$ ratios from deep-sea cores (Savin, Douglas & Stehli, 1975: 1507).

In its generic composition, the Gatuncillo assemblage provides no evidence for an extensive, well-developed dry forest within pollen transport range of the locality. Eight pollen types were recovered that could represent genera occurring in the modern dry forest (Table 6.2), but three of these (*Arrabidaea, Combretum, Eugenia*) have pollen similar to non-dry forest genera (e.g. *Paragonia, Terminalia, Myrcia*), and one (*Ficus*) is widespread in moist/wet forest types. The most convincing records for dry forest elements are *Casearia, Coccoloba, Paullinia* and *Serjania*, among those listed by Gentry (Chapter 7; *Cardiospermum* and *cf Tetragastris*, the latter more common in other kinds of habitats, are also listed by Graham (1985: Table 2) as possibly representing dry forest in southern Central America). All of these are present in trace amounts of 1% or less, with the exception of *Coccoloba* which reaches 3.5% and 2% in two of three samples (Graham, 1985, Table 1). The composition and numerical representations in the Gatuncillo flora, the paleobotanical evidence for a low-lying landscape and warm, moist to wet climates in the region, and independent evidence reflecting widespread tropical conditions on a more global scale, collectively argue against the presence of a well-defined tropical dry forest in southern Central America in the Paleogene. Rather, the record suggests that some elements of the dry forest with broader ecological amplitudes than their modern counterparts, or pre-adapted types in local edaphic, slope, and exposure-controlled habitats were present in the region. Widespread tectonic and volcanic activity during the Paleogene is amply demonstrated by extensive faulting, lava flows, volcanic ash, and tuffs (Stewart & Stewart, 1980). These geological conditions are consistent with the presence of temporary, shifting, open habitats allowing for the pre-adaptation of species to drier habitats, and local persistence of those already adapted to these conditions.

In the Late Oligocene and through the Early Miocene, the composition of microfossil floras available from southern México to southern Central America (Table 6.1) shows a similar level of representation for dry forest (Table 6.2). Only pollen records of *Acacia, Bursera* and *Casearia* are suggestive of dry forest, and these are present only in trace amounts.

Table 6.2. *The fossil record of tropical dry forest genera in northern Latin America*

(a) Arranged by taxon

Taxon	Age	Formation or locality	Source
Acacia	Late Oligocene	San Sebastian	8
	Early Miocene	Culebra	5
	Mid Pliocene	Paraje Solo	2
	Mid Pliocene	Gatun	7
Acalypha	Quaternary	Panamá	1
Arrabidaea[a]	Late Eocene	Gatuncillo	3
Bauhinia	Quaternary	Panamá	1
Bursera	Late Oligocene	San Sebastian	8
	Mid Pliocene	Paraje Solo	2
	Mid Pliocene	Gatun	7
	Quaternary	Costa Rica	10
	Quaternary	Panamá	1
	Quaternary	Venezuela (Lake Valencia)	12
Byrsonima	Quaternary	Costa Rica	10
	Quaternary	Panamá	1
Calliandra	Quaternary	Panamá	1
Casearia	Late Eocene	Gatuncillo	3
	Late Oligocene	San Sebastian	8
	Early Miocene	Culebra	5
Celtis	Mid Pliocene	Paraje Solo	2
	Quaternary	Costa Rica	10
	Quaternary[b]	Colombia	9
	Quaternary	Ecuador	13
	Quaternary	Venezuela (Lake Valencia)	12
Coccoloba	Late Eocene	Gatuncillo	3
	Mid Pliocene	Paraje Solo	2
	Quaternary[c]	Panamá	1
Combretum[d]	Late Eocene	Gatuncillo	3
	Early Miocene	Culebra	5
	Mid Pliocene	Paraje Solo	2
	Mid Pliocene	Gatun	7
	Quaternary	Panamá	1
Cordia	Quaternary	Costa Rica	10
	Quaternary	Panamá	1
	Quaternary	Colombia	9
Croton[e]	Quaternary	Costa Rica	10
	Quaternary	Panamá	1
	Quaternary	Colombia	9
	Paleocene[f]	Guyana	11
	Cenozoic	Venezuela	14
	Cenozoic	Venezuela	15
	Quaternary[g]	Venezuela	15
Erythrina	Mid Pliocene	Gatun	7
	Quaternary	Panamá	1

Table 6.2. (*cont.*)

Taxon	Age	Formation or locality	Source
Eugenia[h]	Late Eocene	Gatuncillo	3
	Late Oligocene	San Sebastian	8
	Early Miocene	Uscari sequence	4
	Early Miocene	Culebra	5
	Early Miocene	Cucaracha	6
	Mid Pliocene	Paraje Solo	2
	Mid Pliocene	Gatun	7
	Quaternary	Costa Rica	10
	Quaternary	Colombia	9
	Early Miocene[i]	Venezuela	14
	Pliocene[i]	Venezuela	14
Ficus	Late Eocene	Gatuncillo	3
	Quaternary	Costa Rica	10
	Quaternary	Ecuador	13
Ipomoea	Quaternary	Panamá	1
cf. *Jatropha*	Mid Pliocene	Gatun	7
cf. *Malpighia*	Mid Pliocene	Paraje Solo	2
Mimosa[j]	Mid Pliocene	Paraje Solo	2
	Quaternary	Costa Rica	10
	Quaternary	Panamá	1
	Quaternary	Colombia	9
Paullinia[k]	Late Eocene	Gatuncillo	3
	Mid Pliocene	Paraje Solo	2
	Mid Pliocene	Gatun	7
Pterocarpus	Quaternary	Panamá	1
Randia	Quaternary	Panamá	1
Serjania	Late Eocene	Gatuncillo	3
	Mid Pliocene	Paraje Solo	2
	Mid Pliocene	Gatun	7
Zanthoxylum	Quaternary	Costa Rica	10
	Quaternary	Panamá	1

(b) Arranged by age

Late Eocene
Arrabidaea,[a] *Casearia, Coccoloba, Combretum,*[d] *Eugenia,*[h] *Ficus, Paullinia, Serjania*

Late Oligocene
Acacia, Bursera, Casearia, Eugenia[h]

Early Miocene
Acacia, Casearia, Combretum,[d] *Eugenia*[h]

Mid Pliocene
Acacia, Bursera, Celtis, Coccoloba, Combretum,[d] *Erythrina, Eugenia,*[h] cf. *Jatropha,* cf. *Malpighia, Mimosa, Paullinia, Serjania*

Table 6.2. (*cont.*)

Quaternary *Acalypha, Bauhinia, Bursera, Byrsonima, Calliandra, Celtis, Coccoloba, Combretum, Cordia, Croton, Erythrina, Eugenia,*[h] *Ficus, Ipomoea, Mimosa,*[j] *Pterocarpus, Randia, Zanthoxylum*

[a] *Paragonia/Arrabidaea.*
[b] cf. *Celtis.*
[c] *Coccoloba*-type.
[d] Including *Combretum/Terminalia.*
[e] Including Crotonoideae.
[f] *Crototricolpites.*
[g] *Crototricolpites.* group.
[h] Including *Eugenia/Myrcia.*
[i] *Myrtaceidites.*
[j] Including Mimosoideae.
[k] Including cf. *Paullinia.*

Sources: 1. Bartlett & Barghoorn (1973); 2. Graham (1976a); 3. Graham (1985); 4. Graham (1987); 5. Graham (1988a); 6. Graham (1988b); 7. Graham (1991b); 8. Graham & Jarzen (1969); 9. Hooghiemstra (1984); 10. Horn (1985); 11. Leidelmeyer (1966); 12. Leyden (1985); 13. Liu & Colinvaux (1988); 14. Lorente (1986); 15. Muller, Di Giacamo & Van Erve (1987).

Late Tertiary: coalescence of elements

In the Pliocene significant changes become evident in the paleovegetation of Central America. Pollen of the Gramineae, which was virtually absent in older deposits, increases to a maximum of 7.5% in the Gatun Formation, suggesting a trend from closed tropical forests to locally open vegetation. By this time, *c.* 4 Ma (Ma = million years ago), uplift and compression forces had resulted in more numerous, larger islands and peninsulas in the isthmian region. Shortly thereafter (between *c.* 3 and 1.8 Ma), continuous emergent land surfaces connected North and South America (Stehli & Webb, 1985; Keller, Zenker & Stone, 1989), although the extent of the land connection varied with sea level fluctuations associated with the Pleistocene glaciations. The emergent land and greater elevations increasingly deflected strong northeast to southwest-trending winds (*nortes*), initiating differentiation of a wetter Atlantic and a drier Pacific side. Maximum elevations increased from an estimated 1200–1400 m in the Early Miocene, to about 1700 m in Gatun (Pliocene) time (Graham, 1989b). Correlated with this environmental and habitat diversity was greater biotic diversity. Individual taxa increase from 44, 55, 21 and 54 types in the Lower Miocene Uscari, Culebra, Cucaracha and

La Boca floras (for references see Table 6.2), to 110 in the Mid Pliocene Gatun Formation. In terms of vegetation, higher elevation (premontane) and drier habitat types are significantly better represented. The Gatun flora contains seven forest types with pollen of 11 or more taxa that can occur in each community: tropical moist forest (38 taxa), tropical wet (31), premontane wet (27), premontane moist (21), lower montane moist (12), premontane rain forest (11), and tropical dry forest (11). With reference to dry forest, the 11 pollen types listed by Graham (1991c: Table 2) are Gramineae (at 7.5%), *Acacia, Allophylus, Bursera, Cedrela, Ceiba, Combretum* (as *Combretum/Terminalia*), cf. *Jatropha, Posoqueria, Pseudobombax* and *Serjania*. Based on the listing in Gentry (Chapter 7), 12 kinds of pollen were recovered that potentially represent this community, compared with about four from the older floras (Table 6.2). Independent geological evidence suggests that conditions favorable to the expansion and coalescence of drier forest elements had developed by Gatun time, and the composition of the plant microfossil flora suggests that the dry forest, as a distinct, recognizable, and at least moderately extensive plant formation, began to appear in the region in Gatun time (about the Mid Pliocene, *c.* 4 Ma).

Identifying the approximate time of appearance of dry forest in southern Central America is, in turn, useful in refining certain aspects of the paleoclimate. Based on the total Gatun assemblage, and its similarity to the modern vegetation of lowland to moderate elevation habitats in northern (Atlantic side) Panamá, rainfall was probably similar to that at present on adjacent Barro Colorado Island (1900–3600 mm y^{-1}, average 2750 mm). With increasing elevations and continuous land connection, drier seasonal conditions became more pronounced on the Pacific side beginning in about the Mid Pliocene. The critical annual rainfall values for sustaining the modern dry forest in northern Latin America is about 1600 mm, with 5–6 months receiving less than about 100 mm (Gentry, Chapter 7; Murphy & Lugo, Chapter 2). If less than 100 mm of precipitation falls in more than 5–6 months, the vegetation trends toward dry evergreen thickets and seasonal deciduous forests; if less than 100 mm of precipitation falls in fewer than 5–6 months, moist forest types develop. This close association of the present dry forest with amounts and distribution of annual rainfall suggests that similar regimes characterized the drier parts of Pacific southern Central America as this vegetation type initially appeared and began its expansion.

To the north, data on the Cenozoic history of the vegetation is limited to the Oligo-Miocene Simojovel microfossil flora from Chiapas, México

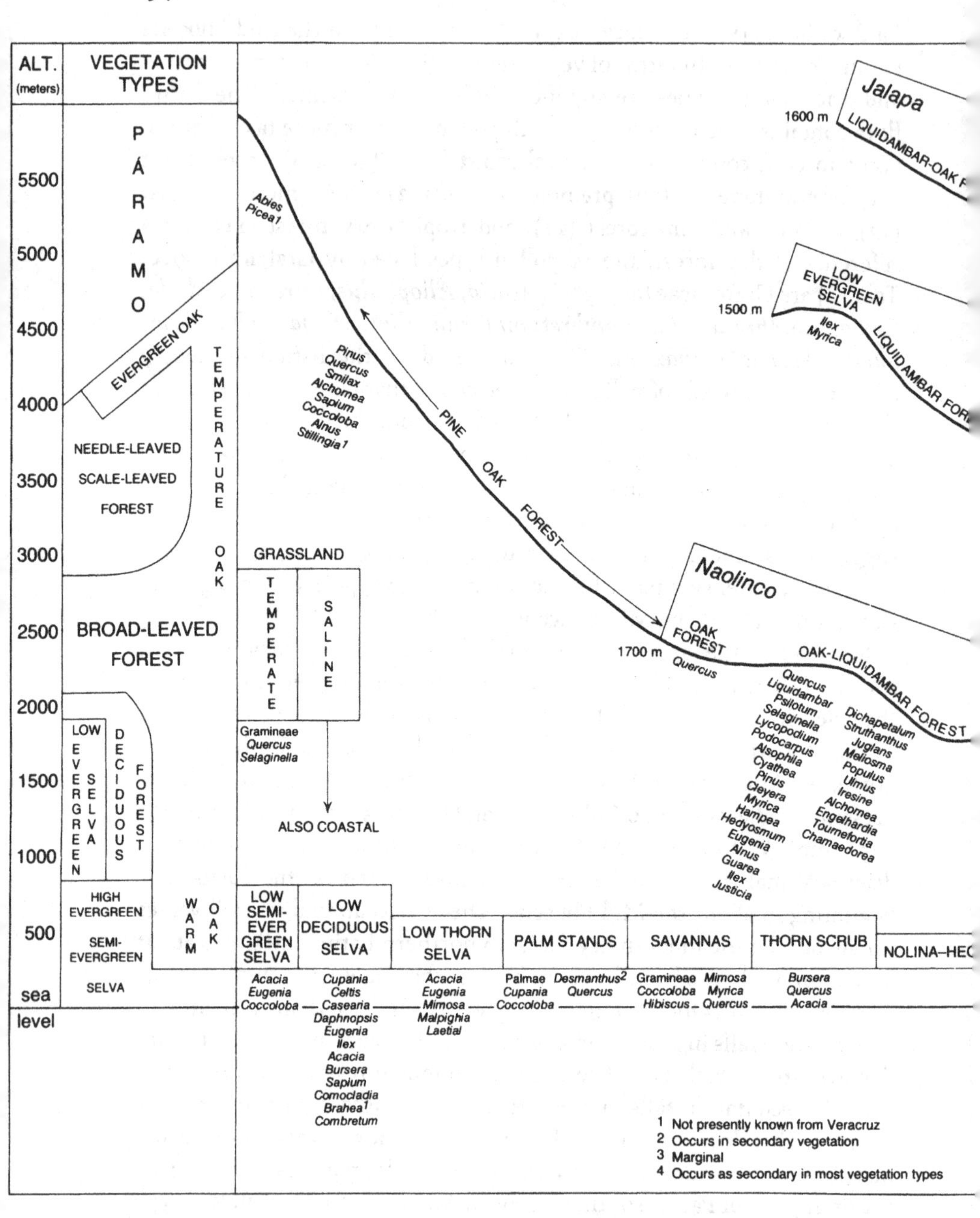

Figure 6.3. Zonation of modern communities along the eastern escarpment of the Mexican Plateau. Generic names represent pollen and spore types recovered from the Mid Pliocene Paraje Solo Formation, southern Veracruz state, México. 'Low deciduous selva' approximately equals tropical dry forest.

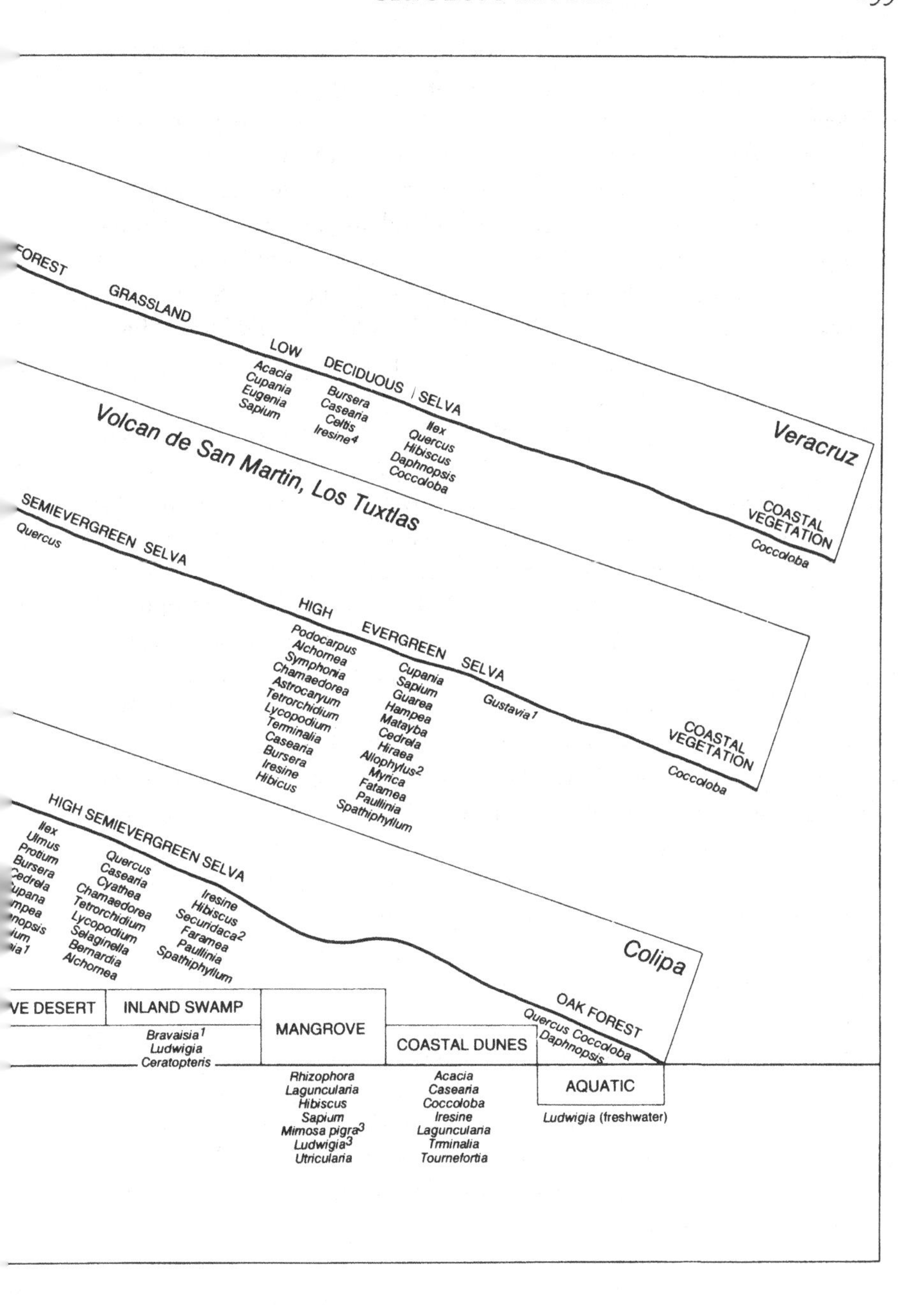
FOREST
GRASSLAND
LOW DECIDUOUS SELVA
Acacia
Cupania
Eugenia
Sapium
Bursera
Casearia
Celtis
Iresine4
Ilex
Quercus
Hibiscus
Daphnopsis
Coccoloba
Veracruz
COASTAL VEGETATION
Coccoloba
Volcan de San Martin, Los Tuxtlas
SEMIEVERGREEN SELVA
Quercus
HIGH EVERGREEN SELVA
Podocarpus
Alchornea
Symphonia
Chamaedorea
Astrocaryum
Tetrorchidium
Lycopodium
Terminalia
Casearia
Bursera
Iresine
Hibicus
Cupania
Sapium
Guarea
Hampea
Matayba
Cedrela
Hiraea
Allophylus2
Myrica
Fatamea
Paullinia
Spathiphyllum
Gustavia1
COASTAL VEGETATION
Coccoloba
HIGH SEMIEVERGREEN SELVA
Ilex
Ulmus
Protium
Bursera
Cedrela
Quercus
Casearia
Cyathea
Chamaedorea
Tetrorchidium
Lycopodium
Selaginella
Bernardia
Alchornea
Iresine
Hibiscus
Securidaca2
Faramea
Paullinia
Spathiphyllum
Colipa
OAK FOREST
Quercus Coccoloba
Daphnopsis
VE DESERT
INLAND SWAMP
Bravaisia1
Ludwigia
Ceratopteris
MANGROVE
Rhizophora
Laguncularia
Hibiscus
Sapium
Mimosa pigra3
Ludwigia3
Utricularia
COASTAL DUNES
Acacia
Casearia
Coccoloba
Iresine
Laguncularia
Trminalia
Tournefortia
AQUATIC
Ludwigia (freshwater)

(Langenheim, Hackner & Bartlett, 1967), and the Mid Pliocene Paraje Solo flora from Veracruz, México (Graham, 1976a). The Simojovel flora has been studied with reference only to environments of deposition for amber produced by *Hymenaea courbaril*. The microfossils identified by Langenheim *et al.* (1967) were from mangrove vegetation (*Rhizophora, Pelliciera, Pachira*), and from an upland temperate community (*Podocarpus, Engelhardtia* now = *Alfaroa/Oreomunnea*). A re-examination of this material is under way (A. Graham, personal communication), and *Ceratopteris, Pteris, Crudia, Quercus* and *Ulmus* have been added to the assemblage. The study is still in progress, but as yet no evidence for a well-defined dry forest is evident.

The younger Paraje Solo flora is about the same age as the Gatun flora of Panamá (Mid Pliocene, *c.* 4 Ma: Akers, 1979, 1981, 1984; Machain-Castillo, 1985). Physiographic diversity was extensive by that time because the region (near Coatzacoalcos, southeastern Veracruz state) is near the confluence of the eastern Transvolcanic Belt and the southern terminus of the Sierra Madre Oriental, both of which had been uplifted by the Mid Pliocene. As a result, a diverse and complex set of paleocommunities was present (Figure 6.3), arranged along elevational gradients extending from sea level to an estimated 2500 m (Graham, 1989b). It is interesting that the same number of possible dry forest genera (12) was independently suggested for the Paraje Solo paleocommunity (Graham, 1976a, as 'low deciduous selva') as for the Gatun flora (Graham, 1991c). Even though the number is coincidental, it is evident that several elements of the dry forest were present in the region by the Mid Pliocene. It is not possible to determine from the available fossil record if these had coalesced into at least local stands of dry forest, or if some of the pollen had blown in from the adjacent Yucatán Peninsula where dry forest presently occurs and may have been more extensively developed in the Pliocene; no direct paleobotanical evidence is available from the area. Nonetheless, elements or forests are better represented in the Mid Pliocene Paraje Solo flora than in the Oligo-Miocene Simojovel flora, consistent with estimates from southern Central America that the dry forest as a forest type began its principal development in northern Latin America during the Pliocene.

Quaternary: dynamics of the tropical dry forest

Older models of climatic change and glacial response for the Quaternary Period traditionally depict four major glacial advances (Wisconsin, Illinoian, Kansan, Nebraskan) separated by four interglacials (Aftonian, Yarmouth, Sangamon, and the present interglacial or Holocene). The duration of each phase was considered about equal through the 1.6 million year span of the Quaternary (Harland *et al.*, 1990), giving a relatively steady pace of about 225,000 years per glacial–interglacial cycle. Recent evidence based on oxygen isotope ratios in deep-sea cores (Johnson, 1982) reveals 18–20 glacial/interglacial cycles, as many as nine of them within the last 800,000 years. Each interglacial lasted about 10,000–15,000 years, while glacial climates characterized 80–90% of Quaternary time, and both were characterized by numerous lesser regressions and advances (Figure 6.4). The work of Davis (e.g. 1983) and others presents convincing evidence that associations within temperate forests did not retain the same composition after each climatic reshuffling. Wolfe (1973, 1979) suggests that the same applies to Tertiary vegetation, and there is no reason to believe that dry forest was any less fluid and dynamic with respect to changes in its range and composition.

In Guatemala, Leyden (1984) has shown that the transition between the Late Glacial and the Holocene at about 11,000 years ago was characterized by a period of aridity that affected the forests. The Late Glacial vegetation consisted of semiarid savanna and juniper scrub, subsequently replaced by a temperate forest, and then by a semi-evergreen seasonal forest with *Brosimum*, as presently characterizes the region. In addition to *Brosimum**, the modern version of this forest includes *Callophyllum*, *Pouteria*, *Cecropia**, *Bursera**, *Spondias*, *Cryosophila** and *Ficus**. Evidence for similar periods of aridity at the Late Glacial/Holocene boundary has been reported from other sites in the neotropics (see references in Leyden, 1984). Since dry forest is transitional between drier scrub communities and more moist forests, it is difficult to determine from the limited available pollen evidence alone how close these Late Glacial/Holocene communities from Guatemala compare with dry forest as presently defined and constituted. It seems clear, however, that between the dry savanna/juniper scrub and the evergreen seasonal forest described for the Late Glacial/Holocene boundary in Guatemala, conditions existed that were favorable for the development of one of several possible versions of dry forest.

When Pleistocene climatic history was earlier viewed as consisting of only four major transition periods between glacial and non-glacial conditions since 1.6 Ma, this and other factors allowed for a view of the modern dry forest, and other neotropical communities, as relatively stable and of great antiquity (Graham, 1976b). The new models, however, suggest a much more dynamic Quaternary environment. If the point designated **t** on Figure 6.4 indicates Quaternary climates compatible with the presence of a dry forest, as reflected by the pollen data from Guatemala, then each of the comparable 18–20 positions on the graph indicates a time of potential dry forest, with the slopes on either side reflecting more arid (savanna, juniper scrub), temperate, or moist/wet vegetation.

This scenario provides the first tentative model of dry forest history based on available paleobotanical data, formulated in association with general paleoenvironmental conditions as reconstructed from several independent approaches. This initial view, to be tested as new information becomes available, is that (a) elements of the dry forest have been present within the vegetation of northern Latin America since at least the Late Eocene in local edaphically and physiographically controlled dry sites, and persisted, as elements, through the earliest Pliocene; (b) these elements began coalescing into a community recognizable as one of several possible versions of dry forest in the Pliocene in northern Latin America; and (c) since that time dry forest has had a dynamic history,

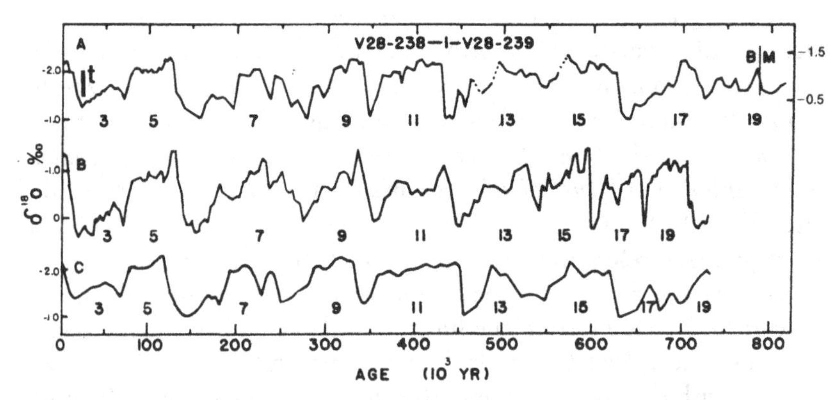

Figure 6.4. Oxygen isotope ratios in three deep-sea cores plotted against time. Low points on the curves represent ^{18}O enrichment, cooler temperatures, and periods of glacial advance; high points are warmer intervals. Isotopic time scale sources: ***A***, Johnson (1982); ***B***, Emiliani (1978); ***C***, Kominz *et al.* (1979). The bar labelled **t** shows the approximate position of Pleistocene aridity phase noted in the Guatemalan study (Leyden, 1984). From Johnson (1982); see also a modified version in Davis (1983: Fig. 1).

experiencing changes in range and composition with each new set of fluctuating environmental conditions.

Southern United States

Both microfossil and macrofossil floras are available from the southeastern and southwestern United States, allowing the composition of the assemblages and paleoenvironments to be estimated from the foliar physiognomy and modern analog methods. Foliar physiognomy of macrofossils can provide a relative guide to the moisture and temperature of paleoenvironments under which fossil leaves grew. Attempts to provide an understanding of both the moisture and temperature limits of the Claiborne, Mid Eocene, floras from western Kentucky and Tennessee were made by Dilcher (1973). Based upon the foliar physiognomy of the fossil leaves from these sediments he suggested that this 45 Ma flora was similar to the seasonally dry tropical forests of the Pacific coasts of Central America which extend from México to Panamá. There are also numerous individual taxa in the Eocene Claiborne floras from southeastern United States that are common to México, Central America and South America. These fossils demonstrate that there have been many successful ancient plant migrations that extended over long periods from the Late Cretaceous through the Eocene, in addition to those known from the Late Oligocene onwards, before continuous land connection was established in the Pliocene (Sharp 1953, 1966; Simpson & Neff, 1985; Taylor, 1988; Webb, 1985, 1991). These taxa lived in western Tennessee, Kentucky and Mississippi near a northward extension of the Gulf of México known as the Mississippi Embayment as well as along the western side of the embayment in Arkansas and Texas. Based upon the foliar physiognomy of the floras discussed above, and the taxa they contain, seasonally dry tropical environments were already existing in areas adjacent to northern México early in the Tertiary.

However, the fossil floras consist of a variety of taxa living today in diverse environments. The dominance of certain families suggests an association that still continues today in seasonally dry tropical forests. The most abundant family in the fossil floras is the Leguminosae, especially the tribes Caesalpiniae, Mimoseae, Swartzieae and Sophoreae (Herendeen, 1990). The second most abundant family is the Lauraceae. Other families, based on both the microfossil and macrofossil record, included the Euphorbiaceae, Sapotaceae, Moraceae, Sapindaceae, Bombacaceae, Anacardiaceae, Arecaceae (Palmae), Araceae, Malpighiaceae, Juglandaceae,

Oleaceae, Fagaceae, Nyctaginaceae, Annonaceae, Myrtaceae, Zingiberaceae, Gentianaceae, Loranthaceae, Hamamelidaceae, Magnoliaceae, Theaceae, Cornaceae and Salicaceae (Grote, 1989; Taylor, 1988). Several of the specific taxa known in these families, such as *Philodendron*, are found today in more moist tropical environments.

It has been suggested that most of the macrofossil remains are derived from plants living in river valley environments and could be considered members of a fossil gallery forest community. Some taxa such as *Nypa* and *Hippomane* represent plants from special environments such as coastal mangroves and swamps. Most macrofossils are preserved in clay deposits that trapped the local vegetation living in the flood plains. Thus, the fossil associations from the Mississippi Embayment floras represent a mixture of plants living in different tropical environments today, only some of which are found in moderately dry habitats. The same applies to elements in the microfossil assemblages. Most of the elements from Eocene sediments in southeastern United States are related to taxa found in more humid/mesic areas today, and very few of the macrofossils at the generic or species level, or microfossils at the generic level, can be related to extant taxa of dry forest.

There is no evidence for an extensive or well-developed dry forest in western United States or Baja California during the Cenozoic (D. Axelrod, personal communication, 1992). However, in contrast to the southeastern United States, the Early Tertiary plant fossil record from the West includes a larger number of individual elements that are members of the dry forest today, or represent lineages that could have given rise to members of this vegetation. The late Mid Eocene Green River flora of Colorado and Utah (MacGinitie, 1969) contains macrofossils identified as *Bursera*, *Caesalpinia*, *Celtis*, *Erythrina* and *Eugenia*. These and other records have been reviewed by Axelrod (1975, 1989), Axelrod & Raven (1985), and Raven & Axelrod (1978). By the Mid Eocene subhumid elements were present that had been derived from older (Cretaceous and Paleogene) laurophyllous forests adapted to spreading dry climates. In North America this vegetation occupied parts of southwestern United States and is thought to have extended into northern and western México. Unfortunately there are no Tertiary floras (micro- or macrofossil) known from western México and, indeed, none from the entire Pacific side through Central America. With cooler and drying climates after the Late Eocene these subhumid elements spread and are represented in Mid Tertiary and later floras in western United States. Raven & Axelrod (1978: 21) write, 'These Miocene floras from southeastern California also have numerous taxa representing thorn scrub vegetation, distributed in *Acacia, Bursera, Caesalpinia, Cardiospermum, Colubrina, Dodonaea, Ficus, Randia, Zizyphus* and others.' They also report (1978: 45) that 'A number of genera that are recorded in Late Cretaceous to Eocene floras, where they are in tropical savanna to thorn forest vegetation, may well have been present over the area of southeastern California by the Eocene, including *Ficus* and *Morus* (Moraceae), *Acacia, Cassia, Caesalpinia,* and *Pithecellobium* (Leguminosae); *Colubrina* and *Zizyphus* (Rhamnaceae); *Cardiospermum* and *Sapindus* (Sapindaceae); *Bumelia* and *Sideroxylon* (Sapotaceae); *Bombax* (Bombacaceae); and others.' These reflect a community more arid than dry forest, but several of the same elements are present. Their coalescence in the north was in the form of chaparral and sclerophyll woodlands adapted to both summer-wet and summer-dry rainfall regimes. It seems likely that a more humid and tropical version of this Tethyan vegetation extended southward along the western side of northern Latin America, augmented by local derivatives of the moist/wet forests, to constitute an early form of dry forest. From the meager Cenozoic paleobotanical record available for northern Latin America, it appears at present that the dry forest as a recognizable vegetation type appeared there in the middle to Late Pliocene (Graham, 1976a, 1991c) and especially in the Plio-Pleistocene (Leyden, 1984). This is consistent with Axelrod's (1975) estimate of the recent origin of the modern subhumid communities. In the absence of an adequate fossil record from the Pacific slope from México through Panamá, where dry forest is now extensively developed, a principal region of origin for dry forest goes, in part by default, to the Madrean area of northern and central western México, from early lineages and pre-adapted elements in the Paleogene, to a recognizable community in the Pliocene and Plio-Pleistocene.

Summary

1. Evidence for the tropical dry forest in northern Latin America presently must be read from the plant microfossil record; there are no extensive, recently studied macrofossil floras from the region.
2. The oldest occurrence (Mid to Late Eocene, Gatuncillo Formation, Panamá) is in the form of individual elements or small populations, probably growing in edaphically controlled habitats within an otherwise mesic environment. These gradually increase in kind and

diversity through the Mid Tertiary, but the first evidence of coalescence into a recognizable community, from the meager data base available, is in the Miocene/Pliocene (Gatun Formation, Panamá).

3. In the southeastern United States, macrofossil evidence indicate that a seasonally dry tropical forest developed in the Mid Eocene, but was gradually eliminated thereafter by increasingly cooler temperatures and other factors. In the southwestern United States elements of a dry savanna–woodland–scrubland were also present by the Mid Eocene, but later tended toward more xeric communities.
4. The chronological resolution is poor, but the sequence seems to be development from early lineages in the Eocene to coalescence in the Mio-Pliocene. In the absence of adequate data from México, one likely site of early occurrence goes temporarily by default to the Madrean area of northern and central western México where tropical dry forest is now extensively developed.

Acknowledgements

The authors gratefully acknowledge comments and suggestions provided by Daniel Axelrod. Research was supported by NSF grant BSR 8819771 (A. G.).

References

Akers, W. H. (1979). Planktic foraminifera and calcareous nannoplankton biostratigraphy of the Neogene of México. *Tulane Studies in Geology and Paleontology* **15**: 1–32.

Akers, W. H. (1981). Planktic foraminifera and calcareous nannoplankton biostratigraphy of the Neogene of México. *Tulane Studies in Geology and Paleontology* **16**: 145–8.

Akers, W. H. (1984). Planktic foraminifera and calcareous nannoplankton biostratigraphy of the Neogene of México. *Tulane Studies in Geology and Paleontology* **18**: 21–36.

Axelrod, D. I. (1975). Evolution and biogeography of Madrean-Tethyan sclerophyll vegetation. *Annals of the Missouri Botanical Garden* **62**: 280–334.

Axelrod, D. I. (1989). Age and origin of chaparral. In *The California Chaparral: paradigms reexamined*, ed. S. C. Keeley, pp. 7–19. Natural History Museum of Los Angeles County, Los Angeles.

Axelrod, D. I. & Raven, P. H. (1985). Origins of the cordilleran flora. *Journal of Biogeography* **12**: 21–47.

Bartlett, A. S. & Barghoorn, E. S. (1973). Phytogeographic history of the Isthmus of Panamá during the past 12,000 years (a history of vegetation, climate and sea-level change). In *Vegetation and Vegetational History of Northern Latin America*, ed. A. Graham, pp. 203–99. Elsevier, Amsterdam.

Davis, M. B. (1983). Quaternary history of deciduous forests of eastern North America and Europe. *Annals of the Missouri Botanical Garden* **70**: 550–63.
Dilcher, D. L. (1973). A paleoclimatic interpretation of the Eocene floras of southeastern North America. In *Vegetation and Vegetational History of Northern Latin America*, ed. A. Graham, pp. 39–59. Elsevier, Amsterdam.
Emiliani, C. (1978). The cause of the ice ages. *Earth and Planetary Science Letters* **37**: 349–52.
Graham, A. (1976a). Studies in neotropical paleobotany. II. The Miocene communities of Veracruz, México. *Annals of the Missouri Botanical Garden* **63**: 787–842.
Graham, A. (1976b). Late Cenozoic evolution of tropical lowland vegetation in Veracruz, México. *Evolution* **29**: 723–35.
Graham, A. (1985). Studies in neotropical paleobotany. IV. The Eocene communities of Panamá. *Annals of the Missouri Botanical Garden* **72**: 504–34.
Graham, A. (1987). Miocene communities and paleoenvironments of southern Costa Rica. *American Journal of Botany* **74**: 1501–18.
Graham, A. (1988a). Studies in neotropical paleobotany. V. The lower Miocene communities of Panamá – the Culebra Formation. *Annals of the Missouri Botanical Garden* **75**: 1440–66.
Graham, A. (1988b). Studies in neotropical paleobotany. VI. The lower Miocene communities of Panamá – the Cucaracha Formation. *Annals of the Missouri Botanical Garden* **75**: 1467–79.
Graham, A. (1989a). Studies in neotropical paleobotany. VII. The lower Miocene communities of Panamá – the La Boca Formation. *Annals of the Missouri Botanical Garden* **76**: 50–66.
Graham, A. (1989b). Late Tertiary paleoaltitudes and vegetational zonation in México and Central America. *Acta Botanica Neerlandica* **38**: 417–24.
Graham, A. (1991a). Studies in neotropical paleobotany. VIII. The Pliocene communities of Panamá – introduction and ferns, gymnosperms, angiosperms (monocots). *Annals of the Missouri Botanical Garden* **78**: 190–200.
Graham, A. (1991b). Studies in neotropical paleobotany. IX. The Pliocene communities of Panamá – angiosperms (dicots). *Annals of the Missouri Botanical Garden* **78**: 201–23.
Graham, A. (1991c). Studies in neotropical paleobotany. X. The Pliocene communities of Panamá – composition, numerical representations, and paleocommunity paleoenvironmental reconstructions. *Annals of the Missouri Botanical Garden* **78**: 465–75.
Graham, A. & Jarzen, D. M. (1969). Studies in neotropical paleobotany. I. The Oligocene communities of Puerto Rico. *Annals of the Missouri Botanical Garden* **56**: 308–57.
Grote, P. (1989). Selected fruits and seeds from the middle Eocene Claiborne formation of southeastern North America. Ph.D. thesis, Indiana University, Bloomington.
Harland, W. B., Armstrong, R. L., Craig, L. E., Smith, A. G. & Smith, D. G. (1990). *A Geologic Time Scale* (text and wall chart). Cambridge University Press, Cambridge.
Herendeen, P. S. (1990). Fossil history of the Leguminosae from the Eocene of Southeastern North America. Ph.D. thesis, Indiana University, Bloomington.
Holdridge, L. R. (1947). Determination of world plant formations from simple climatic data. *Science* **105**: 367–68.
Holdridge, L. R., Grenke, W. C., Hatheway, W. H., Liang, T. & Tosi, J. A., Jr (1971). *Forest Environments in Tropical Life Zones: a pilot study*. Pergamon Press, New York.
Hollick, A. (1928). Paleobotany of Porto Rico. *Scientific Survey of Porto Rico and the Virgin Islands* **7**: 177–393.

Hooghiemstra, H. (1984). Vegetational and climatic history of the High Plain of Bogotá, Colombia: a continuous record of the last 3.5 million years. *Dissertationes Botanicae* **79**: 1–368.

Horn, S. P. (1985). Preliminary pollen analysis of Quaternary sediments from Deep Sea Drilling Project site 565, western Costa Rica. *Initial Reports. Deep Sea Drilling Project* **84**: 533–47.

Johnson, R. G. (1982). Brunhes–Matuyama magnetic reversal dated at 790,000 year B.P. by marine–astronomical correlations. *Quaternary Research* **17**: 135–47.

Keller, G., Zenker, C. E. & Stone, S. M. (1989). Late Neogene history of the Pacific-Caribbean gateway. *Journal of South American Earth Sciences* **2**: 73–108.

Kominz, M. A., Heath, G. R., Ku, T. L. & Pisias, N. G. (1979). Brunhes time scales and the interpretation of climatic change. *Earth and Planetary Science Letters* **45**: 394–410.

Langenheim, J. H., Hackner, B. L. & Bartlett, A. S. (1967). Mangrove pollen at the depositional site of Oligo-Miocene amber from Chiapas, México. *Botanical Museum Leaflets, Harvard University* **21**: 289–324.

Leidelmeyer, P. (1966). The Paleocene and lower Eocene pollen flora of Guyana. *Leidse Geologische Mededelingen* **38**: 49–70.

Leyden, B. W. (1984). Guatemalan forest synthesis after Pleistocene aridity. *Proceedings of the National Academy of Sciences USA* **81**: 4856–59.

Leyden, B. W. (1985). Late Quaternary aridity and Holocene moisture fluctuations in the Lake Valencia Basin, Venezuela. *Ecology* **66**: 1279–95.

Liu, Kam-biu & Colinvaux, P. A. (1988). A 5200-year history of Amazon rain forest. *Journal of Biogeography* **15**: 231–48.

Lorente, M. A. (1986). Palynology and palynofacies of the upper Tertiary in Venezuela. *Dissertationes Botanicae* **99**: 1–222.

MacGinitie, H. D. (1969). The Eocene Green River flora of northwestern Colorado and northeastern Utah. *University of California Publications in Geological Sciences* **83**: 1–202.

Machain-Castillo, M. (1985). Ostracode biostratigraphy and paleoecology of the Pliocene of the Isthmian salt basin, Veracruz, México. *Tulane Studies in Geology and Paleontology* **19**: 123–39.

Muller, J., Di Giacomo, E. de, & Van Erve, A. W. (1987). A palynological zonation for the Cretaceous, Tertiary, and Quaternary of northern South America. *American Association of Stratigraphic Palynologists Contribution Series* No. **19**: 7–76.

Raven, P. H. & Axelrod, D. I. (1978). Origin and relationships of the California flora. *University of California Publications in Botany* **72**: 1–134.

Savin, S. M., Douglas, R. G. & Stehli, F. G. (1975). Tertiary marine paleotemperatures. *Geological Society of America Bulletin* **86**: 1499–1510.

Sharp, A. J. (1953). Notes on the flora of México: world distribution of the woody dicotyledonous families and the origin of the modern vegetation. *Journal of Ecology* **41**: 374–80.

Sharp, A. J. (1966). Some aspects of Mexican phytogeography. *Ciencia Mexicana* **24**: 229–32.

Simpson, B. B. & Neff, J. L. (1985). Plants, their pollinating bees, and the great American interchange. In *The Great American Biotic Interchange*, ed. F. G. Stehli and S. D. Webb, pp. 427–52. Plenum Press, New York.

Stehli, F. G. & Webb, S. D. (ed.) (1985). *The Great American Biotic Interchange*. Plenum Press, New York.

Stewart, R. H. & Stewart, J. L. (with the collaboration of W. P. Woodring) (1980). Geologic Map of the Panamá Canal and Vicinity, Republic of Panamá. Scale 1:100,000. *US Geological Survey Miscellaneous Investigations Map* **I-232**.

Taylor, D. W. (1988). Paleobiogeographic relationships of the Paleogene flora from the southeastern USA: implications for west Gondwanaland affinities. *Palaeogeography, Palaeoclimatology, Palaeoecology* **66**: 265–75.

Webb, S. D. (1985). Late Cenozoic mammal dispersals between the Americas. In *The Great American Biotic Interchange*, ed. F. G. Stehli and S. D. Webb, pp. 201–18. Plenum Press, New York.

Webb, S. D. (1991). Ecogeography and the great American interchange. *Paleobiology* **17**: 266–80.

Wolfe, J. A. (1973). An interpretation of Alaskan Tertiary floras. In *Floristics and Paleofloristics of Asia and Eastern North America*, ed. A. Graham, pp. 201–33. Elsevier, Amsterdam.

Wolfe, J. A. (1979). Temperature parameters of humid to mesic forests of eastern Asia and relation of forests of other regions of the northern hemisphere and Australasia. *US Geological Survey Professional Paper* **1106**.

7

Diversity and floristic composition of neotropical dry forests

ALWYN H. GENTRY

Introduction

Many studies of neotropical dry forest have tended to treat them in a very broad context, typically focusing on how they relate to or are different from moist or wet forests (e.g. Holdridge *et al.*, 1971; Rzedowski, 1978; Gentry, 1982a, 1988; Hartshorn, 1983). Others have taken them as a relatively tractable surrogate for the more diverse moist or wet forests (e.g. Janzen, 1983, 1984, 1988; Hubbell, 1979) where taxonomy often poses severe limitations for the resolution of biologically interesting questions. Other authors have concentrated on the interesting physiological adaptations of dry forest organisms to seasonal water stress (e.g. Medina, Chapter 9; Holbrook, Whitbeck & Mooney, Chapter 10 and included references) or on various aspects of nutrient flow and biomass (e.g. Lugo *et al.* 1978; Murphy & Lugo 1986a, b). In addition there have been floristic and community ecological studies of individual dry forests (e.g. Troth, 1979; Valverde *et al.*, 1979; Thien *et al.*, 1982; Hartshorn, 1983; Heybrock, 1984; Lott, Bullock & Solís, 1987; Kelly *et al.*, 1988; Rico-Gray *et al.*, 1988; Arriaga & León, 1989; Cuadros, 1990; Saldias, 1991; Dodson & Gentry, 1992; see also summaries for México in Rzedowski, 1991a, b, and for the chaco in Prado, 1993). However, there have been remarkably few attempts to focus on the distinctive floristic composition of dry forests as a whole or on how different dry forest plant communities differ from each other.

This review of dry forest floristics and diversity is based largely on the plants >2.5 cm diameter in a series of 28 samples of 0.1 ha. In part this is a subset of the data analysed in a previous more broadly focused review paper (Gentry, 1988); these data also overlap with the dry forest portion of the liana data set analysed in a recent overview of liana distribution (Gentry, 1991), and conclusions reached here about dry forest liana communities are largely anticipated in that summary. However, data for several additional dry forest sites are now available, making possible a more in-depth analysis of dry forest floristic composition and diversity. A few supplementary comments on the structure and reproductive biology of dry forests are also included in this review. In addition, this chapter will attempt to compare trends in endemism for different dry forests, a kind of analysis that has not previously been attempted with this data set.

Methods and study sites

Each forest was sampled with a series of ten 2 × 50 m transects, totalling 0.1 ha of sample area in which all plants >2.5 cm dbh (>2.5 cm maximum diameter in the case of lianas) were identified and recorded. Details of the sampling procedure are given elsewhere (Gentry, 1982a). These dry forest samples (Figure 7.1) thus form a subset of a data base of nearly 200 similar samples worldwide that can be directly compared with moist or wet forests or temperate zone forests sampled by the same technique and reported elsewhere (Gentry, 1982a, 1988, 1991; Lott *et al.*, 1987), as well as with each other. Several important data sets from southern Bolivia (Santa Cruz, Quiapaca, Curuyuqui, Yanaigua) and Paraguayan chaco (A. Taber, personal communication) were obtained after this manuscript was submitted and are not included in all of the analyses.

Since I have limited Antillean samples, data obtained by slightly different methodologies are also included for two Jamaican dry forest samples (Kapos, 1982 and personal communication) and one Puerto Rican one (Murphy & Lugo, 1986a). Both of these sites used large plots divided into 10 × 10 m subplots. The original Jamaican data from Round Hill used a 3 cm lower diameter and did not include lianas; in one sample area I added in the few smaller stems and lianas, but in the other sample area (Round Hill slope) only trees >3 cm dbh are included in the data analysed here.

Supplementary data on the floristic composition of several dry forest field stations are also available from local florulas or checklists, including Santa Rosa National Park, Costa Rica (Janzen & Liesner, 1980), Capeira, Ecuador (Dodson & Gentry, 1991), Chamela, México (Lott, 1985), Sian Ka'an, México (Durán & Olmsted, 1987), the Estación Biológica de Los Llanos, Venezuela (Aristeguieta, 1966), Hato Masaguaral, Venezuela (Troth, 1979), and Round Hill, Jamaica (Kelly *et al.*, 1988). (A revised checklist for Chamela (Lott, 1993) has been published since this manuscript was written.)

Categorization of most of the low rainfall sites of Gentry (1988, 1991; Table 7.1) as 'dry forest' is straightforward, with annual precipitation of <1600 mm closely correlated with physiognomically and floristically recognizable dry forest. However, classification of several sample sites

Figure 7.1. Distribution of neotropical dry forests (shaded), indicating 0.1 ha samples (circles) and local florula sites (arrows).

with 1400–1800 mm annual rainfall is somewhat problematic. This problem is most acute in coastal Brazil where most of lowland Espírito Santo and Rio de Janeiro states have relatively low annual precipitation which is evenly distributed throughout the year. The study sites from Linhares, Espírito Santo (Peixoto & Gentry, 1990) and Jacarepagua, near Rio de Janeiro, with 1400–1500 mm annual rainfall, are here classified as moist forest and excluded from this analysis. A study site from 1000 m elevation at Parque El Rey, Salta, Argentina, with 1500 mm of annual precipitation is also rejected from these analyses except as otherwise indicated. Similar problems arise in Africa where a study site at Makokou, Gabon, has only 1755 mm of annual rainfall but a diverse high-canopy forest better classified as 'moist' rather than 'dry'.

There is a different kind of problem in site classification in northern Colombia where the Loma de los Colorados and Coloso sites are squarely in the middle of a low-rainfall area mapped as Tropical Dry Forest on the Holdridge Life Zone map of Colombia. Nevertheless both forests include patches of large trees belonging to moist forest floristic elements that seem highly anomalous for dry forest, and both have greater species richness than other 'dry' forests. These tiny isolated patches are the last forest remnants in the entire region, so their original status is difficult to evaluate. It is possible that the relatively strong 'double' wet season of this region, unusual in the context of the neotropics, ameliorates some of the expected drought stress, as Terborgh (1986) has implied may occur in similar climatic regimes in the Old World. However, it seems equally likely that what we see here represents an early example of the kind of climatic change associated with massive deforestation that is now being predicted for much of the Earth's tropics (e.g. Salati, Marques & Molion, 1978). The original rainfall of this area may have been significantly greater than the present rainfall and classification of the remnants of semievergreen vegetation as 'dry forest' is somewhat tenuous; these two sites are omitted from most of the analyses.

The samples here recognized as dry forest represent a broad geographic cross section of the neotropics (Figure 7.1). They include sites from 11 different countries and from all of the main dry forest regions except those of the Brazilian shield (see Eiten, 1972, 1978; Sampaio, Chapter 3). All have between 700 and 1600 mm of annual precipitation and a strong dry season. All of the continental sites have several months a year when the woody vegetation is largely or completely deciduous; the Antillean dry forests differ dramatically in having more sclerophyllous leaves and being mostly evergreen or at least semievergreen (Kapos, 1982; Murphy & Lugo, 1986a; Kelly *et al.*, 1988).

Another complication in comparing different dry forests comes from the fact that edaphic factors often confound the climatic ones. For example, forests growing on limestone in relatively moist climates often resemble true dry forests both physiognomically and floristically. Many of the peculiarities of West Indian dry forests are due to edaphic factors associated with prevalence of limestone. I have inadequate data to address this problem adequately and the largely limestone-induced dry forests (e.g. Mogotes de Nevarez, Coloso) are here lumped together with climatically induced ones. (See Murphy & Lugo, Chapter 2, for a more complete discussion.)

Excluded from this analysis are sites from above 1000 m elevation (except in a few cases where data for a 1300 m site at Salta, Argentina are included) and data from drier thorn scrub vegetation which is characterized by shorter stature, more open, scrubby 'forest' (e.g. most of the Península de Guajira of Colombia, most of Falcón and Lara states in northwestern Venezuela, the Península de Santa Elena of Ecuador) as well as the sometimes nearly closed canopy but single-species dominated forests that frequently occur in the intermediate precipitation zone (*c.* 500 mm y^{-1}) between the thorn scrub and closed-canopy dry forest vegetation. Aside from having a different structure (generally with the trees shrubbier and farther apart), not amenable to the sampling technique employed here, most of these drier sites have been decimated by goats, and all have much lower woody plant diversity than the closed-canopy dry forests to which this analysis is restricted. While the chaco might be classified as thorn scrub, it is typically much denser, with a taller canopy, and is here considered as a type of dry forest; my chaco samples are in part from unpublished data of A. Taber (personal communication). Savannas (Huber, 1987) are also excluded from this analysis, e.g. most of the Brazilian cerrado (Eiten, 1972, 1978; Pinto, 1990), most of the Venezuelan/Colombian llanos (Sarmiento, 1984), and the Peruvian/Bolivian Pampas del Heath (Denevan, 1980; A. Gentry & P. Núñez, personal communication) and Beni savannas (Beck, 1984; Haase & Beck, 1989).

For floristic and phytogeographical analysis, the 28 dry forest samples have been categorized into six regional groupings. These follow the phytogeographical regions of Gentry (1982b), except that the sites in México and Central America are analysed separately, and the disjunct patch of dry forest near Tarapoto, Perú, is rather arbitrarily classed with the geographically nearby 'northern Andean' coastal Ecuador/Tumbes dry forests rather than with the far-away 'southern Andean' ones of Bolivia and Argentina (Table 7.1). For some analyses the sites in these six geographical regions are variously combined into larger groupings. For example, all dry forest sites from Costa Rica to Venezuela and Perú are sometimes lumped together as 'equatorial continental' sites to form a kind of core data set against which to compare the other geographically more peripheral sites.

Diversity patterns

It is well known that tropical forests are much more diverse than temperate forests (Figure 7.2). Perhaps less appreciated is the tremendous difference in diversity among different tropical forests (Figure 7.2). At least within the neotropics, the relation between plant species richness and rainfall is highly significant (Figure 7.3; Gentry, 1988). In general dry forests are less diverse than wet or moist forests (e.g. Gentry, 1982a, 1988), although they are more diverse than some subtropical moist forest types. Typically lowland dry forests have about 50–70 species >2.5 cm

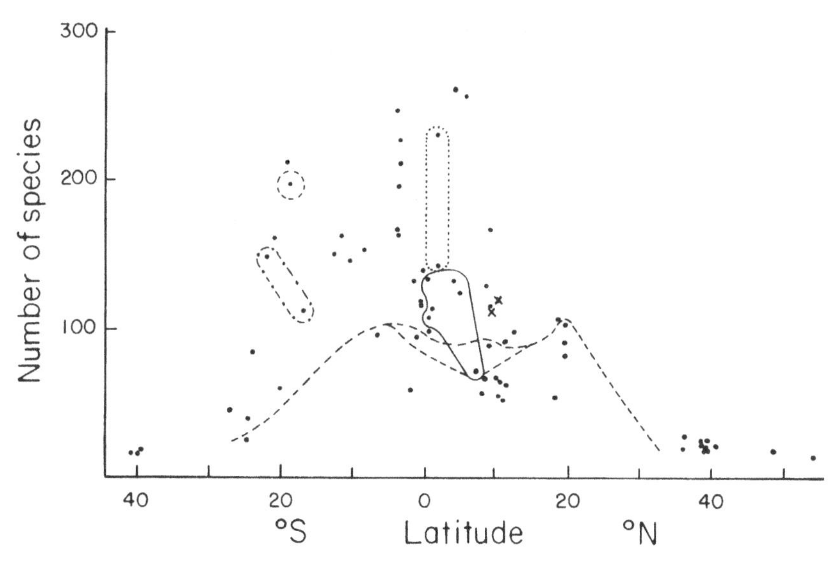

Figure 7.2. Species richness of 0.1 ha samples of lowland (<1000 m) forest as a function of latitude. Dashed line separates dry forest (bottom) from moist and wet forest (top) with intermediate sites (moist forest physiognomy despite relatively strong dry season) indicated by alternate lines. X = the anomalous Coloso and Loma de los Colorados sites in northern Colombia (see text). Data mostly from Gentry (1988).

Table 7.1. *Dry forest study sites*

Site	Coordinates	Elevation (m)	Rainfall (mm)	Temperature (°C)	Dry months
West Indies					
Puerto Rico					
Guánica[1]	17° 57′ N, 65° 52′ W	175	860	25.1	6(–8)
Mogotes de Nevarez[2]	18° 25′ N, 66° 15′ W	50	1500		
Jamaica					
Round Hill (slope)[3a]	17° 50′ N, 77° 15′ W	40	1200	24	
Round Hill (top)[3b]	17° 50′ N, 77° 15′ W	40	1200	24	
México (Jalisco)					
Chamela (upland 1)[4]	19° 30′ N, 105° 3′ W	50	748	25	6–8
Chamela (upland 2)[4]	19° 30′ N, 105° 3′ W	50	748	25	6–8
Chamela (arroyo)[4]	19° 30′ N, 105° 3′ W	50	748	25	6–8
Southern Central America (Costa Rica)					
Guanacaste (upland)[c]	10° 30′ N, 85° 10′ W	100	1600	25	6
Guanacaste (gallery)[c]	10° 30′ N, 85° 10′ W	50	1600	25	6
Southern subtropics					
Argentina					
Salta, Salta[d]	24° 40′ S, 65° 30′ W	1300	712		
Riachuelo, Corrientes	27° 30′ S, 58° 50′ W	60	1200		
Parque El Rey, Salta[5]	24° 45′ S, 64° 40′ W	1000	1500		
Bolivia					
Chaquimayo, La Paz	14° 34′ S, 68° 28′ W	1000	1300		
Santa Cruz, Santa Cruz[6e]	17° 46′ S, 63° 4′ W	375	1171	24.3	6
Quiapaca, Santa Cruz[e]	18° 20′ S, 59° 30′ W	300			
Paraguay					
Fortín Teniente Acosta, Boquerón[7e]	18° 20′ S, 59° 30′ W	300			

Northern South America					
Colombia					
Galerazamba, Bolívar	10° 48′ N, 75° 15′ W	10	500	28	6–7
Tayrona, Magdalena[8]	11° 20′ N, 74° 2′ W	50	1500[f]	27	6–7
Loma de los Colorados, Bolívar	9° 58′ N, 75° 10′ W	240	770[g]	28	
Lomas de Santo Tomás, Tolima	4° 55′ N, 74° 50′ W	320	1500–1800?		
Coloso, Sucre	9° 30′ N, 75° 48′ W	300			
Venezuela					
Boca de Uchire, Anzoategui	10° 9′ N, 65° 25′ W	150	1200		
Estación Biol. Los Llanos, Guarico	8° 56′ N, 67° 25′ W	100	1312	27.5	4
Blohm Ranch, Guarico[g]	8° 34′ N, 67° 35′ W	100	1400		
Pacific coast of South America					
Ecuador					
Capeira, Guayas[10]	2° 0′ N, 79° 58′ W	50	804		8
Perro Muerte, Manabi	1° 36′ S, 80° 42′ W	440	1000		
Perú					
Cerros de Amotape, Tumbes[11]	4° 9′ S, 80° 37′ W	830	1430	22.1	
Tarapoto, San Martín	6° 40′ S, 76° 20′ W	500	1400		

[a] Data for plants >3 cm dbh in contiguous 0.1 ha plot.
[b] Data for plants >3 cm dbh in contiguous 0.1 ha plot supplemented by my own data for plants >2.5 cm, <3 cm dbh.
[c] Both Guanacaste data sets are conglomerated from transects from two nearby localities: Hacienda La Pacífica and Palo Verde.
[d] Site above 1000 m elevation excluded from lowland forest analyses.
[e] Data obtained during 1991 subsequent to drafting of this manuscript and not included in most analyses.
[f] Perhaps lower since 1976–7 total at exact study site was only 600 mm.
[g] Weather data from Cartagena, 80 km northwest.
Sources: Data are my own except the following: 1. Murphy & Lugo (1986a, b); 2. G. Proctor & A. Gentry, personal communication; 3. Kapos (1982); Kelly *et al.* (1988); 4. Lott *et al.* (1987), their data set 1 includes subplots scattered over 1600 ha; 5. Brown *et al.* (1985), relatively moist forest excluded from most analyses; 6. Saldias (1991); 7. A. Taber (personal communication); 8. Heybrock (1984); 9. Troth (1979); 10. Dodson & Gentry (1992); 11. C. Díaz (personal communication).

diameter in 0.1 ha, moist semievergreen forests 100–150 species, wet evergreen forests 150–200 species, and pluvial forests 200–250 species in equivalent samples. The average species richness for the 23 neotropical dry forest samples is 64.9 (69 if the two dubiously dry forest northern Colombian sites are included), compared with the lowland wet and moist forest average of 152 species (Gentry, 1986). Thus from a global perspective dry forests are distinctly unremarkable in their α-diversity.

We might further anticipate from the overall correlation between diversity and precipitation that the drier a dry forest is, the less speciose (and perhaps less deserving of conservational focus) it will be. Surprisingly, that is not the case. Except for the most arid regions, which tend to have open single-species dominated 'forests' (e.g. *Prosopis*, or *Loxopterygium*: Figure 7.4), there is no significant change in diversity of dry forest communities with precipitation (Figure 7.5). Apparently, once the critical rainfall threshold needed to maintain a closed-canopy dry forest is achieved, increases in the amount of precipitation have a negligible effect on species richness until rainfall values high enough to

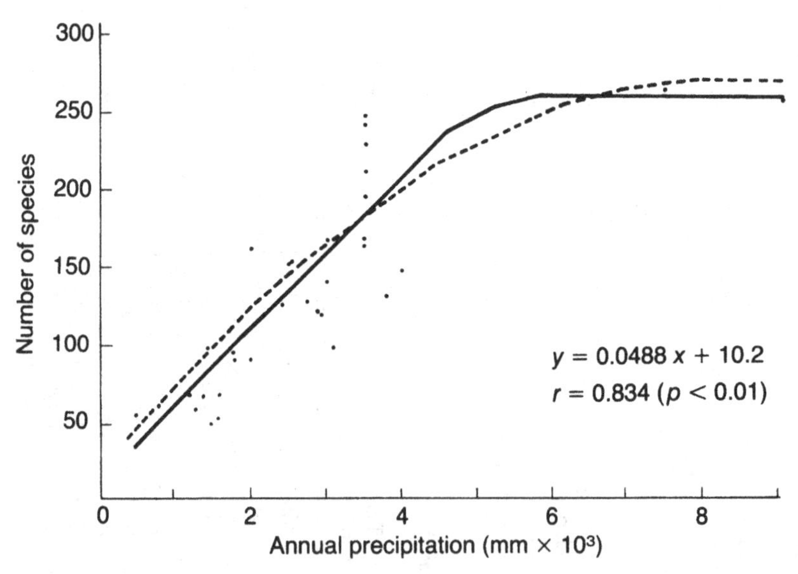

Figure 7.3. Numbers of species in 0.1 ha samples of lowland neotropical forest vs annual precipitation. Solid line: linear regression for sites with <5000 mm annual precipitation with visually estimated asymptote. Dashed line: computer generated curve: $y = 12.37 + 0.0613x - 0.000003598x^2$. The curve is displaced slightly upward from the data points, since questionable morphospecies and lost specimens are treated as distinct species by the computer while the data points represent best estimates of species numbers. Modified from Gentry (1988).

maintain the more diverse moist forest are attained. This lack of a significant correlation between species richness and precipitation **within** dry forest vegetation also holds for different subsets of the community: trees >2.5 cm dbh, trees >10 cm dbh, and lianas. Only dry forest liana diversity comes close ($r^2 = 0.15$, $p = 0.06$) to a significant positive correlation with precipitation (Figure 7.6).

There are significant differences in diversity between different dry forest sites, but these seem to be related more to biogeography than to environment. The most significant differences in diversity relate to the Chamela samples from western México (Lott *et al.*, 1987) which are more species-rich than samples from other dry forest areas. The three Chamela samples average 94.3 species per sample, significantly more than the 67 species average for the more equatorial continental sites. Although the published data are not directly comparable, it appears that Yucatán area dry forests are also less diverse than those from western México. Only 54 species >1 cm dbh (41 of them >5 cm dbh) were found in a 0.96 ha Yucatán dry forest by Rico-Gray *et al.* (1988), suggesting that the diversity of these forests is slightly lower than in most continental dry forests, exactly the opposite of their western México counterparts.

There is also a less pronounced tendency for West Indian dry forests to be less diverse than mainland dry forests. The Antillean dry forest sites average fewer species than the mainland sites (52: Mann–Whitney U, $p < 0.05$).

The chaco and the subtropical Argentinian dry forests have even fewer species, but if the more speciose Bolivian dry forest sample from Chaquimayo is included as 'subtropical' there is no significant difference in species richness from the more equatorial dry forests.

It is noteworthy that the sample from Quiapaca, Bolivia, (collected after this chapter was submitted and not included in most of the analyses) is almost as species-rich (*c.* 86 species) as the Mexican samples. This is the only sample from undisturbed southern dry forest and, if representative, may indicate that the greater diversity of the Mexican dry forests is a general subtropical phenomenon rather than a biogeographic peculiarity of México.

The same trend is apparent at the familial level, with Chamela averaging 39 families per sample compared with 28 for the more equatorial dry forests. As for species richness, familial richness is less than for equivalent moist or wet forest samples which average 46 families per sample. However, despite being less species-rich, the West Indian (28.7

A

B

Figure 7.4. Single-species dominated arid 'forests' excluded from this analysis. ***A***. *Prosopis* woodland, Olmos, Lambayeque, Perú (note heavy disturbance, especially on hillside). ***B***. *Loxopterygium* woodland, Cherrelique, Tumbes, Perú.

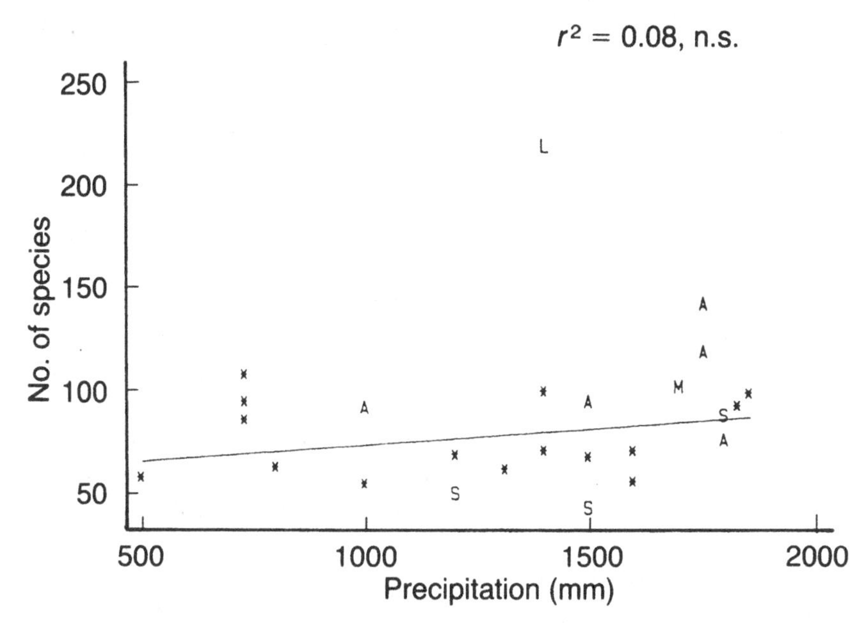

Figure 7.5. Species richness of 0.1 ha samples of plants >2.5 cm dbh for lowland tropical sites with <2000 mm of annual precipitation. *, neotropical dry forests; A, Africa; M, Madagascar; S, subtropical South America; L, Linhares, Espírito Santo, Brazil.

families, 26 if Guánica is included) and Argentinian (27) dry forests have just as many families as do the more equatorial mainland sites (27.9).

If the data are broken down into habit groupings the same general trends are apparent. For trees >10 cm dbh, moist and wet forest sites average nearly twice as many species (42) as do lowland equatorial dry forest sites (24.5). The only significant regional difference in diversity among the different dry forests is that the Chamela forests have more species than do more equatorial sites (36.0 vs 24.5, $p < 0.05$, Mann–Whitney U). Similarly for liana species richness, lowland continental moist and wet forest samples average about twice as many species (35 vs 17) as do equivalent dry forest samples. However, the only significant difference in liana diversity between different dry forest regions is that Antillean communities have fewer species (5.7), in part reflecting the different structure of Antillean dry forests (see below). For trees >2.5 cm in diameter, the lowland equatorial dry forests are less than half as speciose as equivalent moist and wet forests (50 vs 116), with Antillean dry forests just as diverse (46 species) as mainland sites, but with Chamela significantly richer in species (80). Similar trends in diversity are apparent when entire local florulas are compared. Dry forest florulas generally have fewer plant species than do moist or wet forest ones (Gentry, 1988, 1990).

It is possible that parts of the Brazilian cerrado are as rich or richer in species than even the Mexican closed-canopy dry forests, when all habit types are considered. For example, a floristic list of Brazil's Distrito Federal (Filgueiras & Pereira, 1990) includes 2215 native or naturalized vascular plants, over twice as many species as in any of the dry forest local florulas, but the area treated is also far larger and more diverse than any individual field station, so the data are difficult to compare. The cerrado is also conspicuously species-rich in small sample areas. Eiten (1978) found up to 230 species of plants in 0.1 ha samples of cerrado near Brasília, many more than in Lott *et al.*'s (1987) 0.1 ha samples at Chamela. Unfortunately the cerrado data (Eiten, 1978) are not directly comparable with available data for other dry neotropical vegetation which mostly do not include plants <2.5 cm dbh. The only all-plant 0.1 ha sample of (semi-) closed-canopy dry forest (Gentry & Dodson, 1987a) is from Capeira in relatively depauperate western Ecuador, and included only 173 species of vascular plants.

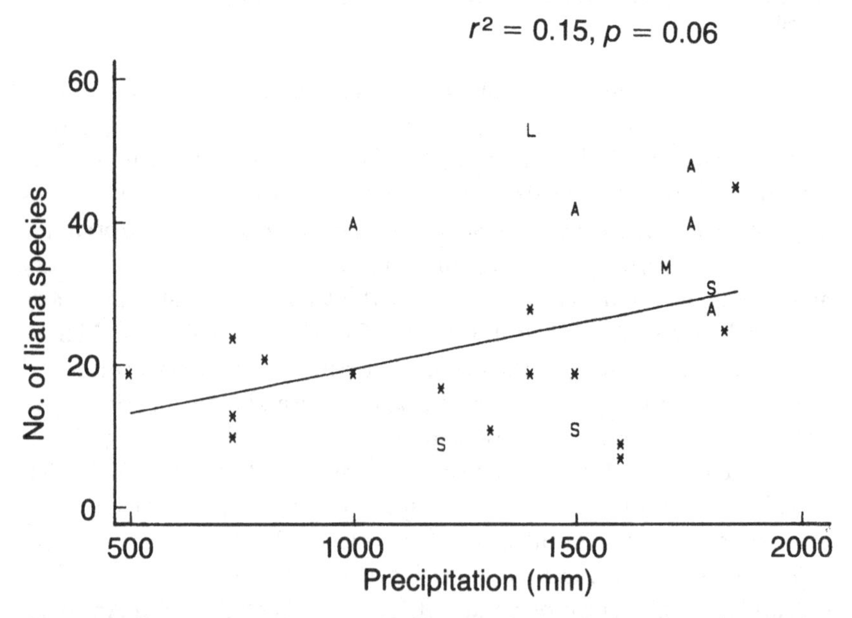

Figure 7.6. Change in species richness of lianas >2.5 cm diameter in 0.1 ha samples of lowland tropical forest with <2000 mm annual precipitation. Abbreviations as in Figure 7.5.

We may conclude that western Mexican dry forests are generally more diverse than other closed-canopy neotropical dry forests for which data are available. Thus dry forests show exactly the opposite trend from moist forests which are significantly less speciose north of 12° latitude (Gentry, 1982b, 1988; but see Wendt, 1993 for an opposing view). There is probably also a parallel increase in dry forest diversity around latitude 19° S to judge from the Quiapaca, Bolivia, data set gathered while this chapter was in press. Except for lianas, which are generally poorly represented on islands (Gentry, 1983), the Antillean dry forests are as diverse as equivalent mainland forests, quite unlike Antillean moist forests which are relatively depauperate (Gentry, 1992; C. Taylor & A. Gentry, personal communication). The cerrado, which is very rich in suffrutices, constitutes a special case; it is structurally more open and only marginally a 'forest', but may be as species-rich as Mexican dry forests.

Structure

In a recent review of dry forest structure, Murphy & Lugo (1986b) focused on structural peculiarities of dry forest as compared to moist or wet forest, suggesting that dry forest canopies and leaf area indices average 50% and basal area 30–75% that of wet forest. However, many of their reported values for dry versus wet forest overlap greatly, and I am more impressed by the structural similarities. Most neotropical forests, including dry forests, tend to be remarkably similar structurally, especially when compared with forests on other continents or in the temperate zone. Although dry forests do differ consistently from moist forests in their lower species richness, they do not differ significantly in most structural attributes, at least for the features measured in Table 7.2. Continental equatorial dry forests have the same average number of plants >2.5 cm diameter in 0.1 ha (370) as do moist and wet forests (373), the same number of trees >10 cm dbh (65 (+2 lianas) vs 64 (+2 lianas)), the same number of lianas >2.5 cm diameter (72 vs 68), and only slightly lower basal area (34.7 m^2 ha^{-1} vs *c.* 40).

An especially good example of this structural homogeneity is provided by lianas (Gentry, 1991). A series of 20 lowland Amazonian moist sites averaged 69 lianas >2.5 cm diameter, just as did a series of nine trans-Andean moist and wet sites, and as 'contrasted' with an average of 70 lianas in 10 dry forest sites and 67 in the Choco pluvial forest (Gentry, 1991). In contrast, African forests may have generally more lianas and

Table 7.2. *Numbers of species and individuals in 0.1 ha samples of lowland neotropical dry forest*

Site	Number of families	Number of species				Number of individuals				Basal area ($m^2 ha^{-1}$)
		Total	Lianas	Trees	Trees + lianas >10 cm dbh	Total	Lianas	Trees	Trees + lianas >10 cm dbh	
West Indies										
Guánica	19	34	0	34	27	1217	0	1217	32	17.8
Mogotes	28	49	11(+1)	37	23	455	37(+1)	418	119	48.1
Round Hill (slope)	26(+)	48(+)	?	48	21	659(+)	?	659(+)	124	28.1
Round Hill (top)	32	58	3(+1)	54	34	566	8(+1)	557	132	36.7
México										
Chamela (upland 1)	37	91	12	79	31	399	42	357	89(+1)	26.3
Chamela (upland 2)	34	89	8	80	28	506	55	451	97	21.8
Chamela (arroyo)	46	103	22	81	49	453	142	311	110	51.0
Southern Central America										
Guanacaste (upland)	22	53	6	47	18	437	81	356	34	*c.* 20.6
Guanacaste (gallery)	35	63	8	55	27	195	24	171	33	*c.* 41.6

Southern subtropics										
Salta	16	25	3	22	11	197	4	193	66	23.6
Riachuelo	27	47	7(+1)	39	24(+2)	451	110(+1)	339	61(+7)	71.2
Parque El Rey	27	40	10	31	20(+1)	190	44	146	39(+1)	33.4
Chaquimayo	29	79	29	50	28(+3)	465	129(+5)	331	77(+3)	47.5
Santa Cruz	30	62	26	36	17(+1)	170	63	107	46(+3)	
Quiapaca	27	86	31	55	27(+s)	395	118	277	66(+2)	
Fortín Acosta (900 m^2)	11(+)	22	0	22	?	141	0	141	?	
Fortín Acosta (600 m^2)	9(+)	*c.* 21	0	*c.* 21	?	428	0	428	?	
Northern Colombia and Venezuela										
Galerazamba	20	55	18	36	18(+5)	396	104	292	44(+6)	29.6
Tayrona	31	67	18	49	34(+1)	337	99	238	71(+1)	36.8
Los Colorados	41	121	37(+3)	81	35(+8)	534	151	383	59(+10)	36.7
Santo Tomás	31	*c.* 71	21	56	23(+1)	*c.* 393	94	*c.* 299	77(+1)	29.0
Coloso	46	113	37(+1)	75	39(+7)	339	100(+1)	238	71(+7)	43.5
Boca de Uchire	20	69	16	53	22	297	75	222	31	13.1
Los Llanos (500 m^2)	21(+)	59	10	49	*c.* 24	330	56	274	44	25.8
Blohm Ranch	31	68	17	51	27	306	77	230	86	31.4
Coastal Ecuador and Perú										
Capeira	27	61	19	42	30(+3)	304	61	243	66(+3)	57.2
Perro Muerte	32(+1)	52	16(+2)	54	17(+2)	325	51(+2)	272	54(+2)	36.4
Cerros de Amotape	29	57	14	43	22(+2)	401	37	377	62(+4)	35.8
Tarapoto	38	102	26(+1)	75	41(+1)	520	86(+1)	434	82(+5)	27.7

some Asian forests fewer than do neotropical forests. Temperate forests are even more distinctive structurally, especially in the more open understory with many fewer small trees (Gentry, 1982a), and in averaging only four lianas >2.5 cm per 0.1 ha (Gentry, 1991).

Most of the structural features suggested by Murphy & Lugo (1986b) as peculiar to dry forest are more pronounced in Antillean dry forests, like the one at Guánica, Puerto Rico that they studied, than in the mainland Central American or South American dry forests I have studied. Perhaps these structural features should be interpreted more as peculiarities of Antillean forests than of dry forests in general.

Thus, although it is not yet clear to what extent dry forests, as a class, differ structurally from other lowland forests, there clearly are some discernible structural differences among different dry forests (Table 7.2). Chamela (and presumably western México dry forests in general) has a denser structure with significantly more individual plants >2.5 cm diameter in 0.1 ha than more equatorial mainland dry forests (594 vs 391). Although this forest has more trees >2.5 cm (373 vs 270), more lianas (79 vs 74), more trees >10 cm dbh (79 vs 59), and a higher basal area (33.0 vs 31.1 m^2 ha^{-1}), none of these individual structural differences is significant on the basis of the currently available data. Nevertheless the cumulative effect of these structural differences is such that the overall density of the Chamela forest is significantly greater.

A more pronounced structural peculiarity distinguishes the Antillean dry forest samples which are characterized by less than a third as many lianas (22) as in the continental sites. The lack of lianas in Antillean dry forests is compensated for by a greater density of both small trees (545 trees >2.5 cm dbh, $p < 0.05$) and medium-sized (especially 7–20 cm dbh) trees (125 trees >10 cm dbh, $p < 0.05$). Since these two opposite trends counteract each other, the overall density of Antillean dry forests, though greater than that of continental forests (560) is not significantly so. The Yucatán Peninsula dry forest studied by Rico-Gray *et al.* (1988), similar to many Antillean dry forests in growing over limestone, is also similar to Antillean forests in the high stem density (619 stems >2.5 cm dbh per 0.1 ha), lack of lianas (at least among the identified species), and high stem density of trees >10 cm dbh (85 per 0.1 ha). In terms of basal area, the Antillean dry forests, like those at Chamela, are almost identical to the average for other dry forests. While more data are needed before hazarding broad generalizations, the structural differences between different neotropical dry forests, and especially between Antillean and continental dry forests, seem at least as great as those between dry and wet forest.

Dry forests also have structural peculiarities that are not associated with the woody flora documented by the 0.1 ha samples. For example, there are many fewer epiphytes in dry forests than in wetter forests. Indeed the near absence of epiphytes in dry forests is one of their most distinctive structural features. Whereas most of the individual plants of a wet forest can be epiphytes (e.g. 4517 epiphytes per 0.1 ha at Río Palenque), epiphytes are insignificant in most dry forests (e.g. 10 epiphytes per 0.1 ha at Capeira, Ecuador) (Gentry & Dodson, 1987b). Epiphytes constitute only 2–4% of the species of a dry forest florula as opposed to about a quarter of the flora of a wet forest site like Río Palenque or La Selva, Costa Rica (Gentry & Dodson, 1987b; Figure 7.7).

Small climbers (vines, *sensu* Gentry, 1990) are another important dry forest component. At a typical neotropical lowland site, vines and lianas together constitute about 20% of the local flora (Gentry, 1991). Although this is true of both moist and dry forests, most climbing species are lianas in moist forests whereas in dry forests they are mostly vines (Gentry & Dodson, 1987a). The western Ecuador dry forests have been thought to be an extreme case. For example, there are half again as many vines species as liana species at Capeira (14% vs 10% of the flora) and small vines literally smother the rest of the forest for part of the year (Figure 7.8). Since all remaining coastal Ecuador dry forests are very

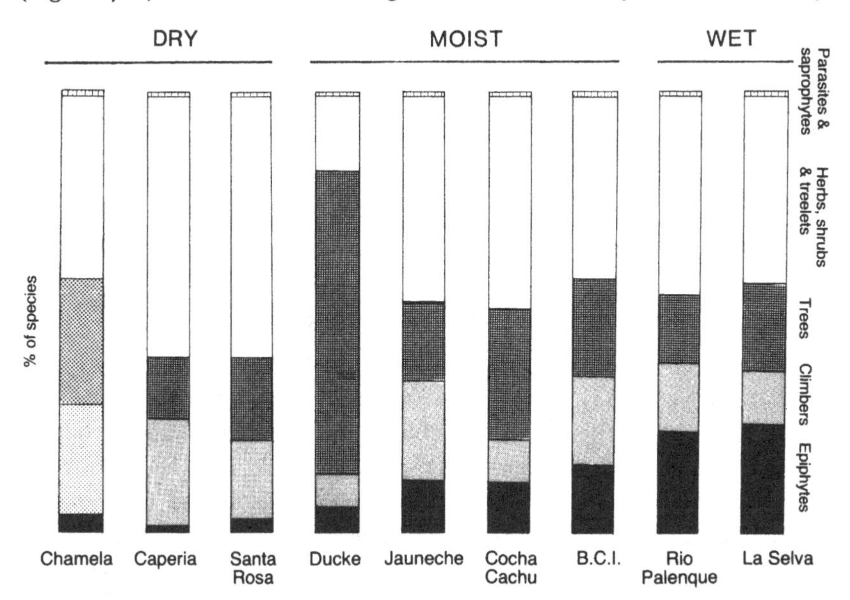

Figure 7.7. Percentage of species in local florulas belonging to different habit groups. Modified from Gentry (1990).

Figure 7.8. Vines smothering trees during wet season in disturbed dry forest at Capeira, Ecuador.

disturbed, the frequency of vines, which are generally rare inside closed-canopy forest (Gentry, 1991) could be due to disturbance regime rather than ecology. However, this prevalence of small vines may be a general feature of the driest dry forests, since it also obtains at Chamela, where there are 131 vine species compared with 61 lianas. In contrast, at the other climatic extreme of dry forest, regions of transition between dry and moist forests seem to have an unusual prevalence of woody lianas (Gentry, 1991).

Dispersal and pollination

Dry forest reproductive biology is summarized elsewhere in this volume (Bullock, Chapter 11). However, recapitulation of a few of the salient points in the context of the data set being analysed here is useful to help set the stage for the dry forest floristic analysis below. The two main trends in dry forest, compared with moist and wet forest, reproductive biology are greater prevalence in dry forests of 'conspicuous'-flowered, 'specialist'-pollinated woody species (as defined by Gentry, 1982a; see also Janzen, 1967; Frankie, 1975) and of wind-dispersed seeds (Gentry, 1982a; Wikander, 1984; Armesto, 1987).

Two-thirds to three-quarters of dry forest woody taxa have conspicuous flowers, generally pollinated by such specialist pollinators as medium-sized to large bees, hummingbirds, or hawkmoths, whereas these figures are reversed for wet forest plants where only about a quarter to a third of the woody species are conspicuous-flowered (Gentry, 1982a). Moist forests are intermediate, with about half the woody species conspicuous and half 'inconspicuous'-flowered. However, since wetter forests are richer in species, the absolute number of conspicuous-flowered species in progressively wetter forests holds approximately constant, with inconspicuous-flowered species accounting for most of the increased species richness of the moist and wet forest plant communities. Conversely, dry forest plant communities might be regarded as consisting mostly of the conspicuous-flowered subset of a more diverse moist or wet forest community.

A similar trend is apparent in seed dispersal, with dry forest woody plants predominantly wind-dispersed (Figure 7.9), while moist and wet-forest ones are mostly bird- or mammal-dispersed. There is also a pronounced dichotomy between trees and lianas with lianas far more prone to be wind-dispersed than are trees, in both dry and moist/wet forests. In dry forests nearly all of the lianas (80%) and a third to a quarter of the trees are wind-dispersed, whereas in wet and especially pluvial forests nearly all are zoochorous: >90% of the trees and about 75% of the lianas (92% in pluvial forest; Gentry, 1982a, 1991).

Floristic composition

The next section of this chapter will attempt to characterize taxonomically the woody flora of neotropical dry forests based on the 350 genera representing 82 different families that are represented in the twenty-five 0.1 ha samples listed in Table 7.1. While this may seem an impressive total, it is a small fraction of the 219 families represented in the complete data set including moist and wet forests. Of the families that attain 2.5 cm diameter in at least one data set, 137 are not represented at all in the dry forest samples. Conversely only three families are represented in the dry forest samples but do not occur in any moist or wet forest sample: Zygophyllaceae, Canellaceae and Julianaceae, each with a single sampled species at a single site. Only three other families are conspicuously better represented in dry than in wet forest: Capparidaceae, Cactaceae and Erythroxylaceae. Even at the generic level the great majority of dry forest taxa also occur in moist and wet forest, and most

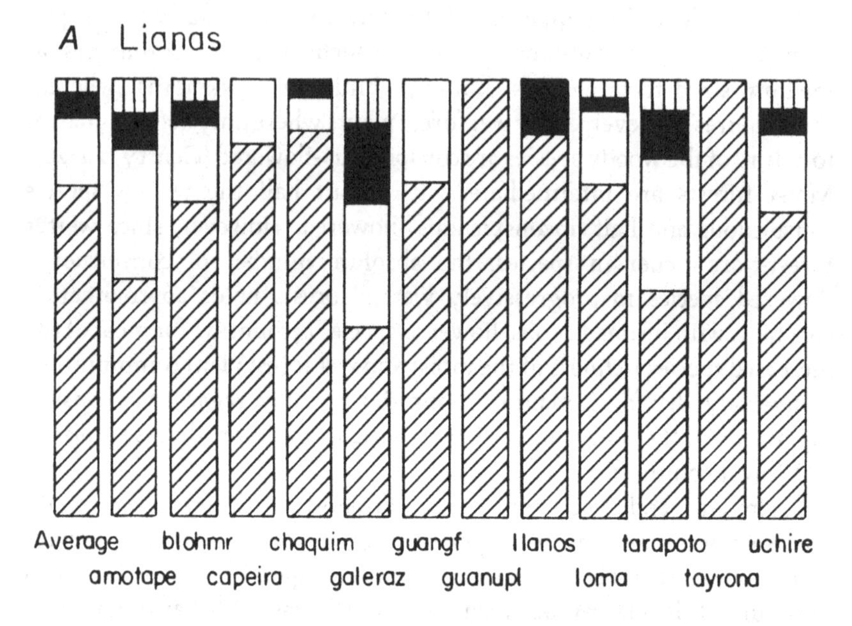

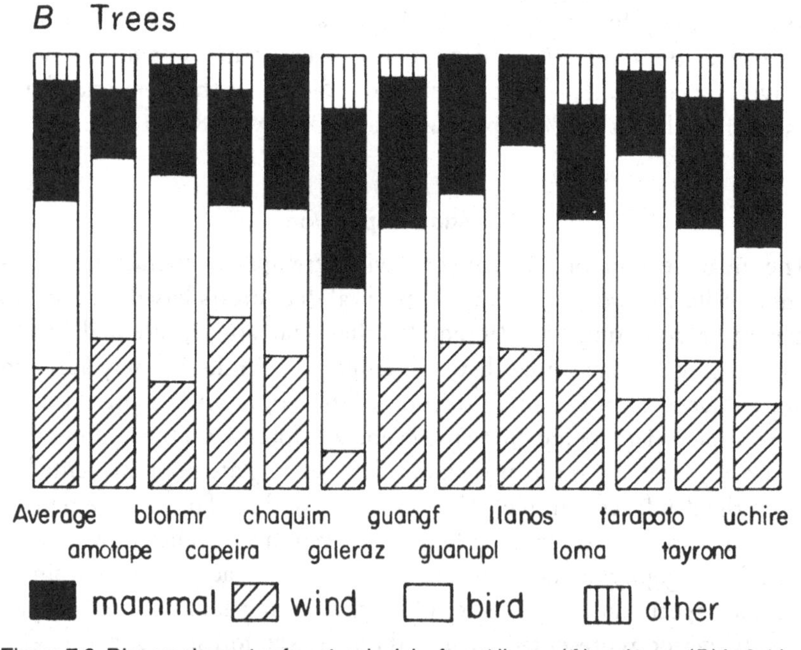

Figure 7.9. Dispersal agents of neotropical dry forest lianas (*A*) and trees (*B*) in 0.1 ha samples. 'Other' includes autochorous and water-dispersed species. First column is average for all the individual sites.

are better represented there. This generalization may be less applicable in México where genera like *Acacia, Bursera, Caesalpinia, Ceiba, Cordia, Lonchocarpus, Lysiloma, Robinsella* and *Zanthoxylum* occur in both dry and moist forests but are better represented in the former, where they probably underwent most of their evolutionary diversification (J. Rzedowski, personal communication). There is also a significant group of genera largely restricted to dry habitats, such as *Cnidoscolus, Forchhammeria, Ziziphus, Guaiacum, Pedilanthus, Podopterus, Thouinia, Exothea,* Burseraceae subfamily Boswellieae, *Ipomoea* subsection *Arborescentes,* and *Jatropha* section *Platyphyllae* (J. Rzedowski, personal communication; see also the section on endemism below), most of which occur both in dry forest and in more xerophytic sites adjacent to the dry forest.

Nevertheless, at the most superficial level the dry forest flora may be characterized as essentially a depauperate subset of the moist/wet forest flora. With the few exceptions noted above, a dry forest milieu does not give rise to a distinctive flora inasmuch as it restricts the occurrence of many of the otherwise typical neotropical families and genera. What characteristics determine which of the normal components of a moist or wet forest plant community will be able to withstand the rigors of a dry forest existence? Before attempting to answer this question it will be instructive to characterize neotropical dry forests floristically, first at the family level then at the generic level.

Families

At the familial level the dry forest communities included in this summary are remarkably consistent in their taxonomic composition (Table 7.3; Figure 7.10). While this is a general property of many different subsets of the neotropics, it is especially pronounced in the relatively depauperate dry forest. To a very large extent, continental dry forests are dominated by two families: Leguminosae and Bignoniaceae. While Leguminosae are usually the dominant arborescent family in moist and wet forests (Gentry, 1988) and Bignoniaceae usually the dominant liana family (Gentry, 1991), their pre-eminent role is accentuated in dry forests. This is not because they are more diverse, however, but rather because so many other families are absent or poorly represented.

Leguminosae average over 12 species per continental dry forest sample, almost twice as many as runner-up Bignoniaceae. They are represented in the overall dry forest data set by at least 48 different

Table 7.3. *Largest families in 0.1 ha samples of plants >2.5 cm dbh in neotropical dry forest*

Bold, most speciose family; underline, second most speciose family; *italic*, third most speciose family.

	Mogotes de Nevarez	Round Hill (top)	Chamela (upland 1)	Chamela (upland 2)	Chamela (arroyo)	Guanacaste (upland)	Guanacaste (gallery)	Salta	Santa Cruz[c]	Chaquimayo	Quipaca[c]	Riachuelo	Galerazamba	Tayrona	Santo Tomás	Boca de Uchire	Est. Los Llanos[a]	Blohm Ranch	Coloso	Loma de los Colorados	Capeira	Perro Muerte	Cerros de Amotape	Tarapoto	Fortín Ten. Acosta[a,b]
Leguminosae	2	2	**21**	**21**	**16**	10	**17**	**7**	6	**17**	**16**	4	**9**	**11**	**10**	**15**	**7**	**13**	**9**	**21**	**13**	5	**7**	**10**	**6**
Bignoniaceae	3	1	2	1	7	**12**	8	2	**12**	10	14	3	6	7	8	12	**7**	6	**9**	13	6	**7**	5	9	–
Rubiaceae	**4**	2	4	6	5	3	4	1	–	2	4	1	1	2	5	4	4	5	2	6	5	1	2	8	–
Sapindaceae	2	3	3	2	5	4	2	2	3	4	6	3	–	4	4	3	1	4	6	8	3	3	1	4	–
Euphorbiaceae	**4**	3	10	8	5	–	–	1	–	3	2	2	4	2	1	4	–	1	4	3	1	3	–	3	–
Capparidaceae	–	–	2	2	4	1	2	–	1	1	2	–	7	6	4	5	–	3	2	6	1	–	–	4	3
Flacourtiaceae	3	3	3	3	2	3	3	–	2	1	2	2	1	2	3	–	3	1	5	2	1	3	3	–	1
Myrtaceae	**4**	**5**	1	1	–	3	–	–	2	–	4	**6**	–	–	4	2	1	1	1	1	2	**5**	1	7	–
Apocynaceae	–	–	3	1	2	1	3	–	3	**5**	3	2	1	2	2	2	–	–	7	6	1	–	1	3	1
Nyctaginaceae	2	2	1	1	2	–	–	1	4	2	3–4	–	3	3	4	4	1	1	2	4	1	1	1	3	–
Malpighiaceae	2	1	2	3	3	4	1	1	–	–	3	–	2	2	–	2	1	1	1	4	1	2	1	3	–
Polygonaceae	2	4	3	2	3	1	–	1	2	1	1	1	2	1	2	1	–	2	1	3	–	2	2	3	1

Anacardiaceae	1	2	–	1	3	3	3	2	–	2	2	1	–	2	2	2	1	1	2	2	1	–	–	1	1
Moraceae	1	2	1	1	1	–	–	–	1	2	–	2	–	–	–	–	–	3	5	3	2	2	2	3	–
Cactaceae	–	–	2	3	–	1	–	–	1	5	–	1	3	1	2	2	1	1	1	2	–	–	2	1	1
Boraginaceae	2	–	3	2	1	2	2	–	1	–	–	2	2	–	–	–	–	1	–	–	3	2	3	3	–
Meliaceae	–	1	1	1	1	2	1	–	–	3	–	1	–	1	3	–	2	1	3	1	–	3	1	3	–
Sapotaceae	2	2	1	–	1	1	1	–	2	2	–	1	–	1	1	–	–	–	6	4	1	1	–	1	1
Rutaceae	1	3	1	2	1	2	–	–	1	–	3–4	2	–	–	–	–	–	1	1	5	–	–	3	3	–
Ulmaceae	–	–	–	–	2	–	–	1	3	4	2	1	2	–	1	–	–	–	2	3	–	2	1	2	1
Verbenaceae	1	1	1	2	1	–	2	–	1	–	–	1	–	1	1	–	–	3	2	1	1	–	1	2	1
Hippocrateaceae	1	–	1	–	1	2	1	–	1	1	2	–	1	1	1	1	–	1	4	1	–	–	–	2	–
Burseraceae	1	1	1	3	1	1	1	–	–	–	–	–	2	2	1	1	–	–	1	2	1	–	–	–	–
Solanaceae	–	–	–	–	–	–	–	1	2	–	–	2	1	1	1	2	1	1	1	1	1	2	1	4	–
Bombacaceae	–	–	–	1	–	–	–	1	1	1	–	–	–	–	1	–	–	1	3	2	4	2	2	1	–
Rhamnaceae	1	2	1	2	2	–	–	–	1	–	2	1	–	1	1	–	–	2	1	–	–	2	1	2	1
Combretaceae	1	–	1	–	1	1	–	–	–	2	2	–	–	–	–	–	–	1	5	1	2	–	1	1	1
Erythroxylaceae	–	1	1	1	1	1	1	–	–	–	1	1	1	1	–	–	1	1	–	1	1	1	1	3	–
Annonaceae	–	–	2	–	2	2	1	–	–	–	1	–	–	–	1	–	2	–	2	1	1	1	–	1	–
Phytolaccaceae	–	–	1	1	1	–	–	1	2	3	3	1	2	–	2	–	–	–	1	1	–	2	–	–	–

[a] From sample <0.1 ha.
[b] Plus several taxa undetermined as to family.
[c] Sample completed subsequent to data tabulation.

genera, again almost twice as many as runner-up Bignoniaceae. Moreover, Leguminosae are the most speciose family in 17 of the 20 continental dry forest data sets and second in the other three. The three exceptions have an unusually diverse sample of Bignoniaceae (two sites) or Myrtaceae (at the most subtropical of all the sampled sites, Riachuelo, Argentina). Leguminosae are poorly represented in the Antilles, averaging only two species per sample.

Bignoniaceae (Figure 7.11) are the undisputed number two family of woody plants of neotropical dry forests, averaging seven species per continental sample, twice as many as third place Rubiaceae. With 26 genera represented in these samples, Bignoniaceae are again second only to Leguminosae. Bignoniaceae are the most speciose or second most speciose family in 16 out of the 20 continental dry forest samples. The four exceptions are Riachuelo, where Bignoniaceae are third after Myrtaceae and Leguminosae; Galerazamba, Colombia, perhaps the most xeric extreme included in the sample, where they are third after

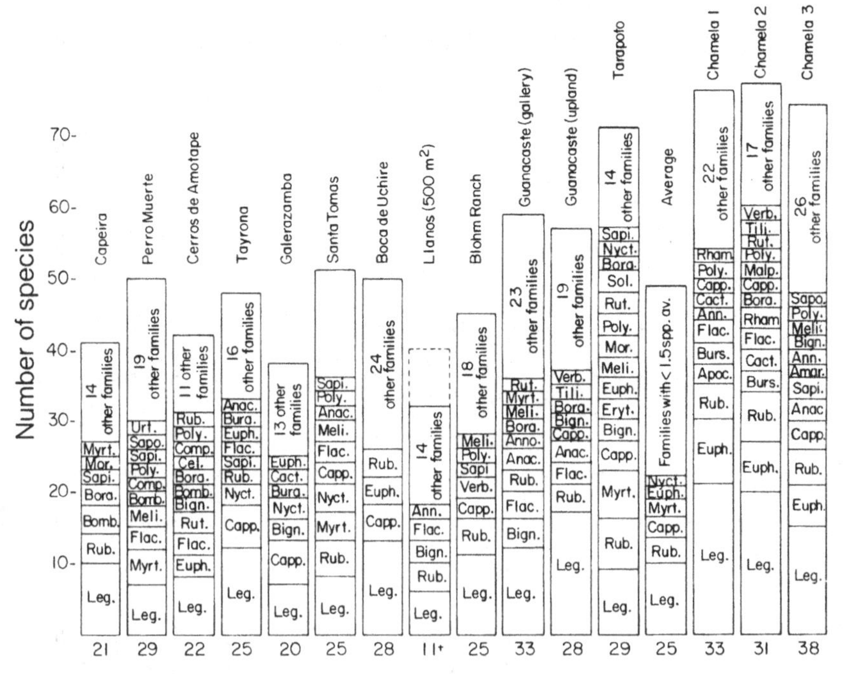

Figure 7.10. Average number of tree species per family for 0.1 ha samples of lowland neotropical dry forest (trees >2.5 cm dbh). Number at bottom of each column is number of families with trees in that sample. 'Average' column excludes samples from Chamela, México.

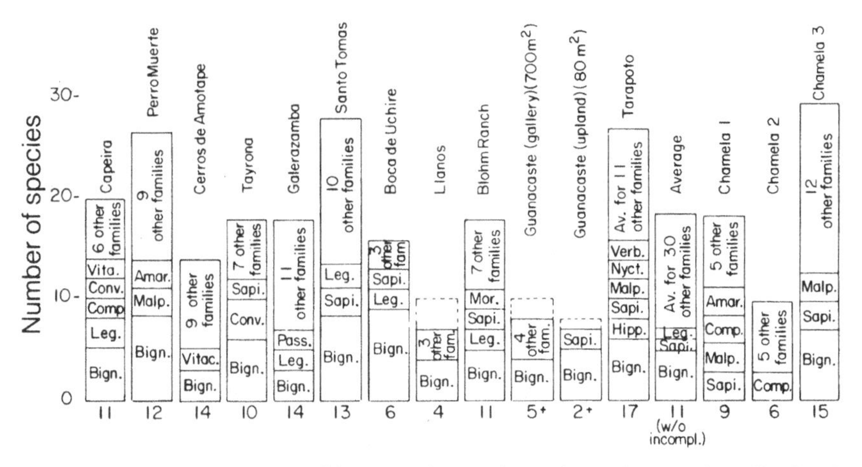

Figure 7.11. Average number of liana species per family for 0.1 ha samples of lowland neotropical dry forest (lianas >2.5 cm maximum diameter). Number at bottom of each column is number of families with lianas in that sample. 'Average' column excludes samples from Chamela, México.

Leguminosae and Capparidaceae; and two of the three samples at Chamela, México, the northernmost site. Apparently Bignoniaceae, more than many other families, are limited in the most extreme sites. Like Leguminosae, Bignoniaceae are conspicuously under-represented in the Antillean dry forests, averaging only 1.5 species per sample.

Myrtaceae is the pre-eminent West Indian dry forest family, averaging four species per Antillean sample. Myrtaceae are the only family beside Leguminosae and Bignoniaceae ever to be the most speciose family in a dry forest sample. Myrtaceae are the most speciose family in both complete Antillean samples (tying with Euphorbiaceae in the Puerto Rican sample) and in subtropical Riachuelo. Abundance of Myrtaceae is also a phenomenon typical of moist and wet forests of coastal Brazil and the Paraná valley (Mori *et al.*, 1983; Peixoto & Gentry, 1990; Spichiger, Bertoni & Loizeau, 1992). With seven species, Myrtaceae are almost the third most speciose family at the geographically anomalous site at Tarapoto, Perú, which would be classed with the southern sites according to the phytogeographic regions of Gentry (1982b). Myrtaceae may tie with Leguminosae as second most speciose family at Perro Muerte, Ecuador, but analysis of this site is not yet complete pending herbarium comparison of the vouchers. Otherwise Myrtaceae are very much an also-ran, and they are only the seventh most speciose dry forest family overall, averaging just 2.2 species per sample (and only 1.8 species per sample in the continental sites).

At least five other families deserve special mention in a summary of dry forest floristics. Rubiaceae, Sapindaceae, Euphorbiaceae, Flacourtiaceae and Capparidaceae are, in order, the next most prevalent dry forest families after Leguminosae and Bignoniaceae. The first three are also the next most prevalent families in number of genera represented in the overall dry forest sample (at least 18 genera of Rubiaceae, 14–15 of Sapindaceae, and 16 of Euphorbiaceae), although Capparidaceae are represented by only six genera and Flacourtiaceae by seven in the overall sample. Rubiaceae have a markedly irregular importance in different samples. They are represented in all the samples, averaging 3.5 species per sample, tie as the most speciose family in one West Indian dry forest, and have more species than any other family in Murphy & Lugo's (1986a) Guánica data set. They are third (after legumes and bignons) in no fewer than nine continental samples. On the other hand, Rubiaceae have only one or two species at many sites.

Sapindaceae, averaging 3.1 species per dry forest sample, are the third most speciose family in five continental dry forest samples and tie for third in one of the Antillean samples. Although many Sapindaceae genera occur sporadically in neotropical dry forests, the majority of its dry forest species belong to the two climbing genera *Paullinia* and *Serjania*. Euphorbiaceae average 2.8 species per sample, making this the fifth most prevalent dry forest family; Euphorbiaceae are also taxonomically diverse, being represented in dry forest samples by 16 different genera. Euphorbiaceae are most prevalent in the northernmost dry forest sites, tying as the most speciose family at one Antillean site, being third at the other, and second only to legumes in two of the Chamela, México samples. Flacourtiaceae tie as the third most speciose family in the Antillean samples, and, like Myrtaceae, average 2.2 species per sample overall. They are represented in these samples mostly by the diverse and ecologically important genus *Casearia*.

Finally, Capparidaceae, averaging 2.1 species per sample, is the seventh most prevalent family for the sampled set of closed-canopy dry forests. Unlike the other important dry forest families, Capparidaceae are not very diverse generically with most of the sampled species belonging to the single large and ecologically important genus *Capparis*. This is one of the very few families that seem to be truly dry forest specialists, being actually better represented in dry than in moist forest. Capparidaceae are especially well represented in coastal Venezuela and Colombia, being the

Figure 7.12. Capparidaceae dominating western Ecuador dry forest. All plants with leaves are *Capparis*. Shrubs along road are *C. crotonoides* HBK.; dark spots are *C. angulata* R. & P., *C. avicennifolia* HBK., and *C.* aff. *flexuosa*.

second most speciose family at Galerazamba and third at Tayrona and Boca de Uchire. They are also the second most speciose family at Fortín Teniente Acosta in the Paraguayan chaco and third in the Bolivian chaco samples.

Capparidaceae, distinctive in being one of the very few evergreen elements of continental dry forests, seem to be generally most prevalent in the very driest sites, including the thorn-scrub vegetation that is not included in my data set. No doubt significantly, Galerazamba is the driest Colombian site and Boca de Uchire the driest Venezuelan one. The chaco sites, which average three Capparidaceae species per 0.1 ha, are even drier. Capparidaceae are also prevalent in similar xeric vegetation in coastal Ecuador (Figure 7.12).

The other most prevalent dry forest woody families are listed in average order of abundance in Table 7.3. The three most diverse of these average 1.8 species per sample. Apocynaceae, represented in dry forest especially by the tree genus *Aspidosperma* and the vine genus *Forsteronia*, average 2.1 species per continental sample and can have as many as six or seven sampled sympatric species in a single forest. Nyctaginaceae and Polygonaceae, averaging 1.8 species per sample, and

Malpighiaceae, averaging 1.6, are also important dry forest elements. Dry forest Nyctaginaceae are mostly trees belonging to the genera *Guapira* and *Neea*, although the most widespread species is the liana *Pisonia aculeata*. Dry forest Polygonaceae are mostly trees belonging to the genera *Ruprechtia*, *Triplaris* and *Coccoloba*, the latter predominantly scandent in moister forests; of these *Ruprechtia* is the only genus largely restricted to dry forest, being especially prevalent in the chaco where *R. triflora* is usually the commonest woody species. Dry forest Malpighiaceae include a variety of genera of wind-dispersed lianas, as well as the shrub genus *Malpighia*, and trees belonging to *Bunchosia* and *Byrsonima*, the latter two, like the lianas, also well represented in moist forest.

Although more prominent in desert situations, Cactaceae are also an important dry forest family. Cactaceae are only the 14th most diverse family as measured by these samples, with an average of 1.3 species per sample, but include at least ten different sampled genera and are the sixth most prevalent dry forest family on that basis. It is noteworthy that Cactaceae are best represented at the latitudinal limits of dry forest, being disproportionately represented at Chamela (2.5 species per sample) and in the southern subtropical sites (2 species per sample).

Cactaceae are even more prevalent in drier forests than any included in these samples; for example, Cactaceae are the second most diverse family (after Leguminosae) in the marginal dry forest of southern Baja California studied by Arriaga & León (1989). Similarly in two sites from the very dry Bolivian chaco, sampled subsequent to preparation of this manuscript, Cactaceae are the most speciose family (in one sample tied with legumes) with seven species per 0.1 ha.

Additional families well represented in one or two dry forest areas include Rutaceae and Celastraceae in the Antilles (respectively averaging 2.3 and 2 species per sample), Boraginaceae (2.8 species per sample, mostly of *Cordia*) and Bombacaceae (2 species per sample) in coastal Ecuador.

The 0.1 ha samples analysed above probably capture most of the dry forest dominants. For example, all of the genera listed by Sarmiento (1975) as characteristic of the South American dry areas he compared (excluding those areas that do not overlap with my purview) are included in my samples except for a few genera of more arid areas (*Loxopterygium*, *Fourcroya*, *Puya*, and a few cactus genera (note that his '*Bombax*' = *Pochota*)). However, many dry forest plants are herbs or vines and thus not included in my samples. Another way to evaluate floristic composition is to compare data from local florulas and species lists which also include the non-woody plants. Data on the predominant families and genera from the available local florulas are summarized in Tables 7.4 and 7.5. Although only six of these are available, they are well distributed geographically. A seventh, less complete data set, from a different site in the Venezuelan llanos than Aristeguieta's (1966) has also been included in some of the analyses below (Troth, 1979 and personal communication). A list of the plant species of the Distrito Federal in the Brazilian cerrado (Filgueiras & Pereira, 1990) is also available for comparison although it covers a much larger area and is not strictly comparable to the local florula data.

Except for Leguminosae, which remains the dominant family at all continental dry forest sites (and in the cerrado), the speciose families and genera indicated by an analysis of local florula data are quite different from the important woody taxa indicated by transect data. This largely reflects the disproportionately large role of herbs and shrubs in dry forest communities (Gentry, 1990). However, the florula data also show the same striking between-site consistency in taxonomic composition of continental dry forests.

The similarity of familial roles is spectacular. If the four sites consisting largely of closed-canopy dry forest are compared, we find Leguminosae as the most speciose family in each site; Euphorbiaceae and Gramineae are both in the top four families at each site. Both Convolvulaceae and Compositae are in the top 5–7 families at each continental dry forest site. Also included in the 10 (to 12) most speciose families at each dry forest site are Rubiaceae, Acanthaceae, Malvaceae (19th at Sian Ka'an), Solanaceae (21st at Chamela), Bignoniaceae, Boraginaceae, Verbenaceae, and Cyperaceae (only 9 species at Chamela). The only other families ever to be included on a site's 'top ten' list are Bromeliaceae and Cucurbitaceae (6th and 10th respectively at Chamela), Orchidaceae and Sapindaceae (5th and 10th at Sian Ka'an), Apocynaceae (9th at Capeira), Polygalaceae (10th at the Estación Biológica de Los Llanos) and ferns and their allies (here treated as a single 'family') (5th at Santa Rosa and 6th at Capeira). Moreover, the same additional families are represented in the 30 most speciose at each site listed in Table 7.4; in addition to the above these include Sterculiaceae, Malpighiaceae, Amaranthaceae, Nyctaginaceae, Cactaceae, Capparidaceae and Moraceae. The only exceptions are Moraceae and Cactaceae which are underrepresented at Capeira, Capparidaceae at Sian Ka'an, and Nyctaginaceae at Santa Rosa. Minor anomalies in this pattern of mainland dry forest

Table 7.4. *Largest families in dry forest local florulas*

Chamela[1]		Sian Ka'an[2]		Santa Rosa[3]		Capeira[4]		Los Llanos[5]		Round Hill[6]	
Leguminosae	115	Leguminosae	104	Leguminosae	123	Leguminosae	80	Leguminosae	52	Euphorbiaceae	8
Caesalp.	24	Caesalp.	27	Caesalp.	20	Caesalp.	11	Caesalp.	14	Myrtaceae	7
Mimosoid.	30	Mimosoid.	31	Mimosoid.	38	Mimosoid.	19	Mimosoid.	6	Rubiaceae	7
Papilion.	61	Papilion.	44	Papilion.	65	Papilion.	50	Papilion.	33	Leguminosae	5
Euphorbiaceae	70	Compositae	46	Euphorbiaceae	35	Compositae	36	Gramineae	45	Ferns	4
Convolvulaceae	30	Gramineae	43	Gramineae	33	Gramineae	32	Cyperaceae	13	Polygonaceae	4
Gramineae	29	Euphorbiaceae	41	Rubiaceae	35	Convolvulaceae	25	Rubiaceae	12	Malpighiaceae	4
Compositae	28	Orchidaceae	32	Ferns	25	Euphorbiaceae	25	Compositae	12	Orchidaceae	4
Rubiaceae	22	Verbenaceae	27	Compositae	24	Ferns	19	Euphorbiaceae	11	Sapindaceae	4
Bromeliaceae	22	Rubiaceae	27	Convolvulaceae	22	Acanthaceae	17	Bignoniaceae	9	Anacardiaceae	3
Malvaceae	22	Convolvulaceae	23	Bignoniaceae	21	Bignoniaceae	17	Convolvulaceae	9	Boraginaceae	3
Acanthaceae	21	Solanaceae	23	Malvaceae	17	Apocynaceae	16	Malvaceae	9	Bromeliaceae	3
Cucurbitaceae	19	Sapindaceae	17	Solanaceae	15	Cyperaceae	15	Polygalaceae	9	Capparidaceae	3
Boraginaceae	17	Cyperaceae	17	Boraginaceae	15	Malvaceae	15	Verbenaceae	6	Celastraceae	3
Verbenaceae	16	Bromeliaceae	15	Moraceae	14	Rubiaceae	14	Labiatae	6	Erythroxylaceae	3
Bignoniaceae	16	Bignoniaceae	15	Cyperaceae	14	Verbenaceae	13	Orchidaceae	5	Flacourtiaceae	3
Cactaceae	15	Polygonaceae	14	Sterculiaceae	13	Cucurbitaceae	12	Malpighiaceae	4	Loranthaceae	3
Asclepiadaceae	13	Boraginaceae	13	Acanthaceae	13	Solanaceae	12	Flacourtiaceae	4	Moraceae	3

Capparidaceae	12	Acanthaceae	13	Verbenaceae	12	Boraginaceae	12	Commelinaceae	4	Passifloraceae	3
Malpighiaceae	12	Asclepiadaceae	13	Orchidaceae	11	Sapindaceae	12	Asclepiadaceae	4		
Sapindaceae	11	Malpighiaceae	13	Apocynaceae	11	Amaranthaceae	12	Myrtaceae	4		
Orchidaceae	11	Malvaceae	12	Sapindaceae	11	Orchidaceae	11	Ferns	3		
Apocynaceae	11	Myrtaceae	12	Capparidaceae	11	Sterculiaceae	9	Araceae	3		
Solanaceae	10	Apocynaceae	11	Malpighiaceae	10	Bromeliaceae	8	Moraceae	3		
Polygonaceae	10	Moraceae	10	Amaranthaceae	9	Passifloraceae	8	Solanaceae	3		
Sterculiaceae	10	Arecaceae	9	Meliaceae	7	Bombacaceae	7	Sterculiaceae	3		
Amaranthaceae	10	Piperaceae	8	Cactaceae	7	Malpighiaceae	7	Turneraceae	3		
Rutaceae	9	Cucurbitaceae	8	Cucurbitaceae	7	Nyctaginaceae	7				
Nyctaginaceae	9	Passifloraceae	8	Labiatae	6	Apocynaceae	6				
Cyperaceae	9	Amaranthaceae	7	Tiliaceae	6	Asclepiadaceae	6				
Flacourtiaceae	8	Cactaceae	7	Asclepiadaceae	6	Capparidaceae	6				
Moraceae	8	Loranthaceae	7	Piperaceae	6	Labiatae	6				
Loranthaceae	7	Nyctaginaceae	7	Flacourtiaceae	5	Onagraceae	6				
Meliaceae	7	Rhamnaceae	7	Turneraceae	5	Phytolaccaceae	6				
		Sapotaceae	7	Melastomataceae	5	Piperaceae	6				
		Scrophulariaceae	7								
		Sterculiaceae	7								

Sources: 1. Lott (1985); 2. Durán & Olmsted (1987); 3. Janzen & Liesner (1980); 4. Dodson & Gentry (1992); 5. Aristeguieta (1966); 6. Kelly *et al.* (1988).

Table 7.5. *Largest genera in dry forest local florulas*

Chamela[1]		Sian Ka'an[2]		Santa Rosa[3]		Capeira[4]		Los Llanos[5]		Round Hill[6]	
Ipomoea	20	*Cassia* (*s.l.*)	18	*Cassia*	12	*Ipomoea*	14	*Cassia*	9	*Eugenia*	5
Tillandsia	16	*Solanum*	15	*Ipomoea*	11	*Cyperus*	11	*Polygala*	8	*Coccoloba*	4
Euphorbia	11	*Ipomoea*	15	*Mimosa*	10	*Passiflora*	8	*Paspalum*	7	*Tillandsia*	3
Acalypha	10	*Tillandsia*	12	*Sida*	9	*Cassia*	7	*Sida*	6	*Capparis*	3
Croton	10	*Croton*	12	*Desmodium*	8	*Acalypha*	7	*Axonopus*	6	*Erythroxylum*	3
Cassia (*s.l.*)	10	*Coccoloba*	10	*Cyperus*	7	*Cordia*	7	*Andropogon*	6	*Passiflora*	3
Lonchocarpus	8	*Pithecellobium*	10	*Acacia*	7	*Ludwigia*	6	*Eragrostris*	5		
Cordia	8	*Acacia*	9	*Capparis*	7	*Desmodium*	6	*Ipomoea*	5		
Phyllanthus	8	*Passiflora*	8	*Psychotria*	7	*Hyptis*	5	*Hyptis*	5		
Acacia	7	*Panicum*	7	*Ficus*	7	*Solanum*	5	*Randia*	4		
Bursera	6	*Eugenia*	7	*Heliotropium*	6	*Chamaesyce*	5	*Rhynchospora*	4		
Jatropha	6	*Piper*	6	*Cordia*	6	*Capparis*	5	*Desmodium*	4		
Caesalpinia	6	*Rhynchospora*	6	*Lonchocarpus*	6	*Vigna*	5	*Pectis*	4		
Cyperus	6	*Lantana*	6	*Solanum*	6	*Adiantum*	5	*Mimosa*	4		
Mimosa	6	*Epidendrum*	6	*Croton*	6	*Thelypteris*	5	*Arrabidaea*	3		
Randia	6	*Eupatorium*	6	*Crotalaria*	6	*Alternanthera*	5	*Cordia*	3		
Capparis	5	*Acalypha*	5	*Pithecellobium*	5	*Sida*	5	*Evolvulus*	3		
Dioscorea	5	*Encyclia*	5	*Piper*	5	*Tillandsia*	4	*Bulbostylis*	3		
Casearia	5	*Stachytarpheta*	5	*Calliandra*	5	*Coccoloba*	4	*Cyperus*	3		
Sida	5	*Heliotropium*	5	*Phaseolus*	5	*Croton*	4	*Chamaesyce*	3		
Lantana	5	*Caesalpinia*	5	*Trichilia*	5	*Paullinia*	4	*Casearia*	3		

Passiflora	5	*Arrabidaea*	5	*Passiflora*	5	*Euphorbia*	4	*Panicum*	3
Justicia	4	*Ficus*	5	*Bursera*	4	*Serjania*	4	*Trachypogon*	3
Ruellia	4	*Diospyros*	5	*Caesalpinia*	4	*Centrosema*	4	*Eriosema*	3
Matelea	4	*Desmodium*	4	*Chamaesyce*	4	*Crotalaria*	4	*Phaseolus*	3
Aristolochia	4	*Euphorbia*	4	*Adiantum*	4	*Pithecellobium*	4	*Byrsonima*	3
Tabebuia	4	*Lonchocarpus*	4	*Euphorbia*	4	*Dicliptera*	3	*Habenaria*	3
Tournefortia	4	*Phoradendron*	4	*Randia*	4	*Aeschynomene*	3		
Bauhinia	4	*Malpighia*	4	*Casearia*	4	*Arrabidaea*	3		
Pithecellobium	4	*Trichilia*	4	*Physalis*	4	*Tabebuia*	3		
Trichilia	4	*Oncidium*	4	*Hyptis*	4	*Heliotropium*	3		
Ficus	4	*Psychotria*	4	*Rhynchosia*	4	*Verbesina*	3		
Zanthoxylum	4	*Serjania*	4	*Centrosema*	4	*Vernonia*	3		
Paullinia	4	*Annona*	4	*Tabebuia*	3	*Jacquemontia*	3		
Solanum	4	*Calliandra*	4	*Cleome*	3	*Dioscorea*	3		
		Cordia	4	*Eupatorium*	3	*Trichilia*	3		
		Phyllanthus	4	*Evolvulus*	3	*Acacia*	3		
		Dioscorea	4	*Jacquemontia*	3	*Lantana*	3		
		Lasiacis	4	*Merremia*	3	*Marsdenia*	3		
		Echites	3	*Arrabidaea*	3	*Mimosa*	3		
		Randia	3	*Cydista*	3	*Piper*	3		
		Erythroxylum	3	*Jatropha*	3	*Panicum*	3		

Sources: 1. Lott (1985); 2. Duran & Olmsted (1987); 3. Janzen & Liesner (1980), supplemented by Liesner, personal communication; 4. Dodson & Gentry (1992); 5. Aristeguieta (1966); 6. Kelly *et al.* (1988).

sites having a remarkably similar familial representation are Polygonaceae (10 species) and Rutaceae (9) being unusually prevalent at Chamela, Polygonaceae (14), Myrtaceae (12), Palmae (9) and Passifloraceae and Piperaceae (8) at Sian Ka'an, Meliaceae (7) at Santa Rosa, and Passifloraceae (8) and Bombacaceae (7) at Capeira.

The Antilles would appear to have a very different overall floristic composition, just as in the case of the woody plant subset discussed above. Euphorbiaceae, Myrtaceae and Rubiaceae stand out as unusually prevalent in the Jamaican dry forest for which a floristic list is available, just as they do in the ecological data. The Antillean sites also seem markedly depauperate, but this is in part due to a different sampling technique, listing mostly the species of the Round Hill forest interior.

The basically savanna sites also have floristic differences from the more strictly dry forest ones. The Los Llanos florula is peculiar in the excess representation of grasses and sedges (which are second and third respectively both in Aristeguieta's (1966) list and in Troth's unpublished data for Hato Masaguaral), as might be expected for savanna sites. Similarly grasses are third and sedges 11th in the Distrito Federal (Filgueiras & Pereira, 1990). The cerrado list is distinctive compared with the local florulas in the representation of Melastomataceae, Myrtaceae and Malpighiaceae among the ten most speciose families, whereas these families are in the top ten in none of the continental local florulas. These families, as well as Bignoniaceae (13th), Verbenaceae (15th) and Apocynaceae (16th) are especially prone to evolution of the shrubs and suffrutices so characteristic of the cerrado.

The general similarities between different dry forest florulas are even more striking when compared with moist and wet forest local florulas where families predominate that are hardly represented in dry forests, e.g. Araceae, Piperaceae, Melastomataceae, Moraceae, Palmae, Gesneriaceae, Sapotaceae, Annonaceae, Guttiferae and Lauraceae (Gentry, 1990).

Generic analysis

Of the *c.* 350 woody genera represented in at least one dry forest sample, over two-thirds are widespread in moist and wet forest and less than one-third are more or less restricted to, or at least most prevalent in, dry forest.

Eleven genera are represented in the samples from all six phytogeographic regions: *Tabebuia, Cordia, Casearia, Bauhinia, Trichilia,*

Erythroxylum, Randia, Hippocratea, Serjania, Croton and *Zanthoxylum*. Thirteen more are included in the samples from all but one region (in each case absent only as a sampling artifact): *Arrabidaea, Celtis, Capparis, Combretum, Pithecellobium, Pterocarpus* and *Paullinia* are lacking only in the Antillean samples; *Bursera* and *Coccoloba* only in the southern subtropical samples; *Macfadyena, Eugenia* and *Acacia* unsampled only at Chamela, and *Ficus* only in southern Central America. These 24 genera are perhaps the most prevalent dry forest genera.

By far the most widespread dry forest genus would appear to be *Tabebuia* of the Bignoniaceae which is represented in every sample from every phytogeographic region. Runners-up are *Casearia* (Flacourtiaceae) and *Trichilia* (Meliaceae), each absent from five samples (plus Murphy & Lugo's Guánica sample), *Erythroxylum* (Erythroxylaceae) and *Arrabidaea* (Bignoniaceae), absent from six samples, *Randia* (Rubiaceae) and *Capparis* (Capparidaceae) absent from seven samples, and *Bursera* (Burseraceae), *Acacia* (Leguminosae) and *Coccoloba* (Polygonaceae), absent from eight. Other genera present in all six phytogeographic regions and represented in samples from at least four of them include *Astronium, Spondias, Forsteronia, Annona, Clytostoma, Cydista, Ceiba, Diospyros, Caesalpinia, Machaerium, Platymiscium, Erythrina, Lonchocarpus, Psidium, Guapira, Ouratea, Ruprechtia, Guettarda, Manilkara, Pouteria, Allophyllus, Guazuma* and *Urera*. Of the widespread dry forest genera only four are not also widespread in moist and wet forest: *Astronium, Caesalpinia, Ruprechtia* and *Guapira*. There are also a few more widespread genera which are clearly more prevalent in dry than in wet forest, notably *Capparis, Erythroxylum, Randia, Serjania* and *Bursera*. Thus the great majority of the widespread dry forest genera are intrinsically moist/wet forest taxa.

There are also some apparent differences in generic-level floristic composition between different dry forest regions. Several of the common wide-ranging genera that are typical of dry forest communities would seem to be distinctly over-represented in certain phytogeographic regions. Genera over-represented in the Antilles compared with other dry forests include *Coccoloba* (3 species per sample), *Eugenia* (2.5), *Erythroxylum* (1.5), and to a lesser extent *Drypetes* (1.3) and *Casearia* (1.3). At Chamela *Caesalpinia* (5 species/sample), *Croton* (3), *Lonchocarpus* (3), and to a lesser extent *Casearia* (2) and *Bursera* (1.7) are disproportionately represented. Genera over-represented in the coastal Colombia/Venezuela data sets include *Acacia* (5), *Capparis* (3.1) and *Arrabidaea* (2.3). In coastal Ecuador *Cordia*, with an average of 2 species per sample may be somewhat over-represented compared with other sites. Although most of the differences probably reflect sampling artifact, others, like the greater representation of *Caesalpinia* and *Bursera* in Mexican dry forest, or of *Capparis* in northern Venezuela/Colombia clearly reflect real biogeographic differences in floristic composition.

The same genera also tend to be speciose in different dry forests when local florulas are compared (Table 7.5). Again, this is the same pattern shown by moist forest genera, but with a completely different set of taxa. Instead of speciose moist forest genera like *Piper, Ficus, Inga, Ocotea, Psychotria, Philodendron, Anthurium* and *Miconia*, and various ferns and orchids (Gentry, 1990), the most speciose genera in dry forest florulas are *Cassia* and *Ipomoea*. Other speciose dry forest genera are *Tillandsia, Euphorbia, Acalypha, Croton, Mimosa, Cordia, Sida, Capparis, Randia* and *Acacia*. There are also a few genera like *Passiflora* and *Cyperus* that tend to be well represented in both dry and moist forest. Several dry forest genera are of special phytogeographic interest in being disproportionately represented at certain sites, for example *Bursera, Lonchocarpus* and *Caesalpinia* in Central American dry forests, several Euphorbiaceae in western México (e.g. *Acalypha, Phyllanthus, Jatropha*), *Pithecellobium, Coccoloba, Diospyros, Eugenia* and *Malpighia* at Sian Ka'an, *Alternanthera* in western Ecuador, and *Polygala* and *Pectis* in Los Llanos.

The cerrado (Filgueiras & Pereira, 1990) is distinctive in having *Habenaria* and *Vernonia* as the most species-rich genera and *Miconia* as sixth, although the other largest genera (respectively, *Cassia* (*s.l.*), *Hyptis, Paspalum, Mimosa, Ipomoea* and *Panicum*) are well represented in most dry forest florulas. The most speciose cerrado fern genera are *Dryopteris* and *Anemia*, poorly represented elsewhere. *Cyrtopodium* and *Banisteriopsis* are also strikingly more species-rich in the cerrado.

Some of these floristic differences reflect ecological rather than phytogeographic differences, e.g. the prevalence of aquatic and semiaquatic taxa like *Ludwigia* in western Ecuador where the florula site includes a large area of fresh water swamp, or the Los Llanos savannas where nine of the most speciose genera (including four of the top seven) are grasses and sedges. The Guanacaste, Costa Rica site is distinctive in the greater representation of moist forest elements like *Ficus* and *Psychotria*, presumably reflecting the presence of the well-developed gallery forests of this region which provide a moist-forest-like habitat despite the seasonally dry climate.

Floristic summary

With remarkably few exceptions like Cactaceae, Capparidaceae and Zygophyllaceae, the dry forest flora is a relatively depauperate selection of the same families that constitute moist and wet forest plant communities. In some cases, at least, we can begin to speculate about what characteristics enable these taxa to withstand the rigors of a dry forest existence.

Two of the most important features disproportionately associated with dry forest woody plants are conspicuous flowers pollinated by such specialized pollinators as hawkmoths, bats, and especially large and medium-sized bees, and wind-dispersed, and perhaps also autochorous, seeds. Thus it is hardly surprising that Bignoniaceae and Leguminosae, perhaps the two families most characterized by the combination of complex conspicuous zygomorphic flowers and wind-dispersed seeds are respectively the pre-eminent dry forest tree and liana families. Another unusually well represented family in dry forest is Euphorbiaceae, probably the most preponderantly autochorous woody plant family.

Similarly, within a family that includes both anemochorous and zoochorous members, genera with wind dispersal will typically be better represented in dry forests. For example, the characteristic and otherwise omnipresent mammal-dispersed legume genus *Inga* is almost exclusively restricted to moist and wet forests, while wind-dispersed *Lonchocarpus*, one of the most prevalent and speciose dry forest genera, is poorly represented in moist and wet forest. Similarly in Sapindaceae, the two large and closely related vine genera, *Serjania* and *Paullinia* show mirror-image ecological preferences, with wind-dispersed *Serjania* most prevalent in dry forest and bird-dispersed *Paullinia* in wet forest.

Some habit differences may also be related to success in a dry forest milieu. For example, several of the most prevalent dry forest genera are predominantly shrubby members of families that are predominantly arborescent or scandent in other habitats, for example *Acalypha* and *Croton* of the Euphorbiaceae or *Cassia* of the Leguminosae. Similarly there is some indication that among scandent plants, vines as opposed to woody lianas may be favored in at least some types of dry forests, like that of western Ecuador. Thus it is no surprise that families of predominantly herbaceous climbers like Cucurbitaceae, Asclepiadaceae, Passifloraceae, and papilionoid legumes may be disproportionately well represented in dry forests as compared to their role in moist or wet forest. Among the most prevalent and speciose dry forest families, the one that seems the most out of place when compared with similar listings of important families of wetter forests is Convolvulaceae. *Ipomoea*, the most speciose of all dry forest genera, is a genus of largely herbaceous or subwoody climbers, with tuberous roots, as contrasted to the woody Convolvulaceae lianas that are more prevalent in moist forests.

Other habit-related suggestions are not borne out. To judge from the prevalence of small-leafleted mimosoids like *Prosopis* and *Acacia* in very dry sites, one might suspect that dry forest Leguminosae are disproportionately mimosoids. In general there are about twice as many papilionoids as mimosoids in dry forest florulas and somewhat fewer species of caesalpinoids than of mimosoids (Table 7.4); however, mimosoids are actually more prevalent, both relatively and in absolute numbers, in moist and wet forest florulas, with almost as many (and sometimes more) species as papilionoids.

While we are still very far from understanding why dry forests are put together the way they are, we do know enough to suspect strongly that there is some kind of underlying order determining their floristic composition.

Endemism

From the perspective of conservation, endemism may be more important than diversity. Concentrations of locally endemic taxa are likely to be more worthy of special conservational focus than higher diversity communities composed mostly of wide-ranging species. Rzedowski (1962, 1978, 1991b) has emphasized that, at least in the context of México, drier areas have higher percentages of generic endemism: 43% endemic genera in arid areas, 28% in semiarid areas, 11% in semihumid areas, and only 4% in humid areas. On the other hand, Gentry (1982c) has suggested that dry forest species tend to be unusually wide-ranging and little prone to local endemism. In this section the dry forest floristic data presented here will be reanalysed from the viewpoint of pinpointing concentrations of endemism.

In dry forest samples from only a single region 174 genera are represented, many only from a single site. Such restricted distributions could provide an indication of greater endemism in those areas with more uniquely represented taxa. For example, one might anticipate *a priori* that the Antilles would be the most distinctive region taxonomically, but this is not supported by these dry forest data. Only 27 genera are restricted to the West Indian samples, in contrast to 54 genera sampled only in the northern Venezuela/Colombia phytogeographic region, 32 at Chamela, and 31 in coastal Ecuador. Only the southern subtropical and southern Central American regions, respectively with 17 and 13 uniquely sampled genera, are less distinctive than the Antilles at this level of analysis. Most of these differences seem more related to sampling artifact than to biogeography; in general the areas like northern Colombia/Venezuela that have more samples also have more 'unusual' genera included in the data set. Thus the transect data would seem inadequate in themselves for a direct analysis of relative endemism.

However, a different kind of analysis of generic endemism patterns is possible, albeit somewhat subjective. Based on monographs, floristic treatments, herbarium analysis and field experience, I have categorized each genus in the dry forest data set as to its distributional pattern: (1) also widespread in moist and wet forests; (2) 'subendemic', being mostly restricted to dry forest but occurring in several phytogeographical regions; (3) 'endemic', or mostly restricted to a single phytogeographic region (but with Central America plus México treated as a single endemism unit and with slight spillover into one adjacent region allowed, e.g. Antillean endemics allowed to occur in south Florida and the Yucatán Peninsula). The genera classified as endemic or as dry forest restricted subendemics are tabulated in Table 7.6.

According to this analysis, the western Mexican region, represented by the three Chamela samples, has by far the highest concentration of generic endemism with 12 Central American endemic genera and another 25 dry forest restricted genera. Runners-up are the southern subtropical area with eight endemic genera and 14 dry forest restricted ones, and the Antilles whose samples include eight West Indian endemic genera (three of which also occur on the Yucatán Peninsula) plus 12 dry forest restricted taxa. At the opposite extreme, the Guanacaste, Costa Rica sample includes only two Central American endemic genera, *Rehdera* of the Verbenaceae and *Sapranthus* of the Annonaceae (the latter also occurring in the Chamela sample) along with 13 dry area restricted taxa.

It is noteworthy that in the case of México, the relatively high levels of endemism in dry forest counteract the generally higher α-diversity of moist and wet forests so that the total Mexican dry forest flora is as large as or slightly larger than the moist forest one (Rzedowski, 1991b and personal communication).

To judge from these patterns, the western Mexican dry forest seems deserving of special conservational attention. From the standpoint of

Table 7.6. *Dry forest specialized genera. Distributions of sampled endemic (E) and dry-forest-restricted (SE) genera by phytogeographic region where sampled. Dry-forest-restricted includes edaphically dry areas*

Antilles		Chamela		N. Colombia/Venezuela		Southern subtropics	
Metopium	E	*Lagrezia*	E[a]	Basellaceae (indet)	E	*Schinus*	SE
Canella	E	*Alstonia*	SE[a]	*Buxus*	SE	*Patagonula*	E
Gyminda	SE	*Thevetia*	SE	*Belencita*	E	*Cnicothamnus*	E
Bucida	SE	*Pachycereus*	E	*Steriphoma*	SE	*Anadenanthera*	SE
Astrocasia	SE	*Peniocereus*	E	*Curatella*	SE	*Gleditsia*	SE[b]
Thrinax	E	*Stenocereus*	E	*Hecatostemon*	E	*Holocalyx*	E
Krugiodendron	E	*Forchhammeria*	SE	*Bowdichia*	SE	*Parapiptadenia*	E
Antirrhea	SE[a]	*Elaeodendron*	SE	*Coursetia*	SE	*Tipuana*	E
Hypelate	E	*Lagascea*	E	*Geoffroea*	SE	*Hexachlamys*	E
Costa Rica		*Amphipterygium*	E	*Prosopis*	SE	*Bougainvillea*	SE
Hemiangium	SE	*Apoplanesia*	E	*Byrsonima*	SE	*Scutia*	SE[a]
Ateleia	SE	*Podopterus*	E	*Copernicia*	SE	*Vasobia*	E
Myrospermum	SE	*Karwinskia*	E	*Alseis*	SE		
Rehdera	E	*Allenanthus*	E	*Pogonopus*	SE		
		Exostemma	SE	*Dilodendron*	SE		
		Hintonia	E	*Melicocca*	SE		
		Recchia	SE[a]	*Corynostylis*	SE		
				Bulnesia	SE		
Coastal Ecuador/Perú		Chamela + Antilles		N. Colombia/Venezuela + S. subtropics			
Raimondea	SE	*Comocladia*	SE	*Acanthocereus*	SE		
Macranthisiphon	E	*Bourreria*	SE	*Cereus*	SE		
Eriotheca	SE	Chamela + Costa Rica		*Syagrus*	SE		
Rochefortia	SE	*Sapranthus*	E	N. Colombia/Venezuela + Coastal Ecuador/Perú			
Pseudogynoxys	SE	*Sciadodendron*	SE	*Morisonia*	SE		
Turbina	SE	*Thouinidium*	SE				
Actinostemon	SE						

Prockia	SE	Chamela + S. subtropics		Chamela–Costa Rica–N. Colombia/Venezuela	
Yucartonia	E	*Opuntia*	SE	*Jacquinia*	SE
Cathedra	SE	*Phyllostylon*	SE		
Condalia	SE			Chamela–N. Colombia/Venezuela–S. subtropics	
Coutarea	SE			*Achatocarpus*	SE
Antilles + Costa Rica		Costa Rica + Coastal Ecuador/Perú		All but Antilles and Costa Rica	
Piscidia	SE	*Ximenia*	SE	*Caesalpinia*	SE
N. Colombia/Venezuela + Costa Rica		Chamela + N. Colombia/Venezuela		*Ruprechtia*	SE
Crescentia	SE	*Gyrocarpus*	SE	All but Costa Rica and S. subtropics	
Godmania	SE	*Diphysa*	SE	*Guapira*	SE
Calycophyllum	SE	*Cnidoscolus*	SE	Chamela–N. Colombia/Venezuela–Coastal Ecuador/Perú	
Malpighia	SE	*Manihot*	SE	*Plumeria*	SE
		Esenbeckia	SE	*Jacquemontia*	SE
Antilles–N. Colombia/Venezuela–Coastal Ecuador/Perú				All but Southern subtropics	
Amyris	SE			*Bursera*	SE
Zizyphus	SE				
All but Antilles and Coastal Ecuador/Perú					
Astronium	SE				

[a] Also Old World.
[b] Also North America.

conservation of endemic genera, the Mexican dry forest is much more important than the Guanacaste, Costa Rica area where much more conservational effort has been expended. Of the dry-forest-restricted Central American genera represented in these samples, *Myrospermum* may be the only one in the Costa Rican samples which does not also occur in western México, whereas 10 of the equivalent western Mexican dry-forest-restricted genera do not reach Guanacaste: *Amphipterygium, Apoplanesia, Comocladia, Elaeodendron, Hintonia, Lagrezia, Pachycereus, Peniocereus, Recchia* and *Stenocereus*.

In addition to looking at regional endemism, we may also use these data to look for supra-regional patterns of phytogeographical relationships. If genera uniquely shared between pairs of sites are considered, the strongest link is between Northern Colombia/Venezuela and coastal Ecuador (16 uniquely shared genera) followed by Northern Colombia/Venezuela with Chamela (15 uniquely shared genera) and northern Colombia/Venezuela with the southern subtropics (13 uniquely shared genera). Surprisingly Chamela and Guanacaste have only six uniquely shared sampled genera. Although such patterns in large part reflect the incompleteness of the data base (with the more extensive Northern Colombia/Venezuela data base disproportionately represented in the various site pairs), it is also noteworthy that the close phytogeographic relationship between Northern Colombia/Venezuela and coastal Ecuador suggested by these data was independently suggested by Sarmiento's (1975) analysis based on a very different generic data set.

There are also some data available for analysing species-level endemism in dry forest. Here the Antilles are distinctive, with the Round Hill list (Kelly *et al.*, 1988) having a higher percentage of endemics than any of the other florulas. At Round Hill, 30 species (24%) are endemic to Jamaica and 39 (31%) to the Antilles, with another 16 endemic to the Antilles plus the Yucatán for a total of 67% regional endemics (with five more species reaching only adjacent Central America). However, the West Indian data are not strictly comparable, since weedy species and forest margin plants, which tend to be widespread, were not included in the list. It should also be noted that, despite its high percentage of endemism, the Jamaican site has many fewer species than the mainland ones, and in absolute numbers actually has fewer endemic species than several of them.

At least among the more comparable mainland florulas, it is again western México that stands out. Of the Chamela species, 16% are local endemics (Lott *et al.*, 1987), many of them newly described (Lott, 1985;

at least four endemic genera). The Sian Ka'an dry forest also has a significant complement of 36 species strictly endemic to the Yucatán Peninsula, or 5% of the total florula, plus at least 54 additional species restricted to Central America and 27 essentially Antillean endemics (Durán & Olmsted, 1987, but excluding several listed species which are actually more widespread); if these three patterns are combined 15% of the Sian Ka'an species are regional endemics. Nineteen per cent of the species at Capeira are endemic to western Ecuador (including the phytogeographically similar adjacent corner of Perú) (Dodson & Gentry, 1991). These values contrast with 15% endemic-to-western-Ecuador species for the moist forest Jauneche flora (Dodson & Gentry, 1991) and 12% for the moist forest flora of Barro Colorado Island, Panamá (Croat & Busey, 1975).

It is interesting to contrast endemism in southern Central American dry forests, where much conservationist attention has been focused, with the Mexican figures. Only 2.5% (19 species) of the Santa Rosa, Costa Rica species are endemic to Costa Rica (data compiled from Janzen & Liesner, 1980 and Liesner, personal communication). If adjacent Nicaragua is included, this figure rises to 3% endemism, and to 4.5% if both Nicaragua and Panamá are included. Even if all of Central America south of México is considered, only 6% of the Santa Rosa species are regionally endemic as contrasted with the 16% **local** endemism in the Chamela flora. In contrast to México, no genus is endemic to southern Central American dry forests. On the other hand, 93% of the Santa Rosa flora reaches México and almost a quarter (23%) of the Santa Rosa species are endemic to México plus Central America. The center of endemism, as well as diversity, of the Mesoamerican dry forest is in western México, with the flora of the southern Central American dry forests generally consisting of the most widespread of the Mexican species.

Conclusions

We may conclude that neotropical dry forests are intrinsically fascinating ecosystems, perhaps not so much for their diversity as for the coordinated way in which their relatively low species diversity is organized. Although their families and genera mostly are shared with lowland moist forest, their species mostly are not. Many unique species have evolved to survive in each manifestation of this distinctive habitat. Alas, not only these species, but the evolutionary milieu that made them possible, are fast disappearing from the Earth. If we are interested in preserving the world's biodiversity, this exceptionally endangered portion of the world's tropics deserves its own conservational priority.

However, our conservational focus may need a certain amount of redirection. We cannot afford to restrict our dry forest conservational attention to a single area like Guanacaste, Costa Rica. The analyses presented here strongly suggest that each different dry forest region merits its own specific conservational focus. We need to expand conservational efforts to encompass each of these floristically distinct dry forests. It may already be too late for western Ecuador, northern Colombia, and much of Central America. However, there are still significant stands of dry forest in western México and in Santa Cruz, Bolivia. To whatever extent a single neotropical dry forest region is to be given conservational priority, that region should be southwestern México where dry forest diversity and endemism felicitously coincide. If the even more neglected Bolivian dry forests turn out to share this high diversity and endemism, as the very preliminary data suggest, then they too should share in the conservational spotlight.

Summary

Neotropical dry forests are generally less diverse than moist forests, but the most diverse dry forests are not the wettest ones, as might be expected, but rather the deciduous forests of western México, which are among the driest closed-canopy tropical forests. In such structural features as number of lianas, number of trees, and basal area, dry forests as a whole differ relatively little from moist and wet forests, but there are major differences between different dry forests, especially between Antillean and continental dry forests. Antillean forests are denser and have smaller trees and many fewer lianas. Dry forests generally have fewer epiphytes and more vines (but fewer lianas) than moist or wet forests. They are also distinctive in a higher percentage of trees and lianas with wind-dispersed seeds and conspicuous flowers.

Most continental neotropical dry forests share a characteristic suite of families and genera. Leguminosae are consistently the dominant family of trees and Bignoniaceae the dominant lianas in continental dry forests just as they are in moist forests, but their prevalence is accentuated in the floristically impoverished dry forest. Myrtaceae is the most speciose tree family in many Antillean and subtropical South American dry forests. The other most important dry forest woody families are Rubiaceae,

Sapindaceae, Euphorbiaceae, Flacourtiaceae and Capparidaceae, the latter especially prevalent in some of the driest sites. The most prevalent dry forest tree genus is *Tabebuia*; other especially prominent woody genera include *Casearia*, *Trichilia*, *Erythroxylum*, *Randia*, *Capparis*, *Bursera*, *Acacia* and *Coccoloba*, all also occurring in moist forest. If entire floras are compared, Leguminosae is always by far the most prevalent family, followed by Euphorbiaceae, Gramineae and Compositae, all better represented in dry than in moist forest. The largest dry forest genera, mostly repeated with partially overlapping sets of species from site to site, are mostly shrubby *Cassia* (*s.l.*), *Croton* and *Acalypha*, mostly scandent and subwoody *Ipomoea*, and in some forests epiphytic *Tillandsia*.

Endemism is non-randomly distributed among dry forests with the same western Mexican dry forests that are the most species-rich having the highest concentration of endemism at both the specific and generic level. Each of the other major dry forest regions has a significant complement of endemic species and a few endemic genera, with dry forest endemism levels in most regions tending to be even higher than in adjacent moist forest.

Acknowledgements

The studies of neotropical forests that form the basis for this review have been variously supported by a series of grants from the National Geographic Society, National Science Foundation (most recently BSR-8607113), Mellon Foundation, and MacArthur Foundation. I thank E. Lott, B. Boyle, P. Murphy, Charlotte Taylor and J. Rzedowski for review comments, R. Clinebell and B. Boyle for assistance with data entry and analysis, V. Kapos, A. Taber and E. Lott for making available their raw data, and various colleagues for assistance with dry forest field work, including H. Cuadros, A. Repizzo, R. Ortiz, and C. Barbosa in Colombia, R. Troth and P. Berry in Venezuela, L. Woodruff and R. Dirzo in Costa Rica, G. Proctor and C. Taylor in Puerto Rico, C. Joose and C. Dodson in Ecuador, Camilo Díaz and D. Smith in Perú, and R. Neumann, R. Palacios, L. Malmierca, C. Cristobal and A. Schinini in Argentina.

References

Aristeguieta, L. (1966). Flórula de la Estación Biológica de Los Llanos. *Boletín de la Sociedad Venezolana de Ciencias Naturales* **110**: 228–307.

Armesto, J. J. (1987). Mecanismos de diseminación de semillas en el bosque de Chiloé: una comparación con otros bosques templados y tropicales. *Anales del IV Congreso Latinoamericano de Botánica* **2**: 7–12.

Arriaga, L. & León, J. L. (1989). The Mexican tropical deciduous forest of Baja California Sur: a floristic and structural approach. *Vegetatio* **84**: 45–52.
Beck, S. G. (1984). Comunidades vegetales de las sabanas inundadas del NE de Bolivia. *Phytocoenologia* **12**: 321–50.
Brown, A. D., Chalukian, S. C. & Malmierca, L. M. (1985). Estudio florístico-estructural de un sector de selva semidecidua del Noreste Argentina 1. Composición florística, densidad, y diversidad. *Darwiniana* **26**: 27–41.
Croat, T. B. & Busey, P. (1975). Geographical affinities of the Barro Colorado Island flora. *Brittonia* **27**: 127–35.
Cuadros, H. (1990). Vegetación Caribeña. In *Caribe Colombiana*, pp. 67–84. Forado FEN Colombia, Bogotá.
Deneven, W. M. (1980). Field work as exploration: the Rio Heath savannas of southeastern Perú. *Geoscience and Man* **21**: 157–63.
Dodson, C. H. & Gentry, A. H. (1991). Biological extinction in western Ecuador. *Annals of the Missouri Botanical Garden* **78**: 273–95.
Dodson, C. H. & Gentry, A. H. (1992). *Flórula de Capeira*. Banco Nacional de Ecuador, Quito.
Durán G., R. & Olmsted, I. C. (1987). *Listado florístico de la Reserva Sian Ka'an*. Amigos de Sian Ka'an, Puerto Morelos.
Eiten, G. (1972). The cerrado vegetation of Brazil. *Botanical Review* **38**: 201–341.
Eiten, G. (1978). Delimitation of the cerrado concept. *Vegetatio* **36**: 169–78.
Filgueiras, T. & Pereira, R. A. (1990). Flora do Distrito Federal. In *Cerrado: caracterização, ocupação e perspectivas*, ed. M. Novaes Pinto, pp. 331–88. UNB-SEMATEC, Brasília.
Frankie, G. W. (1975). Tropical forest phenology and pollinator plant coevolution. In *Coevolution of Animals and Plants*, ed. L. E. Gilbert and P. H. Raven, pp. 192–209. University of Texas Press, Austin.
Gentry, A. H. (1982a). Patterns of neotropical plant species diversity. *Evolutionary Biology* **15**: 1–84.
Gentry, A. H. (1982b). Neotropical floristic diversity: phytogeographical connections between Central and South America, Pleistocene climatic fluctuations, or an accident of the Andean orogeny? *Annals of the Missouri Botanical Garden* **69**: 557–93.
Gentry, A. H. (1982c). Phytogeographic patterns in northwest South America and southern Central America as evidence for a Choco refugium. In *Biological Diversification in the Tropics*, ed. G. T. Prance, pp. 112–36. Columbia University Press, New York.
Gentry, A. H. (1983). Dispersal ecology and diversity in neotropical forest communities. *Sonderbaende des Naturwissenschaftlichen Vereins in Hamburg* **7**: 303–14.
Gentry, A. H. (1986). Species richness and floristic composition of Choco region plant communities. *Caldasia* **15**: 71–91.
Gentry, A. H. (1988). Changes in plant community diversity and floristic composition on environmental and geographical gradients. *Annals of the Missouri Botanical Garden* **75**: 1–34.
Gentry, A. H. (1990). Floristic similarities and differences between southern Central America and upper and central Amazonia. In *Four Neotropical Rain Forests*, ed. A. H. Gentry, pp. 141–57. Yale University Press, New Haven.
Gentry, A. H. (1991). The distribution and evolution of climbing plants. In *The Biology of Vines*, ed. F. E. Putz and H. A. Mooney, pp. 3–42. Cambridge University Press, Cambridge.
Gentry, A. H. (1992). Tropical forest biodiversity: distributional patterns and their conservational significance. *Oikos* **63**: 19–28.
Gentry, A. H. & Dodson, C. H. (1987a). Contribution of non-trees to species richness of tropical rain forest. *Biotropica* **19**: 149–56.

Gentry, A. H. & Dodson, C. H. (1987b). Diversity and biogeography of neotropical vascular epiphytes. *Annals of the Missouri Botanical Garden* **74**: 205–33.
Haase, R. & Beck, S. (1989). Structure and composition of savanna vegetation in northern Bolivia: a preliminary report. *Brittonia* **41**: 80–100.
Hartshorn, G. S. (1983). Plants. In *Costa Rican Natural History*, ed. D. H. Janzen, pp. 118–57. University of Chicago Press, Chicago.
Heybrock, G. (1984). Der Tayrona-Trockenwald Nord-Kolumbiens. Eine Ökosystemstudie unter besonderer Berucksichtigung von Biomasse und Blattflachenindex (LAI). *Giessener Geographische Schriften* **55**: 1–104.
Holdridge, L. R., Grenke, W. C., Hatheway, W. H., Liang, T. & Tosi, J. A., Jr (1971). *Forest Environments in Tropical Life Zones: a pilot study*. Pergamon Press, Oxford.
Hubbell, S. P. (1979). Tree dispersion, abundance, and diversity in a tropical dry forest. *Science* **203**: 1299–1309.
Huber, O. (1987). Neotropical savannas: their flora and vegetation. *Trends in Ecology and Evolution* **2**: 67–71.
Janzen, D. H. (1967). Synchronization of sexual reproduction of trees within the dry season in Central America. *Evolution* **21**: 620–37.
Janzen, D. H. (1983). Food webs: who eats what, why, how, and with what effects in a tropical forest? In *Tropical Rain Forest Ecosystems: structure and function*, ed. F. B. Golley, pp. 167–82. Elsevier, Amsterdam.
Janzen, D. H. (1984). Two ways to be a tropical big moth: Santa Rosa saturniids and sphingids. *Oxford Surveys in Evolutionary Biology* **1**: 85–140.
Janzen, D. H. (1988). Management of habitat fragments in a tropical dry forest: growth. *Annals of the Missouri Botanical Garden* **75**: 105–16.
Janzen, D. H. & Liesner, R. (1980). Annotated checklist of lowland Guanacaste Province, Costa Rica, exclusive of grasses and nonvascular cryptograms. *Brenesia* **18**: 15–90.
Kapos, V. (1982). An ecological investigation of sclerophylly in two tropical forests. Ph.D. thesis, Washington University, St Louis.
Kelly, D. L., Tanner, E. V., Kapos, V., Dickinson, T., Goodfriend, G. & Fairbairn, P. (1988). Jamaican limestone forests: floristics, structure and environment of three examples along a rainfall gradient. *Journal of Tropical Ecology* **4**: 121–56.
Lott, E. J. (1985). *Listados florísticos de México III. La Estación de Biología Chamela, Jalisco*. Instituto de Biología, Universidad Nacional Autónoma de México, México.
Lott, E. J. (1993). Annotated checklist of the vascular flora of the Chamela Bay region, Jalisco, México. *Occasional Papers of the California Academy of Sciences* **148**: 1–60.
Lott, E. J., Bullock, S. H. & Solís Magallanes, J. A. (1987). Floristic diversity and structure of a tropical deciduous forest of coastal Jalisco. *Biotropica* **19**: 228–35.
Lugo, A. E., Gonzalez-Liboy, J. A., Cintrón, B. & Dugger, K. (1978). Structure, productivity, and transpiration of a subtropical dry forest. *Biotropica* **10**: 278–91.
Mori, S. A., Boom, B., Carvalho, A. M. & dos Santos, T. S. (1983). Ecological importance of Myrtaceae in an eastern Brazilian wet forest. *Biotropica* **15**: 68–70.
Murphy, P. G. & Lugo, A. E. (1986a). Structure and biomass of a subtropical dry forest in Puerto Rico. *Biotropica* **18**: 89–96.
Murphy, P. G. & Lugo, A. E. (1986b). Ecology of tropical dry forest. *Annual Review of Ecology and Systematics* **17**: 67–88.
Peixoto, A. L. & Gentry, A. H. (1990). Diversidade e composição floristica da mata de tabuleiro na Reserva Florestal de Linhares (Espírito Santo, Brasil). *Revista Brasileira de Botanica* **13**: 19–25.
Pinto, M. N. (ed.) (1990). Cerrado: caracterização, ocupação e perspectivas. UNB-SEMATEC, Brasília.
Prado, D. E. (1993). What is the Gran Chaco vegetation in South America? I. A review. Contribution to the study of flora and vegetation of the Chaco. V. *Candollea* **48**: 145–72.

Rico-Gray, V., García-Franco, J. G., Puch, A. & Sima, P. (1988). Composition and structure of a tropical dry forest in Yucatán, México. *International Journal of Ecology and Environmental Sciences* **4**: 21–29.

Rzedowski, J. (1962). Contribuciones a la fitogeografia florística e histórica de México I. Algunas consideraciones acerca del elemento endémico en la flora mexicana. *Boletín de la Sociedad Botánica de México* **27**: 52–65.

Rzedowski, J. (1978). *Vegetación de México*. Editorial LIMUSA, México.

Rzedowski, J. (1991a). Diversidad y orígenes de la flora fanerogámica de México. *Acta Botánica Mexicana* **14**: 3–21.

Rzedowski, J. (1991b). El endemismo en la flora fanerogámica mexicana: una apreciación analítica preliminar. *Acta Botánica Mexicana* **15**: 47–64.

Salati, E., Marques, J. & Molion, L. C. (1978). Origem e distribuição das chuvas na Amazonia. *Interciencia* **3**: 200–5.

Saldias P., M. (1991). Inventario de árboles en el bosque alto del Jardín Botánico de Santa Cruz, Bolivia. *Ecología en Bolivia* **17**: 31–41.

Sarmiento, G. (1975). The dry plant formations of South America and their floristic connections. *Journal of Biogeography* **2**: 233–51.

Sarmiento, G. (1984). *The Ecology of Tropical Savannas* (translated by O. Solbrig). Harvard University Press, Cambridge, Mass.

Spichiger, R., Bertoni, B. S. & Loizeau, P.-A. (1992). The forests of the Paraguayan Alto Paraná. *Candollea* **47**: 219–50.

Terborgh, J. (1986). Community aspects of frugivory in tropical forests. In *Frugivores and Seed Dispersal*, ed. A. Estrada and T. H. Fleming, pp. 371–84. Dr W. Junk, Dordrecht.

Thien, L. B., Bradburn, A. S. & Welden, A. L. (1982). The woody vegetation of Dzibilchaltún: a Maya archaeological site in northwest Yucatán, México. *Middle American Research Institute Occasional Paper* **5**. Tulane University, New Orleans.

Troth, R. G. (1979). Vegetational types on a ranch in the central Llanos of Venezuela. In *Vertebrate Ecology in the Northern Neotropics*, ed. J. F. Eisenberg, pp. 17–30. Smithsonian Institution Press, Washington.

Valverde, F. M., de Tazan, G. R. & Rizzo, C. G. (1979). Cubierto vegetal de la Península de Santa Elena. Facultad de Ciencias Naturales, Universidad de Guayaquil, Guayaquil.

Wendt, T. (1993). Composition, floristic affinities, and origins of the canopy tree flora of the Mexican Atlantic slope rain forests. In *Biological Diversity of México: origins and distribution*, ed. T. P. Ramamoorthy, R. Bye, A. Lot and J. Fa, pp. 595–680. Oxford University Press, Oxford.

Wikander, T. (1984). Mecanismos de dispersión de diásporas de una selva decídua en Venezuela. *Biotropica* **16**: 276–83.

8

Vertebrate diversity, ecology, and conservation in neotropical dry forests

GERARDO CEBALLOS

Introduction

The neotropics may sustain the largest number of living species of plants and animals on earth (Wilson, 1988; McNeely *et al.*, 1990). In the vast neotropical region, however, there are extensive regions such as the Amazonian rain forests, the Venezuelan 'tepuis', and the dry forests, where there is little biological knowledge of ecosystem composition and ecological interactions.

Research in neotropical dry forests during the last two decades has provided the scientific community with surprises such as the discovery of an 'extinct' peccary, *Catagonus wagneri*, in the Paraguayan chaco in 1974 (Wetzel *et al.*, 1975). It is, therefore, important to recognize patterns of biological diversity in dry forest to gain insights into causal processes in biogeography and ecosystem function, and to assign them appropriate conservation values.

In this chapter I describe the general patterns of terrestrial vertebrate diversity and conservation in dry forests, and contrast them with patterns in adjacent moist/wet forests. The chapter is divided into five sections. In the first section, a general description of the major neotropical dry forests is given, emphasizing size and degree of perturbation. The following section presents a detailed analysis of patterns of species richness and diversity. The third section is dedicated to a description of community structure and ecological responses to climate seasonality. In the fourth section the origins of the dry forest vertebrate faunas are discussed. Finally, the last section is devoted to conservation problems.

Neotropical dry forests

Neotropical dry forests occur from México to northern Argentina. The main ecological differences between tropical evergreen or semideciduous forests and dry forests are related to differences in the amount and seasonality of annual precipitation (Frankie, Baker & Opler, 1974; Rzedowski, 1978; Hartshorn, 1983; Medina, Chapter 9). Dry forests have a marked seasonality and lower amounts of rainfall. During the dry season, which lasts from four to eight months, almost all plants shed their leaves, and many disperse their seeds (e.g. Bullock & Solís, 1990). The effects of such phenological changes on microclimatic conditions are profound. Comparing Costa Rican dry forests and adjacent semi-deciduous forests at the end of the dry season has shown that semi-deciduous forests have lower soil and air temperatures (6.5 and 5.5 °C cooler, respectively), have a 20% higher relative humidity (Janzen, 1976), and the litter layer has much higher humidity (Duellman, 1965). However, such microclimatic differences tend to disappear during the rainy season. Animal species show diverse ecological, behavioral and physiological responses to cope with the seasonality in climate (see below).

On a geographic basis, dry forest can be divided into two groups: Mesoamerican (México and Central America) and South American. Prehistorically, Mesoamerican dry forest may have covered 6.5×10^5 km^2, distributed in the Yucatán Peninsula and along the Pacific coast from southern Sonora, México (latitude 25° N) to the Península de Nicoya, Costa Rica (10° N).

On the basis of both biological and physical environmental features, the Mesoamerican dry forest can be divided into three main regions: Yucatán, western México, and Central America. In Yucatán, dry forest is restricted to the northern part of the peninsula, and it is in contact with wetter forests to the west and south. It has a floristic composition different from western México (Rzedowski, 1978). In western México, dry forest covers vast areas from southern Sonora to Chiapas, that are geographically and ecologically isolated from other tropical forests. North of the Istmo de Tehauntepec, Mexican dry forest is tenuously connected to more southerly tropical moist forests and are usually in contact with oak forest, and sometimes pine forests. Such isolation has constrained its species richness, but also has facilitated speciation in many groups of plants and animals (e.g. Duellman, 1965; Rzedowski, 1978; Ceballos & Navarro, 1991; Flores, 1991; Ceballos & Rodríguez, 1993).

Relatively large conserved tracts of dry forest are still found in Sinaloa, Nayarit, Jalisco and Michoacán (G. Ceballos, personal observation).

In Central America dry forests were found along a narrow strip on the Pacific coast from Guatemala to Costa Rica (Janzen, 1988). However, these forests have been highly fragmented and destroyed, and only 2% are considered to be still pristine (Janzen, 1988).

Costa Rican and South American dry forests are intermingled with or extensively connected to wet forests (Janzen & Wilson, 1983; Mares, Willig & Lacher, 1985). In South America, the largest areas of dry forest are (or were) found in northern Colombia and Venezuela, in northeastern Brazil ('caatinga'), and in Paraguay, Argentina, Bolivia and Brazil ('chaco'). Dry forests in Colombia and Venezuela are patchily distributed within an area of 6×10^4 km^2 (Eisenberg & Redford, 1979), but most have been degraded, transformed, or destroyed (A. Gentry, personal communication). These forests are (or were) in contact with wet and moist forests, and with savannas ('llanos').

Caatinga dry forests, which cover about 6.5×10^5 km^2 (Mares *et al.*, 1981; Sampaio, Chapter 3) are subject to very severe, prolonged droughts, and extensive areas have been profoundly disturbed by humans (Sampaio, Chapter 3). Enclaves of wetter forest exist, and there are broad transitional areas to wetter vegetation to the west and south (Mares *et al.*, 1981, 1985; Willig, 1983).

Finally, the chaco is an enormous mosaic of grasslands and woodlands that grades into wetter forests to the north and desert scrub to the south (e.g. Mares, 1985; Mares *et al.*, 1985). Large forest tracts have been severely disturbed by logging and cattle raising (Mares, 1985; Saravia-Toledo, 1985; Roig, 1991).

Biodiversity

Species richness

In the neotropics there are at least 1100 species of mammals, 3000 species of birds, and 1700 species of reptiles and amphibians (McNeely *et al.*, 1990; Mares & Schmidly, 1991; Ceballos & Sánchez, 1994). Insect species are largely unknown, but the level of richness is astounding. Nearly 550 butterfly species have been described in Costa Rica alone, and a single square kilometer of Central American forest can hold thousands of insect species (Wilson, 1988).

Tropical wet and moist forests have more species richness and

Table 8.1. *Comparisons of vertebrate species in neotropical dry and rain forests*

Locality	Latitude	Forest type	Number of species		
			Mammals	Birds	Herptiles
Chamela	19° 30′ N	Deciduous	70	270	85
Los Tuxtlas	18° 34′ N	Wet	90	315	65
Costa Grande	18° 30′ N	Deciduous	79	–	55
Lacandona	16° 30′ N	Wet	113	300	109
Guanacaste	10° 50′ N	Deciduous	136	269	70
La Selva	10° 26′ N	Wet	138	410	134
Barro Colorado	9° 9′ N	Moist	76	443	121
Caatinga	8° 0′ S	Deciduous	46	200	47
Manu	11° 45′ S	Wet	150[a]	554	133
Chaco	34° 30′ S	Deciduous	110	409	80[a]

[a] Estimated.
Sources: Duellman (1965, 1990); Sick (1965); Ramírez-Pulido *et al.* (1977); Vanzolini *et al.* (1980); Mares *et al.* (1981, 1985); Janzen (1983); Stiles (1983); Ceballos & Miranda (1986); Karr *et al.* (1990); Glanz (1990); Guyer (1990); Janson & Emmons (1990); Karr *et al.* (1990); Rand & Myers (1990); Redford *et al.* (1990); Rodríguez & Cadle (1990); Wilson (1990); Arizmendi *et al.* (1990); Medellín (1992); García-Aguayo & Ceballos (1994).

community diversity than dry forests (Table 8.1) (e.g. Duellman, 1960, 1965, 1990; Sick, 1965; Ramírez-Pulido, Martínez & Urbano, 1977; Vanzolini, Ramos-Acosta & Vitt, 1980; Mares *et al.*, 1981, 1985; Janzen, 1983; Stiles, 1983; Ceballos & Miranda, 1986; Glanz, 1990; Guyer, 1990; Janson & Emmons, 1990; Karr *et al.*, 1990; Rand & Myers, 1990; Redford, Taber & Simonetti, 1990; Rodríguez & Cadle, 1990; Wilson, 1990; Arizmendi *et al.*, 1990; Medellín, 1992; García-Aguayo & Ceballos, 1994).

Superficially, the fauna (and the flora: Gentry, Chapter 7) can be characterized as a depauperate subset of the moist/wet forest fauna. However, dry forests are special repositories of vertebrate diversity, because of their endemic species (Mares, 1992; Ceballos & Rodríguez, 1993), and the dry forest populations of many widespread species show unique physiological and ecological adaptations to cope with climatic seasonality (e.g. Janzen & Wilson, 1983).

Latitudinal trends in species diversity

The best established geographic pattern of species diversity in the Americas is the decrease in species richness with increasing latitude; i.e.

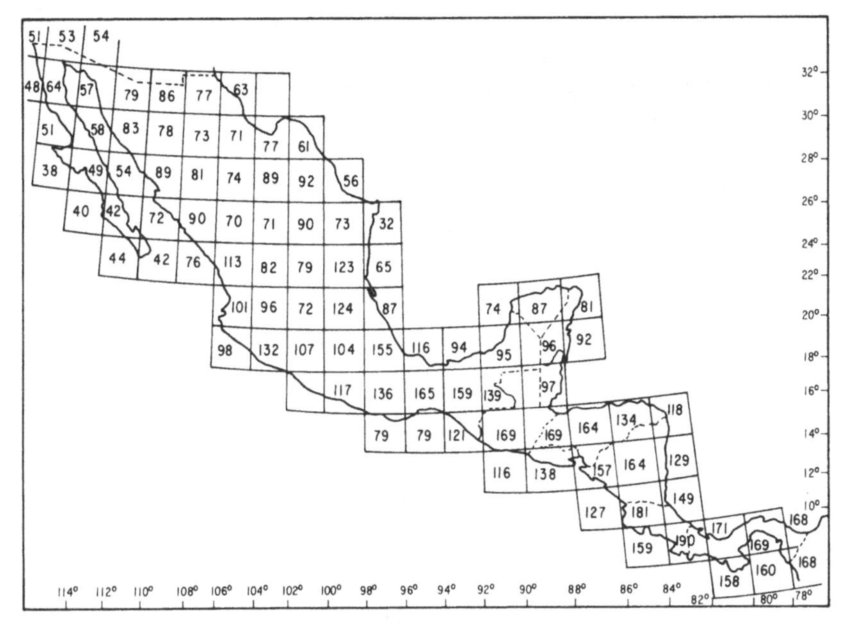

Figure 8.1. Latitudinal trends of mammal species richness in México and Central America, using 2 × 2° quadrants.

the number of vertebrate species increases from the poles towards the tropics (Figure 8.1). This pattern has been well established for birds and mammals in both North and South America (MacArthur & Wilson, 1967; Fleming, 1973; Wilson, 1974; McCoy & Connor, 1980; Mares & Ojeda, 1982; Willig & Selcer, 1989; Ceballos & Navarro, 1991; Willig & Sadlin, 1991).

The increase in mammalian species is attributed mainly to an increase in the number of bat species (Fleming, 1973; Wilson, 1974; Willig & Selcer, 1989; Willig & Sadlin, 1991). Considering only different groups of bats, species richness of phyllostomid (leaf nose) bats increases faster with decreasing latitude than species richness in non-phyllostomid bats (Willig & Selcer, 1989; Willig & Sadlin, 1991).

Some vertebrate groups with nearctic zoogeographical affinities show an opposite trend, declining in diversity with decreasing latitude, such as the families Geomyidae (pocket gophers), Heteromyidae (pocket mice and kangaroo rats), Talpidae (moles), Soricidae (shrews), Sciuridae (squirrels), Anatidae (ducks), Alcidae (auks), Procellaridae (petrels) and Phasianidae (quails).

Latitudinal trends in the diversity of dry forest vertebrates show interesting similarities and contrasts to the overall patterns. Mammal species in dry forest increase with decreasing latitude, but on any given latitude species diversity is relatively higher in wet and moist forests (Figure 8.2). The lower species richness of dry forest is strongly related

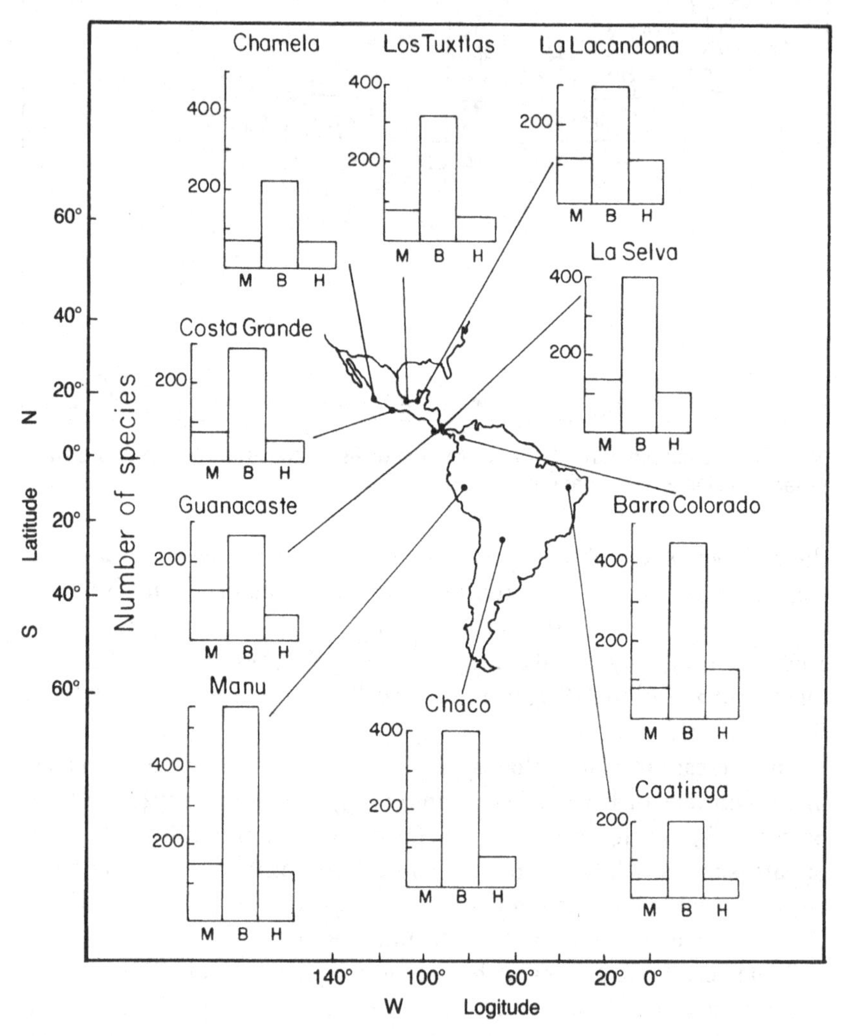

Figure 8.2. Comparisons of vertebrate species in neotropical deciduous and rain forests. M, mammals; B, birds; H, reptiles and amphibians. Deciduous forests are Chamela and Costa Grande in México, Guanacaste in Costa Rica, caatinga in Brazil, and chaco in Paraguay and Argentina. Rain forests are Los Tuxtlas and La Lacandona in México, La Selva in Costa Rica, Barro Colorado Island in Panamá, and Manu in Perú.

to the absence of many specialized species of carnivores, frugivores and semiaquatic mammals.

Bird species richness in dry forest does not increase dramatically with decreasing latitude and the contrast with moist forests is greater than that for mammals (Figure 8.2). The number of herptiles in dry forest does not show a marked increase with decreasing latitude; indeed, reptiles are more diverse in Mexican than in other dry forests (e.g. Casas, 1982; Janzen, 1983; Flores, 1991; Figure 8.2). This is certainly influenced by the fact that México has the highest number of reptile species in the world (McNeely *et al.*, 1990).

Endemism

Of all neotropical dry forests, western México and the chaco support the highest number of endemic terrestrial vertebrate genera and species (Table 8.2). There are few endemic species in other dry forests, suggesting scenarios of either extensive fragmentation and/or little isolation in geological times, and that such forests were readily accessible to animals from surrounding communities (e.g. Short, 1974).

There is a relatively low number of endemic herpetological species in dry forest. Reptiles in the chaco have a low number of endemic species (e.g. Scott & Lovett, 1975) and only one species of lizard is endemic to the caatinga (Vanzolini *et al.*, 1980; Sampaio, Chapter 3). However, western México supports a large number of endemic reptiles and amphibians (173 species, 43% restricted to the dry forest). In the dry forests of the Jalisco lowlands, there are 85 species of reptiles and amphibians, of which 42 (49%) are endemic (García-Aguayo & Ceballos, 1994). Similar percentages of endemicity have been recorded in other localities in western México (e.g. Duellman, 1958, 1960, 1965; Flores, 1991).

Bird endemicity is low in all but Mexican dry forest, which supports at least 25 endemic species. Bird communities in dry forest of Costa Rica are similar to those of northern Central America and México (Slud, 1964; Stiles, 1983). There are only two endemic species in caatinga (Sick, 1965; Sampaio, Chapter 3) and around four in the chaco (Short, 1974).

Among mammals (Table 8.2), relatively high numbers of endemic species are found in western México (26 species; Ceballos & Miranda, 1986; Ceballos & Rodríguez, 1993) and the chaco (22 species; Mares, Ojeda & Barquez, 1989; Redford & Eisenberg, 1992). In northern Colombia and Venezuela only three species of mammals are endemic to dry forest (Eisenberg & Redford, 1979; Eisenberg, 1989; Willig & Mares,

Table 8.2. *Mammals endemic to dry forests in the neotropics. Some species are associated with grasslands or moist/wet habitats within the dry forest ecosystem*

Classification	México	Colombia & Venezuela	Caatinga	Chaco
Marsupialia: Didelphidae (opossums)				
Marmosa canescens	×			
Marmosa xerophila		×		
Insectivora: Soricidae (shrews)				
Megasorex gigas	×			
Chiroptera: Phyllostomidae				
Glossophaga morenoi	×			
Musonycteris harrisoni	×			
Chiroptera: Vespertilionidae				
Myotis carteri	×			
Myotis findleyi	×			
Myotis fortidens	×			
Myotis nesopolus		×		
Rhogeessa genowaysi	×			
Rhogeessa minutilla		×		
Rhogeessa mira	×			
Lagomorpha: Leporidae (hares)				
Lepus flavigularis	×			
Xenarthra: Dasypodidae (armadillos)				
Cabassous chacoensis				×
Chaetophractus villosus				×
Chlamyphorus retusus				×
Rodentia: Sciuridae (squirrels)				
Spermophilus annulatus	×			
Spermophilus adocetus	×			
Sciurus colliaei	×			
Rodentia: Geomyidae (pocket gophers)				
Orthogeomys grandis	×			
Orthogeomys hispidus	×			
Rodentia: Heteromyidae (spiny pocket mice)				
Liomys pictus	×			
Liomys spectabilis	×			
Rodentia: Muridae (rats and mice)				
Akodon dolores				×
Akodon toba				×
Andalgalomys pearsoni				×
Bibimys chacoensis				×
Graomys domorum				×
Graomys edithae				×
Hodomys alleni	×			
Neotoma phenax	×			

Table 8.2. (*cont.*)

Classification	México	Colombia & Venezuela	Caatinga	Chaco
Oryzomys chacoensis				×
Osgoodomys banderanus	×			
Peromyscus perfulvus	×			
Pseudoryzomys wavrini				×
Sigmodon mascotensis	×			
Sigmodon alleni	×			
Tylomys bullaris	×			
Xenomys nelsoni	×			
Rodentia: Caviidae (cavies)				
Kerodon rupestris			×	
Pediolagus salinicola				×
Rodentia: Ctenomyidae (tuco-tucos)				
Ctenomys conoveri				×
Ctenomys dorsalis				×
Ctenomys occultus				×
Ctenomys argentinus				×
Ctenomys bonettoi				×
Ctenomys d'orbigni				×
Ctenomys pundti				×
Ctenomys fochi				×
Ctenomys juri				×
Carnivora: Mustelidae				
Spilogale pygmaea	×			
Artiodactyla: Tayassuidae (peccaries)				
Catagonus wagneri				×
Total species	26	3	1	22

Sources: for México: Ceballos & Miranda (1986), Ceballos & Rodríguez (1993); for Venezuela and Colombia: Eisenberg (1989); for caatinga: Mares *et al.* (1981); for chaco: Myers (1988), Mares *et al.* (1990), Redford & Eisenberg (1992).

1989). Caatinga supports the most depauperate mammalian fauna in the neotropics (Mares *et al.*, 1981, 1985; Streilein, 1982; Willig, 1983; Sampaio, Chapter 3). Its only endemic mammal is a rock-dwelling caviid rodent (*Kerodon rupestris*). Interestingly, the bat communities in the caatinga show a low degree of similarity and are more diverse than nearby moister cerrado habitats (Willig, 1983).

Most endemic dry forest mammals, such as shrews, armadillos, mice, and fossorial species (e.g. tuco-tucos), have low body masses, poor vagilities and short generation times. Such characteristics strongly indicate that speciation events were promoted by habitat fragmentation and isolation of small habitat patches during the Plio-Pleistocene (e.g. Redford & Eisenberg, 1992; G. Ceballos & J. Arroyo, unpublished data). Presently, several species of small mammals such as *Peromyscus perfulvus* in México or *Andalgalomys pearsoni* in the chaco, survive in 'islands' of moist forests or grasslands, respectively, surrounded by 'seas' of dry forest (Ceballos & Miranda, 1986; Redford & Eisenberg, 1992). The lack of larger endemic species may be explained in two ways: they became extinct either as a result of stochastic events associated with small population sizes or because they were unable to move across habitat patches.

Migratory birds

Migratory non-passerine (e.g. *Buteo* hawks) and passerine (e.g. flycatchers, swallows, warblers, vireos) land birds from central and eastern North America annually disperse during the winter into the tropical forests of México, Central America and northern South America (e.g. Stiles, 1983; Hutto, 1986). Both dry forest and moist/wet forests are used as habitats for such species, which spend up to seven months on their tropical wintering grounds, before migrating to the north for the summer breeding season (e.g. Stiles, 1983).

Migratory passerine land birds from western North America behave differently, because their wintering grounds are located exclusively in desert scrubs, dry forest, and pine oak forests in western México (Figure 8.3) (Hutto, 1986; Arizmendi *et al.*, 1990). At least 109 species of warblers, redstarts, vireos, yellowthroats, titmice, flycatchers, gnatcatchers and other birds spend from 8–9 months in their Mexican wintering grounds (Hutto, 1986). Most of these species are more or less restricted to one habitat type, and comprise on average 45% of the species and 55% of the individuals in dry forest during the winter (Hutto, 1986). Foraging migrant and resident birds form mixed-species flocks. The world's largest and most diverse flocks have been recorded in western México (Hutto, 1986).

Relationships of wet and deciduous forest faunas

Vertebrate communities in dry forest are generally represented by groups of typical dry forest species and a subset of species from nearby rain forest faunas. At any given latitude, the number of vertebrates in a dry forest is apparently correlated with its degree of isolation from other tropical forests.

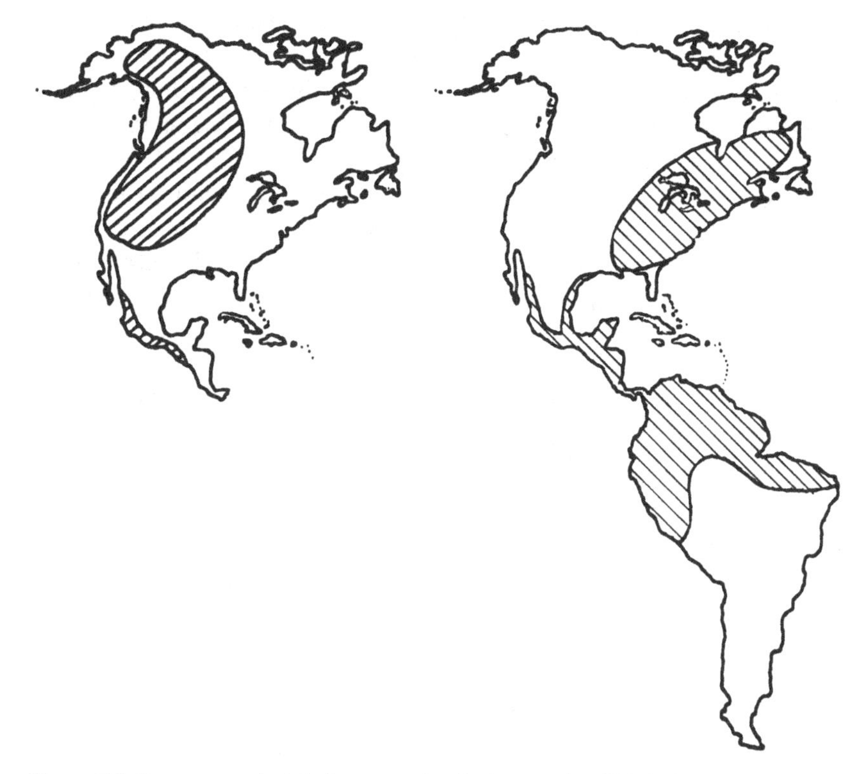

Figure 8.3. Summer and wintering grounds of migratory land birds from western and eastern North America. Modified from Hutto (1986).

Corridors of semideciduous (gallery) forests or riparian vegetation help to increase the diversity of dry forest, because many species widespread in dry forest during the wet season, retreat to those habitats to find refuge during the dry season (Janzen & Wilson, 1983; Stiles, 1983; Eisenberg, 1989; Ceballos, 1989, 1990; Willig & Selcer, 1989; Willig & Mares, 1989). In isolated dry forest, such species tend to be absent because of seasonal food and water scarcity (Duellman, 1965; Janzen, 1983; Ceballos, 1989).

Semideciduous forest patch size and quantity of riparian vegetation appear to be strongly correlated with the number and size of the vertebrates species supported during the dry season. In Costa Rica, Venezuela, and the chaco, dry forests traverse large rivers associated with extensive tracts of moister habitat, and they support populations of large or very specialized birds and mammals such as scarlet macaws (*Ara macao*), toucans (genus *Rhamphastos*), tapirs (genus *Tapirus*) and white-lipped peccaries (*Tayassu tajacu*). In contrast, those species are absent from the small (<200 ha) semideciduous forests and riparian vegetation adjacent to dry forests in western México, although fossil evidence from that region indicates that some of those species were present during the Late Pleistocene (e.g. Shaw & McDonald, 1986; G. Ceballos & J. Arroyo, unpublished data). Notwithstanding, the small tracts of moister habitat in western México are a permanent or seasonal refuge for small-sized frogs such as *Tripion spatulatus*, lizards such as *Basiliscus vittatus*, and mammals such as the pocket gophers (*Pappogeomys bulleri*) and shrews (*Megasorex gigas*) (Duellman, 1958, 1965; Ceballos, 1989, 1990).

Among the families of birds and mammals that tend to be absent from isolated dry forests are: Rhamphastidae (toucans; mostly frugivorous), Galbulidae (jacamars; insectivorous), Cotingidae (cotingas; mostly frugivorous), Tapiridae (tapirs; herbivorous), Dasyproctidae (agoutis; frugivorous), Cebidae (monkeys; frugivorous), Thyropteridae (disk-winged bats; insectivorous) and Myrmecophagidae (anteaters; insectivorous).

Animal responses to climatic seasonality

The marked phenological seasonality of dry forest is related to the time of year and amount of rainfall. The effects of such phenological changes on microclimatic conditions and availability of food resources are profound. In the dry season, the environment desiccates by wind and insolation, and the soil becomes dry (Duellman, 1960; Janzen, 1976). Plant productivity is greatly reduced; for instance, a comparison of litter mass production of dry forest and adjacent semideciduous forest in the Jalisco lowlands, México, showed that litter production is higher (4 vs 7 Mg ha^{-1} y^{-1}) and more stable year-round in the semideciduous forest (Ceballos, 1989). Among all tropical sites, the largest seasonal fluctuation in arthropod number and biomass (5 to 50-fold increase) has been recorded in the same area (Lister & García-Aguayo, 1992). Arthropods reach their lowest abundances during the dry season, and their highest ones in early to mid wet season (e.g. Janzen & Schoener, 1968; Lister & García-Aguayo, 1992).

Animal species show diverse means of coping with the environmental seasonality, including local and regional movements, changes in activity patterns, shifts in diet, seasonal accumulation of fat or food resources, and physiological adaptations to cope with lack of water (e.g. Wilson,

1971; Frankie, 1975; Fleming, 1977; Janzen & Wilson, 1983; Stiles, 1983; Ceballos & Miranda, 1986; Janzen, 1986; Des Granges, 1987; Beck & Lowe, 1991; Lister & García-Aguayo, 1992).

Many species of vertebrates have the vagility to move locally, regionally or geographically during the dry season, searching for food and refuge. Some species such as nectar feeding bats (genus *Leptonycteris*) and passerine migratory land birds, move hundreds or thousands of kilometers. Other species carry out regional movements, either latitudinal or elevational, to other habitats. In western México, hummingbirds move from riparian and pine–oak forests to dry forest during massive blossom of shrubs and trees (Des Granges, 1987), and in Costa Rica bats 'move up to tens of kilometers to stands or populations' of flowering and fruiting trees (Janzen & Wilson, 1983). Howler monkeys (genus *Alouatta*) in Costa Rica (Janzen & Wilson, 1983), and several species of small mammals such as *Peromyscus perfulvus* and *Oryzomys melanotis* in México (Ceballos, 1989, 1990), spread over the dry forest during the rainy season and concentrate in the riparian forest during the dry season.

Changes in patterns of activity are also a common response to seasonality. Some species of amphibians and reptiles remain inactive during the dry season, burrowed in the soil in a dormant state, with lower metabolic rates and energy requirements. For example, the beaded lizard (*Heloderma horridum*) from western México is only active, on average, for three months per year (Beck & Lowe, 1991). Many species of frogs, lizards, tortoises, and snakes are mostly active during the rainy season (e.g. Duellman, 1965). Dormancy is not an alternative for dry forest mammals because of the high energetic cost of maintaining body temperature at tropical temperatures, predation pressures, and lack of adequate food supplies in the right season to accumulate fat reserves (Janzen & Wilson, 1983).

Other species are active year-round, but their activity undergoes striking seasonal changes (e.g. Fleming & Hooker, 1975; Beck & Lowe, 1991; Lister & García-Aguayo, 1992). Dry forest lizards of the genus *Anolis* become inactive during periods of high wind or elevated temperature; they are 2–10 times more active during the rainy season, when they shift from ground to arboreal foraging (Fleming & Hooker, 1975; Lister & García-Aguayo, 1992). Both the agouti (*Dasyprocta punctata*) and paca (*Cuniculus paca*) forage for longer periods of time during the dry season (Smythe, 1983).

Shift in diet is a response to seasonality documented in several species of birds and mammals in dry forest. The broad-billed hummingbird

(*Cynanthus latirostris*) and nectar feeding bats of the genus *Glossophaga* adjust their diet to feed almost exclusively on insects during periods of flower scarcity (Heithaus, Fleming & Opler, 1975; Howell, 1983; Des Granges, 1987), and a foliage-gleaning bat (*Micronycteris hirsuta*) switches from insects to fruits during the dry season (Wilson, 1971). The tamandua (*Tamandua mexicana*) eats ants in the rainy season, but switches to termites, which have a higher moisture content, during the dry season (Lubin & Montgomery, 1981).

Some species have physiological and behavioral specializations that enable them to survive in the during the dry season. The spiny pocket mouse (genus *Liomys*) is able to maintain populations in dry forest throughout the year because it has a granivorous diet, hoarding behavior, and a physiological ability to survive for months on a diet of dry seeds without drinking water (Fleming, 1971, 1974, 1977; Janzen & Wilson, 1983; Janzen, 1986; Ceballos, 1990). During the dry season, the agouti depends on buried seed caches, and the paca relies on stored fat (Smythe, 1983). However, hoarding behavior is an uncommon characteristic in dry forest rodents, particularly when compared with temperate species (Janzen & Wilson, 1983).

Perhaps the most striking vertebrate response to dry forest seasonality is the synchronization of reproduction of many species with periods of high food abundance. For that reason, population densities of vertebrates undergo parallel increases at the end of the rainy season and early dry season, coinciding with periods of high availability of food resources (e.g. Fleming, 1971, 1974; Fleming & Hooker, 1975; Streilein, 1982; Stiles, 1983; Janzen & Wilson, 1983; Ceballos, 1989). In western México, population densities of spiny pocket mice (*Liomys pictus*) increased from 2 to 71 ha^{-1} within two months of the start of the rainy season (Ceballos, 1989, 1990). The tremendous population explosion of dry forest anurans during the wet season provides a noticeable example of this event (Duellman, 1960, 1965).

Some species, such as mammalian or reptile predators, very opportunistic rodents and marsupials, or riparian anurans, may reproduce throughout the year. Species of caatinga small mammals apparently reproduce in any season, but they experience heavy mortality rates during periods of prolonged water scarcity. Such strategy may be a response to very unpredictable rainy patterns (Streilein, 1982). On the other hand, some lizard species such as *Ameiva ameiva* can reproduce year-round in the same habitat by tracking patches of high food resources (Vitt, 1982).

Historical biogeography

Among the most important factors that have shaped the diversity and composition of vertebrate faunas in the neotropics are plate tectonics and faunal interchanges (e.g. Webb & Marshall, 1982; Stehli & Webb, 1985).

In the Late Permian all of the earth's land mass comprised a single continent called Pangaea (Brown & Gibson, 1983). By the Early Cretaceous, the precursors of North and South America were isolated from each other by the splitting of Pangaea into two continents, Gondwanaland and Laurasia (Brown & Gibson, 1983). By the Late Cretaceous all major vertebrate groups had already evolved. Further break-up of the continents continued, more rapidly for Gondwanaland and the future southern continents. South America became an isolated continent tens of millions of years before North America separated from Laurasia (Webb & Marshall, 1982; Eisenberg, 1989). The geographic isolation of North and South America promoted contrasting evolutionary histories: it is certain that post-Jurassic faunal interchanges were very limited, although there is no general agreement about how often such interchanges occurred before the Pliocene (Webb & Marshall, 1982; Stehli & Webb, 1985).

As result of the formation of the Panamanian land bridge the two continents were linked, allowing a large interchange; many South American forms dispersed into North America and vice versa. The faunal interchange has profoundly changed the diversity and composition of both South and North American faunas since the Late Miocene, around 7 million years ago (Webb & Marshall, 1982; Stehli & Webb, 1985). During the interchange South American elements that successfully invaded North America included the Orders Marsupialia (opossums), Edentata (ground sloths), and Rodentia (caviomorph rodents). However, North American elements were more successful (Figures 8.4 and 8.5), and whole South American groups such as Notoungulates, Litopternans, and large herbivorous Edentates (Xenarthrans) became extinct. North American elements that invaded South America included species of the orders Carnivora (Procyonids), Lagomorpha (rabbits), Artiodactyla (deer) and Rodentia (cricetid mice).

During the Pleistocene, repeated climatic fluctuations affected both hemispheres. In the North, there were at least seven glacial periods alternating with periods of warmer climate (Brown & Gibson, 1983). Thus, the geography of climatic zones has changed repeatedly, and as a consequence species and communities have been physically displaced and fragmented (e.g. Leyden, 1984; Davis, 1986; Graham, 1986; Van Devender, 1986; Webb & Barnosky, 1989). Unfortunately, direct evidence from the tropics is very scarce on either plant (Graham & Dilcher, Chapter 6) or animal distributions (e.g. Graham & Lundelius, 1984).

Before the interchange, dry forest and other xeric habitats in South America (and presumably in Central America and México too) supported a high diversity of mammals, including insectivores, specialized desert forms, carnivorous marsupials, browsing and grazing herbivores, and armored omnivores (e.g. Mares, 1985). A dramatic Pleistocene event was the extinction of many mammals on both continents, particularly of large herbivores and carnivores (Martin & Klein, 1984; Webb & Barnosky, 1989). Whole mammalian families (e.g. Glyptodontidae, Megatheriidae and Mylodontidae) completely vanished (Webb & Marshall, 1982; Martin & Klein, 1984). The causes are the subject of continuing controversy (e.g. Martin & Klein, 1984).

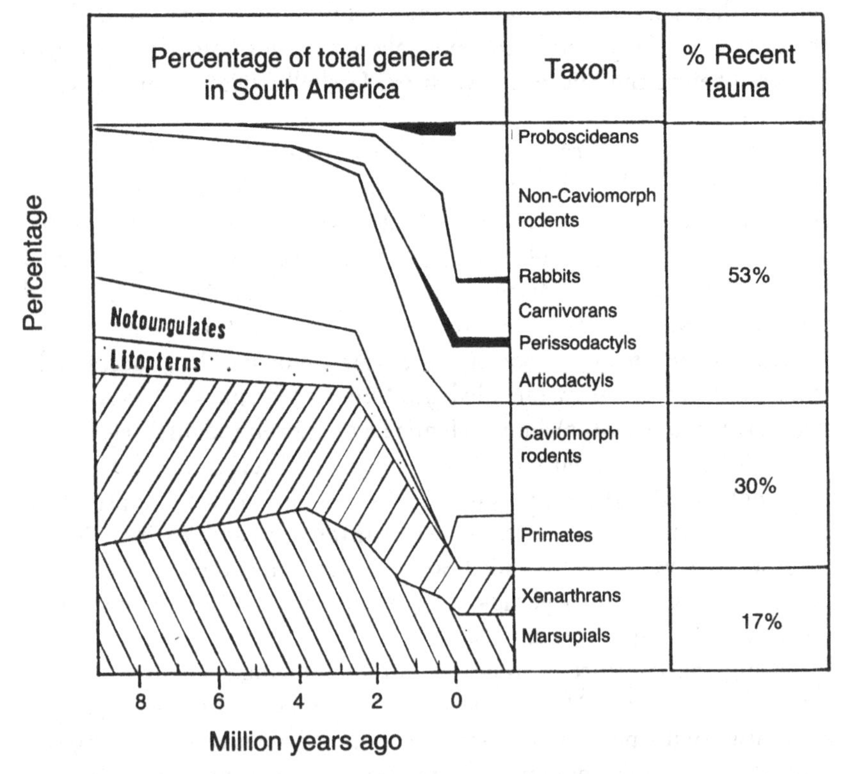

Figure 8.4. Taxonomic composition of South America mammals during the last 10 million years. Modified from Webb & Marshall (1982).

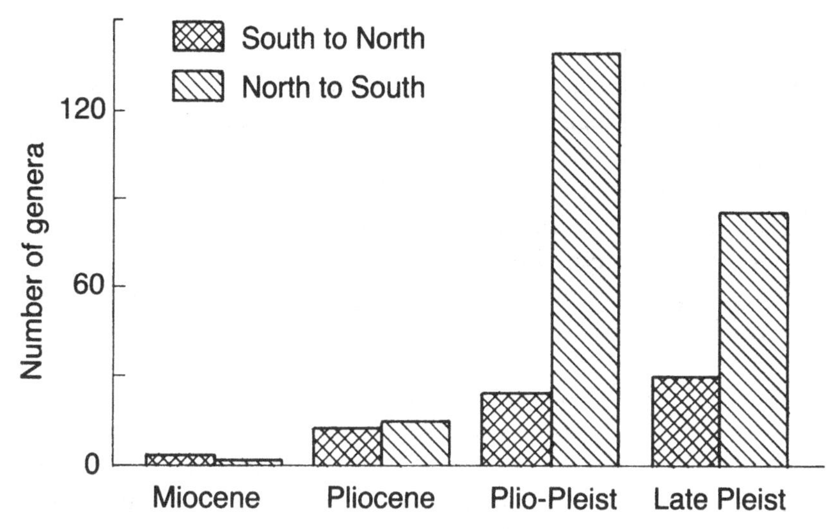

Figure 8.5. Number of genera from South America that invaded North America, and vice versa, since the Miocene. Modified from Webb & Barnosky (1989).

The extensive climatic changes occurring during the Pleistocene expanded, fragmented, and sometimes eliminated dry forests (Duellman, 1965; Sarmiento, 1975; Andrade-Lima, 1982; Graham & Dilcher, Chapter 6). The fragmentation of species' geographic ranges has been argued to be a major mechanism promoting speciation in the neotropics (Prance, 1982), but the debate is still hot (Simpson & Haffer, 1978). Most dry forests lack or have few endemic vertebrate species, suggesting either that isolation mechanisms were not very strong or that putative refugia were too small to support viable populations.

A very general reconstruction of Pleistocene climatic effects indicates that during xeric periods, dry forest expanded greatly at the expense of moister habitats, which were reduced to mesic enclaves along permanent watercourses or windward slopes. Mesic vegetation was used as refuges and corridors by vertebrate species, allowing them to move among dry forests and colonize other habitats, limiting their isolation, and rendering unnecessary any adaptations to xeric conditions (Mares *et al.*, 1985; Redford & Fonseca, 1986).

There is paleoecological evidence that during mesic periods dry forests were greatly fragmented and isolated. This scenario is reflected in the similarity of separated dry forest floras (e.g. caatinga–chaco), plant and animal species with disjunct distribution, and a rich endemic flora with pronounced xeric adaptations (Sarmiento, 1975; Rzedowski, 1978;

Andrade-Lima, 1982; Redford & Fonseca, 1986). During such mesic periods, dry forests were connected by corridors of savanna vegetation, that were used as dispersal routes by vertebrates. Indeed, a 'savanna corridor' has been suggested for the invasion of South American fauna (e.g. ground sloths and camelids) into Yucatán and Florida (Webb, 1974). The presence of fossil records of locally extinct species (e.g. Shaw & McDonald, 1986), and isolated populations of mesic tropical vertebrate species such as the tayra (*Eira barbara*) in dry forests of western México, strongly suggests the presence of a savanna corridor with moist forests enclaves during the Late Pleistocene.

In summary, the effect of the Pleistocene environmental fluctuations in the dry forest biota promoted the following: (1) the maintenance and promotion of large-scale speciation events in the dry forest floras, that resulted in a high endemicity and pronounced adaptation to xeric conditions, (2) the extinction of many xeric adapted species of vertebrates, mainly mammals, and (3) the isolation and speciation of many small-sized vertebrates of dry forests in the chaco and México. Mares (1985) has elegantly summarized the effect of Pleistocene environmental changes on the faunal composition of South American xeric habitats, including dry forest, as follows. 'The great extinctions of the Pleistocene, however, left a legacy of a depauperate fauna in the continent's xeric regions. Most species that inhabit these areas today are descendants of the Late Cenozoic immigrants and are basically eurytopic species.' The majority of the xeric-adapted species, probably including those of the dry forest, failed to survive.

Endangered species and conservation

One of the most pressing global environmental problems is the elevated rate of species extinctions, which is particularly acute in the tropics (e.g. Wilson, 1988). There are no thorough data to evaluate contemporary vertebrate extinctions in the neotropics, but it is known that at least 21 species of mammals and 30 species of birds have disappeared from this region in historic times (Ceballos & Sánchez, 1994). There is an almost complete lack of inventories of threatened or endangered species in dry forest. However, a detailed study in the Chamela–Cuixmala region, Jalisco, México, has shown that at least 40 vertebrate species (excluding fishes), 15% of the regional species richness, are at risk of extinction (Ceballos, García-Aguayo & Rodríguez, 1993).

Wildlife trade is one of the principal industries in South America

(Mares, 1986; Redford & Robinson, 1991). Millions of vertebrates are captured every year for their meat, skins, or for live trade (Mares, 1986; Redford & Robinson, 1991; Ceballos & Sánchez, 1994). For example, at least 5.4 million terrestrial vertebrates of 25 species are legally exported every year from Argentina (Mares & Ojeda, 1982); an additional 6 million hares are killed every year for their meat, valued at an average $24 million (Redford & Robinson, 1991). Hunting is one of the main causes of the decline of several threatened or endangered mammal species such as the chacoan peccary and the jaguar. Among the dry forest species important for the wildlife trade are the three species of peccaries (genus *Tayassu*), spotted cats (e.g. *Felis wiedii, F. pardalis*), foxes (genus *Dusicyon*), parrots and macaws (*Amazona orathrix, A. finschii, Ara macao, A. militaris, A. ambigua*), lizards (*Tupinambis*), and iguanas (*Iguana* and *Ctenosaura*).

There is no doubt that major conservation problems in dry forest are related to habitat destruction. Estimates of deforestation in the neotropics vary extensively, but most tend to indicate that rates are high (Lugo, 1988). There is a general awareness of the conservation problems of the wet forests but not of the present conditions of other tropical ecosystems, especially dry forests (Janzen, 1986; Redford *et al.*, 1990). Deforestation rates are apparently higher than those in rain forests; dry forests are being destroyed mainly to support agriculture and cattle raising activities. Indeed, Central American, Venezuelan, Brazilian and Andean dry forests have almost completely disappeared, and Mexican and chaco dry forests have been reduced (Janzen, 1988; Redford *et al.*, 1990; Gentry, Chapter 7; Sampaio, Chapter 3).

The conservation of large tracts of dry forest is essential for the conservation of many species, including many endemic ones, all western North America land bird migratory species, and many populations of other tropical widespread species. Among the endemic species at risk of extinction are the chacoan peccary, fairy armadillo (*Chlamyphorus retusus*), Chamela wood rat (*Xenomys nelsoni*) and yellow-headed parrot (*Amazona orathrix*). Widespread species that are locally or globally at risk include the jaguar (*Felis onca*), ocelot (*F. pardalis*), tapir (*Tapirus bairdii*), and green macaw (*Ara militaris*) (e.g. Ceballos & Navarro, 1991; Roig, 1991).

Finally, it is important to emphasize that it is essential that we preserve the unique genetic diversity and the ecological processes of the dry forest for both the maintenance of hemispheric biodiversity and the sustainable use of species and systems upon which humans depend. The creation of protected areas that integrate conservation with local socioeconomic issues will be a major step in this direction. Relatively large conserved tracts of dry forest are still found in western México and the chaco. Presently, the proposed Chamela–Cuixmala biosphere reserve in Jalisco, México (Ceballos *et al.*, 1993), and the Santa Rosa National Park are some examples of natural reserves protecting dry forest (Janzen, 1988). However, many more reserves are needed to ensure long-term preservation. If we fail to establish patterns of sustainable use, we will see a continued decline of natural resources, impoverishment of people that depend on these systems, and increased costs to society.

Summary

Neotropical dry forests generally have lower species diversity than tropical moist or wet forests, mainly because of the absence of many specialized species of carnivores, frugivores and semiaquatic mammals. However, dry forests are special respositories of vertebrate diversity because of their endemic species. Dry forests in western México and the chaco support the highest number of endemic terrestrial vertebrate genera and species, suggesting scenarios of geological and ecological isolation.

At any given latitude, the number of vertebrates in dry forest is apparently correlated with its degree of isolation from other tropical forests. Corridors of semideciduous (gallery) forests or riparian vegetation help to increase diversity in dry forest, because many species widespread in dry forest during the wet season retreat to those habitats to find refuge during the dry season.

Among the most important factors that have shaped the diversity and composition of vertebrate faunas in the neotropics are faunal interchange (related to plate tectonics) and Pleistocene climatic changes. As a result of these factors, many dry forest animals became extinct, particularly large herbivores and carnivores. The extensive climatic changes of the Pleistocene expanded, fragmented and sometimes eliminated dry forests. Those changes resulted in large-scale speciation events in dry forest floras, the extinction of many xeric-adapted species of vertebrates, and the isolation and speciation of many small-sized vertebrates in the dry forests of México and the chaco.

Unfortunately, dry forests are being destroyed throughout the neotropics, apparently at a higher rate than in wet forests. Indeed, Central American, Venezuelan, Brazilian and Andean dry forests have almost completely disappeared, and the Mexican and chaco forests have been reduced. The conservation of large tracts of dry forest is essential for the conservation of many endemic species, as well as for all western North American migratory land birds and populations of widespread tropical species. A major effort should be made to protect dry forests in Latin America, with their unique genetic diversity and ecological processes. This is essential for maintaining the hemisphere's biological diversity and the sustainable use of the species and biological systems on which human well-being depends.

Acknowledgements

I thank the Centro de Ecología of the Universidad Nacional Autónoma de México, and the Fundación Ecológica de Cuixmala for financial support to carry out my research in western México. Coro Arizmendi, Alejandro Espinosa, Andres García, and Guadalupe Téllez kindly helped gather information. S. Bullock, M. Mares, B. Miller and H. Mooney read the manuscript and made many helpful suggestions for its improvement.

References

Andrade-Lima, D. de (1982). Present day forest refuges in northeastern Brazil. In *Biological Diversification in the Neotropics*, ed. G. T. Prance, pp. 245–51. Columbia University Press, New York.

Arizmendi, C., Berlanga, H., Márquez, L., Navarijo, L. & Ornelas, F. (1990). *Avifauna de la Región de Chamela, Jalisco*. Universidad Nacional Autónoma de México, México.

Beck, D. D. & Lowe, C. H. (1991). Ecology of the beaded lizard *Heloderma horridum* in coastal México. *Journal of Herpetology* **25**: 395–406.

Brown, J. H., & Gibson, A. C. (1983). *Biogeography*. C. V. Mosby, St Louis, Mo.

Bullock, S. H. & Solís Magallanes, J. A. (1990). Phenology of canopy trees of a tropical deciduous forest in México. *Biotropica* **22**: 22–35.

Casas, G. (1982). Anfibios y Reptiles de la Costa Suroeste del Estado de Jalisco. Tesis Doctoral, Facultad de Ciencias, Universidad Nacional Autónoma de México, México.

Ceballos, G. (1989). Population and community ecology of small mammals in tropical forests from western México. Ph. D. Thesis, University of Arizona, Tucson.

Ceballos, G. (1990). Comparative natural history of small mammals from tropical forests in western México. *Journal of Mammalogy* **71**: 263–6.

Ceballos, G., García-Aguayo, A. & Rodríguez, P. (1993). *Plan de Manejo de la Reserva Ecológica de Chamela–Cuixmala*. Fundación Ecológica de Cuixmala, A.C., México.

Ceballos, G. & Miranda, A. (1986). *Los Mamíferos de Chamela, Jalisco*. Universidad Nacional Autónoma de México, México.

Ceballos, G. & Navarro, D. (1991). Diversity and conservation of Mexican mammals. In *Latin American Mammalogy: history, biodiversity, and conservation*, ed. M. A. Mares and D. J. Schmidly, pp. 167–98. University of Oklahoma Press, Norman.

Ceballos, G. & Rodríguez, P. (1993). Patrones de endemicidad en los mamíferos de México. In *Avances en el Estudio de los Mamíferos de México*, ed. R. A. Medellín and G. Ceballos, pp. 75–99. Asociación Mexicana de Mastozoología, México.

Ceballos, G. & Sánchez, O. (1994). Wildlife diversity and conservation in tropical America. In *Tropical Ecosystems*, ed. M. Balakrishnan, S. W. Bie and R. Borgstrom, pp. 255–84. Oxford and IBH Publishing Co., New Delhi.

Davis, M. B. (1986). Climatic instability, time lags, and community disequilibrium. In *Community Ecology*, ed. J. Diamond and T. J. Case, pp. 269–84. Harper & Row, New York.

Des Granges, J. L. (1987). Organization of a tropical nectar feeding bird guild in a variable environment. *Living Bird* **17**: 199–236.

Duellman, W. E. (1958). A preliminary analysis of the herpetofauna of Colima, México. *University of Michigan. Occasional Papers of the Museum of Zoology* **589**: 1–22.

Duellman, W. E. (1960). A distributional study of the amphibians of the Isthmus of Tehuantepec, México. *University of Kansas Publications. Museum of Natural History* **13**: 19–72.

Duellman, W. E. (1965). A biogeographic account of the herpetofauna of Michoacán, México. *University of Kansas Publications. Museum of Natural History* **15**: 627–709.

Duellman, W. E. (1990). Herpetofaunas in Neotropical rain forests: comparative composition, history, and resource use. In *Four Neotropical Rain Forests*, ed. A. H. Gentry, pp. 455–505. Yale University Press, New Haven.

Eisenberg, J.F. (1989). *Mammals of the Neotropics, Vol. I. The northern neotropics.* University of Chicago Press, Chicago.

Eisenberg, J. F. & Redford, K. H. (1979). A biogeographical analysis of the mammalian fauna of Venezuela. In *Vertebrate Ecology in the Northern Neotropics*, ed. J. F. Eisenberg, pp. 31–8. Smithsonian Institution Press, Washington.

Fleming, T. H. (1971). Population ecology of three species of neotropical rodents. *University of Michigan. Miscellaneous Publications of the Museum of Zoology* **143**: 1–77.

Fleming, T. H. (1973). Numbers of mammal species in North and Central American forest communities. *Ecology* **54**: 555–63.

Fleming, T. H. (1974). The population ecology of two species of Costa Rican heteromyid rodents. *Ecology* **55**: 493–510.

Fleming, T. H. (1977). Response of two species of tropical heteromyid rodents to reduced food and water availability. *Journal of Mammalogy* **58**: 102–6.

Fleming, T. H. & Hooker, R. S. (1975). *Anolis cupressus*: the response of a lizard to tropical seasonality. *Ecology* **56**: 1243–61.

Flores, O. (1991). Análisis de la distribución de la herpetofauna de México. Tesis Doctoral, Facultad de Ciencias, Universidad Nacional Autónoma de México, México.

Frankie, G. W. (1975). Tropical forest phenology and pollinator plant coevolution. In *Coevolution of Plants and Animals*, ed. L. E. Gilbert and P. H. Raven, pp. 192–209. University of Texas Press, Austin.

Frankie, G. W., Baker, H. G. & Opler, P. A. (1974). Comparative phenological studies of trees in tropical dry and wet forests in the lowlands of Costa Rica. *Journal of Ecology* **62**: 881–919.

García-Aguayo, A. & Ceballos, G. (1994). *Guía de Campo de los Reptiles y Anfibios de la Costa de Jalisco.* Fundación Ecológica de Cuixmala and Instituto de Biología, Universidad Nacional Autónoma de México, México.

Glanz, W. E. (1990). Neotropical mammal densities: How unusual is the community on Barro Colorado Island, Panamá. In *Four Neotropical Rain Forests*, ed. A. H. Gentry, pp. 287–313. Yale University Press, New Haven.

Graham, R. W. (1986). Response of mammalian communities to environmental changes during the late Quaternary. In *Community Ecology*, ed. J. Diamond and T. J. Case, pp. 300–13. Harper & Row, New York.

Graham, R. W. & Lundelius, E. L., Jr (1984). Coevolutionary disequilibrium and Pleistocene extinctions. In *Quaternary Extinctions*, ed. P. S. Martin and R. G. Klein, pp. 223–49. University of Arizona Press, Tucson.

Guyer, C. (1990). The herpetofauna of La Selva, Costa Rica. In *Four Neotropical Rain Forests*, ed. A. H. Gentry, pp. 371–85. Yale University Press, New Haven, Conn.

Hartshorn, G. S. (1983). Plants. In *Costa Rican Natural History*, ed. D. H. Janzen, pp. 118–57. University of Chicago Press, Chicago.

Heithaus, E. R., Fleming, T. H. & Opler, P. A. (1975). Foraging patterns and resource utilization in seven species of bats in a seasonal tropical forest. *Ecology* **56**: 841–54.

Howell, D. J. (1983). *Glossophaga soricina* (Murciélago lengua larga, Nectar bat). In *Costa Rican Natural History*, ed. D. H. Janzen, pp. 472–4. University of Chicago Press, Chicago.

Hutto, R. L. (1986). Migratory land birds in western México: a vanishing habitat. *Western Wildlands* **11**: 12–16.

Janson, C. H. & Emmons, L. H. (1990). Ecological structure of the nonflying mammal community at Cocha Cashu biological station, Manu National Park, Perú. In *Four Neotropical Rain Forests*, ed. A. H. Gentry, pp. 314–38. Yale University Press, New Haven.

Janzen, D. H. (1976). The microclimate differences between a deciduous forest and an adjacent riparian forest in Guanacaste province. *Brenesia* **8**: 29–33.

Janzen, D. H. (ed.) (1983). *Costa Rican Natural History*. University of Chicago Press, Chicago.

Janzen, D. H. (1986). Mice, big mammals, and seeds: it matters who defecates what where. In *Frugivores and Seed Dispersal*, ed. A. Estrada and T. H. Fleming, pp. 251–71. Dr W. Junk, Dordrecht.

Janzen, D. H. (1988). Tropical dry forests: the most endangered major tropical ecosystem. In *Biodiversity*, ed. E. O. Wilson, pp. 130–7. National Academy Press, Washington.

Janzen, D. H. & Schoener, T. W. (1968). Differences in insect abundance and diversity between wetter and drier sites during a tropical dry season. *Ecology* **49**: 96–110.

Janzen, D. H. & Wilson, D. E. (1983). Mammals. In *Costa Rican Natural History*, ed. D. H. Janzen, pp. 426–42. University of Chicago Press, Chicago.

Karr, J. R., Robinson, S. K., Blake, J. G. & Bierregaard, R. O., Jr (1990). Birds of four Neotropical forests. In *Four Neotropical Rain Forests*, ed. A. H. Gentry, pp. 237–69. Yale University Press, New Haven.

Leyden, B. W. (1984). Guatemalan forest synthesis after Pleistocene aridity. *Proceedings of the National Academy of Sciences USA* **81**: 4856–9.

Lister, B. & García-Aguayo, A. (1992). Seasonality, predation, and the behaviour of a tropical mainland anole. *Journal of Animal Ecology* **61**: 717–33.

Lubin, Y. D. & Montgomery, G. G. (1981). Defenses of *Nasutitermes* termites (Isoptera, Termitidae) against *Tamandua* anteaters (Edentata, Myrmecophagidae). *Biotropica* **13**: 66–76.

Lugo, A. E. (1988). Estimating reductions in the diversity of tropical forests species. In *Biodiversity*, ed. E. O. Wilson, pp. 51–7. National Academy Press, Washington.

MacArthur, R. & Wilson, E. O. (1967). *The Theory of Island Biogeography*. Princeton University Press, Princeton.

McCoy, E. D. & Connor, E. F. (1980). Latitudinal gradients in species diversity in North American mammals. *Evolution* **34**: 193–203.

McNeely, J. A., Miller, K. R., Reid, W. V., Mittermeier, R. A. & Werner, T. B. (1990). *Conserving the World's Biological Diversity*. International Union for Conservation of Nature and Natural Resources, Gland, Switzerland.

Mares, M. A. (1985). Mammal faunas of xeric habitats and the great American interchange. In *The Great American Biotic Interchange*, ed. F. G. Stehli and S. D. Webb, pp. 489–520. Plenum Publishing Corporation, New York.
Mares, M. A. (1986). Conservation in South America: problems, consequences, and solutions. *Science* **233**: 734–9.
Mares, M. A. (1992). Neotropical mammals and the myth of Amazonian diversity. *Science* **255**: 976–9.
Mares, M. A. & Ojeda, R. A. (1982). Patterns of diversity and adaptation in South American hystricognath rodents. In *Mammalian Biology in South America*, ed. M. A. Mares and H. H. Genoways, pp. 393–432. Pymatuning Laboratory of Ecology, University of Pittsburgh, Linesville.
Mares, M. A. & Schmidly, D. J. (ed.) (1991). *Latin American Mammalogy: history, biodiversity, and conservation*. University of Oklahoma Press, Norman.
Mares, M. A., Willig, M. R. & Lacher, T. E., Jr (1985). The Brazilian caatinga in South American zoogeography: tropical mammals in a dry region. *Journal of Biogeography* **12**: 57–69.
Mares, M. A., Willig, M. R., Streilein, K. E. & Lacher, T. E., Jr (1981). The mammals of northeastern Brazil: a preliminary assessment. *Annals of the Carnegie Museum* **50**: 81–135.
Mares, M. A., Ojeda, R. A. & Barquez, R. (1989). *Guide to the Mammals of Salta Province, Argentina*. University of Oklahoma Press, Norman.
Martin, P. S., & Klein, R. G. (ed.) (1984). *Quaternary Extinctions*. University of Arizona Press, Tucson.
Medellín, R. (1992). Community structure and conservation of mammals in a Mayan tropical rain forest and abandoned agricultural fields. Ph.D. thesis, University of Florida, Gainesville.
Myers, N. (1988). Tropical forests and their species: going, going...? In *Biodiversity*, ed. E. O. Wilson, pp. 28–35. National Academy Press, Washington.
Prance, G. T. (ed.) (1982). *Biological Diversification in the Neotropics*. Columbia University Press, New York.
Ramírez-Pulido, J., Martínez, A. & Urbano, G. (1977). Mamíferos de la Costa Grande de Guerrero, México. *Anales del Instituto de Biología, Universidad Nacional Autónoma de México, Serie Zoología* **48**: 243–92.
Rand, A. S. & Myers, C. W. (1990). The herpetofauna of Barro Colorado Island, Panamá: an ecological summary. In *Four Neotropical Rain Forests*, ed. A. H. Gentry, pp. 386–409. Yale University Press, New Haven.
Redford, K. H. & Eisenberg, J. F. (1992). *Mammals of the Neotropics, Vol. II, The southern cone*. University of Chicago Press, Chicago.
Redford, K. H. & Fonseca, G. A. B. da (1986). The role of gallery forests in the zoogeography of the cerrado's non-volant mammals. *Biotropica* **18**: 126–35.
Redford, K. H. & Robinson, J. H. (1991). Subsistence and commercial uses of wildlife in Latin America. In *Neotropical Wildlife Use and Conservation*, ed. J. G. Robinson and K. H. Redford, pp. 6–23. University of Chicago Press, Chicago.
Redford, K. H., Taber, A. & Simonetti, J. A. (1990). There is more to biodiversity than the tropical rain forests. *Conservation Biology* **4**: 328–330.
Rodríguez, L. B. & Cadle, J. E. (1990). A preliminary overview of the herpetofauna of Cocha Cashu, Manu National Park, Perú. In *Four Neotropical Rain Forests*, ed. A. H. Gentry, pp. 410–25. Yale University Press, New Haven.
Roig, V. (1991). Desertification and distribution of mammals in the southern cone of South America. In *Latin American Mammalogy: history, biodiversity, and conservation*, ed. M. A. Mares and D. J. Schmidly, pp. 239–79. University of Oklahoma Press, Norman.
Rzedowski, J. (1978). *Vegetación de México*. Editorial LIMUSA, México.
Saravia-Toledo, C. (1985). La tierra pública en el desarrollo futuro de las zonas áridas: estado actual y perspectivas. In *IV Reunión de Intercambio Tecnológico en Zonas*

Aridas y Semiáridas, I, pp. 115–40. Centro Argentino de Ingenieros Agrónomos, Buenos Aires.
Sarmiento, G. (1975). The dry plant formations of South America and their floristic connections. *Journal of Biogeography* **2**: 233–51.
Scott, N. J. & Lovett, J. W. (1975). A collection of reptiles and amphibians from the Chaco of Paraguay. *University of Connecticut. Occasional Papers* **3**: 257–66.
Shaw, C. A. & McDonald, H. G. (1986). First record of the giant anteater (Xenarthra, Myrmecophagidae) in North America. *Science* **236**: 186–8.
Short, L. L. (1974). A zoogeographical analysis of the South American Chaco avifauna. *Bulletin of the American Museum of Natural History* **154**: 165–352.
Sick, H. (1965). A fauna do Cerrado. *Arquivos de Zoologia, São Paulo* **12**: 71–93.
Simpson, B. B. & Haffer, J. (1978). Speciation patterns in Amazonian forest biota. *Annual Review of Ecology and Systematics* **9**: 497–518.
Slud, P. (1964). The birds of Costa Rica. *Bulletin of the American Museum of Natural History* **128**: 1–430.
Smythe, N. (1983). *Dasyprocta punctata* and *Agouti paca*. In *Costa Rican Natural History*, ed. D. H. Janzen, pp. 463–5. University of Chicago Press, Chicago.
Stehli, F. G. & Webb, S. D. (ed.) (1985). *The Great American Biotic Interchange*. Plenum Publishing Corporation, New York.
Stiles, F. G. (1983). Birds. In *Costa Rican Natural History*, ed. D. H. Janzen, pp. 502–30. University of Chicago Press, Chicago.
Streilein, K. E. (1982). The ecology of small mammals in semiarid Brazilian caatinga. III. Reproductive biology and population ecology. *Annals of the Carnegie Museum* **51**: 251–69.
Van Devender, T. R. (1986). Climatic cadences and the composition of Chihuahuan desert communities: the late Pleistocene packrat midden record. In *Community Ecology*, ed. J. Diamond and T. J. Case, pp. 285–99. Harper & Row, New York.
Vanzolini, P. E., Ramos-Acosta, A. M. M. & Vitt, L. J. (1980). *Repteis das Caatingas*. Academia Brasileira de Ciencias, Rio de Janeiro.
Vitt, L. J. (1982). Reproductive tactics of *Ameiva ameiva* (Lacertilia: Teiidae) in a seasonally fluctuating tropical habitat. *Canadian Journal of Zoology* **60**: 3113–20.
Webb, S. D. (ed.) (1974). *Pleistocene Mammals of Florida*. University Presses of Florida, Gainesville.
Webb, S. D. & Barnosky, A. D. (1989). Faunal dynamics of Pleistocene mammals. *Annual Review of Earth and Planetary Sciences* **17**: 413–38.
Webb, S. D. & Marshall, L. G. (1982). Historical biogeography of recent South American land mammals. In *Mammalian Biology in South America*, ed. M. A. Mares and H. H. Genoways, pp. 39–52. Pymatuning Laboratory of Ecology, University of Pittsburgh, Linesville.
Wetzel, R. M., Dubos, R. E., Martin, R. L. & Myers, P. (1975). *Catagonus*, an 'extinct' peccary, alive in Paraguay. *Science* **189**: 379–81.
Willig, M. R. (1983). Composition, microgeographic variation, and sexual dimorphism in caatinga and cerrado bat communities from northeast Brazil. *Bulletin of the Carnegie Museum of Natural History* **23**: 1–131.
Willig, M. R. & Mares, M. A. (1989). A comparison of bat assemblages from phytogeographic zones of Venezuela. In *Patterns in the Structure of Mammalian Communities*, ed. D. W. Morris, Z. Abramski, B. J. Fox, and M. R. Willig, pp. 59–67. Texas Tech University, Lubbock.
Willig, M. R. & Sadlin, E. A. (1991). Gradients of species density and turnover in New World bats: a comparison of quadrats and band methodologies. In *Latin American Mammalogy: history, biodiversity, and conservation*, ed. M. A. Mares and D. J. Schmidly, pp. 81–96. University of Oklahoma Press, Norman.
Willig, M. R. & Selcer, K. L. (1989). Bat density gradients in the New World: a statistical assessment. *Journal of Biogeography* **16**: 189–95.

Wilson, D. E. (1971). Food habits of *Micronycteris hirsuta* (Chiroptera: Phyllostomidae). *Mammalia* **35**: 107–10.
Wilson, D. E. (1990). Mammals of La Selva, Costa Rica. In *Four Neotropical Rain Forests*, ed. A. H. Gentry, pp. 273–86. Yale University Press, New Haven.
Wilson, E. O. (ed.) (1988). *Biodiversity*. National Academy Press, Washington.
Wilson, J. W., III. (1974). Analytical zoogeography of North American mammals. *Evolution* **28**: 124–40.

Note added in proof For the caatinga, a second mammal species has been reported as endemic, the armadillo *Tolypeutes tricinctus*; see Cardoso da Silva, J. M. & Oren, D. C. (1993). Observations on the habitat and distribution of the Brazilian three-banded armadillo *Tolypeutes tricinctus*, a threatened Caatinga endemic. *Mammalia* **57**: 149–51.

9

Diversity of life forms of higher plants in neotropical dry forests

ERNESTO MEDINA

Introduction

Tropical dry forests constitute a large set of plant communities occurring under climates characterized by highly seasonal distribution of rainfall (Murphy & Lugo, 1986). The actual percentage of deciduous woody components varies from 100% to 40% depending on the specific forest type and its location within the rainfall gradient (Beard, 1955; Sarmiento, 1972; Hegner, 1979; Mateucci, 1987). In general it is assumed that along a gradient of rainfall in the lowland tropics, under similar temperature conditions, the proportion of deciduous woody components increases more or less linearly as the amount of rainfall received per year decreases below about 2000 mm (Walter, 1973). The structure of the community changes along these rainfall gradients in terms of community height, density of ground cover, proportion of trees and shrubs, and the occurrence of lianas, epiphytes and hemiparasites. The functional attributes of the structural components also change more or less monotonously along the gradients, thus the proportion of deciduous trees and shrubs tends to increase, the presence of epiphytes and hemiparasites is reduced, and at certain levels of the gradient the lianas (woody climbing plants) are important. Along the same gradients the proportion of succulent plants, including Crassulacean acid metabolism (CAM) performing cacti and stem succulent, drought deciduous trees, increases. A growth form showing dominance toward both extremes of the rainfall gradient is the evergreen type of woody plants. The evergreens at each extreme, however, may be separated according to their leaf structure and drought resistance. In this review I will discuss the differentiation of plant components in neotropical dry forests according to morphological and eco-physiological features. The aim is to identify the ecological constraints which determine the dominance of a given set of morpho-physiological properties.

Climatic conditions

There is a continuous variation of both vegetation structure and, frequently, soil properties along aridity gradients in the tropics. This continuity is best assessed by looking at plant available moisture. This parameter may be estimated directly in a given vegetation as the amount of soil water available at the root level, or as the water stress of plant tissues. The latter, however, may be not precise enough to assess drought stress in succulent plants.

The meteorological indices devised to estimate the water stress in a given region are mostly based on the calculation of evaporation using solely air temperature (Holdridge, 1959; Bailey, 1979). Evaporation, however, is a function of both air temperature and humidity. In the tropics average daily or monthly temperature varies little while air humidity oscillates strongly during the year following the pattern of rainfall. Therefore, in spite of the nearly constant temperatures, potential evaporation is invariably higher in the dry season than in the rainy season. Nevertheless, those approaches to calculate evapotranspiration are useful as a first approximation to differentiate vegetation types. For example, monthly Bailey's indices for climatic stations located within well-defined climatic–vegetation types in Venezuela (as described by Ewel, Madriz & Tossi, 1976) show clearly that degree of aridity of vegetation is related to the number of dry months (rainfall < potential evaporation) (Figure 9.1). The number of dry months (months with Bailey's indices below 0.53) decreases across the sequence of vegetation types, from *tropical thorn woodland*, to *very dry tropical forest*, *dry tropical forest* and *humid tropical forest*. A similar differentiation between ecological life zones *sensu* Holdridge (1959) in the lowland tropics was made by Veillon (1963) using the concept of dry months as defined by Gaussen (the number of months with average rainfall smaller than temperature × 2, see Walter, 1973). Veillon found that the wood volume of trees above 20 cm dbh decreases following a quadratic polynomial as the number of dry months increases.

Life forms and growth forms

The concepts of life forms and growth forms used here follow the original propositions of Clements (1920) as discussed recently by Schulze (1982). Life forms are those morphological features of a species which are insensitive to environmental changes, that is trees, shrubs, epiphytes,

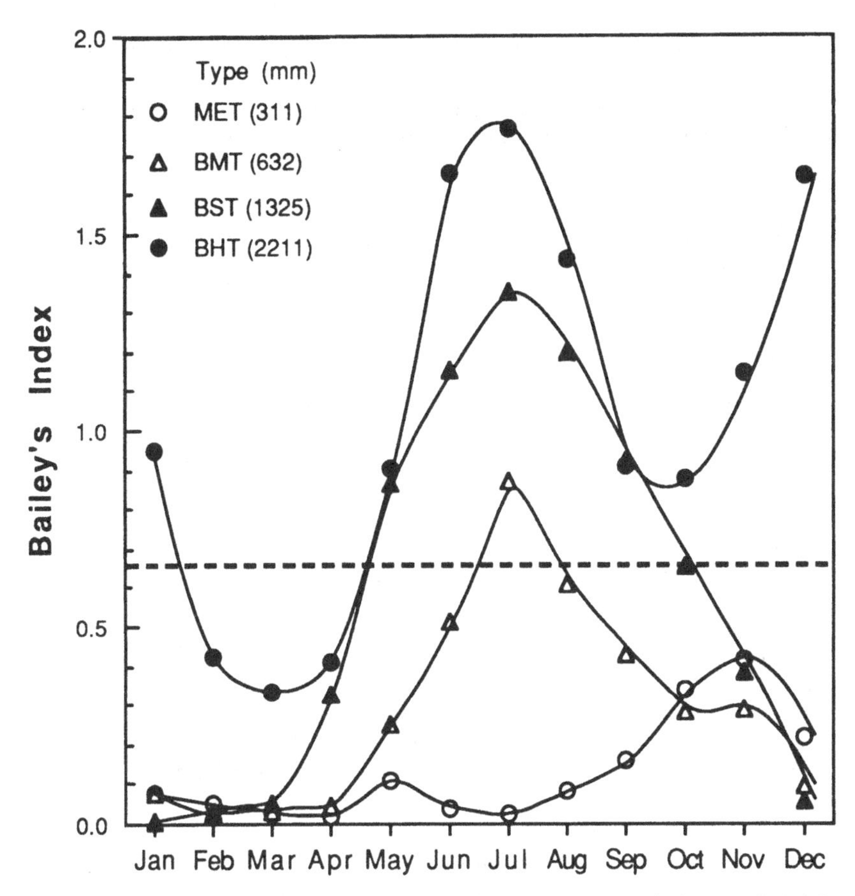

Figure 9.1. Annual course of the monthly Bailey's index (=[0.018 Rainfall/ $0.041^{\text{Temperature}}$]: Bailey, 1979) of meteorological stations corresponding to a series of increasingly drier forest types in northern Venezuela. BHT, humid forest (Caucagua, Miranda state); BST, dry forest (Calabozo, Guárico state); BMT, very dry forest (Barcelona, Anzoátegui state); MET, thorn scrub (Las Piedras, Falcón state). Data from Ewel, Madriz & Tossi (1976).

and so on. On the other hand, growth forms are the direct and quantitative responses made by a plant to different habitats and conditions (Schulze, 1982).

The main life and growth forms identifiable in tropical dry forests are listed in Table 9.1. We differentiate among woody plants (trees and shrubs), herbs, lianas and creeping plants, and epiphytes. The woody deciduous components always constitute the dominant type determining forest physiognomy. A number of genera exemplifying the main life and

Table 9.1. *Dominant life forms and functional attributes of higher plants from tropical dry forests*

Life form and morphological features	Functional attributes
Trees and shrubs	
Mesophyllous	Evergreen
Sclerophyllous	Deciduous
Green-stemmed	C_3 and CAM photosynthesis
Succulent	
Herbs	
Grasses	Annual
Dicots	Perennial
Succulents	Evergreen
	Deciduous
	C_3, C_4 and CAM photosynthesis
Vines	
Herbaceous	
Woody	
Mesophyllous	Evergreen
Sclerophyllous	Deciduous
Succulent	C_3 and CAM photosynthesis
Epiphytes	
Mesophyllous	Evergreen
Sclerophyllous	Deciduous
Succulent	C_3 and CAM photosynthesis
Parasites	Xylem-tapping mistletoes

growth forms found as components of dry forests are given in Table 9.2. Further examples may be found in Sarmiento (1972).

Among woody plants we find succulent and non-succulent stemmed species, the former becoming more frequent in drier areas. These succulent stemmed species include the cacti, and several genera within the families Anacardiaceae, Bombacaceae, Burseraceae, Caricaceae, Cochlospermaceae, Convolvulaceae, Euphorbiaceae and Leguminosae. This group includes the trees of low density wood, water storage trees, and arborescent succulents described by Schulze *et al.* (1988). Succulent stemmed woody plants are characterized by very stable water relations and are all drought deciduous. All the succulent stemmed woody plants either have green stems with CAM metabolism (the large majority of cacti) or possess a chlorophyll-containing bark (*Bursera, Manihot, Pereskia*). Among the non-succulent stemmed species the genus *Cercidium* (Caesalpinoideae) contains species with photosynthetic bark (Adams & Strain, 1968).

Table 9.2. *Examples of life and growth forms in neotropical dry forests (including deciduous forest, thorn forest and thorn scrub)*

Trees and shrubs (woody plants)		
Evergreen		*Capparis, Casearia, Castela, Coccoloba, Guaiacum, Haematoxylon, Jacquinia* (some species deciduous)
Deciduous (obligate or facultative)		*Acacia, Albizzia, Bumelia, Pisonia, Pithecellobium, Prosopis, Swietenia, Tabebuia*
Green-stemmed		*Cercidium*
Succulent deciduous		*Bursera, Cochlospermum, Erythrina, Ipomoea, Jacaratia, Jatropha, Manihot, Pereskia, Spondias*
CAM-type (cacti)		*Lemaireocereus, Opuntia, Pilosocereus, Ritterocereus*
Herbs and rosettes		
Dicots	C_3	*Bastardia, Croton, Melochia, Wedelia*
	C_4	*Alternanthera, Atriplex, Euphorbia, Portulaca, Trianthema*
Grasses	C_4	*Anthephora, Aristida, Cenchrus, Setaria, Sporobolus*
Bromeliads	CAM	*Aechmea, Bromelia*
Lianas and creeping plants		
	C_3	*Arrabidaea, Mansoa, Macfadyena, Cydista, Cissus* (one species CAM-inducible)
	CAM	*Acanthocereus, Selenicereus*
Epiphytes		
	C_3	*Peperomia, Philodendron*
	CAM	*Brassavola, Schomburgkia, Tillandsia*
Hemiparasites		*Ixocactus, Phoradendron, Phtyrusa*

Sources: Aristeguieta (1968); Wikander (1984); Bullock (1985); Ponce & Trujillo (1985).

Within each life form several behavior and photosynthetic types may be found. The main growth forms are the evergreen and deciduous. The former includes plants which are never leafless, although the leaves are generally turned over every year. This may contrast with the concept of evergreens in mediterranean and temperate climates where evergreen plants produce leaves lasting several growing seasons (Chabot & Hicks, 1982). This situation is seldom found in the tropics within dicotyledonous woody plants (Medina, 1984).

C_3 and CAM photosynthetic types may be differentiated. The latter correspond mainly to the columnar and shrubby cacti, while the rest of the woody plants belong to the C_3-photosynthesis type. C_4 elements are found only among the herbaceous life forms, both grasses and dicots.

Vines may be very frequent in dry forests, even reaching levels of codominance. Among vines, woody (lianas) and herbaceous, evergreen and deciduous, fibrous and tuberous rooted, C_3 and CAM types may be found (Wikander, 1980; Ponce & Trujillo, 1985; Castellanos *et al.*, 1989).

The frequency of epiphytes seems to be associated with atmospheric humidity and incidence of dew. In dry forests near coastal locations, where these factors are favorable, CAM epiphyte biomass and diversity can reach a significant level. Within CAM epiphytes, bromeliads with water- and nutrient-absorbing trichomes in their leaves appear to be more abundant and diversified than orchids with a water- and nutrient-absorbing velamen covering their roots. A relatively minor but ever recurring life form in dry forests are the hemiparasites, which are frequently leafless in the cases of extreme dry forests (Rizzini, 1982).

Changes in the proportion of life forms along aridity gradients

The most striking features changing in the life form composition of tropical forests along an aridity gradient are related to tree height and degree of deciduousness (Beard, 1955; Veillon, 1963). Both characteristics are most probably associated with reduced water availability at the root level and particularly increased duration of the dry period (Veillon, 1963; Reich & Borchert, 1984; Lieberman & Lieberman, 1984; Mateucci, 1987; Olivares & Medina, 1992). However, it is noteworthy that the reduction in the proportion of evergreen components is not a monotonous one. At certain levels within the aridity gradient forests in the neotropics are characterized by the conspicuous presence of small

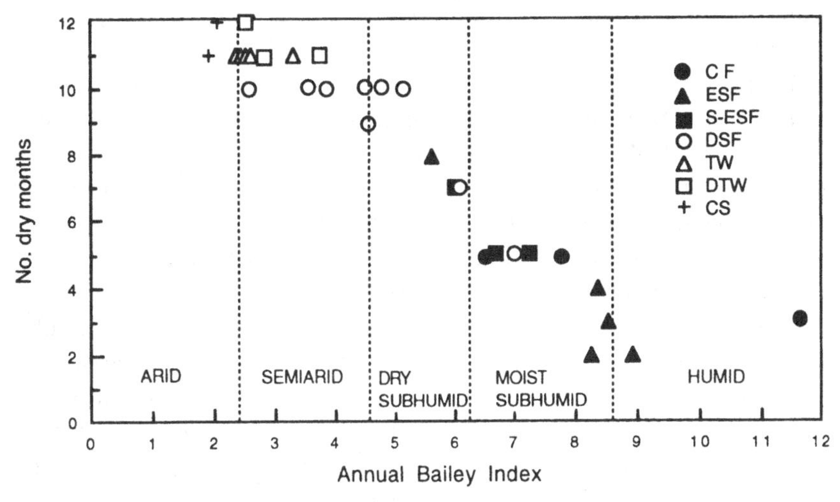

Figure 9.2. Ordination of dry forest sites in the Falcón State in northern Venezuela using temperature and rainfall data. CF, cloud forest; ESF, evergreen seasonal forest; S-ESF, semievergreen seasonal forest; DSF, deciduous seasonal forest; TW, thorn woodland; DTW, deciduous thorn woodland; CS, cactus scrub. Modified from Mateucci (1987).

evergreen trees, particularly in families such as Anacardiaceae (*Astronium graveolens*), Capparidaceae (several species of *Capparis*, *Morisonia americana*), Malphigiaceae (some *Malpighia* species), Sapindaceae (*Talisia*) and Theophrastaceae (several species of *Jacquinia*). Therefore the driest forests in the neotropics are not dominated completely by deciduous woody life forms (Beard, 1955).

Sarmiento (1972) indicates that the evergreens at the humid extreme of the gradient are coriaceous but not sclerophyllous. Sclerophyllous woody plants in the dry tropics are distinguished by their tough consistency, high leaf weight:area ratios, and their resistance to drought stress (Sobrado & Cuenca, 1979; Marín & Medina, 1981; Sobrado, 1986; Medina, García & Cuevas, 1990). These relationships between life forms and climate were quantitatively described for the plant communities in northwestern Venezuela by Mateucci (1987). She showed that the plant communities differentiated in this area could be arranged along aridity gradients as determined using synthetic meteorological indices (Bailey, 1979) (Figure 9.2). It was shown also that the proportion of deciduous woody plants increases steadily from the cloud forest formation (water availability influenced by orographic rains) to the deciduous seasonal forests where they reach a relative cover of 51% (Figure 9.3). The drier plant communities have again a larger cover of evergreen woody plants (sclerophylls).

Tropical rain forests contain the highest diversity measured in terrestrial ecosystems (expressed as number of plant species per unit area) (Gentry, 1988). Woody plants in rain forests, however, tend to converge to a relatively small number of life forms, probably as a result of homogeneous growing conditions during the year leading to a more dense space occupation and higher interspecific competition (see Figure 9.4). It seems that the life-form diversity may be significantly higher in dry forests as a result of increased heterogeneity of habitats and substrates caused by the strong seasonality of rainfall. An expression of these larger variations of life forms in dry forests can be found in the range of wood specific gravity. Dry forests in México are characterized by a higher average of wood specific gravity and also a larger number of classes compared with trees of humid forests (Barajas-Morales, 1985, 1987) (Figure 9.5). The higher specific gravity is probably associated with

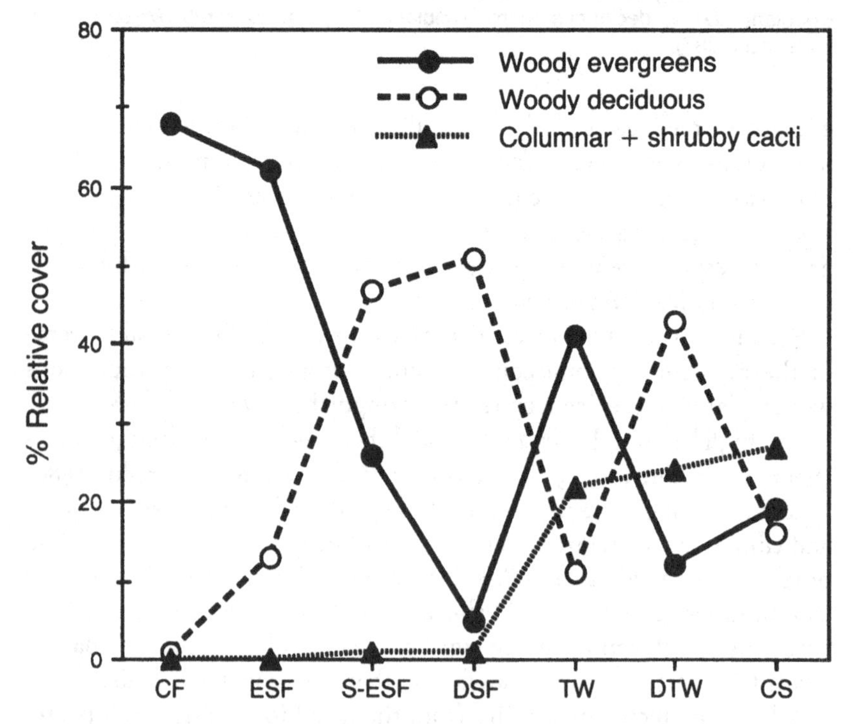

Figure 9.3. Changes in the relative cover of the main growth forms along aridity gradients in Falcón state (Venezuela) as determined by Mateucci (1987). Abbreviations as in Figure 9.2.

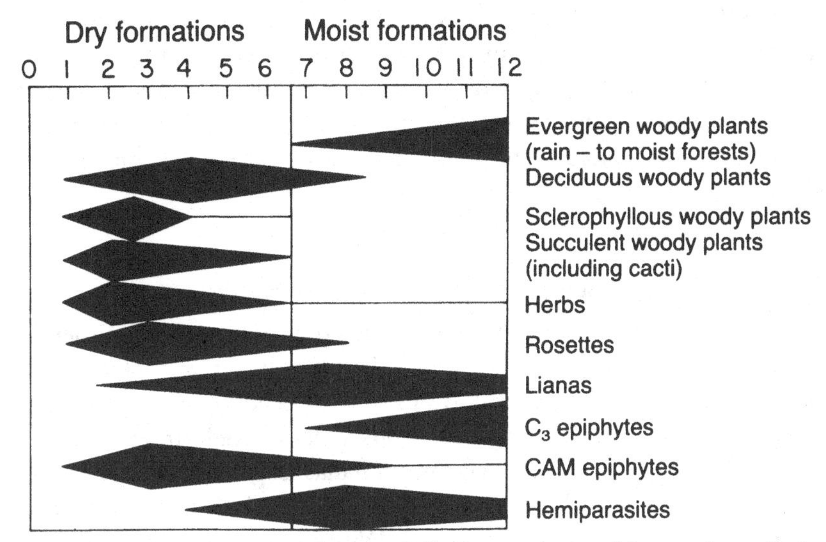

Figure 9.4. Hypothetical distribution of life forms along aridity gradients in the neotropics.

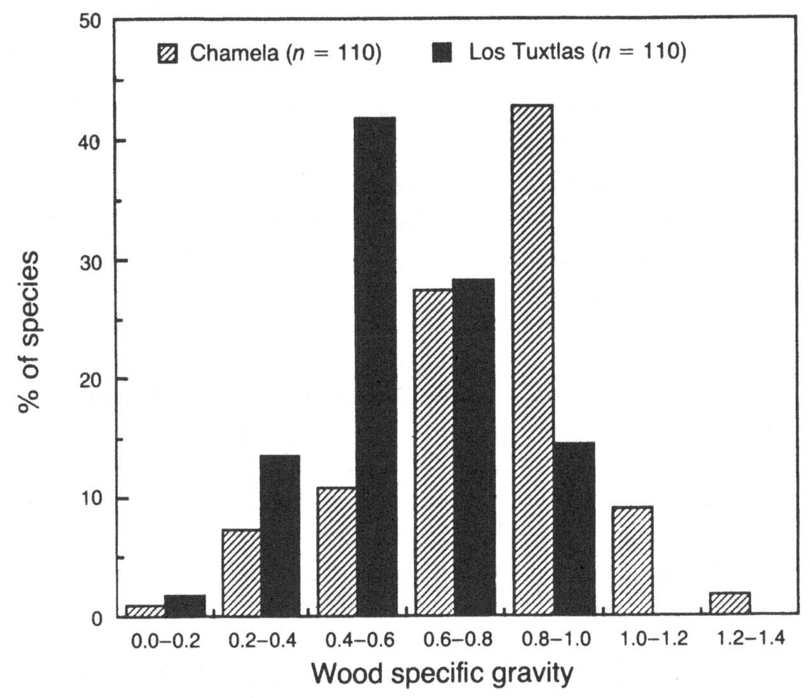

Figure 9.5. Distribution of wood specific gravity observed in a dry forest (Chamela) and in a humid forest (Los Tuxtlas) in México. Modified from Barajas-Morales (1987).

lower efficiency of water transport but less sensitivity to cavitation (Barajas-Morales, 1985). However, succulent trees in the Chamela forest, with wood of low specific gravity, constitute a significant fraction of the population and reveal that the strategy of large stem water capacitance and stable water relations are of selective value in this ecosystem.

Dispersal mechanisms and life forms

Dry forests are characterized by a higher proportion of wind dispersed species compared with moist forests (Frankie, Baker & Opler, 1974; Bullock, Chapter 11). The proportions of anemochorous species in a dry forests, however, are not homogeneously distributed among the different life forms. In typical dry forests lianas in general have a higher proportion of anemochore species than trees and shrubs (Wikander, 1984; Bullock, Chapter 11) (Table 9.3). Wikander found in a dry forest in Charallave, Venezuela that 30.5% of the phanerophytes (trees and shrubs) are typical anemochore species, while among the climbers more than 75% produced wind-dispersed fruits or seeds.

Ecophysiological performance under natural conditions

The diversity of life forms in dry forests may be associated with habitat heterogeneity regarding water availability. However, they do not constitute physiologically homogeneous groups. We will show here examples of the fine subdivision of growth forms based particularly on water relation parameters.

Evergreen (sclerophyllous), deciduous, and succulent woody plants

Evergreen and deciduous growth forms coexist in many types of dry forests (Sarmiento, 1972; Mateucci, 1987). These groups are differentiated according to leaf duration. The sclerophyllous woody plants are never leafless, and frequently change their leaf canopy during the rainy season (Marín & Medina, 1981), thus contrasting with the deciduous species which shed their leaves at the beginning of the dry season. However, leaf life span in deciduous plants is strongly related to the amount of rainfall and duration of the rainy season (Reich & Borchert, 1984). Leaf flushing in the deciduous groups also takes place either shortly before or immediately after the beginning of rains. Differences

Table 9.3. *Morphological characteristics associated with wind dispersion of reproductive units in trees, shrubs and climbers in a dry forest*

Diaspore	Trees and shrubs	Lianas
Feathery seeds	*Ceiba pentandra* *Cochlospermum vitifolium*	*Marsdenia condensiflora* *Matelea maritima* *Matelea planiflora* *Matelea urceolata* *Metastelma* sp.
Winged seeds	*Tabebuia ochracea*	*Cydista equinoctialis* *Macfadyena unguis-cati* *Mansoa verrucifera* *Phryganocidia corymbosa* *Pithecoctenium crucigerum* *Prestonia exerta*
Very small seeds		*Pogonopus speciosus*
Syncarpic fruits with one wing	*Centrolobium paraense* *Fissicalix fendlerii* *Lonchocarpus dipteroneurus* *Machaerium latialatum* *Machaerium robiniaefolium* *Myriospermum frutescens*	*Heteropterys purpurea* *Machaerium moritzianum* *Nissolia fruticosa* *Securidaca diversifolia* *Securidaca* sp.
Schizocarpic, winged fruits	*Beureria cumanensis* *Bulnesia arborea*	*Serjania* sp.
Multi-winged fruits	*Piscidia carthaginensis* *Terminalia* sp.	*Combretum fruticosum* *Paullinia* sp.

Source: Wikander (1980).

are due to the availability of soil water allowing recharging of water reserves in the tree stems (Reich & Borchert, 1984) and the degree of rain seasonality. However, bud-break without leaf expansion appears to be a common phenomenon in some dry forests, particularly in tuberous vines (Bullock & Solís, 1990). In sharply seasonal climates these trees shed their leaves at the onset of drought and do not flush until after the beginning of the rainy season (Lieberman & Lieberman, 1984).

In general, sclerophyllous evergreens produce thicker, heavier leaves (lower specific leaf areas) with relatively lower nutrient content per unit leaf weight (Marín & Medina, 1981). Leaf life span is correlated with the amount of organic matter accumulated in the leaves, particularly in the

Table 9.4. *Values of osmotic pressure (π in bars, average and S.D.) in a gradient from rain forest to very dry forests in Jamaica*

	Rain forest	Moist forest	Slope forest	Slope dry forest	Coastal desert	Sandy beach
Trees and shrubs						
π	8.6 ± 1.9	11.0 ± 2.5	15.3 ± 3.9	30.1 ± 9.0	38.0 ± 13.2	29.6 ± 5.7
n	21	16	16	19	13	4
Hemiparasites						
π	14.8	16.3 ± 1.2	16.8 ± 2.2			
n	1	6	21			
CAM succulents						
π				8.1 ± 1.4	7.5 ± 1.7	
n				11	16	

Source: Harris (1934).

form of non-metabolically active compounds (Chabot & Hicks, 1982; Medina, 1984). However, the thicker leaves of the sclerophylls of dry forests appear to be also more drought tolerant. This drought tolerance is related to the higher osmotic pressure in vacuolar saps of sclerophyll trees compared with deciduous woody plants. On the other hand, the latter produce leaves with more elastic cell walls. These relationships can be observed by measuring the leaf sap osmotic pressure along aridity gradients. Harris (1934) provided one of the earliest sets of data on this subject when studying the vegetation of Jamaica (Table 9.4). There is a clear tendency to more concentrated leaf sap with increased aridity. The CAM succulents measured (cacti) were characterized by significantly lower osmotic pressure of their cell sap.

Actual osmotic pressure of leaf cell sap measured in the field varies significantly among tree growth forms. The evergreen sclerophylls normally show higher osmotic pressure in their leaf sap during the rainy season, while in the dry season even deciduous trees can have quite high values (Table 9.5). The leaves of woody plants with succulent stems, however, always have relatively diluted leaf sap. Succulent stemmed trees are also drought deciduous and due to their low osmotic pressure reach zero turgor at relatively high water potential (Nilsen *et al.*, 1990; Olivares & Medina, 1992). Their cell elasticity has been reported as low in arid communities in California (Nilsen *et al.*, 1990); however, in tropical dry forests cell wall elasticity has been shown to be higher in deciduous than in evergreen trees. Sobrado (1986), using the pressure chamber technique in a lower montane dry forest, showed that deciduous trees produce in general leaves with more elastic cells but lower osmotic concentrations of their leaf sap than those of the evergreen sclerophylls.

Succulent woody plants performing CAM

Activity of CAM succulents varies during the year according to the water availability. These plants possess superficial root systems, therefore the onset of the dry season means the death of the fine root system (Kausch, 1965). Plants become increasingly isolated from the environment, opening their stomata only at night. During this period they fix external CO_2, which is accumulated as malic acid in the vacuoles. The malic acid is decarboxylated during the following day, the resulting CO_2 being incorporated into carbohydrates through photosynthesis. As the dry season proceeds and tissue dehydration increases stomata fail to open even at night and the plant stabilizes at high water potential, effectively recycling

Table 9.5. *Osmotic pressure (MPa) of leaf sap and xylem tension measured in the field during the middle of the dry and rainy seasons of 1984 in a dry forest of the Península de Paraguaná in Falcón State, Venezuela*

	Osmotic pressure		Maximum xylem tension	
Attributes and species	Dry	Rain	Dry	Rain
Evergreens				
Capparis odoratissima	4.31	2.96	5.49	3.57
Capparis linearis	3.66	2.95	5.99	4.50
Jacquinia aristata	4.09	1.51	4.96	2.76
Deciduous				
Casearia tremula	4.55	1.81	6.55	0.78
Pithecellobium dulcis	4.26	1.82	5.76	3.19
Prosopis juliflora	3.13	2.10	4.17	3.11
Succulent woody plants (deciduous)				
Jatropha gossypifolia	1.45	1.19	1.38	0.17
Pereskia guamacho	1.27	1.09	1.33	0.97
Malacophyllous shrub				
Croton cf. *flavens*	4.82	1.28	7.41	0.40

Source: Medina, Olivares & Marin (1985).

the respiratory CO_2 (Szarek & Ting, 1974). The activity of CAM results in high water use efficiency. This behavior has been documented in columnar and shrubby cacti in dry forests in northern Venezuela (Díaz & Medina, 1984; Figure 9.6). The highest diurnal oscillations in titratable acidity (reflecting the carboxylating activity of PEP-carboxylase at night, and the decarboxylation of malate the following day) is measured during periods of higher rainfall. Lüttge *et al.* (1989) report similar observations and show low osmotic pressure of the CAM performing tissues in the cacti in the same range as observed earlier by Harris (1934) in Jamaica (Table 9.4).

Some primitive cacti of the genus *Pereskia* possess non-CAM succulent stems similar to the succulent stemmed tree species described above. Species of this genus produce leaves which become succulent with age. During the rainy season these leaves are characterized by high sensitivity to atmospheric drought and low conductance. At the beginning of the dry season a certain CAM activity is induced, probably extending the leaf life span (Figure 9.6).

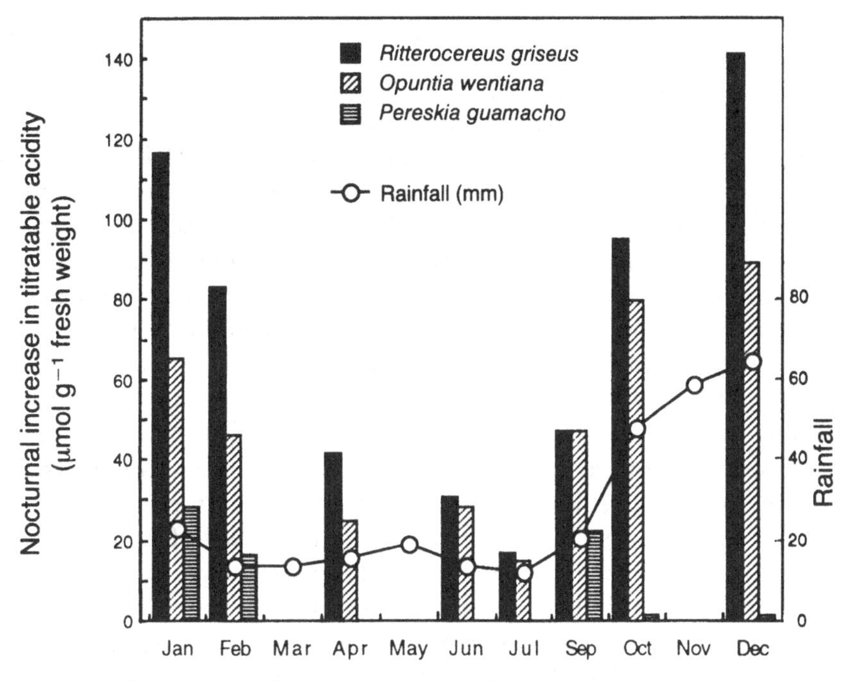

Figure 9.6. Seasonal variation in daily oscillations in titratable acidity in cactus species from dry forests. Modified from Díaz & Medina (1984).

Epiphytes and terrestrial rosettes

The species belonging to these groups are all CAM plants (Medina, Olivares & Díaz, 1986; Mooney, Bullock & Ehleringer, 1989) and present essentially the same seasonal behavior regarding photosynthetic productivity and water availability as that described for the woody succulents performing CAM. These species are characterized by low osmolality of their leaf cell sap, high water potentials throughout the year, and flowering activity concentrated in the dry season, while main growing activity is restricted to the rainy season (Medina *et al.*, 1986; Griffiths *et al.*, 1989; Lee *et al.*, 1989).

Rosette plants such as *Bromelia humilis*, *B. chrysantha* and *Aechmea aquilega* may densely cover the forest floor in many dry forests of northern South America (Medina *et al.*, 1986). It is not known, however, if the present distribution and density of this life form is associated with grazing pressure in dry forests in the neotropics. These rosette plants are normally very thorny being therefore well defended against large herbivores.

Lianas and hemiparasites

Lianas are frequently an important component of many dry forests (Wikander, 1980; Gentry, 1983; Ponce & Trujillo, 1985; Bullock, 1985, 1990; Gentry & Dodson, 1987; see Swaine, Lieberman & Hall, 1990, for an African evergreen dry forest) (Table 9.6). It has been shown that in several dry forests in northern South America and Central America the number of liana individuals can surpass the number of tree individuals, and therefore they may contribute substantially to the maintenance of a large leaf area index, without a corresponding proportional increase in the amount of supporting stem biomass (Gentry, 1983). It seems that the proportion of climber species increases from wet, through moist to dry forests, reaching in the latter up to 34% of all species and 16% of all individuals (Gentry & Dodson, 1987). The canopy of lianas in dry forests may be also highly efficient photosynthetically because they are characterized by high nitrogen content and relatively low leaf weight:area ratios (Castellanos *et al.*, 1989). Their leaf water relations, however, are very little known. In humid forests lianas show relatively low osmotic pressures of leaf cell sap: that is, they do not appear to be under water stress in spite of the relatively small vessel area per unit of leaf area (Walter, 1973: 223). These balanced water relations seem to be associated with their transpiration rates, which are lower than those of the trees supporting them (Coutinho, 1962), but also with the higher hydraulic conductivity of their stems (Gessner, 1956). Higher hydraulic conductivities of liana stems, compared with trees, have been also demonstrated in a dry forest (Gartner *et al.*, 1990). These authors showed that the hydraulic conductivity tends to be higher in wetter environments, or in lianas active during wet seasons, a fact without a clear ecological explanation yet.

Lianas of dry forests are common within families such as Bignoniaceae, Asclepiadaceae, Leguminosae and Combretaceae, and show a variety of growth forms (Wikander, 1984; Ponce & Trujillo, 1985; Castellanos *et al.*, 1989). In a montane dry forest (Ponce & Trujillo, 1985) from a total of 68 woody species 38% were trees while 31% were classified as climbers or subclimbers. Sixty per cent of the tree species were classified as deciduous. Among the climbers the authors differentiated (a) herbaceous or sub-woody annuals (*Gronovia scandens, Dalechampia scandens*), (b) herbaceous perennials with deciduous leaves (*Cerathosanthes palmata, Matelea maritima*), (c) woody, tall climbing deciduous (*Arrabidaea mollisima, Mansoa verrucifera*) and (d) woody evergreens (*Cydista*

Table 9.6. *Life-form distribution counted from florulas of deciduous forests*

	Chamela	Charallave	Caracas
Temp./Rainfall × 100	3.33	2.57	2.45
Number of species	725	166	424
Number of families	100	54	81
% Herbs	28.5	18.7	32.5
% Shrubs	21.7	53.0[a]	23.6
% Trees	24.0		19.8
% Vines	20.8	24.6	16.3
% Epiphytes	3.8	4.2	5.0
% Hemiparasites	0.9	0.6	1.6
% CAM species	6.5	4.8	5.9

[a] includes shrubs and trees.
Sources: Chamela, México: Bullock (1985); Caracas, Venezuela: Berry & Steyermark (1983); Charallave, Venezuela: Wikander (1984).

diversifolia). In another dry forest in northern Venezuela, Wikander (1984) had already reported that from all species registered 20% were classified as woody and 5% as herbaceous lianas. In the dry forests near Caracas, and in Chamela, Mexico, 16 and 21% respectively, of the whole reported flora belongs to the vine life form (Berry & Steyermark, 1983; Bullock, 1985) (Table 9.6). Similar predominance of climbers was reported for dry forests in Ecuador by Gentry & Dodson (1987).

The proportion of hemiparasite species in dry forests is probably always around 2% or less. They behave as evergreens, maintaining lower water potentials than the hosts and transpiring throughout the dry season (Hollinger, 1983; Schulze, Turner & Glatzel, 1984; Goldstein *et al.*, 1989). In general, the leaf conductance of the hemiparasites is higher than that of the host leaves, therefore they have comparatively lower water use efficiencies. This fact is reflected in the relatively more negative $\delta^{13}C$ values of their leaves (Ehleringer, Cook & Tieszen, 1986). Hemiparasites of the family Loranthaceae obtain not only water from their hosts, but also inorganic nutrients and other organic compounds (Ehleringer *et al.*, 1985; Marshall & Ehleringer, 1990). Their abundance and influence on forest productivity have not been assessed yet.

Conclusions

This chapter has presented some relevant data on the diversity of life forms in tropical dry forests, pointing out some of its morpho-physiological properties. It is clear that present knowledge on the role of life form diversity in ecosystem function is incomplete and, at least, heterogeneous. Phenology and water relations of woody components have been better studied, but the actual mechanisms are still in dispute, particularly the aspects related to soil water availability (see Holbrook, Whitbeck & Mooney, Chapter 10). Epiphytes have been studied more or less intensively because of the large number of morpho-physiological characteristics having adaptive value to drought generally found in species belonging to this life form. Woody and herbaceous vines have been receiving more attention recently and the relatively few modern studies are revealing a large number of eco-physiologically interesting properties which will lead to a better understanding of their frequency in seasonal forests. Much more information is required, however, to reach a comprehensive understanding of dry forest function, to understand succession, and reveal the competitive relationships between and within different life forms. Some particularly relevant questions are those related to: (a) the eco-physiology of germination and seed dispersal of all life forms; (b) the clarification of the physiological significance for drought tolerance and resistance of the occurrence of succulent organs, both above and below ground, commonly found in dry forest trees, (c) the quantitative analysis of resource partition and the carbon balance during the dry season, particularly of those trees flowering and maintaining high water potential during the dry season. This understanding is necessary to develop coherent programs for conservation and restoration of dry forests.

Summary

The diversity of life forms in tropical dry forests appears to be larger than in other tropical ecosystems, probably as a consequence of habitat heterogeneity regarding availability of water. This heterogeneity is associated with forest stratification, patchy distribution of soil water reservoirs, and seasonal distribution of rainfall. All neotropical dry forests are characterized by the predominance of deciduous life forms (> 40% of the woody species), although evergreen species, particularly towards the dry end of the humidity gradient, might constitute a large fraction of the community. Structural analyses of dry forests throughout the neotropics have shown the importance of liana species. Other life forms, such as succulent stemmed species (including arborescent cacti), constitute a characteristic set of all neotropical dry forests. Epiphytes, such as bromeliads and orchids, are abundant particularly in dry coastal forests where dew deposition provides a small but continuous source of water. Hemiparasites constitute a small and widespread group in dry forests, but their role in forest dynamics has not been studied. The large majority of woody components of dry forests are wind-dispersed; however, the occurrence of anemochore species is not homogeneous within the different life forms, lianas apparently possess a larger proportion of wind-dispersed species than trees and shrubs. Life-form diversity in dry forests is expressed both on a structural (e.g. specific gravity of wood, plant habit, leaf size) and a physiological basis (stability of water relations as revealed by photosynthetic tissue sap osmolality, drought tolerance and growth seasonality).

References

Adams, M. S. & Strain, B. R. (1968). Photosynthesis in stem and leaves of *Cercidium floridum*: spring and summer diurnal field response to temperature. *Oecologia Plantarum* **3**: 285–7.

Aristeguieta, L. (1968). El bosque seco caducifolio seco de los llanos altos centrales. *Boletín de la Sociedad Venezolana de Ciencias Naturales* **27** (113–114): 395–438.

Bailey, H. P. (1979). Semi-arid climates: their definition and distribution. In *Agriculture in Semi-Arid Environments*, ed. A. E. Hall, G. H. Cannell and H. W. Lawton, pp. 73–97. Ecological Studies Vol. 34. Springer–Verlag, Berlin.

Barajas-Morales, J. (1985) Wood structural differences between trees of two tropical forests in Mexico. *IAWA Bulletin n.s.* **6**: 355–64.

Barajas-Morales, J. (1987). Wood specific gravity in species from two tropical forests in Mexico. *IAWA Bulletin n.s.* **8**: 143–8.

Beard, J. (1955). The classification of tropical American vegetation types. *Ecology* **36**: 89–100.

Berry, P. E. & Steyermark, J. (1983). Flórula de los bosques decíduos de Caracas. *Memorias de la Sociedad de Ciencias Naturales La Salle* **53** (120): 157–214.

Bullock, S. H. (1985). Breeding systems in the flora of a tropical deciduous forest in Mexico. *Biotropica* **17**: 287–301.

Bullock, S. H. (1990). Abundance and allometrics of vines and self-supporting plants in a tropical deciduous forest. *Biotropica* **22**: 106–9.

Bullock, S. H. & Solís Magallanes, J. A. (1990). Phenology of canopy trees of a tropical deciduous forest in Mexico. *Biotropica* **22**: 22–35.

Castellanos, A. E., Mooney, H. A., Bullock, S. H., Jones, C. & Robichaux, R. (1989). Leaf, stem, and metamer characteristics of vines in a tropical deciduous forest in Jalisco, Mexico. *Biotropica* **21**: 41–9.

Chabot, B. F. & Hicks, D. J. (1982). The ecology of leaf life spans. *Annual Review of Ecology and Systematics* **13**: 229–59.

Clements, F. E. (1920). *Plant Indicators: the relations of plant communities to process and practice*. Publication 290. Carnegie Institution of Washington, Washington.

Coutinho, L. M. (1962). Contribuição ao conhecimento da ecologia da mata pluvial tropical. *Faculdade de Filosofia, Ciencias e Letras da Universidade de São Paulo* Boletim No. **257**.

Díaz, M. & Medina, E. (1984). Actividad CAM de Cactáceas en condiciones naturales. In *Physiological Ecology of CAM plants*, ed. E. Medina, pp. 98–113. CIET. Instituto Venezolano de Investigaciones Científicas, Caracas.

Ehleringer, J. R., Cook, C. S. & Tieszen, L. L. (1986). Comparative water use and nitrogen relationships in a mistletoe and its host. *Oecologia* **68**: 279–84.

Ehleringer, J. R., Schulze, E.-D., Ziegler, H., Lange, O. L., Farquhar, G. D. & Cowan, I. R. (1985). Xylem-tapping mistletoes: water or nutrient parasites? *Science* **227**: 1479–81.

Ewel, J. J., Madriz, A. & Tossi J. A. (1976). *Zonas de Vida de Venezuela. Memoria Explicativa sobre el Mapa Ecológico. 2a* edición. Fondo Nacional de Investigaciones Agropecuarias, Ministerio de Agricultura y Cría, Caracas.

Frankie, G. W., Baker, H. H. & Opler, P. A. (1974). Comparative phenological studies of trees in tropical wet and dry forests in the lowlands of Costa Rica. *Journal of Ecology* **52**: 881–919.

Gartner, B. L., Bullock, S. H., Mooney, H. A., Brown, V. B. & Whitbeck, J. L. (1990). Water transport properties of vine and tree stems in a tropical deciduous forests. *American Journal of Botany* **77**: 742–9.

Gentry, A. H. (1983). Lianas and the 'paradox' of contrasting latitudinal gradients in wood and litter production. *Tropical Ecology* **24**: 63–7.

Gentry, A. H. (1988). Changes in plant community diversity and floristic composition on environmental and geographical gradients. *Annals of the Missouri Botanical Garden* **75**: 1–34.

Gentry, A. H. & Dodson, C. H. (1987). Diversity and biogeography of neotropical vascular epiphytes. *Annals of the Missouri Botanical Garden* **74**: 205–33.

Gessner, F. (1956). Wasserhaushalt der Epiphyten und Lianen. In *Handbuch der Pflanzenphysiologie*, Vol. III, ed. H. Ruhland, pp. 915–50. Springer–Verlag, Berlin.

Goldstein, G., Rada, F., Sternberg, L. *et al.* (1989). Gas exchange and water balance of a mistletoe species and its mangrove host. *Oecologia* **78**: 176–83.

Griffiths, H., Smith, J. A. C., Lüttge, U. *et al.* (1989). Ecophysiology of xerophytic and halophytic vegetation of a coastal alluvial plain in northern Venezuela IV. *Tillandsia flexuosa* Sw. and *Schomburgkia humboldtiana* Reichb., epiphytic CAM plants. *New Phytologist* **111**: 273–82.

Harris, J. A. (1934). *The physico-chemical properties of plant saps in relation to phytogeography*. University of Minnesota Press, Minneapolis.

Hegner, R. (1979). *Nichtimmergrüne Waldformationen der Tropen*. Kölner Geographische Arbeiten Heft 37. Geographisches Institut der Universität zu Köln.

Holdridge, L. R. (1959). Simple method for determining potential evapotranspiration from temperature data. *Science* **130**: 572.

Hollinger, D. Y. (1983). Photosynthesis and water relations of the mistletoe *Phoradendron villosum* and its host, the California valley oak, *Quercus lobata*. *Oecologia* **60**: 396–400.

Kausch, W. (1965). Beziehungen zwischen Wurzelwachstum, Transpiration und CO_2-Gaswechsel bei einigen Kakteen. *Planta* **123**: 91–6.

Lee, H. S. J., Lüttge, U., Medina, E. *et al.* (1989). Ecophysiology of xerophytic and halophytic vegetation of a coastal alluvial plain in northern Venezuela. III. *Bromelia humilis* Jacq., a terrestrial CAM bromeliad. *New Phytologist* **111**: 253–71.

Lieberman, D. & Lieberman, M. (1984). The causes and consequences of synchronous flushing in a dry tropical forest. *Biotropica* **16**: 193–201.

Lüttge, U., Medina, E., Cram, W. J., Lee, H. S. J., Popp, M., & Smith, J. A. C. (1989). Ecophysiology of xerophytic and halophytic vegetation of a coastal alluvial plain in northern Venezuela II. Cactaceae. *New Phytologist* **111**: 245–51.

Marín, D. & Medina, E. (1981). Duración foliar, contenido de nutrientes, y esclerofilia en árboles de un bosque muy seco tropical. *Acta Científica Venezolana* **32**: 508–14.

Marshall, J. D. & Ehleringer, J. R. (1990). Are xylem-tapping mistletoes partially heterotrophic? *Oecologia* **84**: 244–8.

Mateucci, S. (1987). The vegetation of Falcón State, Venezuela. *Vegetatio* **70**: 67–91.

Medina E. (1984). Nutrient balance and physiological processes at the leaf level. In *Physiological Ecology of Plants of the Wet Tropics*, ed. E. Medina, H. A. Mooney and C. Vázquez-Yanes, pp. 139–54. Dr W. Junk, The Hague.

Medina, E., García, V. & Cuevas, E. (1990). Sclerophylly and oligotrophic environments: relationships between leaf structure, mineral nutrient content and drought resistance in tropical rain forests of the upper Río Negro region. *Biotropica* **22**: 51–64.

Medina, E., Olivares, E. & Díaz, M. (1986). Water stress and light intensity effects on growth and nocturnal acid accumulation in a terrestrial CAM bromeliad (*Bromelia humilis* Jacq.) under natural conditions. *Oecologia* **70**: 441–6.

Medina, E., Olivares, E. & Marín, D. (1985). Eco-physiological adaptations in the use of water and nutrients by woody plants of arid and semi-arid tropical regions. *Medio Ambiente (Valdivia)* **7**: 91–102.

Mooney H. A., Bullock, S. H. & Ehleringer, J. R. (1989). Carbon isotope ratios of plants of a tropical dry forest in Mexico. *Functional Ecology* **3**: 137–42.

Murphy, P. G. & Lugo, A. E. (1986). Ecology of tropical dry forest. *Annual Review of Ecology and Systematics* **17**: 67–88.

Nilsen, E. T., Sharifi, M. R., Rundel, P. W., Forseth, I. N. & Ehleringer, J. R. (1990). Water relations of stem succulent trees in north-central Baja California. *Oecologia* **82**: 299–303.

Olivares, E. & Medina, E. (1992). Water and nutrient relations of woody perennials from tropical dry forests. *Journal of Vegetation Science* **3**: 383–92.

Ponce, M. & Trujillo, B. (1985). Composición florística y vegetacional de la selva decídua montano-baja del Jardín Botánico Universitario, Maracay, Venezuela. *Ernstia* No. **35**: 30–44.

Reich, P. B. & Borchert, R. (1984). Water stress and tree phenology in a tropical dry forest in the lowlands of Costa Rica. *Journal of Ecology* **72**: 61–74.

Rizzini, C.T. (1982). Loranthaceae. In *Flora de Venezuela: familias Loranthaceae, Hernandiaceae y Chrysobalanaceae*, pp. 7–316. Ediciones Fundación de Educación Ambiental, Caracas.

Sarmiento, G. (1972). Ecological and floristic convergences between seasonal plant formations of tropical and subtropical South America. *Journal of Ecology* **60**: 367–410.

Schulze, E.-D. (1982). Plant life forms and their carbon, water and nutrient relations. In *Encyclopedia of Plant Physiology, New Series, Vol. 12B, Physiological Plant Ecology II, Water Relations and Carbon Assimilation*, ed. O. L. Lange, P. S. Nobel, C. B. Osmond and H. Ziegler, pp. 615–76. Springer-Verlag, Berlin.

Schulze, E.-D., Mooney, H. A., Bullock S. H. & Mendoza, A. (1988). Water contents of wood of tropical deciduous forest species during the dry season. *Boletín Sociedad Botánica de México* **48**: 113–18.

Schulze, E.-D., Turner, N. C. & Glatzel, G. (1984). Carbon, water and nutrient relations of two mistletoes and their hosts: a hypothesis. *Plant Cell and Environment* **7**: 293–9.

Sobrado, M. A. (1986). Aspects of tissue water relations and seasonal changes of leaf water potential components of evergreen and deciduous species coexisting in tropical dry forests. *Oecologia* **68**: 413–16.

Sobrado, M. A. & Cuenca, G. (1979). Aspectos del uso del agua de especies deciduas y siempreverdes en un bosque seco tropical de Venezuela. *Acta Científica Venezolana* **30**: 302–8.

Swaine, M. D., Lieberman, D. & Hall, B. J. (1990). Structure and dynamics of a tropical dry forests in Ghana. *Vegetatio* **88**: 31–51.

Szarek, K. & Ting, I. P. (1974). Seasonal patterns of acid metabolism and gas exchange in *Opuntia basilaris*. *Plant Physiology* **54**: 76–81.
Veillon, J. P. (1963). Relación de ciertas características de la masa forestal de unos bosques de las zonas bajas de Venezuela con el factor climático: humedad pluvial. *Acta Científica Venezolana* **14**: 30–41.
Walter, H. (1973). *Die Vegetation der Erde. Band I. Die Tropischen und subtropischen Zonen*. 3rd edition. VEB Gustav Fischer Verlag, Jena.
Wikander, T. (1980). Mecanismos de dispersión de frutos y semillas del bosque tropical decíduo por la sequía de Charallave. Trabajo de Ascenso. Escuela de Biología, Universidad Central de Venezuela, Caracas.
Wikander, T. (1984). Mecanismos de dispersión de diásporas de una selva decídua en Venezuela. *Biotropica* **16**: 276–83.

10

Drought responses of neotropical dry forest trees

N. MICHELE HOLBROOK, JULIE L. WHITBECK & HAROLD A. MOONEY

Introduction

Many neotropical dry forests are dominated by trees that shed their foliage and remain leafless for a substantial period each year. Because the majority of deciduous species drop their leaves during the dry season and renew their canopies with the onset of the rains, the question of how these trees cope with seasonal reductions in soil moisture and increases in evaporative demand is most simply answered by calling them 'drought avoiders' (*sensu* Levitt, 1972). This categorization, however, gives little insight into the conditions, constraints and consequences that accompany the deciduous habit in these forests. Furthermore, the coexistence of even a small number of evergreen species indicates that the deciduous habit is not unconditionally imposed by the environment and that patterns of leaf fall and renewal must be viewed as part of an integrated response to environmental conditions. Seasonality in water availability clearly plays a major role in structuring patterns of activity and growth in this biome, but there have been few studies of the dominant life form (and fewer of life-form diversity; see Medina, Chapter 9). In this chapter, we address characteristics of trees of tropical dry and deciduous forests that influence their patterns of water use.

Plants respond to changes in resource availability on several scales. Our review considers three such levels: 'structure' encompasses features that remain relatively constant throughout the life of a plant, such as rooting patterns or stem hydraulic properties; 'physiology' focuses on parameters that influence diurnal patterns of water use and gas exchange; while 'phenology' considers seasonal patterns of meristem activity. Although drought-deciduous trees form a major component of neotropical dry forests, comparisons with species that maintain leaves throughout the dry season, as well as the rare pattern of bearing leaves primarily during the dry season (wet season deciduous species), may provide a context for understanding characteristics associated with the deciduous habit. We have attempted to distinguish evergreen or brevideciduous species growing near permanent watercourses from ones within the deciduous forest matrix, but recognize that these cases are not always clear in the literature. We also restrict our discussion to studies conducted in tropical America. Most of the studies we survey focus on the characteristics of particular plant organs and address drought responses at only one scale. We hope to stimulate consideration of the integrated responses of dry forest trees to seasonal changes in water availability.

Structure

Morphological and anatomical aspects of dry forest trees play integral roles in determining the timing and rates of water uptake, transport, and loss to the atmosphere. For example, rooting depth may influence the tempo with which seasonal shifts in soil moisture content are experienced, xylem resistances connect leaf water status with soil moisture levels, and foliar characteristics such as reflectance and thickness affect leaf energy balance and hence the relationship between transpirational water loss and carbon gain. Differing construction and maintenance costs are associated with different strategies of acquiring resources from the environment. In this section we focus on aspects of morphology and anatomy that play important roles in the response to seasonal drought.

Roots

Despite the importance of roots for water acquisition, root anatomy, physiology, and root system architecture are essentially unknown for dry forest trees (see also Martínez-Yrízar, Chapter 13; Jaramillo & Sanford, Chapter 14; Cuevas, Chapter 15). Only limited information is available on root biomass and rooting depth. In comparison with plants of moist and wet forests, dry forest species deploy a greater percentage of their root biomass deeper in the soil profile, where moisture availability tends to be greater and of longer duration (Table 10.1; Sobrado & Cuenca, 1979; Kummerow *et al.*, 1990; but see Vance & Nadkarni, 1992). This trend is paralleled in the gradient from lower montane rain forest to semideciduous lowland forest in Panamá (Cavelier, 1992). Root biomass as a proportion of total plant biomass ranges from 8 to 50% for dry forests and from less than 5 to 33% for wet forests, with means of

Table 10.1. *Root depth distribution in tropical forests. Roots were excavated to a depth of 50 cm; roots below this depth are not included in these calculations. Values of 99% are approximate*

Forest type	Per cent root mass		Source
	0–10 cm	0–50 cm	
Wet (latosol)	79	90	1
Wet (podzol)	95	99	1
Lower montane wet	79	99	2
Lowland semideciduous	65	99	2
Deciduous	57	90	3
Deciduous	51	99	4, 5

Sources: 1. Klinge (1973); 2. Cavelier (1992); 3. Murphy & Lugo (1986b); 4. Kummerow *et al.* (1990); 5. Castellanos *et al.* (1991).

33.5% for deciduous and 16.0% for wet forests (Brown & Lugo, 1982; Murphy & Lugo, 1986a, b). Root:shoot ratios are also approximately twice as high for dry (0.42–0.50) as for wet (0.23) forests (Murphy & Lugo, 1986b; Becker & Castillo, 1990; Castellanos, Maass & Kummerow, 1991).

Overall rates of root production appear relatively high in dry forests. One estimate of annual fine root (≤2 mm diameter) production suggested a value of 423 g m^{-2}, equivalent to the aboveground litter production at that site (Kummerow *et al.*, 1990). Fine root production in relatively aseasonal mature tropical forests is reported to range from 129 to 1380 g m^{-2} y^{-1} (Jordan & Escalante, 1980; Cuevas & Medina, 1988; Sanford, 1989, 1990; Cavelier, 1992; Vance & Nadkarni, 1992), with most values falling between 129 and 228 g m^{-2} y^{-1}, although different methods complicate the interpretation of these results. Increased allocation to roots in dry forest plants is not surprising given the primary role of roots in water and nutrient uptake, and the correlation between nutrient availability and soil water status.

Rooting patterns appear to differ among dry forest trees that exhibit different leaf phenologies. Differences in pre-dawn leaf water potentials at the very beginning of the rainy season suggest differences in either the root system architecture or the responsiveness of drought-deciduous and evergreen tree roots to changes in soil moisture status. During this period the drought-deciduous trees produced new leaves with pre-dawn leaf water potentials that reflected the recently hydrated soil conditions (−0.5 MPa), while evergreen species maintained low pre-dawn water potentials

(−1.6 MPa) that only gradually increased to values measured in the drought-deciduous trees (Sobrado & Cuenca, 1979). The two drought-deciduous species excavated had a fibrous root mass relatively close to the soil surface, and the three evergreen species had single or slightly ramified tap-roots extending deep into the soil. Studies of evergreen and deciduous savanna trees also indicate that evergreen species have larger root systems that allow them access to water throughout the rainless period (Medina, 1984; Sarmiento, Goldstein & Meinzer, 1985), but that do not respond to shallow soil wetting by the first rains. Roots of the wet season deciduous species *Jacquinia pungens* branch only below 0.5–1.5 m depth and the taproots of mature plants extend to at least 3 m in depth (Janzen, 1970).

Roots do not function in isolation, but in close association with the belowground microbial community. While the effects of plant–microbe relationships on the water relations of dry forest trees have not been examined, in other ecosystems these associations can substantially alter root morphology and influence uptake of both water and nutrients (Gianinazzi-Pearson & Gianinazzi, 1983; Harley & Smith, 1983; Janos, 1983). Leguminous trees are dominant in many dry forests (see Gentry, Chapter 7), and association with the nitrogen-fixing bacteria *Rhizobium* is almost universal among species of the Mimosoideae and Papilionoideae and occurs in about 30% of the Caesalpinoideae (Postgate, 1982). Another bacterial symbiotic nitrogen-fixer, *Frankia*, is present in Costa Rican dry forest soils (Paschke & Dawson, 1992). Mycorrhizal associations are also common. The majority of dry forest trees appear to form vesicular-arbuscular mycorrhizal symbioses, although a few maintain ectomycorrhizal associations (Cuenca & Lovera, 1990; E. Rincón & M. Gavito, personal communication). Mycorrhizae not only enhance plant access to soil nutrients, they can also translocate water from soil to roots (Nelsen, 1987; Bethlenfalvay *et al.*, 1988). Whether such associations form a significant component of the water relations of dry forest plants remains unknown, and even descriptive information on the frequency or type of association is minimal.

Leaves

The impression made by deciduousness may overwhelm appreciation of the functional significance of other foliar characters. In contrast to research on desert plants, studies of dry forest vegetation have largely ignored leaf structural characters influencing leaf energy and water balance. Leaf characteristics of drought-deciduous species are clearly distinct from those of evergreen and wet season deciduous species. Several authors describe the leaves of evergreen species in dry forests as sclerophyllous and characterize those of deciduous plants as mesophytic (Sobrado, 1986, 1991; Goldstein *et al.*, 1989; Medina, Chapter 9). Leaves of evergreen species are noticeably thicker than those of deciduous species, and have thick cuticles and a leathery texture (Marín & Medina, 1981). As is characteristic of most ecosystems, the leaves of evergreen species have higher leaf specific mass (g m^{-2}) and lower nitrogen and phosphorus contents on a mass basis than deciduous species (Table 10.2; Marín & Medina, 1981; Medina, 1984; Mooney, Field & Vázquez-Yanes, 1984; Castellanos *et al.*, 1989; Sobrado, 1991; J. L. Whitbeck *et al.*, personal communication). For example, in a Venezuelan dry forest, leaf specific mass is greater in evergreen (131.6 g m^{-2}) than in deciduous (54.4 g m^{-2}) species, and N content is lower in leaves of evergreens (1.79 mmol N g^{-1}) than in those of deciduous (2.38 mmol N g^{-1}) species (Sobrado, 1991).

Estimated construction costs and maintenance costs, calculated on a leaf area basis, are greater for evergreen (mean of four species: 200 g glucose m^{-2}, 2.8 g glucose m^{-2} d^{-1}) than for deciduous leaves (mean of six species: 87 g glucose m^{-2}, 1.3 g glucose m^{-2} d^{-1}: Sobrado, 1991), although there were no differences when calculated in terms of biomass. Janzen (1970) describes the leaves of the wet season deciduous species *Jacquinia pungens* as stiff and sclerophyllous, much like evergreen leaves. Differences in leaf lifespan may confound comparisons between evergreen and deciduous leaves because structural differences can be attributed not only to moisture availability, but also to other selection pressures such as nutrient status, herbivory and mechanical damage (Chabot & Hicks, 1982). In dry forests, the coexistence of species that support foliage exclusively in either the wet or the dry season permits comparisons between two groups of deciduous species possessing similar leaf life spans but differing leaf structure (see also Medina, Chapter 9).

Structural differences at the cellular level can strongly influence leaf physiological processes. Because both cell size and wall thickness directly influence the mechanical properties of cells in terms of providing the physical resistance that enables the generation of hydrostatic pressures (Steudle, Zimmermann & Lüttge, 1977; Oertli, Lips & Agami, 1990), differences in leaf specific weight can constrain leaf water relations. Foliage of dry forests shows a positive correlation between leaf specific

Table 10.2. *Leaf characteristics of tropical dry forest species*

Leaf habit and species	Specific mass (g m^{-2})	Nitrogen (mmol g^{-1})
Drought-deciduous		
Beureria cumanensis	86	2.05
Bulnesia arborea	91	1.41
Cochlospermum vitifolium	54	1.62
Genipa caruto	96	1.29
Godmania macrocarpa	123	1.14
Humboldtiella arborea	37	2.25
Lonchocarpus dipteroneurus	41	2.93
Luehea candida	43	1.31
Mansoa verrucifera (vine)	57	1.94
Pereskia guamacho	54	2.12
Pithecellobium carabobense	85	1.63
Pithecellobium dulce	66	1.85
Pithecellobium ligustrinum	65	2.49
Randia aculeata	33	1.42
Tabebuia billergiana	52	1.65
Mean	65	1.81
Evergreen		
Aristolochia taliscana (vine)	138	–
Byrsonima crassifolia	147	0.57
Capparis aristiguetae (shrub)	145	1.58
Capparis indica	98	–
Capparis linearis	485	0.90
Capparis odoratissima	276	0.82
Capparis pachaca	224	0.96
Capparis verrucosa (shrub)	112	2.25
Curatella americana	135	0.66
Jacquinia revoluta	249	0.69
Memora sp. (vine)	128	1.46
Morisonia americana	147	1.87
Palicourea rigida	154	0.62
Roupala complicata	189	0.60
Vochysia venezuelana	115	0.46
Xylopia aromatica	110	0.76
Mean	177	1.01
Wet season deciduous		
Coccoloba liebmannii	80	–
Forchhammeria pallida	70	–
Jacquinia pungens	100	–
Mean	83	–

Sources: Cuenca (1976, in Medina, 1984); Montes & Medina (1977); Medina (1984); Sobrado (1991); J. Whitbeck *et al.* (personal communication).

weight and estimated turgor pressures at full saturation (Table 10.2; Table 10.4 below). Cell wall elasticity, a measure of a cell's ability to deform reversibly in response to water deficits, spans a broad range among dry forest species (Table 10.4), and different phenological groups appear to have differing capacities to modify their cell wall properties in response to changes in environmental water stress (Sobrado, 1986; Fanjul & Barradas, 1987). Leaf structure and leaf water relations are again interwoven in the support and orientation of the lamina. For example, sclerophyllous leaves are less dependent on turgor pressure for mechanical support than are mesophytic leaves, thus a mesophytic leaf will wilt as mesophyll turgor pressure drops but a sclerophyllous leaf will remain rigid and may overheat. It remains unclear whether there is a direct relationship between construction and maintenance costs of leaves and the mechanical properties of their cell walls.

Stems

Although the mechanisms linking forest stature to rainfall patterns are unexplored, we note that the canopy trees of dry forests are generally about half as tall as those of wet forests (Murphy & Lugo, 1986a). Among neotropical lowland forests, average canopy height is inversely correlated with the number of months receiving less than 200 mm precipitation (Figure 10.1). Differences in root:shoot ratio between deciduous and wet forest trees may thus be due, in part, to differences in aboveground stature. Like forest stature, annual tree diameter growth rates, basal area of trees, and leaf area index of dry forests are reported to be approximately half of wet forest values (Veillon, 1963; Lugo *et al.*, 1978; Murphy & Lugo, 1986a; Bullock, 1990; Martínez-Yrízar *et al.*, 1992).

Seasonal drought appears to exert a strong influence on the structure of vascular conduits. Trees of dry forest have denser wood than trees of continually moist forest (Table 10.3; de Paula & de Hamburgo Alves, 1980; Barajas-Morales, 1985, 1987). Wood of high specific gravity indicates a high cell wall to cell volume ratio, a characteristic of drought-stressed tissues (Carlquist, 1977). Trees of the deciduous forest at Chamela, México, have shorter and narrower vessel elements, greater vessel wall thickness, shorter fibers and rays, and greater abundance (number per cross-sectional area) of these elements than trees of the wet forest at Los Tuxtlas, México (Barajas-Morales, 1985). Because flow through cylindrical pipes increases with the radius to the fourth power, narrower vessels have substantially lower conductance than wide vessels

Table 10.3. *Wood and xylem characteristics of tropical forest species*

Habitat and life form	Stem specific gravity (g cm^{-3})	Vessel diameter (μm)	Vessel wall thickness (μm)	Pore density (mm^{-2})	Source
Wet forest trees					
Random sample	0.57	159	4.1	6.7	1, 2
Drought-deciduous forest trees					
Random sample	0.72	104	6.6	16.0	1, 3, 4
Evergreen					
Hardwood	0.65				3
Lightwood	0.57				5
Deciduous					
Hardwood	0.90				5
Stem succulent	0.67				5
Lightwood	0.40				5
Vines	0.49	155			3, 4
Caatinga					
Trees	0.75	47–127		10.1	6

Sources: 1. Barajas-Morales (1985); 2. Barajas-Morales (1987); 3. Schulze *et al.* (1988); 4. Gartner *et al.* (1990); 5. Borchert (1994b); 6. de Paula & de Hamburgo Alves (1980).

(e.g. Zimmermann, 1978, 1983; Tyree & Ewers, 1991). A greater number of vessels, however, may provide insurance against catastrophic xylem dysfunction in the case of gas embolism blocking water flow in some vessels. Thus some dry forest plants may compromise efficiency of conductance for safety in maintaining an intact water column (Zimmermann, 1978; Sperry, Tyree & Donnelly, 1988).

A recent study addresses the issue of balancing efficiency of water transport with maintenance of an intact water column and demonstrates the value of integrating phenology into any study in a seasonal environment. Examining co-occurring drought-deciduous and evergreen species in a Venezuelan dry forest, Sobrado (1993) observed striking differences in seasonal patterns of stem specific gravity and water content. While both measures remained constant throughout the year in evergreen species, for drought deciduous species water content decreased and specific gravity increased as the dry season progressed such that measurements made in the two seasons differed by as much as 50% of the maximum value. Sobrado suggests that drought-deciduous species experience increased cavitation in the transition from wet to dry seasons. This assertion is reinforced by her measurement of hydraulic conduc-

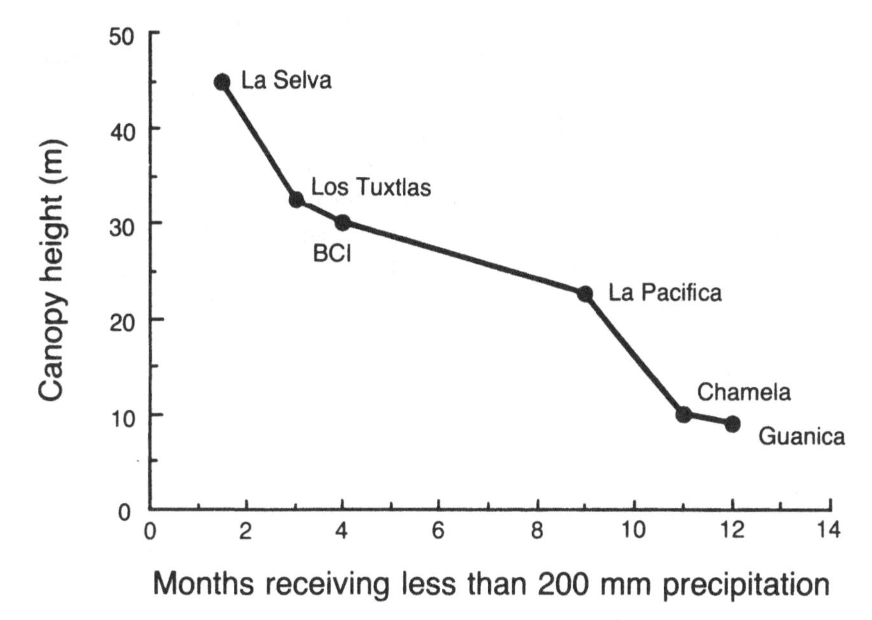

Figure 10.1. Relationship between canopy height and months with less than 200 mm precipitation. Precipitation data are from Frankie *et al.* (1974), Lugo *et al.* (1978), Carabias-Lillo & Guevara Sada (1985), Bullock (1986) and Windsor (1990). Canopy height data are from Holdridge *et al.* (1971), Foster & Brokaw (1982), Murphy & Lugo (1986b), Lott *et al.* (1987) and Bongers *et al.* (1988).

tivity: in the deciduous species hydraulic conductivity is much lower at the time of leaf abscission than during the wet season, suggesting that xylem embolism increases with progressive drought and diminishes the stem cross-sectional area available for water transport.

Differences in wood structure of dry forest trees occur not only on a temporal scale, but also among co-occurring species. Borchert (1994b) reports a complex interaction between leafing phenology and wood specific gravity among trees of a Costa Rican deciduous forest (Table 10.3). While deciduous hardwood species that rehydrate and break bud only after the first heavy rains have wood specific gravities exceeding 0.9 g cm^{-3}, deciduous trees with high stem water contents that typically break bud and/or flower before the heavy rains have wood specific gravities in the range of most evergreen trees sampled (means of 0.49 and 0.57 g cm^{-3} respectively: Table 10.3). Comparing co-occurring drought-deciduous and evergreen trees in a Venezuelan dry forest, Sobrado (1993) found that vessel length was longer in stems of the drought-deciduous species.

Physiological measurements can demonstrate constraints imposed by stem structure to transport through the xylem. Measuring terminal branches of co-occurring drought-deciduous and evergreen trees, Sobrado (1993) found that, although their transpiration rates were equivalent during the wet season, early morning leaf water potentials were slightly lower (more negative) and measures of both hydraulic conductivity and leaf-specific conductivity were greater for the deciduous species than for the evergreens. While the leaf area supported by terminal branches of the deciduous trees was greater than that of the evergreen trees their branches had similar cross-sectional areas; thus, the ratio of conducting tissue area to leaf area supported (the Huber value) was lower for deciduous species. Sobrado (1993) concludes that the deciduous species have more efficient water transport during the wet season than the evergreens. In a study comparing evergreen savanna trees with deciduous species growing in nearby forest patches, Goldstein and co-workers (1989) also found that the Huber values of deciduous trees were lower than those of evergreen trees. One of the evergreen species examined had a slightly lower conductance than any of the other species yet its hydraulic conductivity was similar to those of the deciduous species. In this case, its greater leaf specific conductivity may represent greater investment in vascular structure that could allow it to access and supply sufficient water to its leaves during periods of drought.

Stems constitute large parts of both the volume and the biomass of trees, and many species utilize parts of the stem for water storage (Daubenmire, 1972; Hinckley, Lassoie & Running, 1978; Waring & Running, 1978; Schulze *et al.*, 1988; Holbrook & Sinclair, 1992). Roots also comprise a substantial portion of tree biomass, but data are scarce on root water storage capacity or structural specialization enabling such capacity. The arborescent cactus *Opuntia excelsa*, common at Chamela, draws upon water stored within cladode parenchyma to maintain its photosynthetic tissue fully hydrated throughout the dry season (Lerdau *et al.*, 1992). Buffering of leaf water status also occurs in woody dicotyledons; stem succulent trees of a desert scrubland in Baja California use water stored in their trunks to moderate diurnal fluctuations in leaf turgor (Nilsen *et al.*, 1990). Stem water storage may also serve to buffer stem water potentials during periods of drought and thus possibly avoid cavitation (Tyree *et al.*, 1991). In northwestern Costa Rica, Reich and Borchert (1982, 1984) measured diminishing trunk girth throughout drought.

Many stem succulents are extremely conservative in their water use through the use of Crassulacean acid metabolism (CAM), such that they are able to rely on stored water during periods of low soil water availability. However, the transpirational demands of trees that support leaves are generally large compared to their stem water storage capacity (Nilsen *et al.*, 1990), and species that have high stem water contents are often strictly deciduous. Although stems may not provide viable reservoirs for actively photosynthesizing leaves throughout the dry season, water storage may allow meristem activity during periods of low soil moisture. Shoot activity (i.e. flowering and/or leaf flushing) prior to the onset of the rainy season has been reported in some dry forests species (see Phenology, below), and in some cases, the stem water contents of these trees are higher than those of arborescent cacti growing in the same forest (Schulze *et al.*, 1988; Bullock & Solís-Magallanes, 1990). Similarly, fine root proliferation into dry soil has been observed before the rainy season begins (J. Kummerow, personal communication). Structural features that permit water storage may enable the metabolism and turgor generation necessary for such new tissue production (Reich & Borchert, 1984). Finally, belowground investments appear to be limited in dry forest trees with high stem water contents (Olivares & Medina, 1992).

Although trees dominate dry forests, they mingle with a variety of other plant growth forms. By examining the structural differences and similarities among these growth forms we may improve our understanding of tree species' responses to drought. Epiphytic bromeliads and vines are locally abundant and diverse in dry forests (Lott, Bullock & Solís, 1987; Bullock, 1990; Gentry, Chapter 7). While the bromeliads employ a unique suite of drought-related characteristics (see Medina, Chapter 9), at first glance vine leaves appear structurally, physiologically and phenologically similar to those of adjacent trees. Like trees, vines possess a broad range of leaf morphologies, ranging from large mesic deciduous leaves that wilt under midday drought stress (Fichtner & Schulze, 1990) to small coriaceous evergreen leaves and leafless succulents (Castellanos *et al.*, 1989; J. L. Whitbeck *et al.*, personal communication).

Although vines have less water-conducting tissue per stem cross-sectional area than self-supporting plants of the same habitat (Putz, 1983; Gartner, 1991; but see Bullock, 1990), they typically have larger diameter vessels (Haberlandt, 1914; Gartner *et al.*, 1990) and thus may be able to supply water to the leaves at relatively high rates. Maximum stomatal conductance measured in the annual vine *Cyclanthera multi-*

foliolata at Chamela range from 700 to 1265 mmol m^{-2} s^{-1} (Fichtner & Schulze, 1990; J. L. Whitbeck *et al.*, personal communication). Despite generally slender stems, some vine species store water in stems or roots (Mooney *et al.*, 1992). Because vines need not allocate many resources to support functions, their root systems may face fewer mechanical constraints and be relatively free to locate and exploit soil resources (Putz, 1992). Effective and rapid mining of the soil for water, combined with rapid transport to leaves, distinguish vine species' means of survival from those of tree species in dry forests.

Physiology

The ability to maintain a positive net carbon balance during periods of reduced water availability depends upon interactions between leaf water relations and rates of leaf gas exchange. Resistance to water movement in the xylem results in a conflict between minimizing leaf water deficits and establishing an adequate driving gradient for the transport of water from soil to leaves (Jarvis, 1975). Foliar characteristics that influence the interaction between leaf water content and leaf water potential include the concentration of the cell sap (osmotic potential) and the ability of the cell walls to deform physically, and thus maintain turgor, as the cell volume changes (cell wall elasticity). In this section we focus on water loss rates and leaf water relations of four groups of dry forest trees that differ in phenological and/or structural attributes. Because of the limited data available we discuss drought-deciduous, evergreen, wet season deciduous, and stem succulent trees as discrete categories, but we recognize that dry forest trees exhibit a greater diversity in their behavior, structure and physiology. It is also important to note that studies of physiological responses to drought have focused exclusively on the transition from rainy to dry season. To date, no study has considered the physiological responses of dry forest trees to periods of low precipitation that occur frequently within the rainy season.

Drought-deciduous and evergreen trees

During the rainy season water use patterns of drought-deciduous and evergreen trees of dry forests appear comparable in terms of rates of gas exchange. For example, daily courses of leaf water potential and stomatal conductance were similar between evergreen and deciduous species growing in close association in a predominantly deciduous forest in

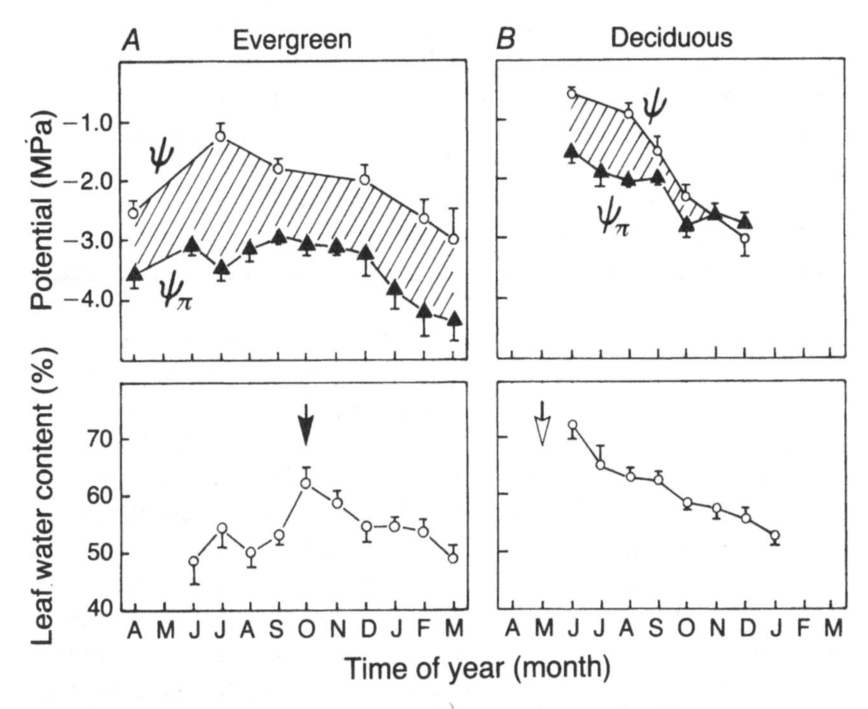

Figure 10.2. Predawn leaf water potential (○), osmotic potential (▲), turgor pressure (shaded area), and percentage leaf water content per unit dry weight as a function in time in evergreen (*A*) and deciduous (*B*) species. Each point is the overall mean of 12 observations (four per species). The evergreen species are *Morisonia americana*, *Capparis verrucosa*, and *C. aristiguetae* (all Capparidaceae); deciduous species include *Lonchocarpus dipteroneurus*, *Humboldtiella arborea* (both Leguminosae), and *Mansoa verrucifera* (Bignoniaceae). The two *Capparis* species are shrubs, while *M. verrucifera* is a vine. All other species are trees. Bars represent 2 standard errors. Arrows mark the production of new leaves. Modified from Sobrado (1986).

Venezuela (Sobrado & Cuenca, 1979). Pre-dawn leaf water potentials of the deciduous trees, however, were higher (less negative) than in the evergreen species (Figure 10.2; Sobrado, 1986). The ranges of other parameters overlap between deciduous and evergreen species: e.g. A_{max} (μmol m^{-2} s^{-1}) respectively, 6.9–9.9 and 6.9–7.9; stomatal conductance (mmol m^{-2} s^{-1}) 140.7–274.8 and 153.2–242.3; and transpiration (mmol m^{-2} s^{-1}) 5.2–7.7 and 5.7–8.0 (Sobrado, 1991). Instantaneous water use-efficiency (μmol CO_2 per mmol H_2O), however, was higher in the deciduous (1.29–1.49) than in the evergreen trees (0.99–1.21). This does not necessarily reflect the water use-efficiency averaged over the course of a day because stomatal conductance and transpiration were measured at A_{max}. Similarities between evergreen and drought-deciduous trees species in stomatal conductance and assimilation rates during the rainy season have been reported in several other forests (Medina & Cuevas, 1990; Olivares & Medina, 1992). These measurements also fall within the range of values observed for forest trees of wetter tropical regions (Fetcher, 1979; Robichaux, Hollsinger & Morse, 1986).

Many drought-deciduous species retain their leaves for a substantial period after the rains cease, although comparisons with evergreen species at this time may be confounded by differences in leaf life span and changes associated with leaf senescence. With the onset of the dry season, decreasing soil water availability and higher vapor pressure deficits are reflected in both the stomatal conductance and leaf water potentials of drought-deciduous trees. For example, despite substantial stomatal closure, minimum leaf water potential of the deciduous tree *Lonchocarpus dipteroneurus* (Leguminosae) declined from approximately −1.5 MPa during the rainy season to <-5.0 MPa before the leaves were shed (Sobrado & Cuenca, 1979). Sobrado (1986) reports a decline of approximately 1.5 MPa in pre-dawn leaf water potential during the first three months of the dry season for two drought-deciduous trees in central Venezuela (Figure 10.2). There is a simultaneous decrease in the osmotic potential (−2.0 to −2.7 MPa), although this is not sufficient to maintain turgor and pre-dawn leaf turgor potentials of zero, and leaf water content declines shortly before the leaves abscise (Figure 10.2).

Substantial declines in the ability of terminal branches and twigs to conduct water may also occur prior to leaf fall (Sobrado, 1993). These findings contrast markedly with evergreen species growing in the same area of central Venezuela (Figure 10.2). Some evergreen species show no seasonal decline in leaf water potential; in other species declines are mirrored by changes in osmotic potential such that turgor pressure during the dry season equals or even exceeds wet season values (Sobrado, 1986). Leaf water content (Figure 10.2), stomatal conductance and branch hydraulic conductivity of the evergreen species also remain stable throughout the year (Sobrado & Cuenca, 1979; Sobrado, 1993).

Wet season deciduous trees

Studies of two wet season deciduous species growing in a dry forest in México demonstrate increasingly conservative patterns of water use as the dry season progresses (Fanjul & Barradas, 1985, 1987). Towards the beginning of the dry season stomata remain open throughout the day (maximum stomatal conductance of 80–160 mmol m^{-2} s^{-1}) with midday leaf water potentials between −1.0 and −2.7 MPa. Later in the dry season these same trees exhibit pronounced midday stomatal closure and minimum leaf water potentials of −3.0 to −3.8 MPa. Stomatal sensitivity to water stress increases as the dry season progresses as indicated by negative correlations of stomatal conductance (g_{st}) and leaf-to-air vapor pressure difference (vpd), and of g_{st} with leaf water potential at the end of the dry season, in contrast to either no relationship (vpd) or a positive one (leaf water potential) at the beginning of the dry season (Fanjul & Barradas, 1985). Mooney, Bullock & Ehleringer (1989) also report increases in the $\delta^{13}C$ ratio of wet season deciduous species during the dry season, consistent with increasing stomatal control.

Stem succulent trees

The water relations of leaf-bearing, stem succulent trees have received little attention. We have already mentioned that dry forest trees with high stem water contents are typically not only leafless during the dry season, but tend to lose their leaves relatively soon following the cessation of the rains (Olivares & Medina, 1992). Sensitivity to drought is indicated by a strong stomatal closure in response to increases in vpd (Olivares & Medina, 1992), which probably accounts for the low midday values of stomatal conductance (*c.* 40 mmol m^{-2} s^{-1}) that occur even during the rainy season (Olivares & Medina, 1992). The combination of strong stomatal control and the potential contribution of water stored within the stem appears to buffer leaf water status. Stem succulents have relatively high leaf water potentials (typically > −1 MPa), even during the beginning of the dry season (Nilsen *et al.*, 1990; Olivares & Medina, 1992).

Mechanisms of turgor maintenance

In comparing the leaf water relations of drought-deciduous, evergreen, wet season deciduous, and stem succulent trees of dry forests we note the following patterns (Figure 10.3, Table 10.4). Drought-deciduous tree leaves have relatively high osmotic potentials and elastic cell walls during the rainy season (Table 10.4; Sobrado, 1986; Fanjul & Barradas, 1987). After the rains cease, there is a decrease in leaf osmotic potential (although not sufficient to maintain turgor) but no change in cell wall properties, such that greater water deficits occur at the turgor loss point

Table 10.4. *Pressure–volume characteristics of the leaves of dry forest trees. Column headings refer to measurements made during the wet and dry season for drought-deciduous and evergreen species, and at the beginning and the end of the dry season for wet season deciduous plants. TLP refers to the turgor loss point*[a]

	Osmotic potential								
	at full saturation (MPa)		at TLP (MPa)		Maximum elastic modulus (MPa)		Relative water content at TLP		
Leaf habit and species	Wet	Dry	Wet	Dry	Wet	Dry	Wet	Dry	Source
Drought-deciduous									
Humboldtiella arborea	−2.0		−2.5		7.3		78–82		1
Lonchocarpus dipteroneurus	−1.8		−2.3		9.5		78–82		1
Trichilia trifolia	−1.6	−2.8	−2.0	−4.9	7.0	11.0	80	75	2
Thouinia paucidentata	−2.6	−2.8	−2.8	−4.1	12.0	7.0	78	77	2
Astronium graveolens	−1.1		−2.1						3
Erythroxylum densum	−2.1		−2.9						3
Erythroxylum orinocense	−1.1		−1.6						3
Tabebuia chrysantha	−1.0		−1.5						3
Mean	−1.7	−2.8	−2.2	−4.5	9.5	9.0	79	76	

Evergreen									
Morisonia americana	−3.3		−4.1		13.9		78–86		1
Capparis verrucosa	−3.0		−3.6		14.4		78–86		1
Capparis aristiguetae	−2.8		−3.3		15.9		78–86		1
Capparis flexuosa	−2.3		−3.8						3
Couepia polyandra	−2.0	−2.0	−2.9	−2.8	12.0	22(12)	85	88	2
Randia armata	−1.7	−2.3	−2.0	−2.6	8.0	23(18)			2
Thouinidium decandrum	−2.3	−3.3	−3.5	−3.8		38(15)	79	75	2
Eugenia casearioides	−2.9		−4.0						3
Mean	−2.5	−2.5	−3.4	−3.1	12.8	28(15)	76	81	
Stem succulent									
Bursera simaruba	−0.5		−0.7						3
Wet season deciduous	Start	End	Start	End	Start	End	Start	End	
Jacquinia pungens	−2.7	−2.9	−4.6	−3.3	50	47	88	93	2
Coccoloba liebmannii	−2.0	−2.6	−3.4	−3.0	25	33	88	91	2
Mean	−2.3	−2.7	−4.0	−3.1	37.5	40	88	92	

[a] Values for dry season elastic moduli of evergreen species in parentheses indicate measurements made at the very end of the dry season, just before leaf turnover. Note that the elastic modulus values are calculated near saturation and thus not necessarily at the same absolute turgor pressure (see text). A larger elastic modulus indicates a more inelastic cell wall.

Sources: 1. Sobrado (1986); 2. Fanjul & Barradas (1987); 3. Olivares & Medina (1992).

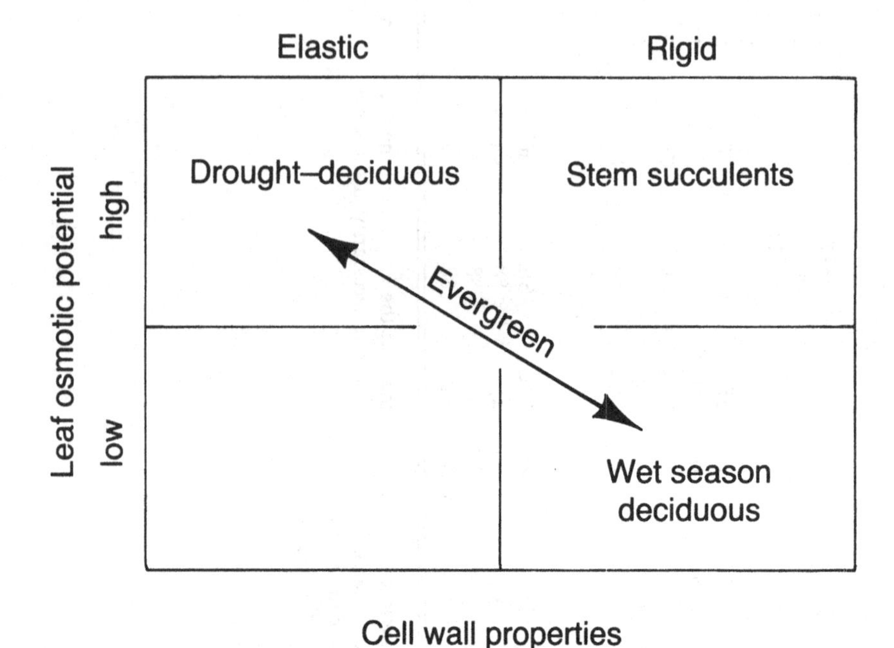

Figure 10.3. Patterns of leaf osmotic potential and cell wall elasticity in dry forest trees.

(Figure 10.2; Table 10.4). In contrast, the two wet season deciduous species studied have low osmotic potentials and extremely inelastic cell walls and demonstrate further decreases in the elasticity of their cell walls as the dry season progresses (Table 10.4; Fanjul & Barradas, 1987). Evergreen species switch between these two strategies, having higher osmotic potentials and more elastic cell walls during the rainy season and lower osmotic potentials and more rigid cell walls during the dry period (Sobrado, 1986; Fanjul & Barradas, 1987). During the rainy season measures of cell wall elasticity of evergreen species are comparable to those in drought-deciduous trees, but the osmotic potentials are lower in the evergreen trees (Table 10.4; Sobrado, 1986; Fanjul & Barradas, 1987). The leaves of stem succulent trees have high osmotic potentials (Olivares & Medina, 1992) and relatively rigid cell walls (Medina & Cuevas, 1990; Nilsen *et al.*, 1990; Olivares & Medina, 1992). They consequently lose turgor at quite small leaf water deficits and have low maximum turgor pressures (Nilsen *et al.*, 1990; Olivares & Medina, 1992).

Although reductions in leaf water potential are typically associated with greater leaf water deficits, low leaf water potentials may permit water movement from deeper roots and drier soils (Turner & Jones, 1980; Kramer, 1983). Concomitant lowering of leaf osmotic potential allows the maintenance of positive turgor pressures. Tropical deciduous forest trees that maintain leaves during the dry season have lower osmotic potentials than those whose leaves are active primarily in the rainy season (Table 10.4, see also Medina, Chapter 9).

Patterns of leaf cell wall elasticity are more difficult to summarize and interpret than leaf osmotic potential (Schulte, 1992). Although increased cell wall elasticity confers the ability to maintain positive turgor pressures to greater leaf water deficits (Robichaux *et al.*, 1986), its role as a drought response is complicated by the influence of cell wall properties on the relationship between water potential and water content. For example, if low leaf water potentials are needed for the substantial movement of water from the soil, then the existence of highly elastic cell walls could lead to excessive leaf water deficits before an adequate soil–leaf gradient was established (note the blank cell in Figure 10.3). Among drought-deciduous and evergreen species, cell wall elasticities are higher during the rainy season than during periods of low soil water availability. Among evergreen species, seasonal changes in cell wall properties and osmotic potentials are such that the relative water content at the turgor loss point remains constant throughout the year. Whether this represents a coordinated response to water availability, however, is not known. The association between stiff cell walls and low osmotic potentials may be purely mechanical insofar as inelastic cell walls may be required to contain the turgor pressures generated by large accumulations of solutes. Alternatively, cell wall properties may be significantly influenced by factors only indirectly associated with leaf water relations (e.g. leaf age: Tyree & Jarvis, 1982; Olivares & Medina, 1992). The occurrence of inelastic cell walls in combination with high osmotic potentials in stem succulent trees awaits further study. Nevertheless, the existence of a high relative water content at the turgor loss point is consistent with conservative water use by plants that rely on water storage (Nilsen *et al.*, 1990; Olivares & Medina, 1992).

Phenology

Differences in the timing of leaf fall, flowering and shoot expansion among deciduous trees are often associated with soil moisture, interannual differences in rainfall patterns, and plant size (Daubenmire, 1972; Frankie, Baker & Opler, 1974; Opler, Frankie & Baker, 1980; Reich &

Borchert, 1982, 1984; Bullock & Solís-Magallanes, 1990; Medina & Cuevas, 1990; Whigham *et al.*, 1990). Such observations strongly support the idea that plant water status plays a dominant role in structuring temporal patterns of the expansion and abscission of leaves and other organs. Nevertheless, in some species bud break occurs during the dry season, leading to the suggestion that factors such as photoperiod and temperature may act as proximal cues (Daubenmire, 1972; Frankie *et al.*, 1974). Despite the primary importance of water stress on cell expansion and growth, in terms of the phenology of dry forest trees its 'mechanism of action' has been described as 'obscure' (Murphy & Lugo, 1986a). Recent studies of the water relations in relation to shoot activity (Borchert, 1994a, b, c) provide a more integrated view of how seasonality in rainfall is related to the variety of phenological behaviors observed among dry forest trees. In this section we examine the relationship between seasonal drought and the phenology of dry forest trees and focus on how plant structure and soil water availability influence patterns of meristem hydration and activity.

Plant water status and phenology

Plant water status is not easily characterized by a single measurement. The movement of water from soil to leaves requires that a gradient in potential exist along this path. Thus, the water status of downstream (distal) organs, such as leaves, may not be the best indicator of the internal conditions most directly influencing bud activity and growth (Borchert, 1994a). Prior to leaf fall, leaf water status may be further uncoupled from that of supporting branches and attached buds. Some earlier studies of shoot activity in dry forest trees emphasized stem diameter fluctuations as an index of plant water status (Daubenmire, 1972; Reich & Borchert, 1984). The dependence of this relationship on stem structure, however, makes these measurements difficult to interpret (Borchert, 1994c).

Leaf abscission

Large increases in evaporative demand and the severe drying of surface soils that characterize extended periods of low rainfall lead the majority of dry forest species to lose their leaves. While the ultimate driving forces for this temporary reduction in surface area include an inability to supply water to the leaves at a rate that permits a net positive carbon balance or to prevent the desiccation and potential damage of more permanent structures, the interactions between water stress and senescence that produce the observed abscission patterns are poorly understood. Virtually all studies of the phenology of dry forest trees report substantial intra- and interspecific variability in the timing of leaf abscission (e.g. Daubenmire, 1972; Frankie *et al.*, 1974; Opler *et al.*, 1980; Reich & Borchert, 1982, 1984; Bullock & Solís-Magallanes, 1990; Borchert, 1994a, b). However, at the spatial scale over which such studies are typically conducted, rainfall – and thus evaporative demand and drying of surface soil layers – remains relatively constant. Differences in the timing of leaf fall therefore reflect the degree to which individual trees are able to access and transport water stored in deeper soil layers and the sensitivity of different species to water stress (Reich & Borchert, 1982, 1984; Bullock & Solís-Magallanes, 1990; Borchert, 1994b).

Studies of the phenology of dry forest trees in Costa Rica report a correspondence between minimum leaf water potential, wood density, species distribution in relation to access to the water table, and patterns of leaf abscission (Table 10.5; Reich & Borchert, 1984; Borchert, 1994a). In areas where the water table is absent or far from the surface, deciduous species that maintain their leaves well into the dry season have extremely low minimum leaf water potentials and dense wood (Table 10.5). Co-occurring species that lose their leaves early in the dry season do so at higher minimum leaf water potentials and often have extremely low density wood due to an abundance of non-lignified, parenchymatous tissues (Borchert, 1994a, b). Leaf structure, on the other hand, is less clearly associated with patterns of leaf fall (Table 10.5). As wood density is inversely related to maximum stem water content (Schulze *et al.*, 1988; Borchert, 1994b), the association between wood density and patterns of leaf abscission may reflect an integration of structural and physiological parameters at the whole plant level.

Substantial variation in the timing, duration and extent of leaf loss exists among dry forests (Daubenmire, 1972; Frankie *et al.*, 1974; Reich & Borchert, 1984; Bullock & Solís-Magallanes, 1990; Borchert, 1994b). Although dry forest regions are defined by total rainfall and length of the dry season, a more important factor in terms of understanding differences in phenology and species composition between forests may be access to ground water (Bullock & Solís-Magallanes, 1990; Borchert, 1994b). Dry forest species differ in their ability to respond to soil moisture in terms of leaf fall (Bullock & Solís-Magallanes, 1990;

Table 10.5. *Relationship between vegetative phenology, habitat, leaf type, wood characteristics, and minimum leaf and branch water potential of dry forest trees in northwestern Costa Rica*[a]

Leaf structure	Wood specific gravity (g cm^{-3})	Maximum stem moisture (% dry mass)	Water potential	
			Leaf (MPa)	Branch (MPa)
Habitat and phenology				
Upland forest: Deciduous; shed leaves mid dry season, bud break and shoot expansion (including flowering) after rains begin; 6 species				
Mesophyll,	1.0	26	—[b]	—[b]
microphyll,	0.91 to 1.2	19 to 31	−3.2 to <−4.0	−3.0 to <−4.0
coriaceous				
Upland forest: Deciduous; shed leaves early in dry season; often flowering while deciduous; bud break may occur at end of dry season; leaf and shoot expansion after rains begin; 3 species				
Mesophyll,	0.45	139	−1.5	−0.8
microphyll	0.39 to 0.49	121 to 171	−0.8 to −2.3	−0.5 to −1.1
Open woodland, scattered trees: Deciduous; shed leaves early to mid dry season; leaf expansion and flowering may occur before rains begin; 7 species				
Mesophyll,	0.70	68	−2.1	−1.8[c]
coriaceous,	0.53 to 0.83	33 to 114	−1.4 to −3.7	−0.9 to <−4.0
sclerophyll				
Moist, lowland sites: Evergreen, leaf changing or briefly deciduous; 7 species (including 3 mesophyll)				
Coriaceous,	0.69	73	−2.9	−1.5
sclerophyll	0.59 to 0.83	41 to 100	−2.2 to −4.0	−0.5 to −2.4
Mesophyll	0.41	168	−1.3	−0.8
	0.32 to 0.49	141 to 196	−1.1 to −1.6	−0.5 to −1.1

[a] Data (means and range) are from Borchert (1994b) but include only species for which all measurements were recorded. Wood specific gravity and maximum (saturated) water content were determined from cores or obtained from Barajas-Morales (1987). Leaf and branch water potentials were measured using a pressure chamber; data presented are the minimum values obtained just prior to leaf fall. Values preceded by < indicate that they exceeded the pressure rating of the chamber.
[b] Mean not calculated due to values exceeding rating of the pressure chamber.
[c] Mean based on six species.

Borchert, 1994b). When species characteristic of upland forests occur in moister soils, those with dense wood experience moderate water stress and leaf fall occurs later in the dry season (Borchert, 1994b). In contrast, species with low density wood lose their leaves as early in the dry season as on upland sites, and with similar leaf water potential, suggesting that they are shallowly rooted or sensitive to increased evaporative demand, or both (Borchert, 1994b).

Details of how plant water status may influence leaf senescence and abscission have received little attention. Periods of low or null precipitation during the rainy season typically do not lead to the same patterns of senescence and abscission as occur at the beginning of the dry season. Are these dry spells insufficient in severity to trigger abscission or are the different responses due to age-related changes in how leaves respond to reduced water availability and increased evaporative demand? Reich and Borchert (1988) report decreased stomatal control in older leaves of dry forest trees. In Ghana, with two periods of drought each year, individual trees lose their leaves only once a year although the forest as a whole has concentrations of leaf fall at the beginning of each dry period (Lieberman, 1982).

Intermittent periods of shoot activity and apparent quiescence are common in trees in aseasonal climates (Borchert, 1973, 1978; Alvim & Alvim, 1978). Comparison of the same species between wet and dry forest areas in Costa Rica showed similar patterning of shoot activity in nine of 11 species; only the two secondary forest species exhibited seasonal patterns of leaf production in the dry forest and continual leaf production in wet areas (Frankie *et al.*, 1974). If the inherent pattern of shoot development arises from the periodic replacement of cohorts of leaves with a specific life span, then any environmental stress which accelerates leaf aging or delays shoot expansion will tend to synchronize the phenology of the community as a whole (Borchert, 1992). Correspondence between seasonal changes in the environment and patterns of plant activity may reflect the entrainment of endogenous rhythms rather than the action of external factors triggering specific developmental steps (Borchert, 1980, 1991, 1992; Reich & Borchert, 1984).

Hydration, bud break and leaf renewal

Dry forest trees exhibit a variety of leaf expansion patterns. In some species bud break and shoot expansion occur only after the first substantial rains of the wet season; others renew their entire canopy during the dry season (Daubenmire, 1972; Frankie *et al.*, 1974; Bullock & Solís-Magallanes, 1990; Borchert, 1994a, b). Some even break bud towards the end of the dry season, but delay leaf expansion until the rains begin (Frankie *et al.*, 1974; Bullock & Solís-Magallanes, 1990; Borchert, 1994a, b). Because of the sensitivity of cell expansion to limitations in water supply, dry season shoot expansion has been considered a paradox (Borchert, 1994a). Earlier studies reported that leaf flushing resulted in marked stem shrinkage regardless of whether shoot expansion occurred before or after the onset of the rains (Daubenmire, 1972; Reich & Borchert, 1982, 1984). This suggests that leaf expansion places particularly strong demands on plant water balance. Thus, Daubenmire (1972) notes that because some drought-deciduous species flower and/or break bud during the height of the dry season, deficient soil moisture could not be the 'force necessitating leaf cast earlier in the dry season'. However, shoot expansion appears to require both the release of xylem tension by leaf shedding and full hydration of the terminal branches (Borchert, 1994a, b). The timing of shoot rehydration varies both between species and microsites reflecting differences in stem water storage and the ability to withdraw water from deep in the soil (Borchert, 1994a, b).

Many deciduous forest tree species without access to the water table must wait for the rains to begin before bud break can occur (Bullock & Solís-Magallanes, 1990; Borchert, 1994b, c). Following the first substantial rainfall or irrigation, rehydration (as indicated by branch segment balancing pressures close to zero) occurs within days, followed shortly by bud break and shoot expansion (Borchert, 1994a, b). This indicates that these trees are not innately dormant, but rather in a state of 'drought-imposed rest' (Borchert, 1991, 1994c). In species with substantial stem water contents, leaf abscission is followed by rapid rehydration of the terminal branches and often flowering (Bullock & Solís-Magallanes, 1990; Borchert, 1994a, b). Both branch rehydration and flowering appear to draw on stem water stores as there is a continual decline in stem moisture content throughout the dry season (Reich & Borchert, 1984). Many of these species break bud after flowering, but shoot development and leaf expansion do not occur until after the soil is recharged (Bullock & Solís-Magallanes, 1990; Borchert, 1994b). As bud break occurs without a change or improvement in plant water status, the involvement of a cue, such as photoperiod, has been postulated (Frankie *et al.*, 1974; Bullock & Solís-Magallanes, 1990).

Irrigation of leafless trees during the dry season leads to rapid rehydration of first xylem vessels and then parenchymatous tissues within the stems (Borchert, 1994c). Estimates of rates of water transport in leafless dense-wooded trees following irrigation fall within the range of 0.7–1.0 m h^{-1}, similar to values for phloem movement (Borchert, 1994a, b). Marked seasonal declines in the hydraulic conductivity of drought-deciduous species (Sobrado, 1993) and observations of gas-filled vessels of deciduous vines during the dry season (Mooney *et al.*, 1992; Borchert, 1994c) suggest that rehydration may involve an actual refilling of embolized xylem vessels. Embolism repair is thought to require positive pressure (Sperry *et al.*, 1987) and it is hypothesized that root pressure may play an important role in the ability of drought-deciduous trees to respond rapidly to changes in soil water availability (Braun, 1984; Borchert, 1994c). 'Bleeding' of sap from irrigated trees was not observed (Borchert, 1994c). This does not rule out the existence of root-generated pressures, as the presence of desiccated tissues surrounding the xylem may act as a potent sink for water rising in the xylem (Borchert, 1994c). In stem succulent species, rehydration of terminal branches occurs during the dry season and appears to involve the redistribution of water stored within the main stem. The driving force for this movement is not known although changes in the solute concentration of parenchymatous tissue adjacent to the xylem may provide the necessary potential gradient to move water up into the distal portions of the shoot (Borchert, 1994c). Further studies of the osmotic activities of living cells along the xylem pathway are needed.

Phenology of belowground meristems

Seasonality in the growth of dry forest trees may be as pronounced in below ground meristems as in those above ground. At Chamela, leaf expansion was preceded by the initiation of fine roots, which were observed just three days after the first rains of the wet season (Kummerow *et al.*, 1990). At the end of the dry season, fine root biomass was 90 g m^{-2}, rising to almost 200 g m^{-2} after the first rains, and reaching 284 g m^{-2} towards the end of the wet season, declining thereafter throughout the dry season (Kummerow *et al.*, 1990). Similar patterns of fine root growth have been observed for a dry forest on a sand dune substrate (Kavanagh & Kellman, 1992; also see Martínez-Yrízar, Chapter 13; Cuevas, Chapter 15). Irrigation during the height of the dry season also resulted in new fine root growth within three days, followed within a week by bud break (J. Kummerow, personal communication). Kummerow *et al.* (1990) further report that this seasonality is not apparent in the biomass of small roots (1.1–5.0 mm diameter), although, as live and dead roots of this size could not be distinguished, it is possible that such fluctuations were obscured. Fine root production may also parallel shoot activity without changes in soil water status. Some species that flower during the dry season initiate fine roots into the extremely dry surface soils concomitant with flowering (J. Kummerow, personal communication). Whether these roots change the water status of the plant is not known. As with early bud break, such root growth might be expected in trees with low wood density and high stem water contents. Elucidation of the patterns and mechanisms of belowground response and their integration with aboveground phenology will greatly enhance our understanding of whole plant phenology in dry forest trees.

Flowering

Dry forest trees exhibit a wide variety of flowering patterns in relation to both rainfall and vegetative phenology (e.g. Daubenmire, 1972; Frankie *et al.*, 1974; Borchert, 1983; Bullock & Solís-Magallanes, 1990). The physiological and structural attributes that support these patterns have not been well studied and there is no information on the water relations or costs of floral displays. Most studies focus on anthesis and the induction of flower buds has received little attention (Borchert, 1992). In many species, flowering and shoot expansion occur simultaneously such that the same physiological and environmental conditions promote both growth and reproduction (Borchert, 1992). Other species flower during the rainy season, often producing terminal inflorescences that mark the end of the year's shoot growth (Borchert, 1992). Another common pattern is one in which floral buds are induced during shoot growth, but remain dormant until after the leaves are shed (Daubenmire, 1972; Frankie *et al.*, 1974; Borchert, 1992). Delays in anthesis could be due to an inhibitory influence of the leaves (Borchert, 1992) or a requirement for a period of water stress followed by hydration (Alvim & Alvim, 1978; Crisosto, Grantz & Meinzer, 1992).

Full hydration of terminal branches appears to be a prerequisite of dry season anthesis (Borchert, 1994a, b). In western México, the maximum stem water content was higher in five species that flower in the mid to late dry season than in 19 other tree species (Schulze *et al.*, 1988; Bullock & Solís-Magallanes, 1990). Declines in stem moisture content and/or stem circumference during flowering indicate a net loss of water from the stem (Borchert, 1994a). Because such species tend to lose their leaves early in the dry season (Bullock & Solís-Magallanes, 1990; Olivares & Medina, 1992) they retain enough water within their stems to support flowering during the dry season. Species without substantial stem water stores must utilize soil moisture to hydrate their apical meristems, and to

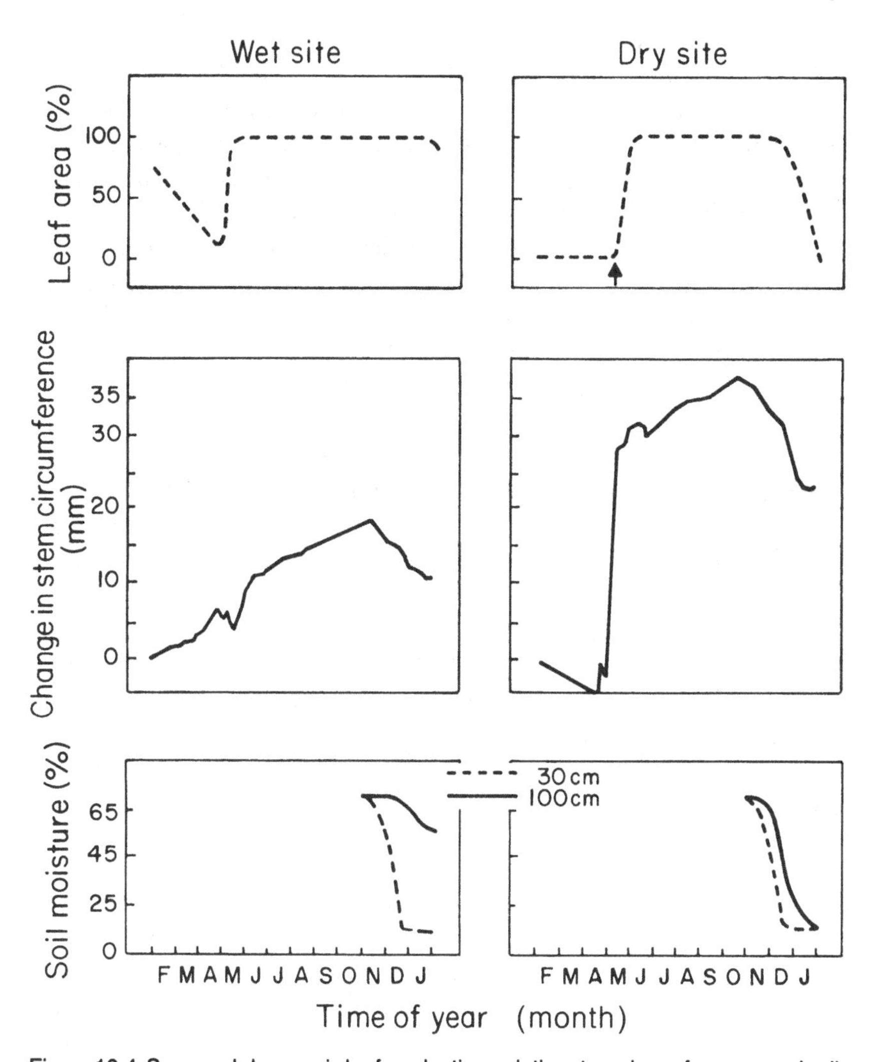

Figure 10.4. Seasonal changes in leaf production, relative stem circumference, and soil moisture of *Tabebuia ochracea neochrysantha* growing in a wet and a dry site. The vertical arrow marks time of flowering. Modified from Reich & Borchert (1982). Increases in stem circumference in this species may reflect changes in hydration of the relatively thick bark, in addition to the net storage of water within the stem (Borchert, 1994c).

support both flower expansion and water loss from floral surfaces (Daubenmire, 1972; Bullock & Solís-Magallanes, 1990; Borchert, 1994a, b). Such species may flower at the beginning of the rainy season, prior to vegetative bud break and leaf expansion, or utilize sporadic dry season rainfalls (Figure 10.4).

A number of evolutionary and ecological pressures that might favor shoot activity during periods of low precipitation and high evaporative demand (e.g. reduced herbivore pressures, increased pollination and/or dispersal services) have been proposed (e.g. Janzen, 1967; Opler, Frankie & Baker, 1976; Augspurger, 1980; Lieberman & Lieberman, 1984). Given our limited knowledge of the water relations of flower development in dry forest trees and the accumulating evidence that a favorable plant water balance is required for anthesis, it is difficult to assess the potential contribution of non-moisture status related factors (Bullock & Solís-Magallanes, 1990; Borchert, 1992).

There remain many unanswered questions regarding how interactions between structural, physiological and environmental factors result in the phenological patterns observed. In dry forests soil moisture appears to determine the timing and rates of leaf fall in species without substantial stem water stores; in wetter regions changes in evaporative demand may be a more important factor. A long-term watering experiment in moist forest in Panamá did not substantially influence patterns of leaf fall (Wright & Cornejo, 1990), whereas irrigation of dry forest trees may lead to marked changes in leafing patterns (Reich & Borchert, 1984; Borchert, 1994c). The present discussion points to the need for more attention directed to rooting patterns and belowground phenology, flowering, hydraulic architecture, leaf energy balance, formation of abscission layers, and the influence of leaf age.

Summary

Many neotropical dry forests are characterized by a predominance of trees that lose their leaves for a substantial portion of the year. Low soil moisture and high evaporative demand during the dry season mean that evergreen species in tropical deciduous forests must make substantial structural investments to supply water at high rates to their leaves. Evergreen species have deeper roots, higher leaf mass per area, and greater investment in xylem transport capacity than co-occurring deciduous species.

Among lowland forests, average canopy height is inversely correlated with the number of months receiving less than 200 mm precipitation. Dry forest trees have denser wood than trees of continually moist forest with shorter and narrower vessel elements and greater vessel wall thickness.

In comparison with plants of moist and wet forests, dry forest species deploy a greater percentage of their root biomass deeper in the soil profile, and their root:shoot ratios are approximately twice as high. Seasonality in growth may be as pronounced below ground as above ground, with fine root production preceding leaf expansion and the initiation of fine roots being observed shortly after the first wet season rainfall.

Leaf water relations and gas exchange parameters of dry forest trees appear not to differ markedly from values reported from wetter regions. Species that maintain leaves during the dry season have lower leaf osmotic potentials during this time than either in the wet season or in drought-deciduous species.

The cumulative transpirational demand imposed by leaves during the dry season far exceeds the storage volume of even a large tree. Dry forest trees with high stem water contents have high leaf water potentials, high osmotic potentials, small leaf water deficits at the turgor loss point, strong stomatal response to vpd and strong sensitivity to drought in terms of leaf fall patterns, but they often flower during the dry season.

Water is the primary limiting resource and leaf abscission the most significant drought response of dry forest trees. Interactions between water availability and structural and physiological characteristics such as rooting depth, stem water storage, hydraulic architecture, and sensitivity to water stress lead to a wide variety of phenological behaviors.

Acknowledgements

We thank P. G. Baer, S. Bigelow, R. Borchert, S. H. Bullock, J. Cavelier, M. Dickenson, J. J. Ewel, K. Fichtner, C. B. Field, B. L. Gartner, G. G. Goldstein, M. L. Goulden, A. A. Held, D. U. Hooper, M. T. Lerdau, E. Medina, E. Nilsen and F. E. Putz for helpful comments, B. Lilley for drafting many of the figures, and R. Borchert for providing information on work in progress.

References

Alvim, P. de T. & Alvim, R. (1978). Relation of climate to growth periodicity in tropical trees. In *Tropical Trees as Living Systems*, ed. P. B. Tomlinson and M. H. Zimmermann, pp. 445–64. Cambridge University Press, Cambridge.

Augspurger, C. K. (1980). Mass flowering of a tropical shrub (*Hybanthus prunifolius*): influence on pollinator attraction and movement. *Evolution* **34**: 475–88.

Barajas-Morales, J. (1985). Wood structural differences between trees of two tropical forests in México. *IAWA Bulletin n.s.* **6**: 355–64.

Barajas-Morales, J. (1987). Wood specific gravity in species from two tropical forests in México. *IAWA Bulletin n.s.* **8**: 143–48.
Becker, P. & Castillo, A. (1990). Root architecture of shrubs and saplings in the understory of a tropical moist forest in lowland Panamá. *Biotropica* **22**: 242–9.
Bethlenfalvay, G. J., Brown, M. S., Ames, R. N. & Thomas, R. S. (1988). Effects of drought on host and endophyte development in mycorrhizal soybeans in relation to water use and phosphate uptake. *Physiologia Plantarum* **72**: 565–71.
Bongers, R., Popma, J., Meave del Castillo, J. & Carabias, J. (1988). Structure and floristic composition of the lowland rain forest at Los Tuxtlas, México. *Vegetatio* **74**: 55–80.
Borchert, R. (1973). Simulation of rhythmic tree growth under constant conditions. *Physiologia Plantarum* **29**: 173–80.
Borchert, R. (1978). Feedback control and age-related changes of shoot growth in seasonal and nonseasonal climates. In *Tropical Trees as Living Systems*, ed. P. B. Tomlinson and M. H. Zimmermann, pp. 497–515. Cambridge University Press, Cambridge.
Borchert, R. (1980). Phenology and ecophysiology of tropical trees: *Erythrina poeppigiana* O. F. Cook. *Ecology* **61**: 1065–74.
Borchert, R. (1983). Phenology and control of flowering in tropical trees. *Biotropica* **15**: 81–9.
Borchert, R. (1991). Growth periodicity and dormancy. In *Physiology of Trees*, ed. A. S. Raghavendra, pp. 221–45. John Wiley & Sons, New York.
Borchert, R. (1992). Computer simulation of tree growth periodicity and climatic hydroperiodicity in tropical forests. *Biotropica* **24**: 385–95.
Borchert, R. (1994a). Water status and development of tropical trees during seasonal drought. *Trees* **8**: 115–25.
Borchert, R. (1994b). Soil and stem water storage determine phenology and distribution of tropical dry forest trees. *Ecology* **75**: 1437–49.
Borchert, R. (1994c). Induction of rehydration and bud break by irrigation or rain in deciduous trees of a tropical dry forest in Costa Rica. *Trees* **8**: 198–204.
Braun, H. J. (1984). The significance of the accessory tissues of the hydrosystem for osmotic water shifting. *IAWA Bulletin n.s.* **5**: 275–94.
Brown, S. & Lugo, A. E. (1982). The storage and production of organic matter in tropical forests and their role in the global carbon cycle. *Biotropica* **14**: 161–87.
Bullock, S. H. (1986). Climate of Chamela, Jalisco, and trends in the south coastal region of México. *Archives for Meteorology, Geophysics and Bioclimatology, Series B* **36**: 297–316.
Bullock, S. H. (1990). Abundance and allometrics of vines and self-supporting plants in a tropical deciduous forest in México. *Biotropica* **22**: 22–35.
Bullock, S. H. & Solís-Magallanes, J. A. (1990). Phenology of canopy trees of a tropical deciduous forest in México. *Biotropica* **22**: 22–35.
Carabias-Lillo, J. & Guevara Sada, S. (1985). Fenología en una selva tropical húmeda y en una comunidad derivada; Los Tuxtlas, Veracruz. In *Investigaciones sobre la Regeneración de Selvas Altas en Veracruz, México, II*, ed. A. Gómez-Pompa and S. Del Amo R., pp. 27–66. Instituto Nacional de Investigación sobre Recursos Bióticos, Xalapa.
Carlquist, S. (1977). Ecological factors in wood evolution: A floristic approach. *American Journal of Botany* **64**: 887–96.
Castellanos, A. E., Mooney, H. A., Bullock, S. H., Jones, C. & Robichaux, R. (1989). Leaf, stem, and metamer characteristics of vines in a tropical deciduous forest in Jalisco, México. *Biotropica* **21**: 41–9.
Castellanos, J., Maass, M. & Kummerow, J. (1991). Root biomass of a dry deciduous tropical forest in México. *Plant and Soil* **131**: 225–8.
Cavelier, J. (1992). Fine-root biomass and soil properties in a semideciduous and a lower montane rain forest in Panamá. *Plant and Soil* **142**: 187–201.

Chabot, B. F. & Hicks, D. J. (1982). The ecology of leaf life spans. *Annual Review of Ecology and Systematics* **13**: 229–60.

Crisosto, C. H., Grantz, D. A. & Meinzer, F. C. (1992). Effects of water deficit on flower opening in coffee (*Coffea arabica* L.). *Tree Physiology* **10**: 127–39.

Cuenca, G. & Lovera, M. (1990). Recolonization by VA-mycorrhizal propagules of severely disturbed areas of the 'Canaima' National Park, in Venezuela. Abstracts of the 8th North American Conference on Mycorrhizae.

Cuevas, E. & Medina, E. (1988). Nutrient dynamics within Amazonian forests. II. Fine root growth, nutrient availability and leaf litter decomposition. *Oecologia* **76**: 222–235.

de Paula, J. E. & de Hamburgo Alves, J. L. (1980). Estudo das estruturas anatómicas e de algunas propriedades físicas da madeira de 14 especies ocurrentes em areas de caatinga. *Brasil Florestal* **10**: 47–58.

Daubenmire, R. (1972). Phenology and other characteristics of tropical semi-deciduous forest in northeastern Costa Rica. *Journal of Ecology* **60**: 147–70.

Fanjul, L. & Barradas, V. L. (1985). Stomatal behavior of two heliophile understory species of a tropical deciduous forest in México. *Journal of Applied Ecology* **22**: 943–54.

Fanjul, L. & Barradas, V. L. (1987). Diurnal and seasonal variation in the water relations of some deciduous and evergreen trees of a deciduous dry forest of the western coast of México. *Journal of Applied Ecology* **24**: 289–303.

Fetcher, N. (1979). Water relations of five tropical tree species on Barro Colorado Island, Panamá. *Oecologia* **40**: 229–33.

Fichtner, K. & Schulze, E.-D. (1990). Xylem water flow in tropical vines as measured by a steady state heating method. *Oecologia* **82**: 355–61.

Foster, R. & Brokaw, N. V. L. (1982). Structure and history of the vegetation of Barro Colorado Island. In *Ecology of a Tropical Forest*, ed. E. G. Leigh, Jr, A. S. Rand and D. Windsor, pp. 67–81. Smithsonian Institution Press, Washington.

Frankie, G. W., Baker, H. G. & Opler, P. A. (1974). Comparative phenological studies of trees in tropical wet and dry forests in the lowlands of Costa Rica. *Journal of Ecology* **62**: 881–919.

Gartner, B. L. (1991). Stem hydraulic properties of vines vs. shrubs of western poison oak, *Toxicodendron diversilobum*. *Oecologia* **87**: 180–9.

Gartner, B. L., Bullock, S. H., Mooney, H. A., Brown, V. & Whitbeck, J. L. (1990). Water transport properties of vine and tree stems in a tropical deciduous forest. *American Journal of Botany* **77**: 742–9.

Gianinazzi-Pearson, V. & Gianinazzi, S. (1983). The physiology of vesicular-arbuscular mycorrhizal roots. *Plant and Soil* **71**: 197–209.

Goldstein, G., Rada, F., Rundel, P., Azocar, A. & Orozco, A. (1989). Gas exchange and water relations of evergreen and deciduous savanna trees. *Annals Sci. For.* **46** (supplement): 448s–453s.

Haberlandt, G. (1914). *Physiological Plant Anatomy*, 4th edition, (Transl. M. Drummond. Reprinted 1965.) Today and Tomorrow's Book Agency, New Delhi.

Harley, S. E. & Smith, J. H. (1983). *Mycorrhizal Symbiosis*. Academic Press, London.

Hinckley, T. M., Lassoie, J. P. & Running, S. W. (1978). Temporal and spatial variations in the water status of forest trees. *Forest Science Monographs* No. 20.

Holdridge, L. R., Grenke, W. C., Hatheway, W. H., Liang, T. & Tosi, J. A., Jr (1971). *Forest Environments in Tropical Life Zones: a pilot study*. Pergamon Press, Oxford.

Holbrook, N. M. & Sinclair, T. R. (1992). Water balance in the arborescent palm, *Sabal palmetto*. Transpiration and stem water storage. *Plant, Cell and Environment* **15**: 401–9.

Janos, D. P. (1983). Tropical mycorrhizas, nutrient cycles and plant growth. In *Tropical Rain Forest: ecology and management*, ed. S.L. Sutton, T. C. Whitmore and A. C. Chadwick, pp. 327–45. Blackwell Scientific, Oxford.

Janzen, D. H. (1967). Synchronization of sexual reproduction of trees within the dry season in Central America. *Evolution* **21**: 620–37.

Janzen, D. H. (1970). *Jacquinia pungens*, an heliophile from the understory of tropical deciduous forests. *Biotropica* **2**: 112–19.

Jarvis, P. G. (1975). Water transfer in plants. In *Heat and Mass Transfer in the Biosphere. I. Transfer Processes in the Plant Environment*, ed. D. A. de Vries and N. H. Afgan, pp. 369–94. John Wiley & Sons, New York.

Jordan, C. F. & Escalante, G. (1980). Root productivity in an Amazonian rain forest. *Ecology* **61**: 14–18.

Kavanagh, T. & Kellman, M. (1992). Seasonal pattern of fine root proliferation in a tropical dry forest. *Biotropica* **24**: 157–65.

Klinge, H. (1973). Root mass estimation in lowland tropical rain forests of central Amazonia, Brazil. I. Fine root masses of a pale yellow latosol and a giant humus podzol. *Tropical Ecology* **14**: 29–38.

Kramer, P. J. (1983). *Water Relations of Plants*. Academic Press, Santa Clara.

Kummerow, J., Castellanos, J., Maass, M. & Larigauderie, A. (1990). Production of fine roots and the seasonality of their growth in a Mexican deciduous dry forest. *Vegetatio* **90**: 73–80.

Lerdau, M. T., Holbrook, N. M., Mooney, H. A., Rich, P. M. & Whitbeck, J. L. (1992). Seasonal patterns of acid fluctuations and resource storage in the arborescent cactus *Opuntia excelsa* in relation to light availability and size. *Oecologia* **92**: 166–71.

Levitt, J. (1972). *Responses of Plants to Environmental Stresses*. Academic Press, New York.

Lieberman, D. (1982). Seasonality and phenology in a dry tropical forest in Ghana. *Journal of Ecology* **70**: 791–806.

Lieberman, D. & Lieberman, M. (1984). The causes and consequences of synchronous flushing in a dry tropical forest. *Biotropica* **16**: 193–201.

Lott, E. J., Bullock, S. H. & Solís-Magallanes, J. A. (1987). Floristic diversity and structure of upland and arroyo forests of coastal Jalisco. *Biotropica* **19**: 228–35.

Lugo, A. E., Gonzales-Liboy, J. A., Cintrón, B. & Dugger, K. (1978). Structure, productivity, and transpiration of a subtropical dry forest in Puerto Rico. *Biotropica* **10**: 278–291.

Marín, D. & Medina, E. (1981). Duración foliar, contenido de nutrientes y esclerofilia en árboles de un bosque muy seco tropical. *Acta Científica Venezolana* **32**: 508–514.

Martínez-Yrízar, A., Sarukhán, J., Pérez-Jiménez, A., Rincón, E., Maass, M., Solís-Magallanes, A. & Cervantes, L. (1992). Above-ground phytomass of a tropical deciduous forest on the coast of Jalisco, México. *Journal of Tropical Ecology* **8**: 87–96.

Medina, E. (1984). Adaptations of tropical trees to moisture stress. In *Tropical Rain Forest Ecosystems: structure and function*, ed. F. B. Golley, pp. 225–37. Elsevier, New York.

Medina, E. & Cuevas, E. (1990). Propiedades fotosintéticas y eficiencia de uso de agua de plantas leñosas del bosque decíduo de Guánica: consideraciones generales y resultados preliminares. *Acta Científica (Puerto Rico)* **4**: 25–36.

Montes, R. & Medina, E. (1977). Seasonal changes in nutrient content of leaves of savanna trees with different ecological behavior. *Geo-Eco-Trop* **1**: 295–307.

Mooney, H. A., Bullock, S. H. & Ehleringer, J. R. (1989). Carbon isotope ratios of plants of a tropical dry forest in México. *Functional Ecology* **3**: 137–42.

Mooney, H. A., Chu, C., Bullock, S. H. & Robichaux, R. (1992). Carbohydrate, water and nitrogen storage in vines of a tropical deciduous forest. *Biotropica* **24**: 134–9.

Mooney, H. A., Field, C. & Vázquez-Yanes, C. (1984). Photosynthetic characteristics of wet tropical forest plants. In *Physiological Ecology of Plants of the Wet Tropics*, ed. E. Medina, H. A. Mooney and C. Vázquez-Yanes, pp. 129–38. Dr W. Junk, The Hague.

Murphy, P. G. & Lugo, A. E. (1986a). Ecology of tropical dry forest. *Annual Review of Ecology and Systematics* **17**: 67–88.

Murphy, P. G. & Lugo, A. E. (1986b). Structure and biomass of a subtropical dry forest in Puerto Rico. *Biotropica* **18**: 89–96.

Nelsen, C. E. (1987). The water relations of vesicular-arbuscular mycorrhizal systems. In *Ecophysiology of VA mycorrhizal plants*, ed. G. R. Safir, pp. 71–91. CRC Press, Boca Raton.

Nilsen, E. T., Sharifi, M. R., Rundel, P. W., Forseth, I. N. & Ehleringer, J. R. (1990). Water relations of stem succulent trees in north-central Baja California. *Oecologia* **82**: 299–303.

Oertli, J. J., Lips, S. H. & Agami, M. (1990). The strength of sclerophyllous cells to resist collapse due to negative turgor. *Acta Oecologica* **11**: 281–90.

Olivares, E. & Medina, E. (1992). Water and nutrient relations of woody perennials from tropical dry forests. *Journal of Vegetation Science* **3**: 383–92.

Opler, P. A., Frankie, G. W. & Baker, H. G. (1976). Rainfall as a factor in the release, timing, and synchronization of anthesis by tropical trees and shrubs. *Journal of Biogeography* **3**: 231–6.

Opler, P. A., Frankie, G. W. & Baker, H. G. (1980). Comparative phenological studies of treelet and shrub species in tropical wet and dry forests in the lowlands of Costa Rica. *Journal of Ecology* **68**: 167–88.

Paschke, M. W. & Dawson, J. O. (1992). The occurrence of *Frankia* in tropical forest soils of Costa Rica. *Plant and Soil* **142**: 63–7.

Postgate, J. R. (1982). *The Fundamentals of Nitrogen Fixation*. Cambridge University Press, Cambridge.

Putz, F. E. (1983). Liana biomass and leaf area of a 'Tierra Firme' forest in the Rio Negro basin, Venezuela. *Biotropica* **15**: 185–9.

Putz, F. E. (1992). Foraging efficiencies of liana and tree roots in heterogeneous soils. *Bulletin of the Ecological Society of America* (Suppl.) **73**: 311.

Reich, P. B. & Borchert, R. (1982). Phenology and ecophysiology of the tropical tree, *Tabebuia neochrysantha* (Bignoniaceae). *Ecology* **63**: 294–9.

Reich, P. B. & Borchert, R. (1984). Water stress and tree phenology in a tropical dry forest in the lowlands of Costa Rica. *Journal of Ecology* **72**: 61–74.

Reich, P. B. & Borchert, R. (1988). Changes with leaf age in stomatal function and water status of several tropical tree species. *Biotropica* **20**: 60–9.

Robichaux, R. H., Holsinger, K. E. & Morse, S. R. (1986). Turgor maintenance in Hawaiian *Dubautia* species: the role of variation in tissue osmotic and elastic properties. In *On the Economy of Plant Form and Function*, ed. T. J. Givnish, pp. 353–80. Cambridge University Press, Cambridge.

Sanford, R. L., Jr (1989). Fine root biomass under a tropical forest light gap opening in Costa Rica. *Journal of Tropical Ecology* **5**: 251–6.

Sanford, R. L., Jr (1990). Fine root biomass under light gap openings in an Amazon rain forest. *Oecologia* **83**: 541–5.

Sarmiento, G., Goldstein, G. & Meinzer, F. (1985). Adaptive strategies of woody species in tropical savannas. *Biological Review* **60**: 315–56.

Schulte, P. J. (1992). The units of currency for plant water status. *Plant, Cell and Environment* **15**: 7–10.

Schulze, E.-D., Mooney, H. A., Bullock, S. H. & Mendoza, A. (1988). Water contents of wood of tropical deciduous forest species during the dry season. *Boletín de la Sociedad Botánica de México* **48**: 113–18.

Sobrado, M. A. (1986). Aspects of tissue water relations and seasonal changes of leaf water potential components of evergreen and deciduous species coexisting in tropical dry forests. *Oecologia* **68**: 413–16.

Sobrado, M. A. (1991). Cost–benefit relationships in deciduous and evergreen leaves of tropical dry forest species. *Functional Ecology* **5**: 608–16.

Sobrado, M. A. (1993). Trade-off between water transport efficiency and leaf life-span in a tropical dry forest. *Oecologia* **96**: 19–23.

Sobrado, M. A. & Cuenca, G. (1979). Aspectos del uso de agua de especies decíduas y siempreverdes en un bosque seco tropical de Venezuela. *Acta Científica Venezolana* **30**: 302–8.

Sperry, J. S., Holbrook, N. M., Zimmermann, M. H. & Tyree, M. T. (1987). Spring filling of xylem vessels in wild grape vine. *Plant Physiology* **83**: 414–17.

Sperry, J. S., Tyree, M. T. & Donnelly, J. R. (1988). Vulnerability of xylem to embolism in a mangrove vs an inland species of Rhizophoraceae. *Physiologia Plantarum* **74**: 276–83.

Steudle, E., Zimmermann, U. & Lüttge, U. (1977). Effect of turgor pressure and cell size on the wall elasticity of plant cells. *Plant Physiology* **59**: 285–9.

Turner, N. C. & Jones, M. M. (1980). Turgor maintenance by osmotic adjustment: a review and evaluation. In *Adaptation of Plants to Water and High Temperature Stress*, ed. N. C. Turner and P. J. Kramer, pp. 87–103. John Wiley & Sons, New York.

Tyree, M. T. & Ewers, F. W. (1991). The hydraulic architecture of trees and other woody plants. *New Phytologist* **119**: 345–60.

Tyree, M. T. & Jarvis, P. G. (1982). Water in tissues and cells. In *Encyclopedia of Plant Physiology, New Series, Vol. 12B, Physiological Plant Ecology II, Water Relations and Carbon Assimilation*, ed. O. L. Lange, P. S. Nobel, C. B. Osmond, and H. Ziegler, pp. 35–77. Springer-Verlag, Berlin.

Tyree, M. T., Snyderman, D. A., Wilmot, T. R. & Machado, J. L. (1991). Water relations and hydraulic architecture of a tropical tree (*Schefflera morototoni*). *Plant Physiology* **96**: 1105–13.

Vance, E. D. & Nadkarni, N. M. (1992). Root biomass distribution in a moist tropical montane forest. *Plant and Soil* **142**: 31–9.

Veillon, J. P. (1963). Relación de ciertas características de la masa forestal de unos bosques de las zonas bajas de Venezuela con el factor climático: humedad pluvial. *Acta Científica Venezolana* **14**: 30–41.

Waring, R. H. & Running, S. W. (1978). Sapwood water storage: its contribution to transpiration and effect upon water conductance through the stems of old-growth Douglas-fir. *Plant, Cell and Environment* **1**: 131–40.

Whigham, D. F., Zugasty Towle, P., Cabrera Cano, E., O'Neill, J. & Ley, E. (1990). The effect of annual variation in precipitation on growth and litter production in a tropical dry forest in the Yucatán of México. *Tropical Ecology* **31**: 23–34.

Windsor, D. (1990). Climate and moisture variability in a tropical forest: long-term records from Barro Colorado Island, Panamá. *Smithsonian Contributions in Earth Science* No. 29.

Wright, S. J. & Cornejo, F. H. (1990). Seasonal drought and leaf fall in a tropical forest. *Ecology* **71**: 1165–75.

Zimmermann, M. H. (1978). Hydraulic architecture of some diffuse-porous trees. *Canadian Journal of Botany* **56**: 2286–95.

Zimmermann, M. H. (1983). *Xylem Structure and the Ascent of Sap*. Springer-Verlag, Berlin.

11

Plant reproduction in neotropical dry forests

STEPHEN H. BULLOCK

Introduction

Temporal and spatial patterns of water availability are perhaps the most widespread determinants of the characteristics of tropical vegetation. Both locally and regionally, floristic diversity and composition, growth forms, phenology and demography are largely functions of hydrology. However, few studies have focused on how the reproduction of tropical forest plants changes along a gradient of water stress, except in regard to phenology, despite extensive field studies (especially of floral biology) and many reviews of the literature (e.g. Baker *et al.*, 1983; Bawa & Hadley, 1990). In this review I emphasize comparisons among species and sites in dry forest and, especially, contrasts between drier and wetter localities, with respect to a series of topics. Do differences in water stress correspond to different patterns of mating and of evolutionary mechanisms, as determined by the frequency of sexual types and of self-incompatibility systems? How do the timing and duration of flowering and fruiting change across a gradient in the duration of water stress, and what determines such changes? What changes occur in the spectra of pollen and seed dispersal systems, and of available vectors? Lastly, extending our perspective to life histories raises questions of how reproductive effort changes due to water stress effects on mortality and growth.

This review is drawn from a heterogeneous literature. There are few comparisons of reproductive biology among localities differing in water stress, or in associated species, which focus on intraspecific or intrageneric variation. Thus many comparisons must be attempted on the basis of dissimilar species lists, and in spite of different methods. In a few cases some attention has been given to local differences between dry and moist sites. Likewise, some comparisons can be made between groups of species identified by function instead of taxonomy, for example between deciduous and evergreen, or animal- and wind-dispersed species.

Another important caveat for this analysis is that the physical and biotic contexts cannot be strictly delimited. For example, most tree species are not restricted to one climatic 'life zone' (Holdridge *et al.*, 1971: 656); also, the biotic environment, the surrounding flora and fauna, varies greatly in composition throughout a species' range, and history. Moreover, it is probably typical of 'tropical dry forest' that a landscape of substantially less than 100 km^2 includes sufficient variation in hydrology that plant phenology and species composition and gross structure differ radically from place to place within that area (Warming, 1909; Peyton, 1980; Lott, Bullock & Solís, 1987; Spichiger & Ramella, 1989; Lewis, 1991; Sampaio, Chapter 3). The distances involved are within the daily foraging or short-term movement range of essentially all pollinating and seed-dispersing animals. To account thoroughly for these complications is difficult, and usually neglected in the primary literature on reproduction, confounding attempts at interpretation or synthesis. Nonetheless, as a guide to conditions at most of the sites mentioned here, some standard features are listed in Table 11.1.

Mating patterns

Genetic recombination systems can strongly affect evolutionary patterns, as well as conservation problems, and thus have been a major focus of research on tropical plants, especially trees. Plant mating patterns are affected most fundamentally by floral sexuality (e.g. hermaphrodite, monoecious and dioecious) and compatibility relations.

Differences among floras in the frequency of sexual types appear to be due largely to different spectra of life forms, and also to certain families or genera dominating regional floras (Baker, 1959; Freeman, Klikoff & Harper, 1980; Fox, 1985; Ramírez & Brito, 1990). Two surveys of floral sexuality which include most or all of a local flora, show remarkable similarities between neotropical moist forest (Barro Colorado) and dry forest (Chamela), and differences between woody and herbaceous plants (Table 11.2). Parallel trends are found in the Shervaroy Hills, India: the frequencies of hermaphrodite, monoecious and dioecious species are distinct between herbaceous and woody plants (Table 11.2) (but are similar among arid scrub, deciduous and semideciduous forests). However, the frequency of dioecy in trees is much lower in India than in the neotropics.

Table 11.1. *Site characteristics*

Site	Rainfall: Annual total (mm)	Rainfall: Months <60 mm	Country	Latitude	Elevation (m)
Galápagos	288[d]	10	Ecuador	0° 54′ S	6
Tapia	450[f]	5	Argentina	26° 50′ S	550
Riacho	639[e]	8	Brazil	8° 27′ S	657
Península de Araya	<500–800	8	Venezuela	10° 29′/39′ N	<50
Chamela[a]	714	7	México	19° 30′ N	100
Colinas de Bello Monte	762[c]	5	Venezuela	10° 30′ N	1100
Maracay	955	5	Venezuela	10° 15′ N	450
Charallave	957	6	Venezuela	10° 16′ N	315
Altos de Pipe	994	5	Venezuela	10° 66′ N	1749
Puerto Morelos	1181[b]	2	México	20° 49′ N	<20
Botucatu	1300	6	Brazil	22° 45′ S	550
Los Llanos[a]	1312	4	Venezuela	8° 56′ N	100
Shervaroys	1500	3	India	11° 50′ N	400–1600
Cañas[a]	1790	5	Costa Rica	10° 32′ N	45
Belém	2277	3	Brazil	1° 18′ S	10
Monteverde	2485	3	Costa Rica	10° 18′ N	1500
Barro Colorado	2600	3	Panamá	9° 9′ N	100
Cerro de la Muerte	2750	3	Costa Rica	9° 34′ N	3100
La Selva	4000	0	Costa Rica	10° 26′ N	50
Los Tuxtlas	4600	0	México	18° 35′ N	150

[a] At 'Cañas' (various sites near Cañas in lowland Guanacaste, e.g. La Pacifica, Comelco, Finca Taboga, Palo Verde) there is a notable diversity of habitats (Holdridge *et al.*, 1971), all disturbed by fire and grazing: moist, dry, riparian and swamp forests, savanna and 'second growth'. Except on limestone hills, species distributions probably relate to complex subsurface drainage patterns. At Los Llanos, savanna trees grow over a water table at 0.8–3.0 m in a typical seasonal fluctuation, while a typical situation for forest trees has a water table at 2.8–5.3 m (Foldats & Rutkis, 1975). At Chamela, there is no evidence of any water table in the greater part of the topography.
[b–f] Rainfall data from nearest available station: [b], Valladolid; [c], Caracas; [d], Puerto Baquerizo; [e], Pesquerias; [f], estimate.

Whether outcrossing is necessary for hermaphroditic trees and shrubs has been examined for many species, and the great majority are considered to require or greatly benefit from cross-pollination (with some marked exceptions, e.g. Gottsberger, 1989). In three dry forests, 74–77% of the species are considered self-incompatible (n = 12 to 36; at

Table 11.2. *Frequency of floral sexual forms. All 'polygamous' forms are included here in monoecy. Data for trees only are in parentheses ('medium to large trees' on Barro Colorado)*

Sexuality	Chamela 708 spp.	(188)	Barro Colorado 1212 spp.	(211)	Shervaroys 814 spp.	(348)
Hermaphrodite	70.2%	(57.9)	76.4	(63.0)	84.2	(80.2)
Monoecious	17.6%	(18.0)	15.2	(16.6)	12.9	(14.6)
Dioecious	12.3%	(23.9)	8.7	(21.3)	2.2	(5.2)

Sources: Chamela: Bullock (1985); Barro Colorado: Croat (1979); Shervaroys: N. Sivaraj & K. V. Krishnamurthy (personal communication).

Cañas, Colinas de Bello Monte and Chamela: see Figure 11.1). With smaller samples, self-incompatibility characterizes 71% of the shrub species tested in Brazilian caatinga ($n = 7$; at Riacho: Machado, 1990), and 62% of shrub and tree species in Argentine chaco ($n = 8$; at Tapia: Aizen & Feinsinger, 1994). For wet forest trees, a figure of 81% is given (at La Selva: Bawa, Perry & Beach, 1985). These figures are generalities and include many species that show low levels of compatibility (Figure 11.1); no differences among these six sites are apparent ($p = 0.58$, Kruskal–Wallis). However, the validity of current attempts to assess compatibility remains open to question; larger samples are clearly a priority. Field tests of self (intra-plant) compatibility often sample less than ten individuals and less than 100 flowers per individual.

In diverse tropical situations, self-compatibility is common, probably because of the scarcity of pollinators, small population sizes, and the frequency of herbaceous species (in Venezuelan cloud forest at Altos de Pipe, Sobrevila & Arroyo, 1982; palm swamp, Ramírez & Brito, 1990; Galápagos Islands, Rick, 1966; McMullen, 1987). Most herbaceous species and many shrubs are self-compatible (Stebbins, 1958), even in tropical forests (Kress, 1983; Bawa *et al.*, 1985; Grove, 1985; McDade, 1985; Montalvo & Ackerman, 1987; Renner, 1989; Machado, 1990).

In contrast to the surveys of taxonomically diverse floras, studies of genera can point to the evolutionary significance of lower compatibility barriers (Figure 11.1), including interspecific crosses (Long, 1975; Daniel, 1983; Dehgan, 1984; D'Arcy, 1986; Neill, 1988). Particularly in herbaceous and shrubby plants in mountainous regions with extreme microgeographic diversity of environmental and disturbance regimes, speciation may commonly involve inbreeding, polyploidy (and

Locality	Number of species tested	0	Index of self-compatibility 0.1–1.0	>1	Sources
Riacho	7	4		1	19
Tapia	8	2		1	20
Chamela	35	13		1	14,17,18
Colinas	12	8		1	5
Cañas	36	20		1	1,2,3,4,7,10,11
Altos	8	0		0	9
La Selva	22	12		1	13
Genus					
Tabebuia	7	7			1,5,14
Cordia	5	2			3,14,19
Capparis	3	1			5,13,14
Casearia	3	1			13,14
Caesalpinia	6	2			1,14
Inga	6	1			12
Lonchocarpus	6	1			1,13,14
Erythrina	4	0		2	16
Hamelia	3	1			13,14
Luehea	3	1			1,7,8,14
Palicourea	3	0			9
Opuntia	3	2			6,14,20
Aphelandra	8	0			15
Angelonia	4	3			19

Figure 11.1. Index of self-compatibility (ratio of fruits set in self- and cross-pollination by hand) in field experiments. Each dot represents the results for one species, but brackets indicate the range of values for species tested at two localities. Note that fruit not seed production is used (except for *Aphelandra*); this increases the data base but introduces systematic bias for higher values. Sources: 1. Bawa (1974); 2. Heithaus, Opler & Baker (1974); 3. Opler *et al.* (1975); 4. Frankie, Opler & Bawa (1976); 5. Ruiz & Arroyo (1978); 6. Grant & Grant (1981); 7. Haber & Frankie (1982); 8. Sazima *et al.* (1982); 9. Sobrevila & Arroyo (1982); 10. Bawa & Webb (1983); 11. Bawa & Webb (1984); 12. Koptur (1984); 13. Bawa *et al.* (1985); 14. Bullock (1985); 15. McDade (1985); 16. Neill (1988); 17. Domínguez & Bullock (1989); 18. Domínguez (1990); 19. Machado (1990); 20. Aizen & Feinsinger (1994).

aneuploidy) and interspecific hybridization (Gentry, 1989). Referring to wetter regions, Gentry postulates that almost half of the neotropical flora may be accounted for by 'explosive and essentially sympatric speciation' (1989: 126). Studies of dry forest plants are relatively scarce and unsynthesized in this context. Vicariance events also merit more attention (Ramamoorthy & Lorence, 1987). Changes in chromosome number and arrangement are important in evolution at the infrageneric level in some common families of dry forests (Holm-Nielsen, 1979; Morawetz,

1986; Piovano & Bernardello, 1991) and there is evidence for a trend to higher ploidy in drier areas (Morawetz, 1986). The focal regions of highest species diversity are in western México and southeast Bolivia (Rzedowski, 1993; Gentry, Chapter 7), far from those of wet forest plants.

At the intraspecific level, partial compatibility has been contrasted between wetter and drier localities for only five species: three show greater self-compatibility in the drier sites (*Ochroma pyramidale*, *Lonchocarpus eriocarinalis* and *Luehea seemannii*), and two show the opposite pattern (*Casearia corymbosa* and *Capparis verrucosa*) (Bawa, 1974; Ruiz & Arroyo, 1978; Bawa *et al.*, 1985; Bullock, 1985). The difference between *L. seemannii* populations is attributed to its lack of an outcrossing pollinator in the drier region (Haber & Frankie, 1982). However, partial self-compatibility occurs in *Luehea speciosa* where its bat pollinator is present in southern Brazil (Sazima, Fabian & Sazima, 1982). Several changes occur in the breeding system of *Metrosideros collina* trees along an elevational gradient in Hawaii (Carpenter, 1976). In the drier lowlands, a self-compatible morph with insect- or self-pollinated yellow flowers is more frequent, but at higher elevations a partially self-incompatible morph with bird-pollinated red flowers is more frequent. Below the population level, differences in self-compatibility among individuals also exist (Bawa, 1974; Opler, Baker & Frankie, 1975; Bullock *et al.*, 1987), but they are not widely documented or well understood.

Phenology

Drier forests are distinctive not only in the degree of leaf deciduousness, but also in temporal characteristics of reproductive activity, including synchrony, duration and timing. The mechanisms are more complex than in vegetative development, but associations among reproductive phenology patterns and other life history features are apparent. However, evaluation of the data from drier sites is difficult for several reasons: unrecorded or unreported variations in subsurface hydrology and climatic anomalies, low temporal resolution, small sample sizes, and lack of both biological and actuarial detail available on individuals and species.

It may be characteristic of drier sites that synchronous flowering within populations is more frequent, and that synchrony among species is greater. These features are pronounced in species which flower during the dry season in response to isolated precipitation events (Gentry, 1974;

Opler, Frankie & Baker, 1976) and in those which flower at the dry–wet transition (e.g. Opler, Frankie & Baker, 1980 near Cañas; Bullock, 1986; Bullock & Solís, 1990, at Chamela).

The duration of flowering periods may be a useful comparative index of adaptation to water stress. In a Costa Rican wet forest (La Selva) 47% of tree species show 'extended' flowering, compared with 12% in a dry forest (near Cañas; Frankie, Baker & Opler, 1974). Average duration of extended flowering is similar between these habitats (23.6 and 24.7 weeks, respectively), as is the duration of 'seasonal' flowering (6.5 and 5.6 weeks: Frankie *et al.*, 1974). However, the frequency distributions of duration may be strongly skewed, and should be compared without prior division into classes. In a Mexican deciduous forest, the modal duration is less than six weeks and accounts for 41% of the canopy tree species studied (Bullock & Solís, 1990).

The period of the year when flowering and fruiting occur is also affected by water stress. However, most diverse dry forest sites present diverse strategies: some species flower while leafless, others while expanding new leaves, in full leaf, or with senescent leaves (e.g. Fenner, 1982; Barbosa *et al.*, 1989). Dry season flowering (and leaf production) is more common on moister sites (Boaler, 1966; Burger, 1974; Opler *et al.*, 1980; S. Bullock & J. Solís, personal communication) and savannas (Monasterio & Sarmiento, 1976 at Los Llanos; Lieberman, 1982; see also Holbrook, Whitbeck & Mooney, Chapter 10). At very dry sites, flowering is commonly associated with the initiation of rains, nearly simultaneous with leaf expansion (Lieberman, 1982; Prasad & Hegde, 1986; Bullock & Solís, 1990; Guevara de Lampe *et al.*, 1992 on the Península de Araya). The strongest pattern is that phenology differs among life forms (Boaler, 1966; Malaisse, 1974; Croat, 1975 at Barro Colorado; Monasterio & Sarmiento, 1976; Carabias-Lillo & Guevara, 1985 at Los Tuxtlas; Sivaraj & Krishnamurthy, 1989). Drought-limited herbaceous plants most often flower during and after the peak of their vegetative development. Many species from various life forms may be notably uncoupled from both soil moisture and precipitation regimes by the use of stored water (Schulze *et al.*, 1988; Medina, Chapter 9), e.g. orchids, cacti, succulent trees and vines. On the other hand, some woody plants are active in the dry season apparently because they tap deeper moisture.

Variations within a single population have received scant attention. The flowering of individuals on microsites with distinct soil moisture regimes may differ in calendar date and in relation to vegetative status of the tree (Reich & Borchert, 1982). Comparing riparian and upland individuals of four species near Cañas, Daubenmire (1972) shows differences in the timing and/or duration of flowering in two species, and changes in the relative timing of flowering and leafing in three. Under constant irrigation, some species will grow, flower and fruit 'continuously', while others only maintain leaves and others continue a calendar of deciduousness and seasonal reproduction (Bullock & Solís, 1990; Wright & Cornejo, 1990, at Barro Colorado).

Borchert (1983) argues that flowering of trees (particularly flowering that uses the apical meristem) is determined by developmental sequences: induction of floral primordia is 'stimulated by conditions that inhibit vegetative growth', and anthesis is released by leaf drop or rehydrating rainfall which alleviate water stress. More detailed descriptive or experimental studies will help to evaluate these ideas. However, they do not apply directly in cases of flowering on lateral inflorescences (nearly) simultaneous with apical growth, and have problems where development and anthesis are long delayed after leaf drop, or occur without rehydration, or occur before rains with vegetative bud break or leaf expansion. The mechanisms controlling both vegetative and sexual development are clearly diverse (Ramia, 1981 at Maracay; Bullock & Solís, 1990; Wright & Cornejo, 1990). Most mechanisms allow for a wide scope in their precise definition and quantitative tuning, and such scope complicates the demonstration and discussion of mechanisms. Moreover, it is apparent that the timing of development evolves through selection on several aspects of the plant's life history (Opler, 1978). Although these patterns operate in both dry and moist forests, it is clear that phenological control by water stress is more prevalent in dry forest.

The timing and duration of reproduction may differ appreciably among groups of plants with different pollination (or dispersal) vectors. The modal duration of flowering near Cañas increases across groups of plants pollinated by 'large' bees, sphingid moths and bats (Figure 11.2*A*). Sphingid-pollinated species have a greater tendency to flower in the wet season (Figure 11.2*B*). The seasonality of flowers visited by all sizes of bees is markedly different from that of large bee flowers near Cañas (Opler, 1983). Thus, the mixture of pollination types present may well affect community phenology; however, analyses comparing sites or habitats by vector type are lacking.

Regarding the phenology of seed dispersal in dry forests, it is apparent that the patterns depend on the stratum or life form considered and soil drought patterns (Monasterio & Sarmiento, 1976; Opler *et al.*, 1980). Fleshy fruits are notoriously concentrated in the wet season and scarce in

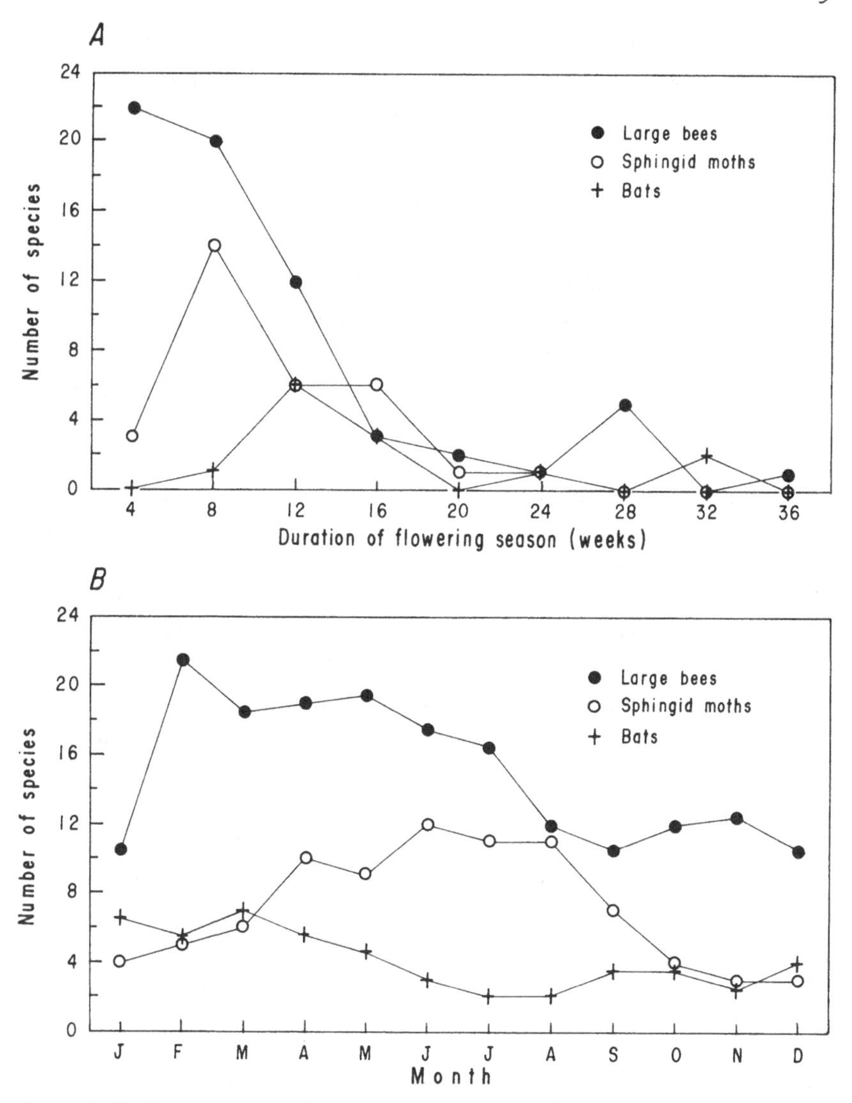

Figure 11.2. Flowering phenology of some plants near Cañas grouped by pollinator type. *A*. Frequency distribution of the duration of flowering. *B*. Number of species per calendar month. Sources: Heithaus, Fleming & Opler (1975); Frankie *et al*. (1983); Opler (1983); Haber & Frankie (1989).

the dry season; anemochores and autochores are absent or inconspicuous in the wet season and abundant in the dry season (Frankie *et al.*, 1974; Bullock & Solís, 1990; Guevara de Lampe *et al.*, 1992). The period between flowering and seed dispersal is so variable that multiple explanations of the determinants of fruiting seasons are needed. Selective factors that are often mentioned include physiological constraints, pre-germination predation, post-germination establishment, and dispersal efficiency (Janzen, 1967; Daubenmire, 1972; Frankie *et al.*, 1974).

Pollination and dispersal spectra

The frequency of different modes of pollination and dispersal differs between wetter and drier sites on both a continental and a local scale, as well as between sympatric evergreen and deciduous trees.

Ranking neotropical forests by increasing precipitation, Gentry (1982, Chapter 7) found an increasing number of tree and liana species with 'inconspicuous' flowers, presumably pollinated by wasps, social bees, short-tongued flies, thrips, etc. In contrast, conspicuous-flowered species remain equally common along this gradient (Gentry, 1982; Bawa *et al.*, 1985). Floristics and systematics provide other suggestive lines of evidence on pollination spectra. For example, a paucity of beetle pollination is apparent from the poor representation of certain plant families which have a high frequency of beetle-pollinated species (Araceae, Cyclanthaceae, Arecaceae, Annonaceae, Lauraceae, Myristicaceae: Schatz, 1990). In some families, pollination systems may differ consistently between drier and wetter regions. In the Andean region and the Brazilian and Guiana shields, Malvaceae is well represented by larger woody forms with hummingbird or even bat pollination, while in drier regions with more open vegetation, smaller species with bee pollination and less outcrossing are prevalent. This evolutionary trend is repeated in several genera (Gottsberger, 1986). Cactaceae is a neotropical family with diversity highest to the north and south of the Equator. In northern subtropical deserts cacti are mostly bee-pollinated (Grant & Grant, 1979) while in the deciduous forest at Chamela, species pollinated by sphingids and bats are more common. Capparidaceae is even more typically a family of semiarid tropics and subtropics, with sphingid-pollinated shrubs and trees in *Capparis* and *Crataeva*, as well as species pollinated by bees (*Capparis*) and wind (*Forchhammeria*).

Regarding dispersal mode in trees, precipitation is correlated positively with the frequency of zoochory ($r = 0.86$, $p = 0.003$, angular transformed data; see Figure 11.3). Parallel differences occur at the local level: topographically drier and moister sites support greater and lesser frequencies of anemochorous tree species, respectively (Gentry, 1982, near Cañas; Gottsberger & Silberbauer-Gottsberger, 1983, near Botucatu;

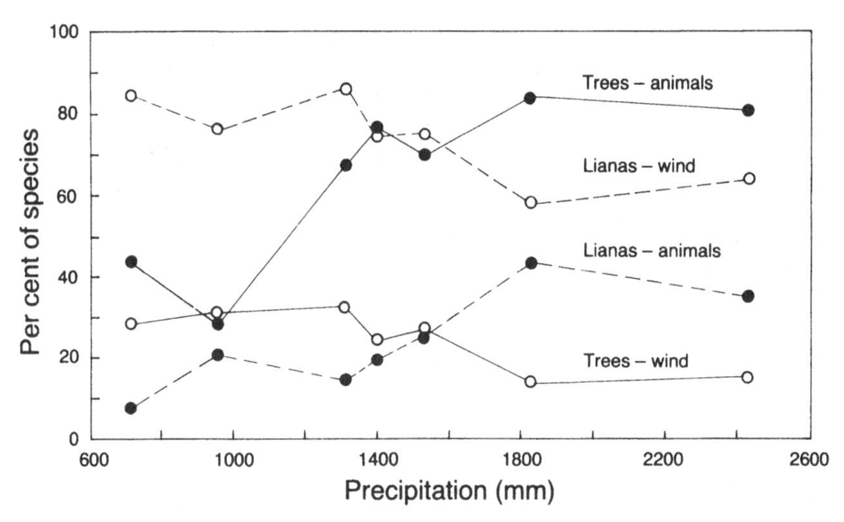

Figure 11.3. Frequency of anemochory and zoochory in trees and lianas in relation to annual precipitation. Sources: Gentry (1982), Wikander (1984), and data of Lott *et al.* (1987).

Wikander, 1984, near Charallave; S. Bullock, personal observation, at Chamela). In lianas, wind is the predominant vector at all sites, but precipitation does correlate with the frequency of zoochory (r = 0.70, p = 0.034, angular transformed data; see Figure 11.3). Dispersal types are known to differ in the spatial patterns of dispersed seed and juvenile plants (Hubbell, 1979; Heithaus & Fleming, 1981; both near Cañas). These patterns have yet be examined in the context of site differences in plant size or species diversity.

Perhaps the most striking association of reproductive characters with drought is at the local level, among trees with contrasting leaf longevities. Near Cañas, the evergreen habit is significantly (not exclusively!) associated with single-seeded, animal-dispersed fruits and small unspecialized flowers, but the deciduous habit is associated with several-seeded fruits, abiotic dispersal and specialized flowers (Table 11.3). At Chamela, the dispersal and seed contrasts are similarly significant between evergreen and deciduous trees, but the floral contrast is not (Table 11.3).

Several neglected aspects of flower and seed biology warrant more attention. Surveys of the rewards for pollinators have concentrated on nectar, and comparisons within and among pollinator types (Baker & Baker, 1983; Opler, 1983); differences among habitats, independent of

Table 11.3. *Numbers of species of trees with particular combinations of leaf habit and reproductive characteristics near Cañas (Opler, 1978) and Chamela (S. H. Bullock, unpublished data)*

	Cañas		Chamela	
Reproductive characteristics	Evergreen	Deciduous	Evergreen	Deciduous
Pollinators				
Specialized	17	58	17	69
Generalized	25	31	18	59
Dispersal vector				
Animals	36	34	29	46
Physical	6	55	6	82
Seeds per fruit				
One	27	26	22	33
More than one	15	63	13	95

pollinator types, have not been surveyed. Some notable vectors have been neglected: pollination by wind (Bullock, 1994) and by flies, and seed dispersal by ants or explosive fruits. Seed size, anatomy, chemistry and germination types provide insightful bases for comparison of habitats and of populations, but there are no surveys that include tropical dry forest.

Animal vectors

Variations in plant reproduction are effected in part by variations in the populations or behavior of animals that serve as pollen and seed dispersal vectors. There are conspicuous differences among sites and habitats in faunal composition (Ceballos, Chapter 8) and in such characteristics of vectors as dependence on floral products or fruit, host specificity, seasonality and length of flight periods, foraging energetics and migratory tendencies. Not surprisingly, the quality and quantity of information are limited by the difficulty of observation and identification.

Most bees are dependent on pollen and floral exudates for larval and adult nutrition. Although the degree to which bees are parasites or commensals of plants is unknown, they are generally considered to pollinate more neotropical plants than do other animals. A major biogeographic trend in bees, and probably in pollination, is the shift from dominance of solitary species in drier regions or higher latitudes to highly social species in wetter or equatorial regions (Michener, 1979;

Roubik, 1989; Ayala, Griswold & Bullock, 1993). Corresponding differences may be expected among habitats in several features important to pollination: length of the flight season, functional response to abundant resources, individual daily limitation to nectar or pollen host species, short-term individual constancy, foraging range (potentially vagrant or necessarily central-place), dispersion of nest sites, and the activity of males.

Comparison of the Cañas and two California bee faunas shows both longer seasons and broader feeding niches in the tropics (Heithaus, 1979a, b). There is a striking range of body size among bee species (Raw, 1985; M. Aizen, personal communication): for example, from 2.7 mg to 1.8 g wet mass at Chamela (S. Bullock, personal observation). However, the effects of size on patterns of flower use, or even on foraging range and speed, have not been explored, and size distributions for local faunas are not available.

Adult sphingid moths often consume large volumes of high quality nectar, but apparently are not constant to particular host species. Except in the caatinga, most neotropical forests support tens of species (Schreiber, 1978), but differences in local diversity are not well understood. Faunal comparisons can reasonably be made on the bases of sizes and seasonality. Mass, wing length and (log) proboscis length are well correlated across many species; comparison among the faunas of Chamela, Cañas, and Belém (in equatorial moist forest) by proboscis length shows no significant difference (43–63 species per site; $p = 0.59$, Kruskal–Wallis; data from Moss, 1920; Bullock & Pescador, 1983; Haber & Frankie, 1989; Pescador, 1994). However, the duration of species' flight seasons as shown by light trap data, differs among these sites with the less seasonal sites having more species with long flight seasons (Figure 11.4*A*, $p = 0.033$, Kolmogorov–Smirnov). There are also differences in the number of species active in each month, and thus in the month-to-month turnover in species composition (Figure 11.4*B*). Seasonal long-distance and interhabitat movements are known in some species of neotropical sphingids (Beebe, 1949; Haber & Frankie, 1989; Powell & Brown, 1990).

The biogeography of some volant vertebrate vectors is well studied (Karr, 1980; Koopman, 1981; Stiles, 1981), but a few comments can be added on simplistic size comparisons. Flower- and fruit-feeding bat faunas show no significant differences in body weight distribution among five sites in dry, wet and montane forests ($p = 0.83$, ANOVA: LaVal & Fitch, 1977, at La Selva, Cañas and Monteverde; Ceballos & Miranda, 1986, at Chamela; Coates-Estrada & Estrada, 1986, at Los Tuxtlas).

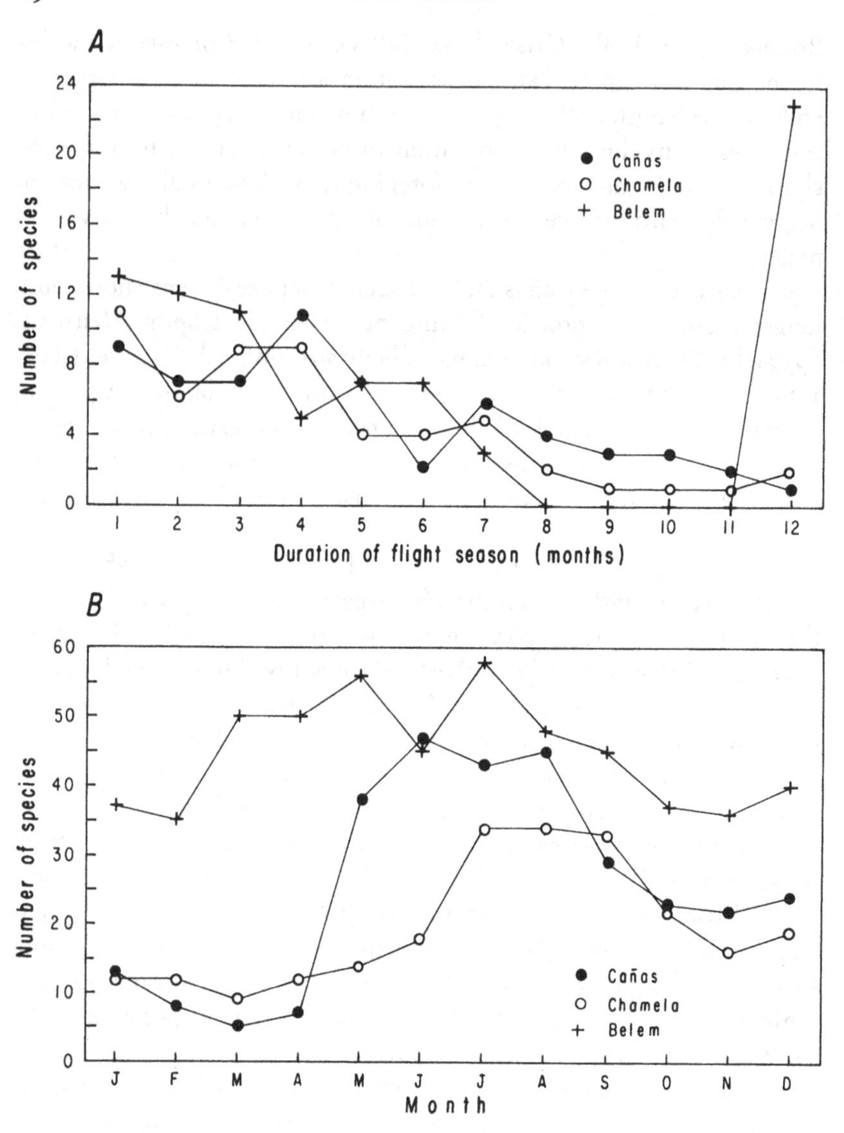

Figure 11.4. Phenology of sphingid moth faunas. *A*. Duration of flight period. *B*. Number of species per calendar month. Sources: Chamela: Pescador (1994); Cañas: Haber & Frankie (1989); Belém: Moss (1920).

A comparison of sizes in hummingbird faunas among six sites in dry, wet and montane forests shows no significant differences regarding body weight and wing loading ($p = 0.15$ and $p = 0.20$, Kruskal–Wallis; dry: near Cañas, Wolf, 1970; Chamela, Arizmendi & Ornelas, 1990; wet:

Trinidad and Tobago, Feinsinger, Swarm & Wolfe, 1985; montane: Monteverde, Feinsinger & Chaplin, 1975, Feinsinger & Colwell, 1978; Cerro de la Muerte, Wolf, Stiles & Hainsworth, 1976). Living close to the limits of their food resources (Montgomerie & Gass, 1981), hummingbirds respond rapidly to changes in food abundance by changing foraging location, or abandoning territories for vagrancy, or by migration (Wolf, 1970; DesGranges, 1978). Dry forests consistently support fewer species than wet or higher elevation forests.

Reproductive effort

The scaling and scheduling of reproduction affect population and ecosystem dynamics, and may differ considerably among sites with different degrees of water stress. A few predictions can be ventured, although even the simplest data are scarce. As the amount of rainfall decreases, its variability increases; this probably increases the variability of mortality in all life stages, which should favor increased reproductive effort (the proportion of total assimilation expended on reproduction). This climatic trend probably also increases the variance in reproductive output and decreases the mean frequency of reproduction (Ridpath, 1985). In trees, reproductive output per individual may be lower on drier sites, because trees there are generally smaller in terms of trunk diameters (Veillon, 1963; Pérez, 1970; Holdridge *et al.*, 1971) and corresponding crown diameters (Pérez, 1970). Also, lower species diversity and smaller size suggest higher density per species, and all of these factors affect relations with pollen and seed dispersal vectors. For example, as flower or fruit crop size decreases, dispersal of pollen or seeds away from the parent by animal vectors should increase.

Some gross indications of reproductive effort and its variability are available from litter trap data, although these have systematic biases and poor resolution. Moist valley and dry hill sites at Chamela, which differ in stem size distribution and floristics (Lott *et al.*, 1987) and phenology (S. Bullock & J. Solís, personal communication), also differ in the contribution of 'reproductive parts' to annual litter, respectively 5.6% and 9.4% (Martínez-Yrízar & Sarukhán, 1990). In the Zambezian *miombo* ecosystem, fruits account for most of the 4.1% reproductive component of litter, but climatic anomalies can provoke an increase of more than an order of magnitude in fruit mass (Malaisse *et al.*, 1975). In the moist forest of Barro Colorado, fruits account for 5.0% and flowers for 2.1% of (non-trash) litter, but the interannual variabilities in raw mass are more than five- and three-fold, respectively, which contrasts with less than 1.5-fold variation in leaf litter (Leigh & Windsor, 1982). In dry forest at Puerto Morelos, reproductive litter did not vary significantly over four years, probably because spatial heterogeneity was greater than for leaf litter: the coefficient of variation for reproductive litter was 2.4 to 5.4 times greater than for leaf litter in the same years (Whigham *et al.*, 1990).

Another parameter available for comparisons among sites is 'reproductive efficacy', the percentage of flowers which produce fruits under natural pollination compared to the percentage under hand cross-pollination (Ruiz & Arroyo, 1978). The data show no significant difference between two dry forest sites, but much higher efficacy in a cloud forest (Figure 11.5; $p = 0.00009$, Kruskal–Wallis). The majority of dry forest examples fall below 30%, suggesting that fecundity may often be limited by inadequate pollination. On the other hand, the high frequency of self-compatibility in the cloud forest site may account for higher efficacy. It would be interesting to test if efficacy varies among years, sites, habitats or pollinator types, although the effects of individual crop size must also be considered.

With reference to fruit or seed production by individual plants, little is known about the relationship of reproduction to age, size or growth, whether constant or decelerating, or whether affected by water relations.

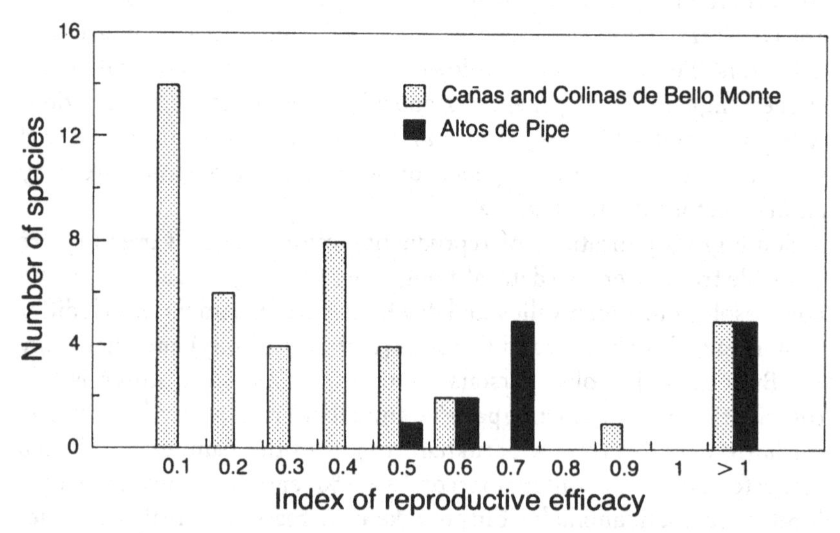

Figure 11.5. Index of Reproductive Efficacy in dry forests (Cañas and Colinas de Bello Monte) and cloud forest (Altos de Pipe). The index is the percentage of fruits from natural pollinations divided by percentage fruits from hand cross-pollination. Sources as in Figure 11.1 and Montalvo & Ackerman (1987).

Table 11.4. *Size–fecundity relationship in trees of tropical dry forests, expressed as* r^2 *of* Log(No. Fruits + 1) *on* Log(Basal Area)

Species	r^2	Period (years)		Source
Stemmadenia donnell-smithii	0.44	1		1
Ateleia herbert-smithii	0.53	1		2
Spondias purpurea	0.49	1		3
Jacaratia mexicana	0.28	13	mean; range 0.13–0.48	4
Cochlospermum vitifolium	0.25	9	mean; range 0.07–0.44	4
Bursera simaruba	0.54	11	mean; range 0.42–0.62	5

Sources: 1. McDiarmid, Ricklefs & Foster (1977) (for 1971); 2. Janzen (1989) (for 1982); 3. Mandujano, Gallina & Bullock (1994); 4. S. H. Bullock (unpublished data); 5. G. Stevens (unpublished data).

For conventional trees, if 'size' (leaf or root area? reserves?) can be acceptably indexed by trunk diameter, data on reproductive effort could be obtained rapidly. Reproductive output has not been extensively sampled, but existing data suggest that its relationship with trunk diameter is not strict (Table 11.4). However, it appears that fruit production does not accelerate or decelerate with tree size. The scatter of untransformed data suggests that the variance in reproduction increases with size. Rather than a correlation, such data suggest that vegetative size sets an approximate upper limit to reproduction. There may be significant habitat differences in the important factors keeping individuals below such a limit. Of course, reproductive effort can be varied by a change in the slope of the size–output curve, or by a shift in the minimum size for reproduction.

Reproductive effort and growth have been compared for *Opuntia* species between sites with (moister) dense forest and (drier) open scrub vegetation. At Chamela, comparisons between coastal scrub and forest populations of *Opuntia excelsa* show that the former has significantly less height growth with increasing size, but the populations are similar in fruits produced per cladode (S. Bullock *et al.*, personal communication). Similarly, in the Galápagos Islands, *Opuntia* species in scrub sites differ from those in forest by showing less height growth and similar reproduction as measured by the number of fruits per cladode. However, reproductive effort is greater in the drier site in terms of fruit cal cm^{-2} cladode (Racine & Downhower, 1974). This shows the importance of identifying the currency of reproductive costs, an unsettled issue, and may support the idea of greater reproductive effort on drier sites.

Non-annual reproduction might be more common in drier forests as a result of the greater climatic variability affecting assimilation, development, pest resistance or pollinator populations. There are scarcely any data on the phenomenon, regardless of causes. In some species, supra-annual schedules may be genetically programmed, but no relation to drought has been suggested (Janzen, 1978). Because most observation of non-reproduction is part of multispecies studies with few individuals per species, and lasting for few years, the sampling biases are more notable than the results. For example, in a two-year study of trees in Costa Rican wet and dry forests, fruits were not observed on 13% and 8% of the species, respectively. However, the numbers of individuals monitored averaged 2.5 and 10.2 per species, respectively (Frankie *et al.*, 1974). At Chamela, 12% of tree species did not fruit during more than three years, but 85% of these were represented by only one individual (Bullock & Solís, 1990). Adequate data are few, but the propensity for variation of frequency among individuals clearly differs among species (Figure 11.6). Quantitative data are important to relate non-reproduction to minor variation around low reproduction. Intraspecific differences between sites (Figure 11.6), especially along a gradient of water stress, would be instructive but are yet to be studied.

The flower and fruit crops of trees are often large compared to the

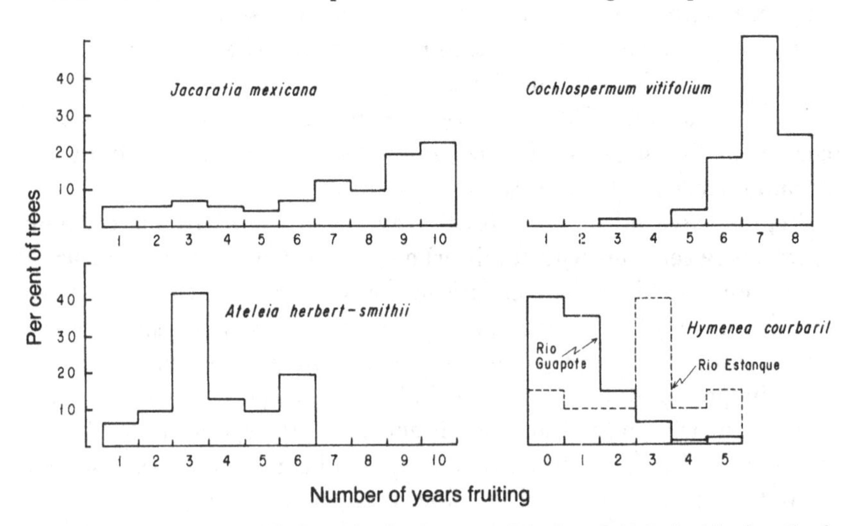

Figure 11.6. Frequency of fruit production in trees of dry forest. Note that the length of the recorded period ranges from 5 to 10 years. Sources: *Hymenea courbaril* ($n = 94$ and 20), Janzen (1978); *Ateleia herbert-smithii* ($n = 31$), Janzen (1989); *Jacaratia mexicana* ($n = 71$) and *Cochlospermum vitifolium* ($n = 46$), S. Bullock (unpublished data).

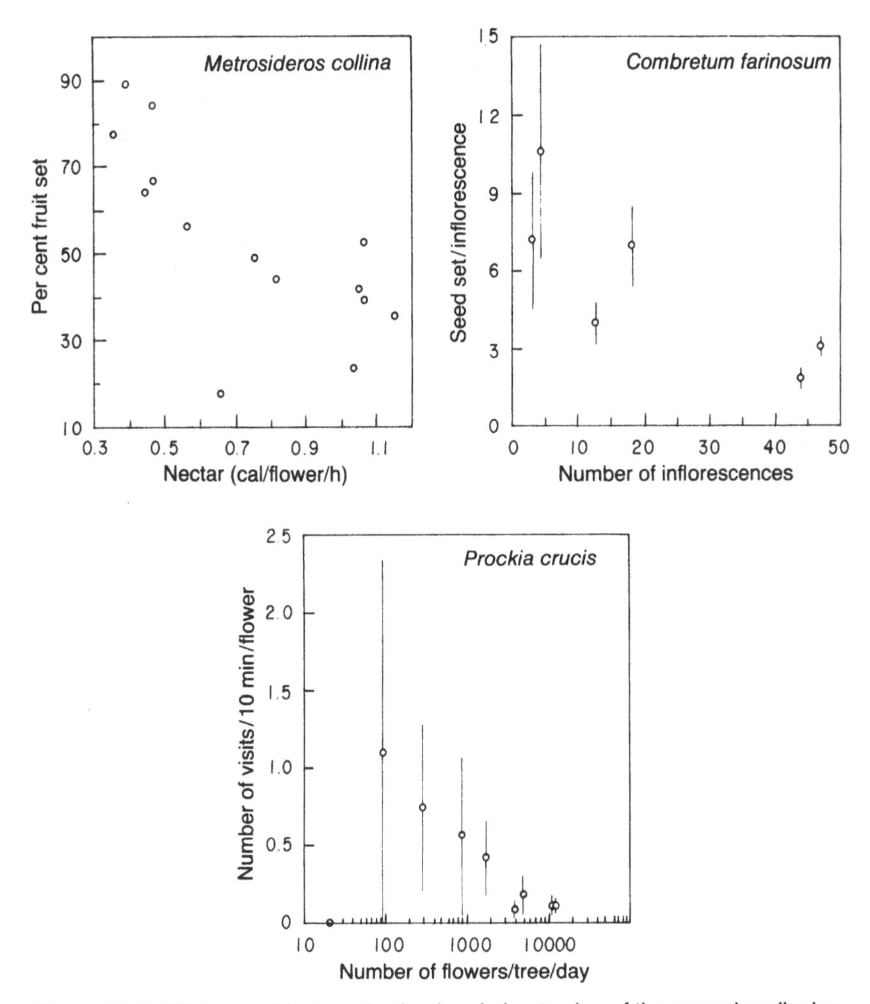

Figure 11.7. Efficiency of fruit production in relation to size of the reward or display. Sources: *Metrosideros collina*, Carpenter (1976); *Combretum farinosum*, Schemske (1980); *Prockia crucis*, Bullock *et al.* (1989).

energetic requirements of individual pollinators or seed dispersers. Thus, increases in plant size may result in increased rates of self-pollination and of deposition of seeds below the mother plant (Arroyo, 1976). Of course, animal foraging is affected by the apparent daily crop size, and thus depends on the phenology of maturation more than on annual reproduction (Gentry, 1974; McDade, 1985) (recall that seasons are generally shorter at drier sites). However, an inverse relation between size and efficiency is probably common in woody plants (Figure 11.7). Fruit set per inflorescence is inversely related to number of inflorescences in *Metrosideros collina* trees, because pollinating birds move less among trees so that outcrossing decreases (Carpenter, 1976). *Combretum farinosum* lianas, pollinated by hummingbirds in lowland Guanacaste, are similarly limited by decreased outcrossing (Schemske, 1980). *Prockia crucis*, a big-bang flowering (*sensu* Gentry, 1974) tree visited only by pollen-collecting bees at Chamela, shows a constant or decreasing rate of visitation per flower as crop size increases over four orders of magnitude; the relationship depends on the neighborhood of conspecifics (Bullock, Martínez & Ayala, 1989). Increasing density of flowering conspecifics may often result in greater outcrossing, as shown in moist forest trees of *Cavanillesia platanifolia* (Murawski *et al.*, 1990). Thus, outcrossing may be greater at drier sites where conspecifics grow closer together and have smaller canopies, which should increase pollinator movement among plants.

Summary

Sexuality and self-compatibility are similar between continental lowland wet and dry forest floras, but differ among life forms in all habitats. Intrageneric studies show that ploidy tends to increase in parallel with water stress. Intraspecific comparisons of breeding systems between sites or habitats are scarce.

The relationship of reproductive to vegetative phenology and to moisture availability depends on life form, with trees and other high-storage plants showing the most diversity. On sites with more prolonged soil water deficit flowering is generally shorter in duration as well as more synchronous within and among species, and fruiting is more common in the dry season.

Floristic differences among forest types bias pollination and dispersal spectra, and some drought-related evolutionary trends have been suggested. Among trees, conspicuous flowers and wind-dispersed seeds are relatively more frequent in drier forests.

Faunas of pollination and seed dispersal vectors differ among and within habitats in taxonomic composition, seasonality and energetics. Bee faunas differ in composition and sociality, and probably in seasonality and body size distribution. Sphingid moth faunas do not differ in body size but in seasonality, particularly between very wet and dry forests. Bat and hummingbird faunas differ in diversity but not in body size between wet and dry forests, and migration may be more common in drier sites.

Female fecundity may often be limited by a lack of effective pollination, but comparisons among habitats are few. Low fecundity and non-annual reproduction are frequent in trees, but their relation to variation in the length or quality of the growing season is unknown. Trunk size is apparently a mediocre predictor of reproduction. Greater density of smaller individuals may favor outcrossing in drier forests.

Acknowledgements

I am grateful for the useful comments and encouragement of T. Bullock, A. Gentry, G. Gottsberger, C. Martínez del Río, R. Primack, I. Silberbauer-Gottsberger, B. Simpson, N. Martijena, E. Medina, H. Mooney and J. Neff.

References

Aizen, M. A. & Feinsinger, P. (1994). Forest fragmentation, pollination, and plant reproduction in a Chaco dry forest, Argentina. *Ecology* **75**: 330–51.

Arizmendi, M. C. & Ornelas, F. (1990). Hummingbirds and their floral resources in the west coast of México. *Biotropica* **22**: 172–80.

Arroyo, M. T. K. (1976). Geitonogamy in animal pollinated tropical angiosperms: a stimulus for the evolution of self-incompatibility. *Taxon* **25**: 543–48.

Ayala, R., Griswold, T. L. & Bullock, S. H. (1993). The native bees of México. In *Biological Diversity of México: origins and distribution*, ed. T. P. Ramamoorthy, R. Bye, A. Lot and J. Fa, pp. 179–227. Oxford University Press, New York.

Baker, H. G. (1959). Reproductive methods as factors in speciation of flowering plants. *Cold Spring Harbor Symposia on Quantitative Biology* **24**: 177–90.

Baker, H. G. & Baker, I. (1983). Floral nectar sugar constituents in relation to pollinator type. In *Handbook of Experimental Pollination Biology*, ed. C. E. Jones and R. J. Little, pp. 117–41. Van Nostrand Reinhold Co., New York.

Baker, H. G., Bawa, K. S., Frankie, G. W. & Opler, P. A. (1983). Reproductive biology of plants in tropical forests. In *Tropical Rain Forest Ecosystems: structure and function*, ed. F. B. Golley, pp. 183–215. Elsevier, Amsterdam.

Barbosa, D. C. A., Alves, J. L. H., Prazeres, S. M. & Paiva, A. M. A. (1989). Dados fenológicos de 10 espécies arbóreas de uma área de caatinga (Alagoinha-PE). *Acta Botanica Brasileira* **3** (Suppl.): 109–17.

Bawa, K. S. (1974). Breeding systems of tree species of a lowland tropical community. *Evolution* **28**: 85–92.

Bawa, K. S., Bullock, S. H., Perry, D. R., Coville, R. E. & Grayum, M. H. (1985). Reproductive biology of tropical lowland rain forest trees. II. Pollination systems. *American Journal of Botany* **72**: 346–56.

Bawa, K. S. & Hadley, M. (ed.) (1990). *Reproductive Ecology of Tropical Forest Plants*. UNESCO, Paris.

Bawa, K. S., Perry, D. R. & Beach, J. H. (1985). Reproductive biology of tropical lowland rain forest trees. I. Sexual systems and incompatibility mechanisms. *American Journal of Botany* **72**: 331–45.
Bawa, K. S. & Webb, C. J. (1983). Floral variation and sexual differentiation in *Muntingia calabura* (Elaeocarpaceae), a species with hermaphroditic flowers. *Evolution* **37**: 1271–82.
Bawa, K. S. & Webb, C. J. (1984). Flower, fruit and seed abortion in tropical forest trees: implications for the evolution of paternal and maternal reproductive patterns. *American Journal of Botany* **71**: 736–51.
Beebe, W. (1949). Insect migration at Rancho Grande in North-Central Venezuela. General Account. *Zoologica* **34**: 107–10.
Boaler, S. B. (1966). Ecology of a miombo site, Lupa North Forest Reserve, Tanzania. II. Plant communities and seasonal variation in the vegetation. *Journal of Ecology* **54**: 465–79.
Borchert, R. (1983). Phenology and control of flowering in tropical trees. *Biotropica* **15**: 81–9.
Bullock, S. H. (1985). Breeding systems in the flora of a tropical deciduous forest in México. *Biotropica* **17**: 287–301.
Bullock, S. H. (1986). Observations and an experiment on synchronous flowering. *Madroño* **33**: 223–4.
Bullock, S. H. (1994). Wind pollination of neotropical dioecious trees. *Biotropica* **26**: 172–79.
Bullock, S. H., Ayala, R., Baker, I. & Baker, H. G. (1987). Reproductive biology of the tree *Ipomoea wolcottiana* (Convolvulaceae). *Madroño* **34**: 304–14.
Bullock, S. H., Martínez del Río, C. & Ayala, R. (1989). Bee visitation rates to trees of *Prockia crucis* differing in flower number. *Oecologia* **78**: 389–93.
Bullock, S. H. & Pescador, A. (1983). Wing and proboscis dimensions in a sphingid fauna from western México. *Biotropica* **15**: 292–4.
Bullock, S. H. & Solís Magallanes, J. A. (1990). Phenology of canopy trees of a tropical deciduous forest in México. *Biotropica* **22**: 22–35.
Burger, W. C. (1974). Flowering periodicity at four altitudinal levels in eastern Ethiopia. *Biotropica* **6**: 38–42.
Carabias-Lillo, J. & Guevara Sada, S. (1985). Fenología en una selva tropical húmeda y en una comunidad derivada; Los Tuxtlas, Veracruz. In *Investigaciones sobre la Regeneración de Selvas II*, ed. A. Gómez-Pompa and S. Del Amo R., pp. 27–66. Instituto Nacional de Investigaciones sobre Recursos Bióticos, Xalapa.
Carpenter, F. L. (1976). Plant-pollinator interactions in Hawaii: pollination energetics of *Metrosideros collina* (Myrtaceae). *Ecology* **57**: 1125–44.
Ceballos, G. & Miranda, A. (1986). *Los Mamíferos de Chamela*. Universidad Nacional Autónoma de México, México.
Coates-Estrada, R. & Estrada, A. (1986). *Manual de Identificación de Campo de los Mamíferos de la Estación de Biología Los Tuxtlas*. Universidad Nacional Autónoma de México, México.
Croat, T. B. (1975). Phenological behavior of habit and habitat classes on Barro Colorado Island (Panama Canal Zone). *Biotropica* **7**: 270–7.
Croat, T. B. (1979). The sexuality of the Barro Colorado Island flora (Panama). *Phytologia* **42**: 319–48.
Daniel, T. F. (1983). *Carlowrightia. Flora Neotropica* **34**: 1–116.
D'Arcy, W. G. (ed.) (1986). *Solanaceae: biology and systematics*. Columbia University Press, New York.
Daubenmire, R. (1972). Phenology and other characteristics of tropical semi-deciduous forest in north-western Costa Rica. *Journal of Ecology* **60**: 147–70.
Dehgan, B. (1984). Phylogenetic significance of interspecific compatibility in *Jatropha* (Euphorbiaceae). *Systematic Botany* **9**: 467–78.

DesGranges, J. L. (1978). Organization of a tropical nectar feeding bird guild in a variable environment. *Living Bird* **17**: 199–236.
Domínguez, C. A. (1990). Consecuencias ecológicas y evolutivas del patrón de floración sincrónico y masivo de *Erythroxylum havanense* Jacq. (Erythroxylaceae). Tesis Doctoral, Centro de Ecología, Universidad Nacional Autónoma de México, México.
Domínguez, C. A. & Bullock, S. H. (1989). La reproducción de *Croton suberosus* (Euphorbiaceae) en luz y sombra. *Revista de Biología Tropical* **37**: 1–10.
Feinsinger, P. & Chaplin, S. B. (1975). On the relationship between wing disc loading and foraging strategies in hummingbirds. *American Naturalist* **109**: 217–24.
Feinsinger, P. & Colwell, R. K. (1978). Community organization among neotropical nectar-feeding birds. *American Zoologist* **18**: 779–95.
Feinsinger, P., Swarm, L. A. & Wolfe, J. A. (1985). Nectar feeding birds on Trinidad and Tobago: a comparison of diverse and depauperate guilds. *Ecological Monographs* **55**: 1–28.
Fenner, M. (1982). Aspects of the ecology of *Acacia-Commiphora* woodland near Kibwezi, Kenya. *Journal of the East Africa Natural History Society and National Museum* **175**: 1–11.
Foldats, E. & Rutkis, E. (1975). Ecological studies of chaparro (*Curatella americana* L.) and manteco (*Byrsonima crassifolia* H.B.K.) in Venezuela. *Journal of Biogeography* **2**: 159–78.
Fox, J. F. (1985). Incidence of dioecy in relation to growth form, pollination and dispersal. *Oecologia* **67**: 244–9.
Frankie, G. W., Baker, H. G. & Opler, P. A. (1974). Comparative phenological studies of trees in wet and dry forests in the lowlands of Costa Rica. *Journal of Ecology* **62**: 881–919.
Frankie, G. W., Haber, W. A., Opler, P. A. & Bawa, K. S. (1983). Characteristics and organization of the large bee pollination system in Costa Rican dry forest. In *Handbook of Experimental Pollination Biology*, ed. C. E. Jones and R. J. Little, pp. 411–47. Van Nostrand Reinhold Co., New York.
Frankie, G. W., Opler, P. A. & Bawa, K. S. (1976). Foraging behavior of solitary bees: implications for outcrossing of a neotropical forest tree species. *Journal of Ecology* **64**: 1049–57.
Freeman, D. C., Klikoff, L. G. & Harper, K. T. (1980). Ecology of dioecy in the intermountain region of western North America. *Oecologia* **44**: 410–17.
Gentry, A. H. (1974). Flowering phenology and diversity in tropical Bignoniaceae. *Biotropica* **6**: 64–8.
Gentry, A. H. (1982). Patterns of neotropical plant species diversity. *Evolutionary Biology* **15**: 1–84.
Gentry, A. H. (1989). Speciation in tropical forests. In *Tropical forests: botanical dynamics, speciation and diversity*, ed. L. B. Holm-Nielsen, I. C. Nielsen and H. Balslev, pp. 113–34. Academic Press, London.
Gottsberger, G. (1986). Some pollination strategies in neotropical savannas and forests. *Plant Systematics and Evolution* **152**: 29–45.
Gottsberger, G. (1989). Comments on flower evolution and beetle pollination in the genera *Annona* and *Rollinia* (Annonaceae). *Plant Systematics and Evolution* **167**: 189–94.
Gottsberger, G. & Silberbauer-Gottsberger, I. (1983). Dispersal and distribution in the cerrado vegetation of Brazil. *Sonderbaende des Naturwissenschaftlichen Vereins in Hamburg* **7**: 315–52.
Grant, B. R. & Grant, P. R. (1981). Exploitation of *Opuntia* cactus by birds on the Galápagos. *Oecologia* **49**: 179–87.
Grant, V. & Grant, K. A. (1979). The pollination spectrum of the Southwestern American cactus flora. *Plant Systematics and Evolution* **133**: 29–37.
Grove, K. F. (1985). Reproductive biology of neotropical wet forest understory plants. Ph.D. thesis, University of Iowa, Iowa City.

Guevara de Lampe, M., Bergeron, Y., McNeil, R. & Leduc, A. (1992). Seasonal flowering and fruiting patterns in tropical semi-arid vegetation of northeastern Venezuela. *Biotropica* **24**: 64–76.
Haber, W. A. & Frankie, G. W. (1982). Pollination of *Luehea* (Tiliaceae) in Costa Rican deciduous forest: significance of adapted and non-adapted pollinators. *Ecology* **63**: 1740–50.
Haber, W. A. & Frankie, G. W. (1989). A tropical hawkmoth community: Costa Rican dry forest Sphingidae. *Biotropica* **21**: 155–72.
Heithaus, E. R. (1979a). Community structure of neotropical flower visiting bees and wasps: diversity and phenology. *Ecology* **60**: 190–202.
Heithaus, E. R. (1979b). Flower-feeding specialization in wild bee and wasp communities in seasonal neotropical habitats. *Oecologia* **42**: 179–94.
Heithaus, E. R. & Fleming, T. H. (1981). Frugivorous bats, seed shadows and the structure of tropical forests. *Biotropica* **13** (Suppl.): 45–53.
Heithaus, E. R., Fleming, T. H. & Opler, P. A. (1975). Foraging patterns and resource utilization in seven species of bats in a seasonal tropical forest. *Ecology* **56**: 841–54.
Heithaus, E. R., Opler, P. A. & Baker, H. G. (1974). Bat activity and pollination of *Bauhinia pauletia*: plant-pollinator coevolution. *Ecology* **55**: 412–19.
Holdridge, L. R., Grenke, W. C., Hatheway, W. H., Liang, T. & Tosi, J. A., Jr (1971). *Forest Environments in Tropical Life Zones*. Pergamon Press, Oxford.
Holm-Nielsen, L. B. (1979). Comments on the distribution and evolution of the genus *Phyllanthus* (Euphorbiaceae). In *Tropical Botany*, ed. K. Larsen and L. B. Holm-Nielsen, pp. 277–90. Academic Press, London.
Hubbell, S. P. (1979). Tree dispersion, abundance, and diversity in a tropical dry forest. *Science* **203**: 1299–1309.
Janzen, D. H. (1967). Synchronization of sexual reproduction of trees within the dry season in Central America. *Evolution* **21**: 620–37.
Janzen, D. H. (1978). Seeding patterns of tropical trees. In *Tropical Trees as Living Systems*, ed. P. B. Tomlinson and M. H. Zimmermann, pp. 83–128. Cambridge University Press, Cambridge.
Janzen, D. H. (1989). Natural history of a wind-pollinated Central American dry forest legume tree (*Ateleia herbert-smithii* Pittier). In *Advances in Legume Biology*, ed. C. H. Stirton and J. L. Zarucchi. *Monographs in Systematic Botany. Missouri Botanical Garden* **29**: 293–376.
Karr, J. R. (1980). Geographic variation in the avifaunas of tropical forest undergrowth. *Auk* **97**: 283–98.
Koopman, K. F. (1981). The distribution patterns of New World nectar-feeding bats. *Annals of the Missouri Botanical Garden* **68**: 352–69.
Koptur, S. (1984). Outcrossing and pollinator limitation of fruit set: breeding systems of neotropical *Inga* trees (Fabaceae: Mimosoideae). *Evolution* **38**: 1130–43.
Kress, W. J. (1983). Self-incompatibility systems in Central American *Heliconia* (Heliconiaceae). *Evolution* **37**: 735–44.
LaVal, R. K. & Fitch, H. S. (1977). Structure, movements and reproduction in three Costa Rican bat communities. *Occasional Papers of the Museum of Natural History. University of Kansas* **69**: 1–28.
Leigh, E. G., Jr & Windsor, D. M. (1982). Forest production and regulation of primary consumers on Barro Colorado Island. In *The Ecology of a Tropical Forest*, ed. E. G. Leigh Jr, A. S. Rand and D. M. Windsor, pp. 111–22. Smithsonian Institution Press, Washington.
Lewis, J. P. (1991). Three levels of floristical variation in the forests of Chaco, Argentina. *Journal of Vegetation Science* **2**: 125–30.
Lieberman, D. (1982). Seasonality and phenology in a dry tropical forest in Ghana. *Journal of Ecology* **70**: 791–806.

Long, R. (1975). Artificial interspecific hybridization in temperate and tropical species of *Ruellia* (Acanthaceae). *Brittonia* **27**: 289–96.
Lott, E. J., Bullock, S. H. & Solís Magallanes, J. A. (1987). Floristic diversity and structure in upland and arroyo forests in coastal Jalisco. *Biotropica* **19**: 228–35.
McDade, L. A. (1985). Breeding systems of Central American *Aphelandra* (Acanthaceae). *American Journal of Botany* **72**: 1515–21.
McDiarmid, R. W., Ricklefs, R. E. & Foster, M. S. (1977). Dispersal of *Stemmadenia donnell-smithii* (Apocynaceae) by birds. *Biotropica* **9**: 9–25.
McMullen, C. K. (1987). Breeding systems of selected Galápagos Islands angiosperms. *American Journal of Botany* **74**: 1694–1705.
Machado, I. C. S. (1990). Biologia floral de espécies de caatinga no município de Alagoinha (PE). Tese de doutoramento, Universidade Estadual de Campinas, Campinas.
Malaisse, F. (1974). Phenology of the Zambezian woodland area with emphasis on the miombo ecosystem. In *Phenology and Seasonality Modeling*, ed. H. Lieth, pp. 269–86. Springer-Verlag, Berlin.
Malaisse, F., Freson, R., Goffinet, G. & Malaisse-Mousset, M. (1975). Litter fall and litter breakdown in miombo. In *Tropical Ecological Systems*, ed. F. B. Golley and E. Medina, pp. 137–52. Springer-Verlag, New York.
Mandujano, S., Gallina, S. & Bullock, S. H. (1994). Frugivory and the dispersal of *Spondias purpurea* in a tropical deciduous forest. *Revista de Biología Tropical* **42**: 107–14.
Martínez-Yrízar, A. & Sarukhán, J. (1990). Litterfall patterns in a tropical deciduous forest in México over a five-year period. *Journal of Tropical Ecology* **6**: 433–44.
Michener, C. D. (1979). The biogeography of bees. *Annals of the Missouri Botanical Garden* **66**: 277–347.
Monasterio, M. & Sarmiento, G. (1976). Phenological strategies of plant species in the tropical savanna and the semi-deciduous forest of the Venezuelan llanos. *Journal of Biogeography* **3**: 325–56.
Montalvo, A. M. & Ackerman, J. D. (1987). Limitations to fruit production in *Ionopsis utricularioides* (Orchidaceae). *Biotropica* **19**: 24–31.
Montgomerie, R. D. & Gass, C. L. (1981). Energy limitation of hummingbird populations in tropical and temperate communities. *Oecologia* **50**: 162–65.
Morawetz, W. (1986). Remarks on karyological differentiation patterns in tropical woody plants. *Plant Systematics and Evolution* **152**: 49–100.
Moss, A. M. (1920). Sphingidae of Para, Brazil. *Novitates Zoologicae* **27**: 333–424.
Murawski, D. A., Hamrick, J. L., Hubbell, S. P. & Foster, R. B. (1990). Mating systems of two Bombacaceous trees of a neotropical moist forest. *Oecologia* **82**: 501–6.
Neill, D. A. (1988). Experimental studies on species relationships in *Erythrina* (Leguminosae: Papilionoideae). *Annals of the Missouri Botanical Garden* **75**: 886–969.
Opler, P. A. (1978). Interaction of plant life history components as related to arboreal herbivory. In *The Ecology of Arboreal Folivores*, ed. G. G. Montgomery, pp. 23–31. Smithsonian Institution Press, Washington.
Opler, P. A. (1983). Nectar production in a tropical ecosystem. In *The Biology of Nectaries*, ed. B. Bentley and T. Elias, pp. 30–79. Columbia University Press, New York.
Opler, P. A., Baker, H. G. & Frankie, G. W. (1975). Reproductive biology of some Costa Rican *Cordia* species (Boraginaceae). *Biotropica* **7**: 234–47.
Opler, P. A., Frankie, G. W. & Baker, H. G. (1976). Rainfall as a factor in the release, timing, and synchronization of anthesis by tropical trees and shrubs. *Journal of Biogeography* **3**: 231–36.
Opler, P. A., Frankie, G. W. & Baker, H. G. (1980). Comparative phenological studies of treelet and shrub species in tropical wet and dry forests in the lowlands of Costa Rica. *Journal of Ecology* **68**: 167–88.

Pérez, J. W. (1970). Relation of crown diameter to stem diameter in forests of Puerto Rico, Dominica and Thailand. In *A Tropical Rain Forest*, ed. H. T. Odum and R. F. Pigeon, pp. B105-B122. US Atomic Energy Commission, Oak Ridge.
Pescador, A. (1994). *Manual de Identificación para las Mariposas de la Familia Sphingidae (Lepidoptera) de la Estación de Biología 'Chamela', Jalisco, México.* Cuadernos 22, Instituto de Biología, Universidad Nacional Autónoma de México, México.
Peyton, B. (1980). Ecology, distribution and food habits of spectacled bears, *Tremarctos ornatus*, in Peru. *Journal of Mammalogy* **61**: 639–52.
Piovano, M. A. & Bernardello, L. M. (1991). Chromosome numbers in Argentinian Acanthaceae. *Systematic Botany* **16**: 89–97.
Powell, J. A. & Brown, J. W. (1990). Concentrations of lowland sphingid and noctuid moths at high mountain passes in eastern México. *Biotropica* **22**: 316–19.
Prasad, S. N. & Hegde, M. (1986). Phenology and seasonality in the tropical deciduous forest of Bandipur, South India. *Proceedings of the Indian Academy of Sciences (Plant Sciences)* **96**: 121–33.
Racine, C. H. & Downhower, J. F. (1974). Vegetative and reproductive strategies of *Opuntia* (Cactaceae) in the Galápagos Islands. *Biotropica* **6**: 175–86.
Ramamoorthy, T. P. & Lorence, D. H. (1987). Species vicariance in the Mexican flora and description of a new species of *Salvia* (Lamiaceae). *Bulletin. Museum National d'Histoire Naturelle Paris, 4°serie. Section B, Adansonia* **9**: 167–75.
Ramia, M. (1981). Fenología de árboles en el bosque deciduo tropical. *Memoria de la Sociedad de Ciencias Naturales La Salle* **41**: 9–33.
Ramírez, N. & Brito, Y. (1990). Reproductive biology of a tropical palm swamp community in the Venezuelan llanos. *American Journal of Botany* **77**: 1260–71.
Raw, A. (1985). The ecology of Jamaican bees (Hymenoptera). *Revista Brasileira de Entomologia* **29**: 1–16.
Reich, P. B. & Borchert, R. (1982). Water stress and tree phenology in a tropical dry forest in the lowlands of Costa Rica. *Journal of Ecology* **72**: 61–74.
Renner, S. (1989). A survey of reproductive biology in neotropical Melastomataceae and Memecylaceae. *Annals of the Missouri Botanical Garden* **76**: 496–518.
Ridpath, M. G. (1985). Ecology of the wet–dry tropics: how different? In *Ecology of the Wet–Dry Tropics*, ed. M. G. Ridpath and L. K. Corbett, pp. 3–20. Ecological Society of Australia, Casuarina.
Rick, C. M. (1966). Some plant-animal relations on the Galápagos Islands. In *The Galápagos*, ed. R. I. Bowman, pp. 215–24. University of California Press, Berkeley.
Roubik, D. W. (1989). *Ecology and Natural History of Tropical Bees.* Cambridge University Press, Cambridge.
Ruiz Zapata, T. & Arroyo, M. T. K. (1978). Plant reproductive ecology of a secondary deciduous tropical forest in Venezuela. *Biotropica* **10**: 221–30.
Rzedowski, J. (1993). Diversity and origins of the phanerogamic flora of México. In *Biological Diversity of México: origins and distribution*, ed. T. P. Ramamoorthy, R. Bye, A. Lot and J. Fa, pp. 129–44. Oxford University Press, New York.
Sazima, M., Fabian, M. E. & Sazima, I. (1982). Polinização de *Luehea speciosa* (Tiliaceae) por *Glossophaga soricina* (Chiroptera, Phyllostomidae). *Revista Brasileira de Biologia* **42**: 505–13.
Schatz, G. E. (1990). Some aspects of pollination biology in Central American forests. In *Reproductive Ecology of Tropical Forest Plants*, ed. K. S. Bawa and M. Hadley, pp. 69–84. UNESCO, Paris.
Schemske, D. W. (1980). Floral ecology and hummingbird pollination of *Combretum farinosum* in Costa Rica. *Biotropica* **12**: 169–81.
Schreiber, H. (1978). Dispersal centres of Sphingidae (Lepidoptera) in the neotropical region. *Biogeographica* **10**: 1–195.

Schulze, E.-D., Mooney, H. A., Bullock, S. H. & Mendoza, A. (1988). Water contents of wood of tropical deciduous forest species during the dry season. *Boletín de la Sociedad Botánica de México* **48**: 113–18.

Sivaraj, N. & Krishnamurthy, K. V. (1989). Flowering phenology in the vegetation of Shervaroys, South India. *Vegetatio* **79**: 85–8.

Sobrevila, C. & Arroyo, M. T. K. (1982). Breeding systems in a montane tropical cloud forest in Venezuela. *Plant Systematics and Evolution* **140**: 19–37.

Spichiger, R. & Ramella, L. (1989). The forests of the Paraguayan Chaco. In *Tropical Forests: botanical dynamics, speciation and diversity*, ed. L. B. Holm-Nielsen, I. C. Nielsen and H. Balslev, pp. 259–70. Academic Press, London.

Stebbins, G. L. (1958). Longevity, habitat and release of genetic variability in higher plants. *Cold Spring Harbor Symposia on Quantitative Biology* **23**: 365–78.

Stiles, F. G. (1981). Geographical aspects of bird-flower coevolution, with particular reference to Central America. *Annals of the Missouri Botanical Garden* **68**: 323–51.

Veillon, J. P. (1963). Relación de ciertas características de la masa forestal de unos bosques de las zonas bajas de Venezuela con el factor climático: humedad pluvial. *Acta Científica Venezolana* **14**: 30–41.

Warming, J. E. B. (1909). *Oecology of Plants: an introduction to the study of plant communities*. Clarendon Press, Oxford.

Whigham, D. F., Zugasty Towle, P., Cabrera Cano, E., O'Neill, J. & Ley, E. (1990). The effect of annual variation in precipitation on growth and litter production in a tropical dry forest in the Yucatán of México. *Tropical Ecology* **31**: 23–34.

Wikander, T. (1984). Mecanismos de dispersión de diasporas de una selva decidua en Venezuela. *Biotropica* **16**: 276–83.

Wolf, L. L. (1970). The impact of seasonal flowering on the biology of some tropical hummingbirds. *Condor* **72**: 1–14.

Wolf, L. L., Stiles, F. G. & Hainsworth, F. R. (1976). Ecological organization of a tropical, highland hummingbird community. *Journal of Animal Ecology* **45**: 349–79.

Wright, S. J. & Cornejo, F. H. (1990). Seasonal drought and leaf fall in a tropical forest. *Ecology* **71**: 1165–75.

12

Plant–herbivore interactions in Mesoamerican tropical dry forests

RODOLFO DIRZO & CÉSAR A. DOMÍNGUEZ

Introduction

The tropical dry forests of Mesoamerica constitute an ecological theater where a plethora of herbivory plays occur with a great variety of forms, intensity and spatio-temporal variation. In contrast to other types of tropical forests, the herbivory events in dry forests are frequently quite apparent or visible to the eyes of the professional or even the casual observer. In these forests, for example, complete defoliation events are not so uncommon, particularly on some deciduous (as opposed to evergreen) species of trees and shrubs, and these occurrences suggest that massive defoliations and insect 'outbreaks' are not exclusively extra-tropical phenomena. Likewise, levels of seed predation for some individuals in some species can sometimes be in the order of 100% (e.g. *Acacia cornigera*: R. Dirzo, personal observation).

From the point of view of the consumers, the protagonists of these ecological plays comprise two major groups.

1 The array of vertebrates spans four orders of magnitude in body weight. In Santa Rosa National Park ('SRNP', Guanacaste Province, Costa Rica), these include forest understory mammals ranging from tapirs (*Tapirus bairdii*) (200,000 g) and white-tailed deer (*Odocoileus virginianus*), which actively consume seeds and foliage, to 30 g seed-eating mice such as *Liomys pictus* or *Oryzomys palustris*, and forest canopy mammals ranging from leaf- and flower-eating howler monkeys (*Alouatta palliata*) (8000 g) to small (40 g) leaf- and seed-eating rats such as *Nyctomys sumichrastii*.
2 The array of invertebrate herbivores is extremely diverse. In SRNP these include not less than 110 species of seed-eating beetles (Janzen, 1980), and 3140 species of folivorous caterpillars (Janzen, 1988).

Likewise, from the point of view of the resource base, the plants also comprise a genetically diverse array. For example, there are not less than 975 species in the Guanacaste lowlands (Janzen & Liesner, 1980), or 1082 species in the Chamela area, state of Jalisco, México (Lott, 1993). This wealth of Latin plant binomials can come, to the eyes of the animals in the next trophic level, as an outstanding variety of consumable items such as young embryos, seeds, herbaceous and woody stems, leaves, buds and saps with an amazing variation in their nutritional and toxic properties and, additionally, highly variable in space and time.

Given this scenario, is it possible to detect some structure or discernible patterns in the trophic interactions between plants and phytophagous animals from dry forest? The answer is 'yes'; but very emphatically, yes in part. As this chapter will show, our knowledge on this theme is still quite preliminary and fragmentary. Nevertheless, we hope to illustrate that some progress has been made.

In two comprehensive reviews of the patterns of herbivory in SRNP, Janzen (1981, 1983a) distinguishes two major levels of analysis: (1) the community-level analysis of the degree of herbivory and its spatio-temporal variation; and (2) the patterns and consequences of herbivory on plant populations and on individual plant or animal performance and fitness. In the present contribution we follow such a sequence, but in the light of some more recent studies and considering the evidence obtained from other Mesoamerican forests, mainly in México. Our review is largely from the point of view of the effect of the animals on the plants.

Community-level analysis of folivory

By far the most evident and correspondingly the most studied type of herbivory on vegetative parts is folivory or leaf consumption. The image of the levels of herbivory that we can get from assessments of folivory is, however, a gross underestimate of the herbivores' impacts on the community, for it ignores the less evident (and difficult to measure) but still omnipresent episodes of phloem sucking, shoot apex removal, branch and twig cutting, leaf and wood mining, and belowground herbivory. Moreover, the measurements of folivory are not completely free from difficulties and inaccuracies of some degree, as we will discuss below. The consumption of leaf tissue is carried out in two contrasting patterns depending on whether the folivores are invertebrates (insects) or vertebrates.

Invertebrate leaf-eaters are fundamentally insects. Mollusks seem to play a minor role at least at Chamela and in other Costa Rican dry forests with which we are familiar. This situation is similar in many wet forests (Dirzo and Domínguez, personal observation). Of the insects, the best documented are leaf-eating caterpillars and these, in association with some beetles (e.g. Chrysomelidae) and leaf-cutting ants (Attini), collectively determine the following patterns.

Levels of folivory

Janzen (1981) provides a qualitative image of the levels of damage in SRNP: 'a few individuals show no damage, many show light to moderate damage and a few are heavily defoliated'. The accuracy of this statement is reflected in a data set gathered in a survey of the standing levels of damage on a sample of 45 trees of 33 species in Guanacaste, Costa Rica (Figure 12.1; R. Dirzo, personal communication; $n = 1$ for 21 species and $n = 2$ for 12 species at the Palo Verde field station). The survey shows that one tree was undamaged, but the vast majority (71%) of the sampled plants had between 1 and 12% leaf area eaten, while 11% of the plants sustained a level of 25–50% defoliation and only one tree (2.2% of the sample) was completely defoliated. This general picture remains relatively similar between years, although the allocation of individual trees and species into the different categories of damage tends to change somewhat. An indication of such changes is provided by a study of between-year variation of the levels of folivory in 16 species at Chamela (Table 12.1; Filip *et al.*, 1995). Measurements of percentage leaf area loss in three consecutive years (1983–5) show a relative similarity in the overall level of damage (6.7, 9.2 and 7.8%). Interestingly, this relative consistency occurs even though 1983 and 1985 were the wettest and driest years in the history of the station. However, six of the 16 species showed at least one significant between-year variation.

Although it would be tempting to try to use these kinds of data to estimate the leaf area consumed by insects in the community, this may not be appropriate because the standing levels of damage (i.e. defined by a single measurement) underestimate the real leaf area eaten to a varying degree. A comparison of single (instantaneous) measurements of insect damage with repeated (long-term) measurements (achieved by labelling individual leaves and drawing their intact and damaged contours repeatedly) for a set of 12 species at Chamela (Table 12.2) provides an indication of the degree of underestimate of the single measurements. Table 12.2 shows that in seven of the 12 species, the single measurements yield

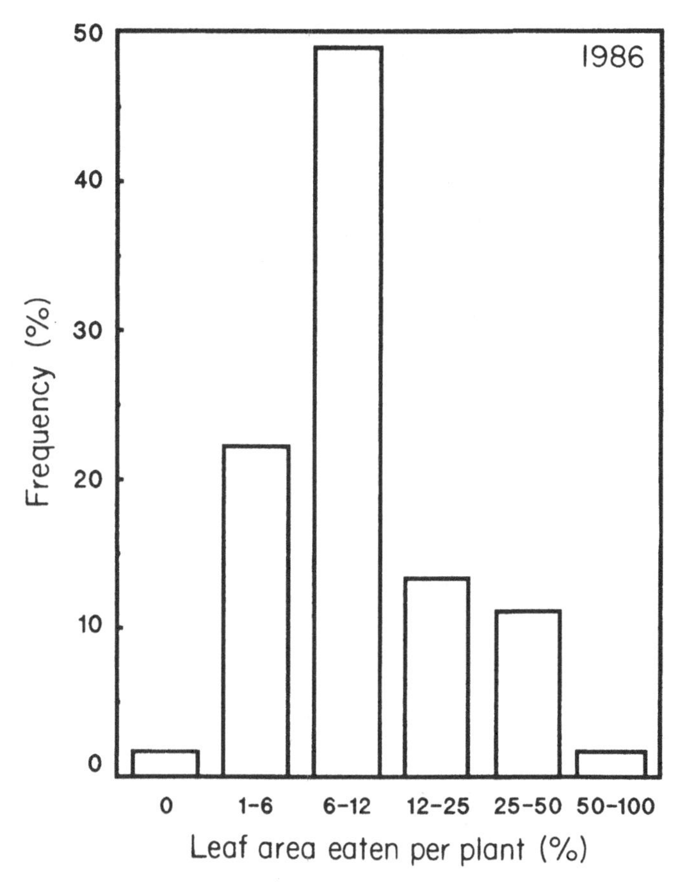

Figure 12.1. Frequency distribution of damage on trees from the Palo Verde field station, Guanacaste Province, Costa Rica.

significantly lower values of herbivory compared with the long-term measurements. Moreover, there is a large interspecific variation in the degree of the underestimate. A discrepancy ratio (long-term: instantaneous values) gives a direct measure of the underestimate. For these data, the ratio ranges from 0.79 to 4.52 and shows that the discrepancy tends to be greater in the more heavily damaged species (Spearman's rank correlation, $r_s = 0.76$, $p = 0.016$). Taking all the species together, the overall mean discrepancy ratio is 1.99. Thus, in general, single measurements underestimate leaf area loss by half, although there is great interspecific variation.

Table 12.1. *Percentage leaf area loss for 16 species at Chamela, Jalisco, México, in three consecutive years (mean ± SD)*

Species	Percentage leaf area loss			Statistical comparison[a]
	1983	1984	1985	
Bursera heteresthes	5.08 ± 1.06	2.72 ± 1.80	1.56 ± 1.43	n.s.
Casearia species	9.67 ± 4.53	12.32 ± 6.10	7.31 ± 4.15	n.s.
Celaenodendron mexicanum	7.06 ± 2.99	12.66 ± 4.18		n.s.
Cochlospermum vitifolium	4.47 ± 2.13	3.89 ± 1.27		n.s.
Cordia alliodora	7.0 ± 2.79	4.73 ± 2.73	2.61 ± 2.32	83 84 85
Cordia elaeagnoides	3.08 ± 2.19	3.90 ± 2.41		n.s.
Croton pseudoniveus	4.13 ± 2.91	19.80 ± 5.22	16.98 ± 10.29	83 84 85
Guapira species	7.56 ± 4.76	10.36 ± 4.74	9.04 ± 2.83	n.s.
Heliocarpus pallidus	11.88 ± 4.45	11.01 ± 4.71	8.40 ± 3.09	n.s.
Ipomoea wolcottiana	11.38 ± 4.75	17.27 ± 6.73	13.35 ± 8.46	n.s.
Jatropha standleyi	9.75 ± 4.13	10.40 ± 5.32	19.53 ± 9.44	83 84 85
Lonchocarpus eriocarinalis	4.71 ± 2.25	5.75 ± 1.61	4.93 ± 2.53	n.s.
Plumeria rubra	5.32 ± 2.80	10.80 ± 4.19	1.92 ± 1.55	83 85 84
Psidium sartorianum	3.26 ± 0.71	2.62 ± 1.67		n.s.
Thouinia paucidentata	7.41 ± 2.75	9.96 ± 5.21	2.86 ± 1.67	83 84 85
Trichilia trifolia	5.89 ± 1.89	9.40 ± 4.21	2.30 ± 1.70	83 84 85
Overall mean	6.73	9.22	7.57	n.s.

[a] Years not jointly underlined are significantly different ($0.05 > p > 0.01$, ANOVA).
Source: Filip *et al.* (1995).

The results of the long-term study indicate that mean leaf area loss for a group of important tree species at Chamela is 17%, with a range of 1.2–73%. This mean value is higher than the upper values of the range commonly reported for some wet forests (*c.* 8–10%) (Dirzo, 1987). This is probably a reflection of the fact that in other studies the levels of herbivory have been obtained from single measurements. It would be extremely interesting to assess the discrepancy ratios for wet forest species. This would allow us to define unequivocally whether the apparently higher levels of herbivory in dry forest are indeed real or a reflection of the fact that damage is more evident in these forests.

Table 12.2. *Estimates of per cent leaf area loss, using single and repeated measurements, and their discrepancy ratio, at Chamela, 1985*

Species	Single	Repeated	Ratio[a]
Jatropha standleyi	19.53	72.66	3.72***
Ipomoea wolcottiana	13.35	29.65	2.22***
Croton pseudoniveus	16.98	19.74	1.16
Guapira species	9.04	26.47	2.93***
Heliocarpus pallidus	8.40	17.80	2.13**
Casearia species	7.31	15.99	2.19**
Plumeria rubra	1.92	8.67	4.52**
Thouinia paucidentata	2.86	4.63	1.62*
Lonchocarpus eriocarinalis	4.93	4.14	0.84
Cordia alliodora	2.61	2.48	0.95
Trichilia trifolia	2.30	1.95	0.85
Bursera heteresthes	1.56	1.24	0.79
Overall mean ± SD	7.57 ± 6.16	17.13 ± 2.01	1.99 ± 1.22

[a] Significance by t-test: *, $p < 0.05$; **, $p < 0.01$; ***, $p < 0.001$.
Source: Filip *et al.* (1995).

Likewise, it would be useful to investigate how species- and site-specific the correction factors are, as well as how consistent they are between years. This might provide a practical tool for estimates of the levels of damage for whole communities, on the basis of single measurements, which are much simpler and less time-consuming, in comparison with the complicated repeated measures.

At this stage, even though we have an idea of the underestimate by instantaneous measurements, it seems likely that the general shape of the frequency distributions of leaf area loss, like the ones obtained at Palo Verde (cf. Figure 12.1), will not change markedly. Also, it appears that the level of folivory at the scale of the whole community does not change greatly between years (e.g. Janzen, 1981; but see Foster, 1990) while particular species and individuals change from year to year, sometimes dramatically. Two clear examples of this are *Cnidoscolus spinosus* and *Erythroxylum havanense* at Chamela (personal observation).

Within-year variation in herbivory

Given the seasonal deciduousness of this kind of forest, it is obvious that folivory will tend to be concentrated in the rainy season. However, temporal variation of folivory within the rainy season also seems to occur. Janzen (1981) and Janzen & Waterman (1984) qualitatively describe a pattern for SRNP. These authors state that there is a very conspicuous peak in the activity of insect defoliators during the first half of the rainy season, followed by a much less populous second generation

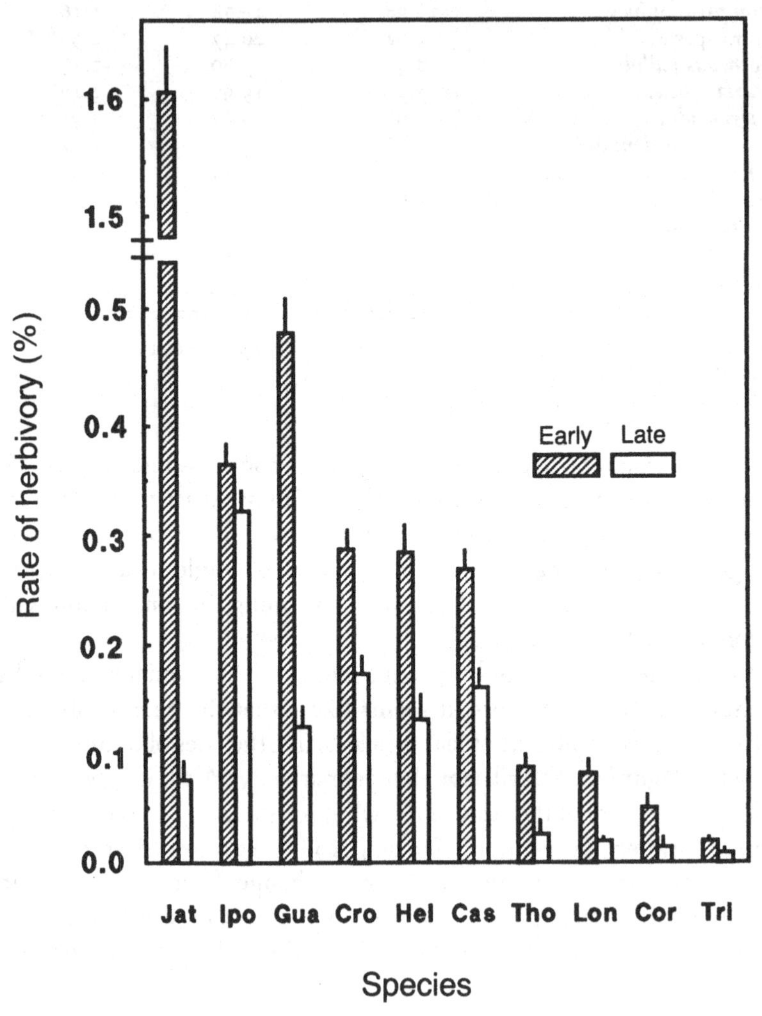

Figure 12.2. Rates of herbivory (mean ± SD) in 10 species at the Chamela field station, state of Jalisco, México, measured in two periods, early and late, of the rainy season of 1985. Species codes: Jat, *Jatropha standleyi*; Ipo, *Ipomoea wolcottiana*; Gua, *Guapira* species; Cro, *Croton pseudoniveus*; Hel, *Heliocarpus pallidus*; Cas, *Casearia* species; Tho, *Thouinia paucidentata*; Lon, *Lonchocarpus eriocarinalis*; Cor, *Cordia alliodora*; Tri, *Trichilia trifolia*. From Filip *et al.* (1995).

during the second part of the rainy season. This early peak of activity should translate into an early peak of folivory. We have some qualitative observations from Palo Verde which lend support to this statement, and a more quantitative study at Chamela also points in the same direction (Filip *et al.*, 1995; Figure 12.2). The rates of herbivory (percentage leaf area eaten per day) were calculated for a set of 10 species and for two periods in the wet season: early (July–early August) and late (August–late September). Figure 12.2 shows that, with the unique exception of *Ipomoea wolcottiana*, all the species had a greater rate of herbivory early in the season and in all of these cases the differences were highly significant. The daily rates of herbivory obtained in these species fall within the range reported in the few other studies (Nuñez-Farfán & Dirzo, 1988; Coley, 1990) where comparable rates have been measured. The only exception is *Jatropha standleyi*, which presents the second highest rate ever reported (cf. 2.2% per day in *Heliocarpus appendiculatus* at Los Tuxtlas, state of Veracruz, México: Nuñez-Farfán & Dirzo, 1988).

Janzen & Waterman (1984) explored the hypothesis that foliage quality, in terms of digestibility and nutritional value, declines with aging of the leaves; this might explain in part the early peak of activity of insect defoliators and the high values of herbivory early in the season. To this effect, they carried out a within-season analysis of chemical defenses and found no significant change in fiber contents, total phenolic equivalents or condensed tannin equivalents in 80 species of woody plants from SRNP. This and some preliminary data from Chamela suggest that the secondary chemistry and nutritional quality of the leaves are not strong determinants of the early wet season peak of herbivory. However, this does not deny that other secondary compounds or nutritional attributes might be important. An obvious alternative for the explanation of this pattern is that parasites and predators of folivores could build up as the season progresses, to levels where they would depress the populations of folivorous insects. Another possibility is that if the emergence of many herbivorous insects is synchronous and occurs early in the season, the low levels of damage characteristic of the late part of the season might be due, at least in part, to the insects' life history patterns. We were not able to find relevant data for the assessment of these two possibilities. Clearly these aspects warrant further study.

Spatial heterogeneity in folivory

One of the most obvious spatial contrasts in Mesoamerican dry forest is given by the occurrence of riparian vegetation. Arroyo flood plains, riparian or gallery forests sustain an evergreen vegetation (in Costa Rica) or semideciduous forests (in México) which contrasts with the rest of the deciduous forest. The possible existence of heterogeneity in herbivory associated with deciduous as opposed to evergreen canopies has been noticed by other workers (Janzen, 1983a; Stanton, 1975). At Palo Verde, a series of consecutive surveys of herbivory in the riparian (Arroyo Colmenar) and deciduous forests has been carried out to explore the possible differences in herbivory. Typical results are given in Figure 12.3. The figure shows the distribution of damage in randomly sampled leaves from 10 species typical of each forest (100 leaves per species). There is a highly significant heterogeneity ($\chi^2 = 466.8$, $p < 0.001$) whereby the leaves of the riparian plants are over-represented in the low categories of damage and under-represented in the categories of intense damage. In contrast, the leaves of the dry forest plants are under-represented in the categories of low damage and overrepresented in the intense damage classes. Similar results have been obtained from the same kind of survey comparing arroyos (semideciduous forest) and hillsides (deciduous forest) at Chamela (R. Dirzo, personal communication).

The consistency and magnitude of these differences calls for an explanation. An evolutionary interpretation for this kind of result has been given by Stanton (1975). She suggests that because dry forest plants drop their leaves during the dry season, these are unavailable to herbivores for a prolonged period. In these circumstances, some of the folivorous insects look for food and shelter in the riparian forest during the dry season and therefore apply a more constant and perhaps intense selective pressure on the evergreen plants. Stanton reasoned that the evergreen riparian plants should be more likely to evolve defense mechanisms that should render them more highly resistant to herbivores. In the chemical survey of Janzen & Waterman (1984) mentioned above, a comparison of evergreen and deciduous plants from Santa Rosa was made (see Table 12.3) and it was found that the evergreen species on average, have leaves with significantly higher fiber content; this was true for leaves in three age categories. In contrast, alkaloids were more common on the foliage of deciduous species. This is still an exciting and open question that deserves further study.

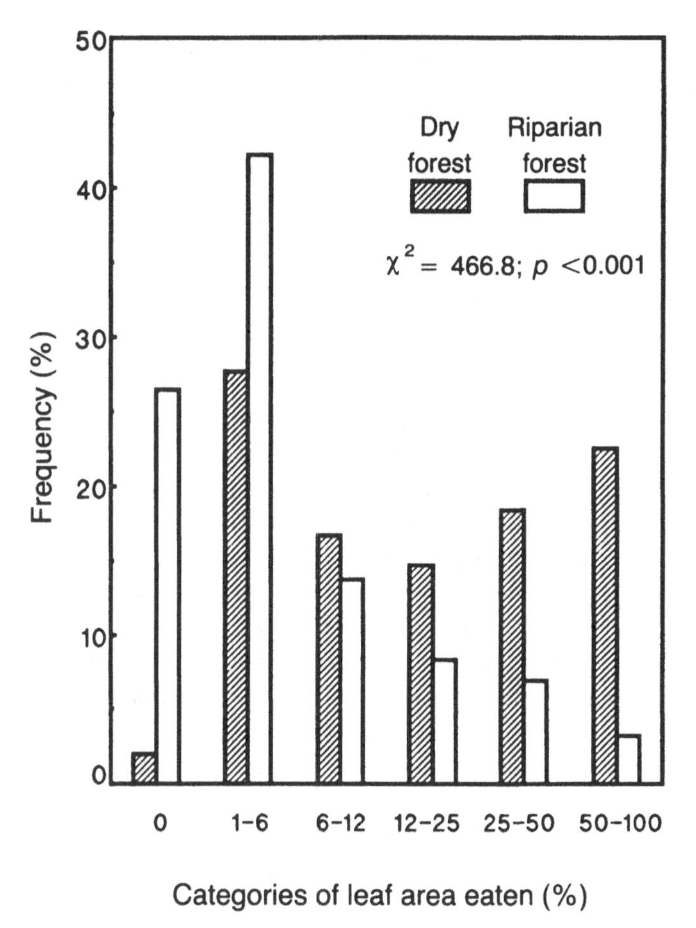

Figure 12.3. Frequency distribution on damage of leaves from riparian and deciduous forest from Palo Verde in 1989. The χ^2 analysis tests the null hypothesis of independence of categories of leaf area eaten and type of forest.

Folivory by vertebrates

In marked contrast, folivory by vertebrates in Mesoamerican dry forest remains practically unstudied. On the basis of his own experience and of the scarce literature available, Janzen (1981) concludes that vertebrate leaf-eaters represent a relatively trivial portion of the overall herbivory threat and impact. Studies on the patterns of herbivory at the Los Tuxtlas

Table 12.3. *Percentage of acid detergent fiber per unit dry weight of leaves (mean ± SD) at three stages of development from evergreen and deciduous species from Santa Rosa National Park, Guanacaste, Costa Rica*

Leaf age	Deciduous	Evergreen	*t*-test
Young	33.6 ± 7.0	44.1 ± 8.1	3.35; $p < 0.01$
Middle-aged	37.9 ± 7.2	44.5 ± 5.1	4.01; $p < 0.001$
Old	38.9 ± 6.3	44.5 ± 8.4	2.59; $p < 0.02$

Source: Janzen & Waterman (1984).

wet forest lead to the same conclusion. However, in the light of some recent studies at Los Tuxtlas (Dirzo & Miranda, 1990), the marked and consistent lack of vertebrate herbivory observed there appears to be a human-made phenomenon due to contemporary defaunation. It would be tempting to suggest that contemporary defaunation might be an explanation for the reduced vertebrate folivory reported for Mesoamerican dry forest. However, at SRNP, where defaunation does not seem to be a major problem and given the low densities of howler, spider and white-faced monkeys, tapirs, peccaries, white-tailed deer, squirrels, porcupines, ctenosaur and iguana lizards, Janzen (1981) still concludes that vertebrates are much less of a threat than are the leaf-eating insects. At first sight, a similar situation seems to occur in Chamela. It is important to bear in mind, however, that no quantitative study of vertebrate folivory has been made, and it may be too soon to attempt conclusions.

Community-level analysis of seed predation

This theme has been extensively studied by Janzen and a review of his copious literature would be beyond the scope of the present chapter. Here we will only mention some salient patterns and refer the reader to the appropriate literature (Janzen, 1980, 1981, 1983a, b, 1985).

In a long-term study of the insects that feed on the seeds of plants of the Guanacaste Province, Janzen (1980, 1981) has found that of the *c.* 975 species of plants in the area, at least 100 regularly have beetle larvae that develop in the seeds and which generally (not always) kill the seeds. Sixty-three per cent of the prey species are Leguminosae although only 17% of the plant species in the Guanacaste lowlands belong to this family. Thus, Leguminosae are used as prey in excess of their taxonomic

Table 12.4. *Number of species of plants whose seeds are attacked by 0, 1, 2, 3, 4 and 5 species of seed-eating beetles at Santa Rosa National Park, Guanacaste, Costa Rica*

No. beetle species per plant species	No. plant species	
	Observed	Expected[a]
0	875	825.0
1	59	138.0
2	25	11.5
3	11	0.6
4	4	0.03
5	1	0.001

[a] from a random distribution ($n = 163$).
Source: Janzen (1981).

abundance. Another pattern in the use of seeds as prey for beetles is shown in Table 12.4. Of the 100 known prey species, 59 were fed on by a single species of beetle; 25 were fed on by two; 11 by three, 4 by four and one by five. There were 163 cases of a beetle attacking a species of prey. If these 163 cases were randomly distributed (i.e. following a Poisson distribution) on the 975 available species of prey, the resulting distribution (see Table 12.4) would be significantly different from the observed one ($\chi^2 = 117$; $p < 0.001$). This result indicates that (1) there are too many unattacked species, (2) too few are preyed on by only one species, and (3) too many are preyed on by 2–5 species. These contrasting distributions bear the implication that over evolutionary time most prey species have been extremely difficult for predators to attack (colonize) while, in contrast, a few have been extremely susceptible to colonization. A possible explanation for the latter is that a prey species susceptible to one predator species may lack defenses against beetle larvae in general (Janzen, 1981).

From the point of view of the species of beetle, 75% of them prey on only one species, 13% prey on two, 8% on three, only 2% prey on four, one species preys on six and one preys on eight. Interestingly, in a similar but not so extensive study, Andrade (1989) found, for the wet forest at Los Tuxtlas, a remarkably similar pattern for a sample of 100 species of woody plants. These taxonomic patterns indicate that seed predation by insects, from the animals' viewpoint, is an interaction of great specificity. From the numerous studies of Janzen on this topic, it appears that the proximal explanations for these highly species-specific patterns of herbivory may be related in considerable degree to the defensive chemistry of the seeds.

In a more ecological vein, it is noted that invertebrates can act as pre-dispersal or post-dispersal predators, or both. Regarding the question of the magnitude of seed predation at the level of the community, data do not seem to be available and this may be due partly to the difficulty of measuring the intensity of seed predation, even within one single species crop. In a subsequent review, Janzen (1983a) indicates that about 20% of all species of plants in the Guanacaste lowlands have pre-dispersal seed predation by insects.

Seed predation by vertebrates

Vertebrate seed predators, like invertebrates, can act as pre-dispersal predators or post-dispersal predators, or both. In contrast to insects, however, vertebrates are highly euryphagous. One major distinction between vertebrate and invertebrate seed predators is that many vertebrates (birds, tapirs, mice, agoutis) combine roles as predators and as dispersal agents. The distinction between these two types of plant–animal interaction (dispersal and predation) is not easy and seems to be very much dependent on the plant and animal species involved, their specific location and the plants' seeding patterns. Some fascinating accounts of these complex interplays have been documented by Janzen (e.g. 1980, 1981, 1983a, b), such as the interaction between the seeds of some trees, beetles, mice, and horses.

Quantitative information on the impact of vertebrate seed predators is markedly lacking and most of the information comes from the detailed natural history account of the interaction between a given plant species and its guild of vertebrate predators. These interactions are commonly mingled in a complex fashion with invertebrate predators and with the role that vertebrates may play, to a varying degree, as dispersal agents.

Janzen (1983a) argues that, taken collectively, both vertebrate and insect seed predators, per biomass unit eaten, have a larger impact on the structure of dry forest than any other animal form. Some isolated studies seem to confirm this statement. At Chamela, for example, between seeds of *Erythroxylum havanense* which were placed on the ground with and without protective exclosures for vertebrates, protected seeds had a chance 3.7 times higher of successful establishment (Gryj, 1990). Although there are some quantitative studies on the density of small mammals (mice) in some sites (e.g. Ceballos, 1989) the lack of information on seed consumption prevents any extrapolations of density data to evaluate the effect on seed banks.

As in the case of folivory by vertebrates (see below), or perhaps more so, contemporary patterns of the interactions between plants and their seed predators are likely to be obscured by the effect of contemporary defaunation (cf. Dirzo & Miranda, 1990) or perhaps the Pleistocene megafaunal extinctions (Janzen & Martin, 1982).

Effects of herbivory on plant populations and fitness

Apart from the documented cases of embryo and seed predation where the effect of the animals on the individual plant is lethal, we have very little evidence of the consequences on individual plant performance of the wide range of levels of herbivory we described above. Likewise, we are not aware of any published studies on the demographic behavior of a plant population under different herbivory regimes. Although there are descriptions of the approximate levels of seed predation for trees of a given population, in the absence of demographic analyses we cannot be sure if those levels of seed mortality have an impact on the population or simply remove individuals that had no future if they were to be killed later by other agents such as, for instance, density-dependent effects at the seedling or juvenile plant stage.

Evidence of the possible consequences of herbivory on plant population dynamics or on plant performance has remained largely as anecdotal accounts and the experimental approach or field manipulations have been very much neglected. Moreover, given the high levels of herbivory (see above) and the currently available instrumentation, it is surprising that no eco-physiological measurements seem to have been made for dry forest plants under different levels of natural or artificial herbivory.

Although it is still premature, the few available studies suggest a potentially negative effect of herbivory on plant performance and fitness. Here we present some representative examples.

Effects on seeds and seedlings

When a seed-eating insect develops inside a seed, it usually kills the seed, but this is not always the case. If the insect does not develop fast enough, or if it does not damage the embryo, or if the seed germinates quickly, the seed may not be killed. Thus, seed predation occasionally may be non-lethal damage. The effect of non-lethal insect damage on seeds has been investigated experimentally by Janzen (1976), by applying three levels of simulated beetle attack on the large (5–10 g) seeds of *Mucuna andreana*. These seeds are not attacked in the wild and appear to be heavily defended by secondary metabolites (L-DOPA). In effect, then, this experiment evaluated how much the fitness of *M. andreana* would be reduced if a mutant seed predator was to appear, with the capability of dealing with the seeds' defensive chemistry. The simulated herbivory consisted of drilling holes into the seeds, to remove 1, 5 and 10% of the seed weight. The resulting seedlings of these treatments were also subjected to artificial herbivory and shading. A summary of the results of this study is given in Table 12.5. The data set documents a conspicuous decline in seed and seedling fitness with an increase in the degree of herbivory. Fitness reduction is apparent in terms of seed germination, seedling survival and seedling biomass. This study provides unequivocal evidence that seed herbivory may dramatically reduce seedling fitness. The study also illustrates the potential adaptive value of effective seed defense that reduces damage. Similar results with the same type of experimental protocol have been obtained with the large seeds of some wet forest species in México (Domínguez & Dirzo, personal communication), indicating that this response is similar in some wet and dry tropical forests.

Effects on reproductive plants

Perhaps the most quoted evidence of the effects of defoliation on seed production for plants from dry forest is Rockwood's (1973) study on artificial defoliation in six plant species from Guanacaste (Table 12.6). Rockwood applied complete hand-defoliations to paired plants of the species listed in Table 12.6. All experimental trees were reproductive, and with the exception of *Cochlospermum vitifolium*, were defoliated twice by hand. For all the species there was a marked negative effect of defoliation on number of fruits and crop weight, but a number of serious limitations of this study are evident from Table 12.6. On the one hand, the statistical analysis is confusing. For example, the tremendous differences between treatments in *Acacia farnesiana*, *Spondias purpurea* and *Crescentia alata* are reported as weakly statistically significant or even not significant (see *S. purpurea*). On the other hand, and perhaps more critical, is the fact that reproductive performance was measured as number of fruits and weight of the whole fruit crop, which probably are not very appropriate indicators of reproductive success and fitness for these species. Another caveat of the experiment refers to the questionable

Table 12.5. *Performance of seedlings of* Mucuna andreana *subjected to three shading and cutting treatments and derived from seeds artificially predated at three levels (plus the control, undamaged)*

Treatment	Germination (%)	Survival age (days)	Biomass (g)
Shaded and cut at 30 cm			
Control	95	90	0.49*
1% loss	80	81	0.35*
5% loss	80	71	0.16
10% loss	65	60	0.16
Shaded and cut at 2–5 cm			
Control	100	116	0.36*
1% loss	90	75	0.12*
5% loss	75	56	0.05
10% loss	45	38	0.03
No shading, cut at 2–5 cm			
Control	95	122*	0.31
1% loss	95	93	0.24*
5% loss	85	58	0.08
10% loss	80	51	0.06

* Significantly different ($p < 0.001$) from the next most severely damaged seed treatment.
Source: Janzen (1976).

Table 12.6. *The effects of artificial defoliation on mean number of fruits and fruit crop weight per individual tree of six species from Guanacaste, Costa Rica*

		Mean no. fruits		Mean weight (g)	
Species	No. pairs	Control	Defoliated	Control	Defoliated
Acacia farnesiana	10	155.4	0.7**	199.9	9.0*
Bauhinia ungulata	10	24.9	1.6**	41.3	2.1**
Cochlospermum vitifolium	10	3.8	1.8	102.0	48.5
Gliricidia sepium	10	419.5	96.0*	3150.5	545.3*
Spondias purpurea	10	1182.0	0	5000.0	0
Crescentia alata	8	36.0	4.6*	11600.0	2200.0*

Significance levels: *, $p < 0.05$; **, $p < 0.001$.
Source: Rockwood (1973).

Table 12.7. *Number of mature fruits (mean ± SD) per plant of* Erythroxylum havanense *subjected to defoliation treatments (1987)*

Treatment	*n*	No. fruits	Comparison[a]
Control	15	136.9 ± 52.8	a
0% defoliation	15	72.2 ± 26.4	a
25% defoliation	15	30.2 ± 9.4	b
100% defoliation	15	42.9 ± 13.9	b

[a] Treatments with the same letter are not significantly different.
Source: Domínguez (1990).

realism of the defoliation treatments. Despite these limitations, the importance of the experimental results has kept this as one of the most cited examples of the effect of defoliation on reproductive performance.

A recent and more detailed experimental study (Domínguez, 1990) shows the complex nature of the consequences of defoliation in some dry forest plants. The study was carried out with *Erythroxylum havanense* at Chamela. This is an abundant shrubby species in dry forest such as at Chamela and Guanacaste. The species is proleptic, i.e. it has an annual resting period between flower induction and anthesis (see Borchert, 1983). Folivory on species such as this could have an effect in at least two consecutive seasons, because flower production in a given year depends on the photosynthates of the preceding season. Thus herbivory could affect the reproductive performance of both the current season (in terms of seed production) as well as the season following the defoliation episode (in terms of flower and seed production). For this study, four defoliation treatments were applied seven days after the natural flowering of 1987: (1) control (no manipulation); (2) 0% defoliation and chemical protection with insecticide; (3) 25% defoliation and protection; and, 4) 100% defoliation with protection to the subsequent leaf crop. Here we describe some salient results for the effects within the season (1987) in which the treatments were applied and in the subsequent season (1988).

Defoliation affected fruit production in 1987 (Table 12.7). In the two groups of plants experimentally defoliated fruit production was about a third of the undefoliated plants (36 vs 104, respectively), while the two groups of defoliated plants were not significantly different.

Another interesting consequence of defoliation (Table 12.8) was on the

Table 12.8. *Mean time to fruit ripening in plants of* Erythroxylum havanense *subjected to defoliation treatments (1987)*

Treatment	n	Mean time (days)	Mean rank	Comparison[a]
Control	14	21.6	18.4	a
0% defoliation	15	22.1	24.5	a
25% defoliation	14	22.3	23.1	b
100% defoliation	13	30.1	49.8	b

[a] Treatments with the same letter are not significantly different.
Source: Domínguez (1990).

number of days required for fruit maturation (days from fruit initiation to the time when fruits turned intense red). This comparison shows that fruits on the plants with 100% defoliation took a considerably longer time to reach maturation. In these plants fruit development was stopped after defoliation and continued only after the plants had produced a new crop of leaves. The delay in fruit maturation also generates a delay in the process of dispersal by frugivorous birds and seedling establishment, thus reducing the time seedlings have for growth before the onset of the dry season.

Regarding effects in the next reproductive season after defoliation, the number of flowers per plant, a critical aspect in the attraction of pollinators and therefore pollen movement, was affected by defoliation. The results (Table 12.9) show a monotonic decrease in the number of flowers per plant with the degree of defoliation, although the most dramatic and only significant reduction was with 100% defoliation. In accordance with this, the production of mature fruits also decreased with the intensity of defoliation (Table 12.9). However, fruit production is highly dependent on the number of flowers available for pollination, plus the process of maturation itself (pollination, fecundity, capability for fruit filling, etc.). A series of analyses using the number of flowers and number of initiated fruits as covariates indicates that defoliation affects fruit production in the next season through its effect on flower production and not through the subsequent process of maturation.

The results of this study show that the effects of defoliation in this plant are evident during two consecutive seasons. Moreover, this study with *E. havanense* seems to be the only one for dry forest plants in which the results of the experimental defoliation collectively indicate that herbivores may affect fitness by reducing the number of descendants

Table 12.9. *Number of flowers and fruits (mean ± SD) produced by plants of* Erythroxylum havanense *one year after they were subjected to defoliation treatments*

Treatment	*n*	Flowers[a]	Fruits[a]
Control	15	1454 ± 467 a	193 ± 53 a
0% defoliation	15	1189 ± 395 a	145 ± 39 a
25% defoliation	15	845 ± 207 a	164 ± 54 a
100% defoliation	15	259 ± 62 b	74 ± 22 b

[a] Means with the same letter are not significantly different.
Source: Domínguez (1990).

produced by the defoliated plants. In plants in general, the number of descendants produced depends on the number of flower buds, pollination efficiency, rate of abortion and the number of mature fruits. In each of these phases complex physiological processes occur and this study suggests that herbivory may disturb several of them, and eventually affect reproductive fitness. Another important aspect of this experiment is the fact that several components of fitness were affected only by complete defoliation. Complete defoliations in this species caused by lycaenid and geometrid larvae, which have been observed at Chamela and Palo Verde (R. Dirzo, personal observation). Although a frequency distribution of damage yields a skewed distribution whereby most of the plants have low to intermediate levels of damage, some individual plants do get heavily defoliated and many do so in years of 'outbreaks' of defoliating insects (see Janzen, 1985).

Finally, it is worth mentioning that we have no demographic information for this species at Chamela so we do not know if reduction in seed production effectively limits recruitment to the population of seedlings or mature plants (see Louda, 1982). Clearly, long-term demographic studies are needed.

Summary

Herbivores apparently constitute an important factor affecting the ecology of tropical dry forest plants, although there is much spatial, temporal and among-species variation. The protagonists in plant–herbivore interactions in any Mesoamerican forest include hundreds of plant species, each with many different tissues, thousands of phytophagous insects and a few dozen vertebrates. Most attention has been given to leaf-eating caterpillars and to seed-eating beetles, to the neglect of herbivores feeding on phloem, wood, shoots and roots.

At the community level, standing levels of herbivory may be generally less than 10%. However, the discrepancy ratio of repeated: instantaneous measurements had an average value of 2 in a study of 12 species, and varied from 0.8 to 4.5 among species. This suggests that total leaf area eaten is about 17%. Herbivory rates are highest early in the wet season, but the evidence suggests that digestibility and nutritional value do not determine this pattern. Preliminary contrasts of instantaneous measures between adjacent deciduous and semideciduous forests show less folivory in the latter.

Seed predation by beetles in one region is concentrated on about 10% of the flora. There are disproportionate numbers of Leguminosae among the prey, and of plants attacked by several beetle species. The distribution of number of hosts per beetle species, greatly biased to one host, is remarkably similar in dry and wet forests. Whether due to invertebrates or vertebrates, seed mortality by predation has not been assessed at the community level; nevertheless, some population studies suggest that it has a major impact on the structure of dry as well as wet forests.

Defoliation experiments in several species have shown negative effects on fruit production. Detailed studies of *Erythroxylum havanense* showed effects on fruit production in the season of defoliation, and on flower production (and hence on fruits) in the following year. Demographic studies of herbivory from phytocentric or zoocentric perspectives have been limited to one or a few life stages, however, and the importance of herbivory as a selective pressure also needs wider study.

Acknowledgements

We are grateful to Germán Avila for his help in the preparation of this manuscript and his valuable suggestions. Alvaro Miranda and E. Roger Guevara made useful comments on earlier drafts. Gustavo Verduzco and María del Carmen Vázquez-Rojas provided valuable help with the field work in the defoliation experiments. This work was supported by the Centro de Ecología, Universidad Nacional Autónoma de México, and by grants from CONACyT and DGAPA.

References

Andrade, L. O. (1989). Cultivo de insectos depredadores de propágulos de árboles tropicales en Los Tuxtlas, Veracruz. Tesis de Licenciatura, Facultad de Ciencias, Universidad Nacional Autónoma de México, México.

Borchert, R. (1983). Phenology and control of flowering in tropical trees. *Biotropica* **15**: 81–9.

Ceballos, G. (1989). Population and community ecology of small mammals from tropical deciduous and arroyo forests in western México. Ph.D. thesis, University of Arizona, Tucson.

Coley, P.D. (1990). Tasas de herbivorismo en diferentes árboles tropicales. In *Ecología de un Bosque Tropical*, ed. E. G. Leigh, Jr, R. A. Stanley and D. M. Windsor, pp. 191–200. Smithsonian Tropical Research Institute, Balboa, Panamá.

Dirzo, R. (1987). Estudios sobre interacciones planta-herbívoro en Los Tuxtlas, Veracruz. *Revista de Biología Tropical* **34**: 119–31.

Dirzo, R., & Miranda, A. (1990). Altered patterns of herbivory and diversity in the forest understory: a case study of the possible consequences of contemporary defaunation. In *Plant–Animal Interactions*, ed. P. W. Price, T. L. Lewinsonn, G. W. Fernández and W. W. Benson, pp. 273–87. Wiley-Interscience, New York.

Domínguez, C. A. (1990). Consecuencias ecológicas y evolutivas del patrón de floración sincrónico y masivo de *Erythroxylum havanense* Jacq. (Erythroxylaceae). Tesis Doctoral, Centro de Ecología, Universidad Nacional Autónoma de México, México.

Filip, V., Dirzo, R., Maass, J. M. & Sarukhán, J. (1995). Within and between year variation in the levels of herbivory on the foliage of trees from a Mexican deciduous forest. *Biotropica* **27**: 78–86.

Foster, R. B. (1990). Hambruna en la Isla de Barro Colorado. In *Ecología de un Bosque Tropical*, ed. E. G. Leigh, Jr, R. A. Stanley, and D. M. Windsor, pp. 271–83. Smithsonian Tropical Research Institute, Balboa, Panamá.

Gryj, E. O. (1990). Dispersión de frutos del arbusto *Erythroxylum havanense* Jacq. en Chamela, Jalisco. Tesis de Licenciatura, Facultad de Ciencias, Universidad Nacional Autónoma de México, México.

Janzen, D. H. (1976). Reduction of *Mucuna andreana* (Leguminosae) seedling fitness by artificial seed damage. *Ecology* **57**: 826–28.

Janzen, D. H. (1980). Specificity of seed-attacking beetles in Costa Rican deciduous forest. *Journal of Ecology* **68**: 929–52.

Janzen, D. H. (1981). Patterns of herbivory in a tropical deciduous forest. *Biotropica* **13**: 271–82.

Janzen, D. H. (1983a). Food webs: who eats what, why, how, and with what effects in a tropical forest? In *Tropical Rain Forest Ecosystems: structure and function*, ed. F. B. Golley, pp. 163–82. Elsevier, Amsterdam.

Janzen, D. H. (1983b). Dispersal of seeds by vertebrate guts. In *Coevolution*, ed. D. J. Futuyma and M. Slatkin, pp. 232–62. Sinauer Associates, Sunderland.

Janzen, D. H. (1985). A host plant is more than its chemistry. *Illinois Natural History Survey Bulletin* **33**: 141–74.

Janzen, D. H. (1988). Ecological characterization of Costa Rican dry forest caterpillar fauna. *Biotropica* **20**: 120–35.

Janzen, D. H. & Liesner, R. (1980). Annotated check-list of plants of lowland Guanacaste Province, Costa Rica, exclusive of grasses and non-vascular cryptogams. *Brenesia* **18**: 15–90.

Janzen, D. H. & Martin, P. S. (1982). Neotropical anachronisms: the fruits the gomphotheres ate. *Science* **215**: 19–27.

Janzen, D. H. & Waterman, P. G. (1984). A seasonal census of phenolics, fibre and alkaloids in foliage of forest trees in Costa Rica: some factors influencing their

distribution and relation to host selection by Sphingidae and Saturniidae. *Biological Journal of the Linnean Society* **21**: 439–54.
Lott, E. J. (1993). Annotated checklist of the vascular flora of the Chamela Bay region, Jalisco, México. *Occasional Papers of the California Academy of Sciences* **148**: 1–60.
Louda, S. M. 1982. Limitation of the recruitment of the shrub *Haploppapus squarrosus* (Asteraceae) by flower and seed feeding insects. *Journal of Ecology* **70**: 43–53.
Nuñez-Farfán, J. & Dirzo, R. (1988). Within-gap spatial heterogeneity and seedling performance in a Mexican tropical forest. *Oikos* **51**: 274–84.
Rockwood, L. L. (1973). The effect of defoliation on seed production of six Costa Rican tree species. *Ecology* **54**: 1363–69.
Stanton, N. (1975). Herbivore pressure on two types of tropical forests. *Biotropica* **7**: 8–11.

13

Biomass distribution and primary productivity of tropical dry forests

ANGELINA MARTÍNEZ-YRÍZAR

Introduction

A large amount of work has been published on various aspects of the functioning at the ecosystem level in forests throughout the world. However, information for tropical dry forests is sparse. In comprehensive reviews of forest productivity studies (Bray & Gorham, 1964; Art & Marks, 1971; Murphy, 1975; Jordan & Murphy, 1978; Brown & Lugo, 1982; Proctor, 1984; Vitousek, 1984; Murphy & Lugo, 1986a; Vogt, Grier & Vogt, 1986), there are no more than ten references regarding studies on primary productivity of tropical dry forests. Subsequently a few more studies have been published, particularly from Brazil, México, Puerto Rico and India (Table 13.1).

This review of biomass distribution and primary productivity of tropical dry forests emphasizes the aboveground portion of the vegetation. There is no consensus on a precise definition of dry forest. Following Murphy & Lugo (1986a), my review includes tree-dominated sites in tropical regions in which at least the dominant trees are drought deciduous. These sites include those with strongly seasonal climate, that is, including a well-defined dry season. Mean annual rainfall is usually below 1500 mm and annual potential evapotranspiration:precipitation ratio is normally greater than one.

After a review of forest phytomass, I survey the process of litterfall and of litter decomposition, treating methodologies reviewed by Proctor (1983) and Wieder & Lang (1982). The significance of litterfall and litter decomposition in nutrient cycling in tropical deciduous forests is presented in detail by Jaramillo & Sanford (Chapter 14). Finally, I examine the few attempts that have been made to estimate net primary productivity (NPP).

A list of the studies conducted in dry forests throughout the world and general descriptive information for each study site is presented in Table 13.1. Each study is labelled by a number which is used as a reference in the rest of the tables.

Biomass distribution

Aboveground phytomass

There are at least 15 dry forest sites, in different parts of the world, for which phytomass measurements are available for comparison (Table 13.2). Aboveground phytomass ranges about one order of magnitude, from 28 to 268 Mg ha^{-1}.

From the communities listed in Table 13.2, the correlation between aboveground phytomass and annual precipitation was statistically significant ($r^2 = 0.34$, $n = 12$, $p < 0.05$). However, the significance of the correlation was increased ($r^2 = 0.68$, $n = 9$, $p < 0.01$) when values from studies Nos. 3, 12 and 19 were excluded from this analysis (Figure 13.1). These results suggest that successional status may have a stronger influence on aboveground phytomass than the dominant climatic variable rainfall. Other important factors influencing forest structure, such as soil type and aspect, or the history of human perturbations, are reported so differently in each study, that the opportunity for comparison is precluded.

The highest aboveground phytomass value (268 Mg ha^{-1}) corresponds to a dry monsoon forest in Ping Kong, northwestern Thailand, which has a continuous canopy at 17–26 m. The lowest phytomass value of 28 Mg ha^{-1} is for a mixed deciduous forests at Udaipur, India. This value may be the result of the combined effects of the semiarid climate and the intense human activity in the forest of the hilly region (Vyas, Garg & Vyas, 1977).

At Chamela, México, Martínez-Yrízar *et al.* (1992b) reported an aboveground phytomass of 85 Mg ha^{-1} (Table 13.2). These authors followed a dimensional analysis approach to develop allometric relations to predict the forest phytomass. Castellanos, Maass & Kummerow (1991) used these relations to estimate the standing crop phytomass of a nearby undisturbed site at 74 Mg ha^{-1}. The difference in phytomass between these two sites probably lies in the larger amount of relatively small stems (dbh <5 cm) found by Castellanos *et al.* (46% vs 23% at the site of Martínez-Yrízar *et al.*, 1992b).

The aboveground phytomass value for the forest at Guánica (Murphy

Table 13.1. *Sites of studies of tropical dry forests throughout the world. Sites are arranged according to decreasing annual precipitation. Number of dry months is the approximate, defined by each author*

Site	Latitude	Elevation (m)	Rainfall (mm)	No. dry months	Temperature (°C)	Source
Ping Kong, Thailand	19° 50′ N	–	1500	5	25	1
Palo Verde, Costa Rica	10° 20′ N	low	1500	5	26	2
Indian Church, Belize	17° 45′ N	–	1480	5	26.4	3
Calabozo, Venezuela	8° 48′ N	100	1330	5	27	4
Lubumbashi, Zaïre	11° 29′ S	1208	1273	5	20	5
Olokemeji, Nigeria	7° 24′ N	250	1232	5	23	6
Olokemeji, Nigeria	7° 24′ N	250	1232	5	23	7
Ibadan, Nigeria	7° 23′ N	250	1200	6	26.5	8
Varanasi, India	25° 18′ N	90–365	1100	>6	26	9
Varanasi, India	25° 18′ N	90–365	1100	>6	26	10
Varanasi, India	25° 18′ N	90–365	1100	>6	26	11
Varanasi, India	25° 18′ N	90–365	1100	>6	26	12
Varanasi, India	25° 18′ N	90–365	1100	>6	26	13
Varanasi, India	25° 18′ N	90–365	1100	>6	26	14
Varanasi, India	25° 18′ N	–	1057	>6	26	15
Chandraprabha, India	25° 52′ N	–	1050	>6	26	16
Guánica, Puerto Rico	17° 57′ N	175	860	5	25.1	17
Guánica, Puerto Rico	17° 57′ N	175	860	5	25.1	18
Guánica, Puerto Rico	17° 57′ N	175	860	5	25.1	19
Guánica, Puerto Rico	17° 57′ N	175	860	5	25.1	20

Araripina, Brazil	7° 32′ S	800	831	7	23.5	21
Serra Talhada, Brazil	7° 59′ S	500	803	7	24.0	22
Chamela, México	19° 30′ N	150	707	6–8	24.9	23
Chamela, México	19° 30′ N	150	707	6–8	24.9	24
Chamela, México	19° 30′ N	150	707	6–8	24.9	25
Chamela, México	19° 30′ N	150	707	6–8	24.9	26
Chamela, México	19° 30′ N	150	707	6–8	24.9	27
Chamela, México	19° 30′ N	150	707	6–8	24.9	28
Chamela, México	19° 30′ N	150	707	6–8	24.9	29
Chamela, México	19° 30′ N	150	707	6–8	24.9	30
Udaipur, India	24° 35′ N	587	603	10	25	31
Udaipur, India	24° 35′ N	587	603	10	25	32
Udaipur, India	24° 35′ N	587	603	10	25	33
Bambey, Senegal	14° 43′ N	20	500	7	–	34
Dehra Dun, India	29° 59′ N	385–610	–	–	–	35
Dehra Dun, India	29° 59′ N	750	–	–	–	36
Sri Lanka	8° N	low	–	–	–	37

Sources: 1. Owaga *et al.* (1965); 2. Gessel *et al.* (1980); 3. Lambert *et al.* (1980); 4. Medina & Zelwer (1972); 5. Malaisse *et al.* (1975); 6. Hopkins (1966); 7. Madge (1969); 8. Madge (1965); 9. Singh (1968); 10. Bandhu (1971, in Ref. 11); 11. Misra (1972); 12. Singh (1975); 13. Gaur & Pandey (1978); 14. Pandey *et al.* (1980); 15. Singh & Singh (1981); 16. Singh (1989); 17. Lugo *et al.* (1978); 18. Lugo & Murphy (1986); 19. Murphy & Lugo (1986b); 20. Cintrón & Lugo (1990); 21. E. Sampaio *et al.* (personal communication); 22. Sampaio *et al.* (1990); 23. Martínez-Yrízar (1980); 24. Vizcaíno (1983); 25. Ceballos (1989); 26. Martínez-Yrízar & Sarukhán (1990); 27. Patiño (1990); 28. Castellanos *et al.* (1991); 29. Martínez-Yrízar *et al.* (1992a); 30. Martínez-Yrízar *et al.* (1992b); 31. Garg and Vyas (1975); 32. Ranawat & Vyas (1975); 33. Vyas *et al.* (1977); 34. Jung (1969); 35. Kumar *et al.* (1984); 36. Soni *et al.* (1984); 37. Hladik (in Hladik, 1978).

Table 13.2. *Dry phytomass (Mg ha^{-1}) of dry forests. ABG, aboveground phytomass; BLG, belowground phytomass.*

Different values for a given study site derive from samples in different community types, topographic situations or watersheds. See Table 13.1 for sources.

Site	Phytomass ABG	Phytomass BLG	Phytomass Total	Ratio BLG:ABG	Source
Ping Kong	144.0	16.0	160.0	0.11	1
Ping Kong	268.0	25.0	293.0	0.09	1
Indian Church	56.0	–	–	–	3
Varanasi	–	7.6[a]	–	–	10
Varanasi	205.5	34.3	239.8	0.17	12
Varanasi	64.3	9.5	73.8	0.15	15
Chandraprabha	95.0	–	–	–	16
Guánica	44.7[b]	45.0	89.7	1.01	19
Araripina	67.3	10.9	78.2	0.16	21
Serra Talhada	84.0	–	–	–	22
Chamela	73.6	30.9	104.5	0.42	28
Chamela	85.0	35.7[c]	120.7	0.42	30
Udaipur	28.2	–	–	–	32
Dehra Dun	61.9[d]	–	–	–	35
Dehra Dun	23.0[d]	–	–	–	36

[a] Includes roots $<$25 mm diameter only.
[b] Does not include 8.3 Mg ha^{-1} of standing dead stems.
[c] Estimated considering BLG:ABG ratio = 0.42.
[d] Includes shrubs, herbs and grasses only.

& Lugo, 1986b) is roughly half the amount reported for Chamela by Martínez-Yrízar *et al.* (1992b). Structural differences based on successional status between the two forests, may explain again why phytomass is so variable from site to site. Although both forests are characterized by a large number of relatively small stems, many trees (85%) in Guánica were $<$5 cm dbh.

In the two localities in the caatinga in Brazil, the aboveground phytomass was very different (67 and 84 Mg ha^{-1}), probably related to differences in soil nutrients. The Araripina site with lower phytomass has soils low in phosphorus (E. Sampaio, personal communication).

Recent work in the southern Sonoran desert in México, has shown that aboveground phytomass on the northern slopes with tropical thorn scrub, and in the arroyos with xero-riparian forests, is 12 and 21 Mg ha^{-1}, respectively. These values are close to the lower limit reported for

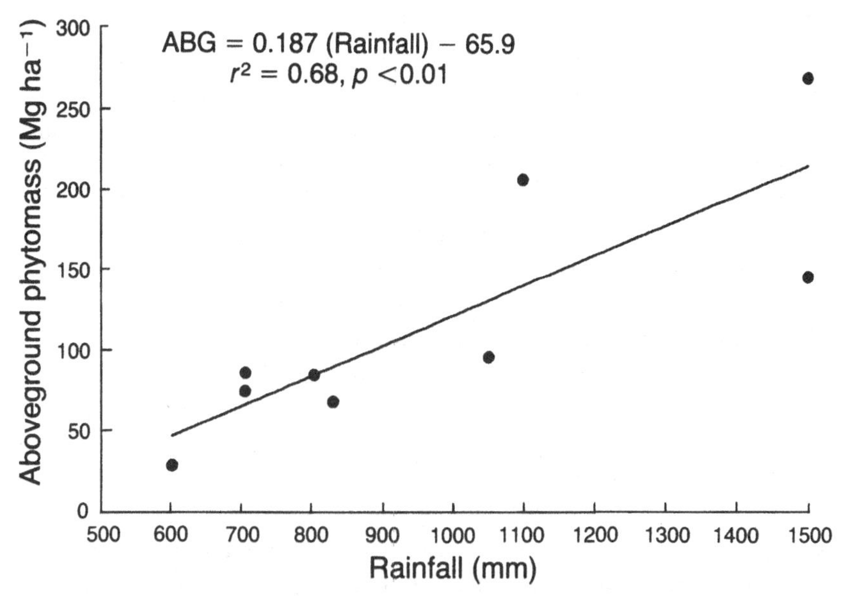

Figure 13.1. Correlation between aboveground phytomass (Mg ha^{-1}) and annual rainfall (mm) in dry forests. Data from Tables 13.1 and 13.2.

tropical dry forests, and confirm the boundaries between true desert (6 Mg ha^{-1}), thorn scrub, and tropical deciduous forests (Búrquez *et al.*, 1992).

Belowground phytomass

It is difficult to estimate belowground phytomass of most forests and such data are few (Table 13.2). Also, these data are difficult to compare since the methods used for soil excavation, root extraction and further processing differ greatly between studies. For example, Singh & Singh (1981) estimated only the phytomass of roots less than 25 mm diameter. Other studies have included excavated roots of all sizes (Murphy & Lugo, 1986b; Sampaio, Salcedo & Kauffman, 1990; Castellanos *et al.*, 1991) and have even made corrections for taproot biomass (Castellanos *et al.*, 1991).

Root mass in dry forests seems to be confined mainly to the superficial soil layers. At Chamela, an analysis of the depth distribution of the different diameter classes revealed that 62% of the root biomass occurs between the soil surface and 20 cm depth, and 91% in the upper 40 cm (Castellanos *et al.*, 1991). Similarly, in the forest at Guánica, about 90%

of the root biomass was concentrated within 40 cm of the soil surface, with 57% in the upper 10 cm (Murphy & Lugo, 1986b). The small and fine root category (diameter <5 mm) may have little influence on total root mass; for example, at Chamela they constitute one-third of all roots (Castellanos *et al.*, 1991). However, small and fine roots are functionally more important than medium-large roots and may be as important for nutrient cycling as aboveground litter production (Singh & Singh, 1981; Kummerow *et al.*, 1990; see Cuevas, Chapter 15).

Root:shoot ratios are highly variable between the sites (Table 13.2). The belowground fraction in the monsoon forest at Ping Kong represented only 9% of the aboveground phytomass. Root:shoot ratios were higher in the other sites, varying from 0.11 in the monsoon forest–savanna forest ecotone at Ping Kong to 1.01 in the Guánica forest. The greater belowground biomass at the Puerto Rican site may be related to forest history. Murphy & Lugo (1986b) reported that 50% of trees at Guánica have their origin from coppicing, and suggested that the prevalence of this mode of growth may be inhibiting the development of an aboveground structure comparable to that which existed prior to cutting.

Litter production

Total litterfall

The information presented in this section refers mainly to estimates of aboveground litter production. Belowground litter production in dry forests is practically unknown (Vogt *et al.*, 1986; Kummerow *et al.*, 1990; Cuevas, Chapter 15).

Total litterfall in dry forests varies over a wide range (Table 13.3). The highest annual litter production, 12.6 Mg ha^{-1} y^{-1}, corresponds to a seasonally dry tropical hardwood forest in Belize. The lowest amount, 1.5 Mg ha^{-1} y^{-1}, is from an Indian deciduous forest at Varanasi, dominated by *Shorea robusta*. In this latter study litter was collected from the ground, not in traps, and a correction for significant decomposition or other loss *in situ* was not included.

Excluding data from studies in which litterfall was estimated based on collections from the forest floor and/or the quantification of litter input was incomplete, a regression analysis showed a positive and statistically significant relationship between total litterfall and mean annual rainfall ($r^2 = 0.61$, $n = 16$, $p < 0.01$; Figure 13.2). In contrast, Spain (1984) found that total litterfall in moist/wet forests (rainfall >1000 mm) was

Table 13.3. *Annual leaf fall and total litterfall (Mg ha^{-1} y^{-1} dry mass) in dry forests. The percentage contribution of leaves to total litterfall is in parentheses*

Different values for a given study site derive from samples in different community types, topographic situations or watersheds. See Table 13.1 for sources.

Site	Litterfall			Source
	Leaves		Total	
Palo Verde	–		8.2	2
Indian Church	–		12.6	3
Calabozo	–		8.2	4
Lubumbashi	2.9	(74)	3.9[a]	5
Olokemeji	4.6		–	6
Olokemeji	5.0		–	7
Ibadan	3.4	(61)	5.6	8
Varanasi	1.0		–	9
Varanasi	3.2		–	9
Varanasi	4.2		–	9
Varanasi	5.0		–	9
Varanasi	6.2		–	9
Varanasi	–		7.7	10
Varanasi	1.3	(87)	1.5	13
Varanasi	1.7	(94)	1.8[ab]	13
Varanasi	–		1.7[ab]	14
Guánica	2.5	(86)	2.9	17
Guánica	4.3	(90)	4.8	18
Chamela	2.4	(70)	3.4[b]	24
Chamela	3.6	(70)	5.1[b]	24
Chamela	–		4.2	25
Chamela	–		6.9	25
Chamela	2.7	(67)	4.0	26
Chamela	4.8	(73)	6.6	26
Chamela	2.6	(74)	3.5	29
Chamela	2.6	(71)	3.6	29
Chamela	3.0	(70)	4.2	29
Chamela	3.1	(69)	4.6	29
Chamela	3.3	(67)	4.9	29
Udaipur	4.0		–	31
Udaipur	4.8		–	33
Bambey	2.5	(22)	11.6[a]	34
Sri Lanka	2.4	(52)	4.6	37
Sri Lanka	–		3.9	37

[a] Litterfall collected from the ground.
[b] Includes leaves and woody material only.

not significantly related to mean annual precipitation. The significance of the correlation in my analysis indicates that litter production in dry forests is strongly limited by water availability, while in moist/wet forests production may be limited by other factors.

Litterfall components

The relative importance of the different litterfall components to total litter production is usually determined by sorting the litter into fractions (e.g. leaves, fine woody material, flowers, etc.). Leaves normally constitute the largest fraction. Less important, in decreasing order, are: fine woody material, reproductive remains and fragmented debris.

The relative contribution of leaf fall to total litter production in dry forests exhibits considerable variation from site to site (Table 13.3). This ratio largely depends on the age of the forest, weather conditions, disturbance, definition of litterfall fractions (i.e. wood size limits), and other factors. Since the contribution of the different litterfall fractions may also vary from year to year, it is extremely important to extend litter production measurements beyond a single year (Proctor, 1983). For example, in the miombo forest at Lubumbashi, Zaïre, leaf fall averaged

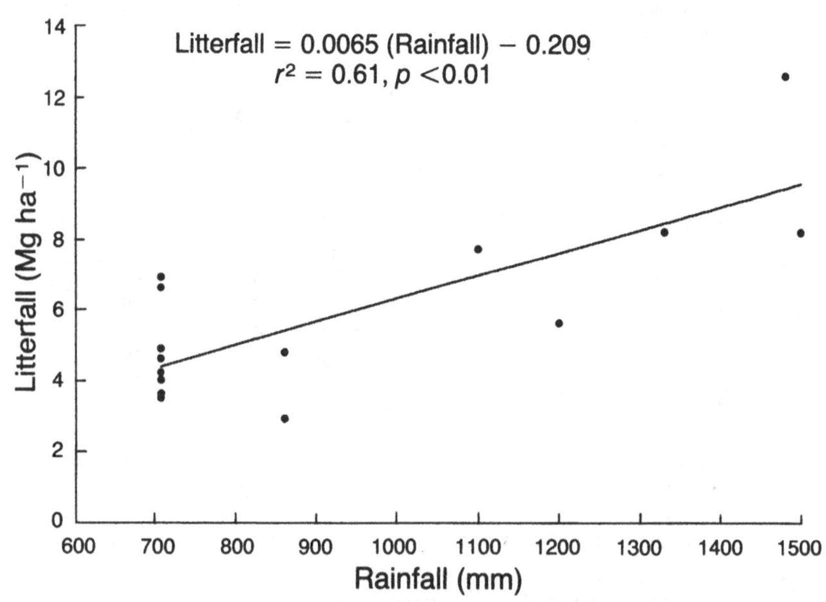

Figure 13.2. Correlation between annual litterfall (Mg ha^{-1} y^{-1}) and annual rainfall (mm) in dry forests. Data from Tables 13.1 and 13.3.

74% of the total litterfall in four consecutive years, but it fell to 47% in the fifth year because of heavy fruiting (Malaisse *et al.*, 1975). In the forest at Bambey, Senegal, leaf-fall contribution was very low (22%). Fruit production of the dominant species (*Acacia albida*) was exceptionally high during the study year (Jung, 1969).

Woody litter

Although litter traps collect small fragments of branches falling from dead trees, larger pieces of wood (>1 cm diameter) are not usually found in the traps (Martínez-Yrízar & Sarukhán, 1990). This problem may be overcome by using transects or permanent quadrats from which the woody litter is periodically collected from the forest floor (Collins, 1977; Shiavini da Silva, 1983). However, most litterfall studies have failed to address this problem and comparisons of litterfall measurements should be made with caution (Proctor, 1983).

When hurricanes or heavy rains occur, mainly in summer, the standing crop of woody litter in the forest floor increases considerably (Lugo *et al.*, 1978; Patiño, 1990). In the forest at Lubumbashi, twig fall was greatest in the late rainy season and related to the high moisture content of rotted wood (Malaisse *et al.*, 1975). In the Guánica forest fine wood fall was seven times the mean value after the passage of a hurricane (Cintrón & Lugo, 1990). Different intensities of wood fall in relation to storm patterns have also been reported in Australian wet forests (Lowman, 1988).

Litterfall seasonality

Litterfall seasonality is a feature common to all dry forests because of the pronounced seasonal changes in rainfall. Maximum rates of litterfall occur during the period of minimum rainfall as reported by all studies listed in Table 13.3. In some cases, there may be a significant time-lagged correlation between minimum rainfall and maximum litterfall. In the Chamela forest, for example, Martínez-Yrízar & Sarukhán (1990) showed that heavy litterfall occurs just at the onset of the dry season in a deciduous forest site, while in an adjacent semideciduous forest (with greater soil moisture storage), maximum litterfall occurred four months after the onset of the dry season.

With the exception of a few species in which leaf drop seems to be regulated by photoperiod (Bullock & Solís, 1990), there is little doubt that

Table 13.4. *Range of total annual litterfall (dry mass) in dry forests. Data from studies lasting for two years or more*

Different values for a given study site derive from samples in different community types, topographic situations or watersheds. See Table 13.1 for sources.

Site	Duration (years)	Litterfall (Mg ha^{-1} y^{-1})	Source
Olokemeji	2	4.5–4.7[a]	6
Varanasi	2	0.9–1.1[a]	9
Varanasi	2	2.8–3.6[a]	9
Varanasi	2	4.2–4.3	9
Varanasi	2	5.0–5.1[a]	9
Varanasi	2	6.1–6.4[a]	9
Chamela	2	3.2–3.5[b]	24
Chamela	2	4.5–5.7[b]	24
Guánica	3	1.6–3.9	20
Chamela	4	3.1–4.0	29
Chamela	4	3.3–4.6	29
Chamela	4	3.4–5.0	29
Chamela	4	4.3–5.3	29
Chamela	4	4.5–5.2	29
Lubumbashi	5	2.5–3.4[a]	5
Chamela	5	3.7–4.3	26
Chamela	5	6.1–7.0	26

[a] Includes leaf litter only.
[b] Includes leaves and woody material only.

water deficit is the major determinant of leaf shedding in dry forests (Lieberman, 1982; Reich & Borchert, 1984; Bullock & Solís, 1990; Holbrook, Whitbeck & Mooney, Chapter 10).

Annual variations of litterfall

There is little published information on annual variations of litterfall in dry forests. Only a few studies have been carried out for more than two years (Table 13.4). Long-term studies are essential to improve litterfall estimates and to determine adequately the extent of the variations of the forest behavior.

Studies with multi-year data have reported mixed results (Table 13.4). The forests studied by Singh (1968) in Varanasi and by Hopkins (1966) at Olokemeji, Nigeria produced almost the same amount of total litterfall in two consecutive years. Martínez-Yrízar & Sarukhán (1990) found no statistically significant annual variations in total litterfall over a five-year period in a deciduous forest at Chamela. However, they did find significant annual variations in a semideciduous forest in the same locality (Table 13.4).

Possible direct response of litterfall to annual variations in rainfall has been suggested by Cintrón & Lugo (1990) from long-term studies at the Guánica forest. By comparing two consecutive studies, they attributed the higher amount of litter produced during one of the studies to better rainfall conditions. Using each of five watersheds as units of research at the ecosystem level at Chamela, Martínez-Yrízar *et al.* (1992a) found a positive and significant correlation between annual total litterfall and annual rainfall in three of the watersheds over a period of six years.

Standing crop litter

Standing crop litter refers to the surface litter or mean annual amount of forest floor mass. There is considerable variability in the mean amount of litter accumulated on the floor of dry forests (Table 13.5). Values range from 1.3 to 10.5 Mg ha^{-1}. Site-to-site variations of litter buildup may be the result of the separate or combined effects of various causes (Vogt *et al.*, 1986): (1) differences in litter quality, e.g. lignin and nutrient content which affects substrate quality for decomposition, (2) differences in soil moisture and nutrient availability at the decomposition site, and (3) incomplete quantification of total litter input.

Only a few attempts have been made to measure the relative contribution of the different litter components to total buildup. At Chamela, woody litter (diameter <5 cm) and the leaf fraction constituted 50% and 42% respectively of the mean standing crop litter (Patiño, 1990). Cintrón & Lugo (1990) found that the leaf and miscellaneous fraction made up 77% of the mean total on a ridgetop deciduous forest at Guánica.

Litter decomposition

The decomposition of litter (k) can be estimated by applying the mean annual standing crop litter and the annual litterfall values to the equation modified by Nye (1961). As both litter standing crop and rates of litter disappearance in dry forests vary markedly between wet and dry seasons (Hopkins, 1966; Madge, 1969), K represents a mean decomposition value for the whole year, and should be used for comparative purposes only (Swaine, Lieberman & Hall, 1990). It must also be assumed that litter

Table 13.5. *Mean annual litter standing crop (LSC, dry mass), annual litterfall (Input, dry mass) and annual turnover rate of total litter in dry forests. k = Input/LSC. See Table 13.1 for sources*

Site	LSC ($Mg\ ha^{-1}$)	Input ($Mg\ ha^{-1}\ y^{-1}$)	Turnover, k^{-1} (years)	Source
Indian Church	7.2	12.6	0.57	3
Lubumbashi	3.3[a]	3.9[a]	0.85[a]	5
Olokemeji	1.7[a]	4.6[a]	0.37[a]	6
Ibadan	2.1	5.6	0.38	8
Varanasi	1.3	1.7	0.76	13
Guánica	8.7	2.9	3.00	17
Guánica	12.3[a]	4.3[a]	2.86[a]	18
Chamela	8.9	4.3	2.07	23
Chamela	10.2	6.4	1.59	23
Chamela	7.6	3.7	2.05	27
Chamela	3.2[a]	2.3[a]	1.39[a]	27
Bambey	10.5	11.9	0.88	34

[a] Leaf-litter fraction only.

accumulation in the ecosystem has reached a steady-state level (Olson, 1963).

Annual turnover rates (k^{-1}) of total litter in dry forests are summarized in Table 13.5. There is a great variation between sites for k^{-1} values. The Guánica forest has the lowest turnover rate (3 years), while at Ibadan, only 0.38 year is needed for the standing crop litter to be recycled. Turnover rates for the leaf fraction of the standing crop are even faster (Table 13.5). Leaves are known to decompose at a higher rate than wood or mixed litter (John, 1973; Malaisse *et al.*, 1975; Swift, Heal & Anderson, 1979). At Chamela the annual turnover rate for leaf litter and total litter was 1.4 and 2.1 years, respectively (Patiño, 1990; Table 13.5). Turnover rates for the leaf fraction of litter are also higher in other types of dry forests (Swaine *et al.*, 1990).

Litter-bag studies

Only a few direct measurements of litter decomposition have been made on dry forests (Madge, 1965, 1969; Singh, 1969; Malaisse *et al.*, 1975; Shukla, 1976 in Gaur & Pandey, 1978; Martínez-Yrízar, 1980; Lugo & Murphy, 1986). Using the litter-bag technique to analyse the process of litter decomposition, Martínez-Yrízar (1980) compared the rates of leaf litter breakdown of selected species in the Chamela forest. This study showed that the decomposition of leaves of locally common species occurred at different rates under the same conditions, probably as a result of differences in leaf toughness. Softer leaves decomposed faster. It was also found that the rate of decomposition was affected by the groups of decomposing organisms involved, as illustrated by experiments with litter bags of different mesh sizes.

Under controlled environmental conditions, Singh (1969) also reported that leaf litter of trees from dry forest decompose at different rates. Fastest decomposition rates were recorded for the species with the lowest leaf C:N ratio and lignin content. Lugo & Murphy (1986) found that the release of nutrients from decomposing leaf litter differed depending on the nutrient and the location of the bag.

Seasonality of litter decomposition

The process of litter decomposition in dry forests takes place mainly during the rainy season, when the conditions for the growth and activity of soil microorganisms are improved (see Cuevas, Chapter 15). Pronounced seasonal changes in standing crop litter in dry forests have been documented by several authors (Madge, 1965, 1969; Hopkins, 1966; Singh, 1969; Malaisse *et al.*, 1975; Lugo *et al.*, 1978; Martínez-Yrízar, 1980; Pandey, Gaur & Singh, 1980; Cintrón & Lugo, 1990). Such temporal fluctuations indicate that the rates of decomposition of surface litter vary seasonally. For example, in the Chamela forest, Martínez-Yrízar (1980) found a 68% reduction in standing crop litter at the end of the rainy season. In the Chakia forest in India, Shukla (1976, in Gaur & Pandey, 1978) showed that with the onset of the rainy season about 60% of the original litter had disappeared after four months. Madge (1969) also found that there was 5–6 times more surface leaf litter during the dry than the wet season in a semideciduous forest at Olokemeji.

Information on soil microflora further demonstrates seasonal rhythms of decomposition in dry forests (Malaisse *et al.*, 1975). There are significant seasonal fluctuations in biomass, composition, relative abundance, diversity and distribution of important consumer groups (i.e. nematodes and psocids) associated with the decomposition of organic matter in tropical forest soils (Madge, 1969; Pradhan & Dash, 1987; García-Aldrete, 1988).

The period taken for a leaf to decay depends on the time of year at which it falls (Hopkins, 1966; Madge, 1969). Using the litter-bag technique, Martínez-Yrízar (1980) demonstrated that leaves of different species lose 20–50% of their original ash-free dry mass within three months of being placed in the field at the beginning of the rainy season. In contrast, in an experiment initiated at the beginning of the dry season the leaves remained almost intact for 3–4 months until the next wet period. Madge (1969) also showed that leaf disks laid out during the wet season became rapidly fragmented and disappeared in about two months. Extrapolation from leaf disks placed on the forest floor at the start of the dry season showed that the discs would completely decomposed by about seven months after starting the experiment.

Net primary productivity

Aboveground productivity

Net primary productivity (NPP) has been estimated by various authors by multiplying annual leaf litter production by three, considering that annual leaf fall represents one-third of total annual NPP (Bray & Gorham, 1964). However, the use of litterfall as an index of NPP should be taken with caution, since corrections are not usually made for mortality of all plant parts and losses of organic matter to herbivores (Medina & Klinge, 1983).

From the limited data set available in the literature, Murphy & Lugo (1986a) reported that the range of total NPP for tropical dry forests is 8–21 Mg ha^{-1} y^{-1} with 6–16 Mg ha^{-1} y^{-1} above ground.

Assuming that NPP is the sum of stem wood biomass production plus litterfall, Lugo & Murphy (1986) calculated an aboveground NPP value of 6.9 Mg ha^{-1} y^{-1} for the forest at Guánica. Total NPP estimate for the dry forest at Varanasi was 15 Mg ha^{-1} y^{-1}. This value was estimated as the sum of the annual increment in the non-green parts and peak leaf biomass (by species), with corrections for fine root growth and various litterfall categories (Singh, 1989).

Regarding gross primary productivity (GPP) measurements in dry forests, Lugo *et al.* (1978) estimated GPP for the Guánica forest based on data from gas-exchange measurements and corrections for seasonal leaf area index changes. Using a correlation between GPP and soil moisture and the annual water budget for the forest, it was possible to calculate the monthly changes of GPP of the forest. The calculated mean daily rate of GPP from this curve was 3.0 g C m^{-2} day^{-1}.

Belowground productivity

Information on root productivity in tropical forests is limited. Only a few studies have tried to estimate belowground productivity in dry forests (Medina & Klinge, 1983; Murphy & Lugo, 1986a; Vogt *et al.*, 1986; Cuevas, Chapter 15). According to Murphy & Lugo (1986a) root productivity of dry forests (root size limit not indicated) ranges from 2 to 5 Mg ha^{-1} y^{-1}.

By summing significant fine root (<1 mm) biomass changes between sampling dates, Kummerow *et al.* (1990) found that annual production in the Chamela forest was 4.2 Mg ha^{-1} y^{-1}. They also reported seasonal changes driven by periodic drought and wet seasons in this forest. Fine roots were shed with the progress of the dry season and their growth was very rapid after the initiation of the rainy season. This result suggests that fine root litter decomposition may be as important for nutrient cycling as aboveground litterfall (Kummerow *et al.*, 1990; Jaramillo & Sanford, Chapter 14).

Following a different methodology, annual net productivity of small roots (<25 mm) at Varanasi was estimated at 2.8 Mg ha^{-1} y^{-1} (Singh & Singh, 1981). This value was computed as the difference between the annual maximum and minimum standing crop of roots.

Summary

Studies on the structure and functional characteristics of tropical dry forests suggest that aboveground phytomass varies from 28 to 268 Mg ha^{-1} with 9–50% of the total phytomass allocated to roots. Root mass changes seasonally and seems to be confined mainly to the superficial soil layers. Annual litter production varies from 2 to 13 Mg ha^{-1} y^{-1}. Leaves constitute on average 70% of total litterfall. Most of the annual litter is produced during the dry season. Annual litterfall is positively correlated with annual rainfall. Litter accumulation on the forest floor varies between sites over a wide range from 1.3 to 12.3 Mg ha^{-1}. Litter turnover rate is low at Chamela and Guánica compared with other locations. Net primary productivity estimates are limited. Information from Guánica and Varanasi indicates that NPP varies from 7 to 15 Mg ha^{-1} y^{-1}. Differences in functional traits from site to site largely depend on successional status and history of human perturbations, among other important factors.

Despite their geographical and cultural importance, dry forests are among the least known tropical ecosystems. Most work on biomass distribution and primary productivity has been done in the American tropics and in India. Large tracts of dry forest occur in Africa and islands in tropical regions, and information on these is scarce. All tropical dry forests are subject to increasing pressures from humanity, and extensive and intensive studies are urgently needed for a better understanding of their long-term behavior and response to human exploitation.

Acknowledgements

I gratefully thank Drs A. Búrquez, M. Maass and V. Jaramillo for their help and valuable discussions during the preparation of this chapter. I also thank Drs P. S. Martin, P. Vitousek and R. Felger for critically reviewing the manuscript.

References

Art, H. W. & Marks, P. L. (1971). A summary table of biomass and net annual primary productivity in forest ecosystems of the world. In *Forest Biomass Studies*, pp. 3–32. Life Sciences and Agriculture Experiment Station Miscellaneous Publication 132, University of Maine, Orono.

Bray, J. R. & Gorham, E. (1964). Litter production in forests of the world. *Advances in Ecological Research* **2**: 101–57.

Brown, S. & Lugo, A. E. (1982). The storage and production of organic matter in tropical forests and their role in the global carbon cycle. *Biotropica* **14**: 161–87.

Bullock, S. H. & Solís Magallanes, J. A. (1990). Phenology of canopy trees of a tropical deciduous forest in México. *Biotropica* **22**: 22–35.

Búrquez, A., Martínez-Yrízar, A., Núñez, S., Quintero, T. & Aparicio, A. (1992). Aboveground phytomass of a Sonoran Desert community. *American Journal of Botany* **79**(6): 186.

Castellanos, J., Maass, M. & Kummerow, J. (1991). Root biomass of a dry deciduous tropical forest in México. *Plant and Soil* **131**: 225–8.

Ceballos, G. J. (1989). Population and community ecology of small mammals from tropical deciduous and arroyo forests in Western México. Ph.D. thesis, University of Arizona, Tucson.

Cintrón, B. B. & Lugo, A. E. (1990). Litterfall in a subtropical dry forest: Guánica, Puerto Rico. *Acta Científica (San Juan)* **4**: 37–49.

Collins, N. M. (1977). Vegetation and litter production in southern Guinea savanna, Nigeria. *Oecologia* **28**: 163–5.

García-Aldrete, A. (1988). The psocids (Psocoptera), of Chamela, Jalisco, México. Species, diversity, abundance, distribution and seasonal changes. *Folia Entomológica Mexicana* **77**: 63–84.

Garg, K. R. & Vyas, N. L. (1975). Litter production in deciduous forest near Udaipur (south Rajasthan), India. In *Tropical Ecological Systems*, ed. F. B. Golley and E. Medina, pp. 132–5. Springer-Verlag, New York.

Gaur, J. P. & Pandey, H. N. (1978). Litter production in two tropical deciduous forest communities at Varanasi, India. *Oikos* **30**: 570–5.

Gessel, S. P., Cole, D. W., Johnson, D. & Turner, J. (1980). The nutrient cycles of two Costa Rican forests. In *Progress in Ecology*, ed. V. P. Agarwal and V. K. Sharma, pp. 23–44. Today & Tomorrow's Printers and Publishers, New Delhi.
Hladik, A. (1978). Phenology of leaf production in rain forests of Gabon: distribution and composition of food for folivores. In *The Ecology of Arboreal Folivores*, ed. G. G. Montgomery, pp. 51–71. Smithsonian Institution Press, Washington.
Hopkins, B. (1966). Vegetation of the Olokemeji forest, Nigeria. IV. The litter and soil with special reference to their seasonal changes. *Journal of Ecology* **54**: 687–703.
John, D. M. (1973). Accumulation and decay of litter and net production of forest in tropical West Africa. *Oikos* **24**: 430–5.
Jordan, C. F. & Murphy, P. G. (1978). A latitudinal gradient of wood and litter production, and its implications regarding competition and species diversity in trees. *American Midland Naturalist* **99**: 415–34.
Jung, G. (1969). Cycles biogéochimiques dans un écosystème de région tropicale sèche *Acacia albida* (Del.) sol ferrugineux tropical peu lessivé (Dior). (Note préliminaire). *Oecologia Plantarum* **4**: 195–210.
Kumar, O. M., Vasistha, H. B. & Soni, P. (1984). Phytosociology and standing crop biomass of five different forest types of east Dehra Dun Forest Division. *Indian Forester* **110**: 791–801.
Kummerow, J., Castellanos, J., Maass, M. & Larigauderie, A. (1990). Production of fine roots and the seasonality of their growth in a Mexican deciduous dry forest. *Vegetatio* **90**: 73–80.
Lambert, D. H., Arnason, J. T. & Gale, J.L. (1980). Leaf-litter and changing nutrient levels in a seasonally dry tropical hardwood forest Belize, C.A. *Plant and Soil* **55**: 429–43.
Lieberman, D. (1982). Seasonality and phenology in a dry tropical forest in Ghana. *Journal of Ecology* **70**: 791–806.
Lowman, M. D. (1988). Litterfall and leaf decay in three Australian rainforest formations. *Journal of Ecology* **76**: 451–65.
Lugo, A. E., Gonzalez-Liboy, J. A., Cintrón, B. & Dugger, K. (1978). Structure, productivity and transpiration of a subtropical dry forest in Puerto Rico. *Biotropica* **10**: 278–91.
Lugo, A. E. & Murphy, P. G. (1986). Nutrient dynamics of a Puerto Rican subtropical dry forest. *Journal of Tropical Ecology* **2**: 55–72.
Madge, D. S. (1965). Leaf fall and litter disappearance in a tropical forest. *Pedobiologia* **5**: 273–88.
Madge, D. S. (1969). Litter disappearance in forest and savanna. *Pedobiologia* **9**: 288–99.
Malaisse, F., Freson, R., Gaffinet, G. & Malaisse-Mousset, M. (1975). Litterfall and litter breakdown in Miombo. In *Tropical Ecological Systems*, ed. F. B. Golley and E. Medina, pp. 137–52. Springer-Verlag, New York.
Martínez-Yrízar, A. (1980). Tasas de descomposición de materia orgánica foliar de especies arbóreas de selvas en clima estacional. Tesis de Licenciatura, Facultad de Ciencias, Universidad Nacional Autónoma de México, México.
Martínez-Yrízar, A. & Sarukhán, J. (1990). Litterfall patterns in a tropical deciduous forest in México over a five-year period. *Journal of Tropical Ecology* **6**: 433–44.
Martínez-Yrízar, A., Sarukhán, J., Maass, M., Patiño, C. & Centeno, P. (1992a). Litter production in a tropical deciduous forest in western México. *American Journal of Botany* **79**(6): 184.
Martínez-Yrízar, A., Sarukhán, J., Pérez-Jiménez, A., Rincón, E., Maass, M., Solís Magallanes, A. & Cervantes, L. (1992b). Aboveground phytomass of a tropical deciduous forest on the coast of Jalisco, México. *Journal of Tropical Ecology* **8**: 87–96.
Medina, E. & Klinge, H. (1983). Productivity of tropical forests and tropical woodlands. In *Encyclopedia of Plant Physiology, New Series, Vol. 12D*,

Physiological Plant Ecology IV, ed. O. L. Lange, P. S. Nobel, C. B. Osmond and H. Ziegler, pp. 281–303. Springer-Verlag, Berlin.

Medina, E. & Zelwer, M. (1972). Soil respiration in tropical plant communities. In *Tropical Ecology, with Emphasis on Organic Matter Production*, ed. P. Golley and F. B. Golley, pp. 245–67. Institute of Ecology, University of Georgia, Athens.

Misra, R. (1972). A comparative study of net primary productivity of dry deciduous forest and grassland of Varanasi. In *Tropical Ecology, with Emphasis on Organic Matter Production*, ed. P. Golley and F. B. Golley, pp. 279–93. Institute of Ecology, University of Georgia, Athens.

Murphy, P. G. (1975). Net primary productivity in tropical terrestrial ecosystems. In *Primary Productivity of the Biosphere*, ed. H. Lieth and R. H. Whittaker, pp. 217–31. Springer-Verlag, New York.

Murphy, P. G. & Lugo, A. E. (1986a). Ecology of tropical dry forest. *Annual Review of Ecology and Systematics* **17**: 67–88.

Murphy, P. G. & Lugo, A. E. (1986b). Structure and biomass of a subtropical dry forest in Puerto Rico. *Biotropica* **18**: 89–96.

Nye, P. H. (1961). Organic matter and nutrient cycles under moist tropical forest. *Plant and Soil* **13**: 333–46.

Ogawa, H., Yoda, K., Ogino, K. & Kira, T. (1965). Comparative ecological studies on three main types of vegetation in Thailand. II. Plant Biomass. *Nature and Life in Southeast Asia* **4**: 49–81.

Olson, J. S. (1963). Energy storage and the balance of producers and decomposers in ecological systems. *Ecology* **4**: 322–31.

Pandey, H. N., Gaur, J. P. & Singh R. N. (1980). Litter input and decomposition in tropical dry deciduous forest, grassland and abandoned crop field communities at Varanasi, India. *Acta Oecologica (Oecologia Plantarum)* **1**: 317–323.

Patiño, M. C. (1990). Variación espacial y temporal de la capa de hojarasca (mantillo) en una selva baja caducifolia en Chamela, Jal. México. Tesis de Licenciatura. Facultad de Ciencias, Universidad Nacional Autónoma de México, México.

Pradhan, G. B. & Dash, M. C. (1987). Distribution and population dynamics of soil nematodes in a tropical forest ecosystem from Sambalpur, India. *Proceedings of the Indian Academy of Sciences (Animal Science)* **96**: 395–402.

Proctor, J. (1983). Tropical forest litterfall I. Problems of data comparison. In *Tropical Rain Forest: ecology and management*, ed. S. L. Sutton, T. C. Whitmore and A. C. Chadwick, pp. 267–73. British Ecological Society and Blackwell Scientific Publications, Oxford.

Proctor, J. (1984). Tropical forest litterfall II. The data set. In *Tropical Rain Forest*, ed. A. C. Chadwick and S. L. Sutton, pp. 83–113. Leeds Philosophical and Literary Society (Science Section), Leeds.

Ranawat, M. P. S. & Vyas, L. N. (1975). Litter production in deciduous forest of Koriyat, Udaipur (south Rajasthan), India. *Biologia* (Bratislava) **30**: 41–7.

Reich, P. B. & Borchert, R. (1984). Water stress and tree phenology in a tropical dry forest in the lowlands of Costa Rica. *Journal of Ecology* **72**: 61–74.

Sampaio, E. V. S. B., Salcedo, I. H. & Kauffman, J. B. (1990). Efeito da intensidade de fogo na rebrota de plantas de caatinga em Serra Talhada, PE. In *Resumos da XVI Reunião Nordestina de Botânica*, p. 55. Sociedade Botânica do Brasil, Recife.

Shiavini da Silva, I. (1983). Alguns aspectos da ciclagem de nutrientes em uma area de cerrado (Brasilia, D.F.): chuva, produção e decomposição de liter. Dissertação de Mestrado em Ecologia, Universidade de Brasilia, Brasilia.

Singh, K. P. (1968). Litter production and nutrient turnover in deciduous forests of Varanasi. In *Proceedings of the Symposium of Recent Advances in Tropical Ecology*, ed. R. Misra and B. Gopal, pp. 655–65. International Society for Tropical Ecology, Varanasi, India.

Singh, K. P. (1969). Studies in decomposition of leaf litter of important trees of tropical deciduous forests at Varanasi. *Tropical Ecology* **10**: 292–311.

Singh, K. P. (1989). Mineral nutrients in tropical deciduous forests and savanna ecosystems in India. In *Mineral Nutrients in Tropical Forest and Savanna Ecosystems*, ed. J. Proctor, pp. 153–8. Blackwell Scientific Publications, Oxford.
Singh, K. P. & Singh R. P. (1981). Seasonal variation in biomass and energy of small roots in tropical dry deciduous forest, Varanasi, India. *Oikos* **37**: 88–92.
Singh, R. P. (1975). Biomass, nutrient and productivity structure of a stand of dry deciduous forest of Varanasi. *Tropical Ecology* **16**: 104–9.
Soni, P., Kumar, O. M. & Vasistha, H. B. (1984). Community structure and biomass production in a northern tropical dry mixed deciduous forest of east Dehra Dun Division. *Indian Forester* **110**: 954–7.
Spain, A. V. (1984). Litterfall and the standing crop of litter in three tropical Australian rainforests. *Journal of Ecology* **72**: 947–61.
Swaine, M. D., Lieberman, D. & Hall, J. B. (1990). Structure and dynamics of a tropical dry forest in Ghana. *Vegetatio* **88**: 31–51.
Swift M. J., Heal, O. W. & Anderson, J. M. (1979). *Decomposition in Terrestrial Ecosystems*. University of California Press, Berkeley.
Vitousek, P. M. (1984). Litterfall, nutrient cycling, and nutrient limitation in tropical forests. *Ecology* **65**: 285–98.
Vizcaíno, M. (1983). Patrones temporales y espaciales de producción de hojarasca en una selva baja caducifolia en la costa de Jalisco, México. Tesis de Maestría, Facultad de Ciencias, Universidad Nacional Autónoma de México. México.
Vogt, K. A., Grier, C. C. & Vogt, D. J. (1986). Production, turnover and nutrient dynamics of above- and belowground detritus of world forests. *Advances in Ecological Research* **15**: 303–77.
Vyas, L. N., Garg, R. K. & Vyas, N. L. (1977). Stand structure and above ground biomass in dry deciduous forests of Aravalli Hills at Udaipur (Rajasthan), India. *Biologia (Bratislava)* **32**: 265–70.
Wieder, R. K. & Lang, G. E. (1982). A critique of the analytical methods used in examining decomposition data obtained from litter bags. *Ecology* **63**: 1636–42.

14

Nutrient cycling in tropical deciduous forests

VICTOR J. JARAMILLO &
ROBERT L. SANFORD, Jr

Introduction

Nutrient cycling processes have been well documented for tropical moist forest (Vitousek & Sanford, 1986; Bruijnzeel, 1991) but few comprehensive syntheses exist for tropical and subtropical dry and deciduous forests (Lugo & Murphy, 1986; Singh, 1989). Tropical dry forests are considered among the most threatened tropical ecosystems (Janzen, 1988) because they experience considerable exploitative pressure (Murphy & Lugo, 1986). In India, such pressures have been responsible for the transformation of vast areas of deciduous forest into savanna (Singh, 1989). The current rate of destruction of deciduous forest makes it imperative that we gain a thorough understanding of nutrient cycling in the remaining intact and successional forests. At present one of the principal agricultural practices relies on forest slash burning, resulting not only in recurrent nutrient losses thereby affecting the long-term productivity of the system but also substantially contributing to emissions of C and N to the atmosphere (Kauffman, Sanford & Sampaio, 1990; Maass, Chapter 17).

Seasonally dry forest production is controlled by the amount and distribution of annual rainfall (Martínez-Yrízar, Chapter 13), and this may explain why nutrients have not been considered in detail (Murphy & Lugo, 1986; Singh, 1989). For example in a recent review of forest nutrient cycling there was only one reference to tropical dry forest (Vogt, Grier & Vogt, 1986). If we assume that water availability alone limits primary production in tropical deciduous forest, we could conclude that nutrient limitation is not important. However, multiple resource limitation of plant growth is common in natural communities (Chapin *et al.*, 1987). Results from experimentation in a semiarid ecosystem where water availability is the main control shows a synergistic response by the plant community to the addition of water and N in the field (Lauenroth, Dodd & Sims, 1978). This suggests a careful evaluation of the assumption that tropical deciduous forests are not nutrient-limited. There is strong evidence that some tropical forests are limited by P availability (Vitousek, 1984). In rigorous terms, nutrient limitation can only be shown through experimental fertilization (Chapin, Vitousek & Van Cleve, 1986). However, no such studies have been reported for tropical deciduous forests.

Immobilization–mineralization phases in the soil may ultimately determine the timing and movement of nutrients within an ecosystem. Microbial processes are key for N transformations (Rosswall, 1982) and for the mineralization of organic P (Adams, Attiwill & Polglase, 1989) which in lowland forests may be the most important pool controlling P availability for plants (Cole & Sanford, 1989; Salcedo, Elliott & Sampaio, 1991). Furthermore, microbial biomass is very important in controlling N and P conservation (Vitousek & Matson, 1984; Wood, Bormann & Voigt, 1984).

The relative importance of nutrients in litterfall vs microbial biomass for tropical deciduous forests has been discussed recently (Singh *et al.*, 1989; Raghubanshi *et al.*, 1990; Kundu, 1990; see below). However, another pathway for nutrient cycling, fine root turnover, may also be crucial (Vogt *et al.*, 1986; Cuevas, Chapter 15). This may be particularly important for deciduous forests because they have a comparatively large proportion of belowground biomass (Castellanos, Maass & Kummerow, 1991; Martínez-Yrízar, Chapter 13).

For this chapter we review all available data on nutrient cycling in tropical deciduous forests. Our objective is the synthesis of this information, the identification of existing knowledge gaps and the generation of key hypotheses to help direct nutrient cycling research in these forests.

Approach

We treat deciduous forest as a subset of the dry forest of Murphy & Lugo (1986). Deciduous forests characteristically have dominant trees that lose their leaves in the dry season creating a gray-colored landscape. Published information for México and India (Rzedowski, 1978; Singh, 1989; Jaramillo-Villalobos, 1994) indicates that deciduous forest occurs in areas with annual rainfall between 600 and 1300–1400 mm. Most of the data on nutrient cycling have been collected only in these two countries so we include data from additional dry forests (*sensu* Murphy & Lugo, 1986) for comparison. General information on the sites considered for this review is in Table 13.1 of Martínez-Yrízar (Chapter 13), except for the subtropical dry forest in Australia. This forest is located at 149° 45′ E and 27° 19′ S, at an elevation of 285 m, and has a mean annual rainfall of 560 mm (Moore, Russell & Coaldrake, 1967). Where possible, we compare the data with that of moist forests (Vitousek & Sanford, 1986). To avoid repetitious citations, unless otherwise indicated, all references for moist forest come from the above-mentioned review.

Nutrient concentration and content in biomass

Leaves

In some cases, leaf nutrient concentration reflects site fertility; however, growth rate variations among species cause tissue concentrations of wild plants to be less sensitive indicators of soil nutrient availability (Chapin, 1980). Nevertheless, the relationship among nutrient supply, leaf nutrient content and physiological performance is useful for understanding production processes of plants in the tropics (Medina, 1984). Unfortunately, the scarcity of information for deciduous forest does not allow for an in-depth analysis of the relationship between soil fertility and leaf nutrient status. The data show that deciduous species tend to have higher N and P concentrations than evergreen or sclerophyllous plants (Table 14.1). This pattern is consistent with that of tundra and temperate ecosystems (Chapin, Johnson & McKendrick, 1980, and references therein).

The data from the hill site at Chamela, México, show the highest N, P, and Ca concentrations and thus seems an outlier compared to the other sites. However, other studies at Chamela, but from a different location, have shown even higher leaf N concentrations (3.64%: mean calculated from Castellanos *et al.*, 1991). Unfortunately, no leaf nutrient data are available for the other Mexican locations referred to in the remainder of this review. In general, N concentrations tend to be lower and P concentrations similar to those from moderately fertile sites in moist forest (Vitousek & Sanford, 1986). The N:P ratios show great variability but are not as high as those from infertile sites in moist forests (range 28–53). Nevertheless, they are generally much higher than those of plants with well-balanced nutrition (*c.* 7:1: Shaver, Fetcher & Chapin, 1986).

Table 14.1. *Nutrient concentration and nitrogen:phosphorus ratios of leaves in tropical deciduous (DEC) and dry (DRY) forests*

Site	Forest type	Nutrient concentration (%)					N:P	Source
		N	P	K	Ca	Mg		
India[a]	DEC	1.95	0.13	0.75			15	1
México	DEC	2.96	0.28	1.52	1.91	0.48	11	2
Venezuela 1								
Deciduous species	DRY	2.10	0.16	1.70	1.58	0.36	13	3
Evergreen species	DRY	1.18	0.07				17	3
Venezuela 2								
Deciduous species	DRY	1.71	0.12	0.65	0.77	0.29	15	4
Evergreen species	DRY	1.21	0.06				21	4
Puerto Rico	DRY	1.64	0.06	0.78			25	5
Belize	DRY	1.50	0.10	1.20	1.00	0.50	15	6
Australia	DRY	1.66	0.06	0.74	0.74	0.19	27	7
Zaïre	DEC	2.20	0.11	0.90	0.53		20	8

[a] All data from India cited in the following tables are from deciduous forests.
Sources: 1. Singh (1989); 2. R. Esteban *et al.* (personal communication; these values are for the Hill Site; see text and Martínez-Yrízar, Chapter 13, Table 13.1); 3. Marín & Medina (1981; cited in Medina, 1984); 4. Cuenca (1976; cited in Medina, 1984); 5. Lugo & Murphy (1986); 6. Lambert *et al.* (1980); 7. Moore *et al.* (1967); 8. Malaisse (1979)

Roots

As is the case for moist forest, root biomass and root nutrient data are very scarce for deciduous forest (Tables 14.2 and 14.3; see also Martínez-Yrízar, Chapter 13). Fine root production in deciduous forest is also very poorly documented despite its ecological significance (Kummerow *et al.*, 1990; Cuevas, Chapter 15). Root N:P ratios of deciduous forest in India suggest that nutrient limitation does not slow root decomposition. Estimates of N and P content (excluding the immature forest site: Table 14.3) are within the range reported for moist forest; however, root N content is consistently higher than reported for temperate and tropical forest (Vogt *et al.*, 1986). Roots from the Indian deciduous forest are the most nutrient rich.

Allocation to roots increases under nutrient or water limitation (Mooney, 1972). On average, allocation to belowground components in deciduous and dry forests as a proportion of total biomass is 27% with above- to belowground ratios as high as 0.85 (see Table 13.2 in Martínez-Yrízar, Chapter 13). These proportions are, in general, substantially

Table 14.2. *Nutrient concentration in roots*

Site	Nutrient concentration (%)					Source
	N	P	K	Ca	Mg	
India						
<10 mm diameter	0.85	0.08	0.58			1
>10 mm diameter	0.50	0.05	0.33			1
Australia[a]	0.88	0.02	0.18	0.81	0.13	2
Puerto Rico[b]						
Live	1.24	0.04	0.25			3
Dead	1.26	0.71	0.08			3

[a] No information on root sizes.
[b] Includes roots <1 mm to >10 mm diameter.
Sources: 1. Singh (1989); 2. Moore *et al.* (1967); 3. Lugo & Murphy (1986).

Table 14.3. *Biomass and nutrient content of roots*

Site	Biomass ($g\ m^{-2}$)	Nutrient content ($g\ m^{-2}$)					Source
		N	P	K	Ca	Mg	
India 1	3430	21.3	5.6	69.1			1
India 2[a]	960	6.2	0.9				2
Puerto Rico	4500	54.6	1.8	11.0			3
Australia	4070	36.0	0.6	32.9	7.2	5.4	4

[a] Immature forest.
Sources: 1. Misra (1972); 2. Singh (1975); 3. Lugo & Murphy (1986); 4. Moore *et al.* (1967).

higher than in moist forest (11%: calculated from Vitousek & Sanford, 1986, excluding the low-statured sclerophyllous forests) and suggest that root nutrient input may be an important pathway for nutrient cycling in deciduous and dry forest (see below).

Nutrient transfer

Litterfall in deciduous forest represents a more synchronized pulse of nutrient input to the soil system than in many other tropical forests. Surprisingly, extremely few studies have measured nutrients in litterfall appropriately (Tables 14.4 and 14.5). Litterfall N and P concentrations at the Mexican site rank among the highest for the tropical forests considered by Vitousek (1984; approximate ranges for N = 0.5–1.9% and

Table 14.4. *Nutrient concentration in litterfall. The fine litterfall fraction includes leaves, flowers, fruits and small twigs*

Site	Nutrient concentration (%)					Source
	N	P	K	Ca	Mg	
México						
Leaf	2.26	0.108	0.77	1.69	0.94	1
Fine	2.47	0.132	0.82	1.16	0.73	1
Puerto Rico						
Leaf	1.01	0.019	0.71			2
Wood	0.9	0.02	0.24			2
Belize						
Total	1.20	0.070	0.50	3.00	0.30	3

Sources: 1. R. Esteban *et al.* (personal communication); 2. Lugo & Murphy (1986); 3. Lambert *et al.* (1980).

Table 14.5. *Dry mass and nutrient content of litterfall*

Site	Biomass ($g\ m^{-2}$)	Nutrient content ($g\ m^{-2}$)					N:P	Source
		N	P	K	Ca	Mg		
México								
Leaf	231	5.2	0.25	1.8	3.9	2.2	21	1
Fine	258	6.4	0.34	2.1	3.0	1.9	19	1
Puerto Rico								
Leaf	430	4.4	0.07	3.5			62	2
Wood	50	0.4	0.01	0.1				2
Total	480	4.8	0.08	3.6			61	2
Belize								
Total	1260	15.6	0.92	5.9	37.3	3.2	17	3

Sources: 1. R. Esteban *et al.* (personal communication); 2. Lugo & Murphy (1986); 3. Lambert *et al.* (1980).

P = 0.01–0.15%). The Guánica site in Puerto Rico has the lowest N and P litterfall concentrations reported for dry forests. Both the Mexican and Puerto Rican sites show similar litterfall mass while the more humid successional forest in Belize, as expected, shows greater productivity as well as higher N and P contents (Table 14.5). Nitrogen and P returns in litterfall at the Chamela and Guánica sites are similar to those of 'montane' and 'spodosol/psamment' forests of the moist tropics. Phosphorus returns at Chamela, however, resemble those of the 'oxisol/ultisol' forests, while in Guánica they are considerably lower than all the

Table 14.6. *Nutrient use-efficiency (litterfall mass:element mass)*

Site	N	P	K	Ca	Mg	Source
México						
Leaf	44	926	130	59	106	1
Fine	40	760	122	86	137	1
Puerto Rico						
Leaf	97	6056	123			2
India[a]	79	1059	167			3

[a] Litterfall was collected every 4 weeks from the ground.
Sources: 1. This study, based on R. Esteban *et al.* (personal communication); 2. Lugo & Murphy (1986); 3. Singh (1989).

moist forest sites. The N:P mass ratio of litterfall at Guánica is very high for tropical forests (Vitousek, 1984) and higher than for any temperate forest (Vogt *et al.*, 1986). The N:P mass ratio at Chamela corresponds to a moderately fertile site of the moist tropics.

Nutrient use-efficiency estimated as the dry mass:element mass ratio of litterfall (Vitousek, 1984; Table 14.6) suggests that the Chamela and the Indian forests (Singh, 1989) are not nutrient-limited. However, Singh *et al.* (1989) reported a deciduous forest in the same Vindhyan region of India as extremely nutrient poor. This indicates there is variability regarding the nutrient status of deciduous forest in that area. Both the high nutrient use-efficiency and N:P ratios (Tables 14.5 and 14.6) show that the neotropical Guánica site may be limited by P. In fact, Lugo & Murphy (1986) estimated that 65% of the vegetation P requirement at their site was obtained through retranslocation.

Phosphorus use-efficiency is highest among the nutrients studied in dry forests (Table 14.6), suggesting its possible key role in these ecosystems. However, as mentioned above, nutrient limitation can be verified only with direct fertilization experiments.

Nutrient release through litter decomposition

Surface litter decomposition has been considered one of the main pathways for nutrient cycling in forest ecosystems. Among documented deciduous forests, those of India show exceptionally high surface litter P concentrations while the Chamela site in México is highest in N (Table 14.7). Interestingly, both sclerophyllous dry forests (Puerto Rico and Australia) show remarkably similar N and P contents in litter (Table 14.8).

Table 14.7. *Nutrient concentration of surface litter*

Site	Nutrient concentration (%)					Source
	N	P	K	Ca	Mg	
México						
Leaf	2.29	0.08	0.21	2.12	0.39	1
Total	1.84	0.07	0.14	2.79	0.35	1
India						
Terminalia–Shorea	0.96	0.39	0.48	2.94	0.45	2
Shorea–Buchanania	0.63	0.23	0.50	1.44	0.46	2
Diospyros–Anogeissus	0.80	0.29	0.72	2.46	0.48	2
Tectona	0.78	0.18	0.24	2.34	0.25	2
Butea	1.43	0.14	0.66	1.98	0.62	2
Puerto Rico						
Leaf	1.34	0.03	0.14			3
Belize	0.90	0.01	0.39	4.00	0.34	4
Australia						
Leaf	1.14	0.03	0.07	0.91	0.09	5
Wood	0.40	0.01	0.04	0.71	0.04	5

Sources: 1. R. Esteban *et al.* (personal communication); 2. Singh (1969); 3. Lugo & Murphy (1986); 4. Lambert *et al.* (1980); 5. Moore *et al.* (1967).

Table 14.8. *Biomass and nutrient content of surface litter*

Site	Biomass ($g\ m^{-2}$)	Nutrient content ($g\ m^{-2}$)					Source
		N	P	K	Ca	Mg	
México							
Leaf	318	7.3	0.2	0.7	6.7	1.2	1
Total	764	14.1	0.5	1.1	21.3	2.7	1
India							
Total	770	12.7	1.0		18.5		2
Total	790	4.6	0.3				3
Total		8.0	0.6	3.8			4
Belize							
Total	720	6.5	0.1	2.8	28.7	2.4	5
Puerto Rico							
Leaf	1230	16.5	0.4	1.7			6
Australia							
Leaf + twigs	1430	16.3	0.4	1.1	13.0	1.3	7

Sources: 1. R. Esteban *et al.* (personal communication); 2. Misra (1972); 3. Singh (1975); 4. Singh (1989); 5. Lambert *et al.* (1980); 6. Lugo & Murphy (1986); 7. Moore *et al.* (1967).

Table 14.9. *Mean residence time for organic matter (OM) and nutrients and decomposition constant (K_L). Mean residence time is surface litter mass divided by litterfall mass for the variable of interest. K_L = litterfall mass/surface litter mass. Values calculated from data in Tables 14.5 and 14.8*

	Mean residence (years)						
Site	OM	N	P	K	Ca	Mg	K_L
México							
Leaf	1.37	1.40	1.00	0.39	1.71	0.55	0.73
Puerto Rico							
Leaf	2.87	3.72	5.21	0.49			0.35
Belize							
Total	0.57	0.42	0.10	0.48	0.77	0.76	1.76

We calculated the mean residence time for organic matter and nutrients and the decomposition constant K_L (Vogt *et al.*, 1986; Anderson & Swift, 1983) for all deciduous forest sites that have sufficient data (Table 14.9). Vogt *et al.* (1986) suggest that the relative importance of nutrient uptake and microbial immobilization can be assessed either by comparing N and P concentration of aboveground litterfall to that of surface litter or by comparing the mean residence time of organic matter in the forest floor to that of the nutrient of interest. If nutrient concentrations in surface litter are lower than in aboveground litterfall then plant nutrient uptake is assumed to prevail over microbial immobilization. If the mean residence time of organic matter is higher than that of the nutrient of interest, plant uptake is assumed to prevail over immobilization. Both comparisons are calculated for three forests and the results are consistent. The more moist, seasonally dry forest site (Belize) has the lowest residence time for organic matter, Ca, P and N, and shows potential immobilization for Ca and Mg. In the Mexican deciduous forest, plant uptake of P, K and Mg prevails and Ca is immobilized. For N there seems to be a balance between uptake and immobilization since residence time for organic matter and N are very similar. The dry forest in Puerto Rico has the slowest organic matter decomposition, which is expected given the prevalence of sclerophyllous species (see Medina, Chapter 9), and shows immobilization of both N and P. This agrees with its high N use-efficiency when compared to the Mexican site. This site has high P immobilization potential which is also consistent with the high P use-efficiency calculated for this forest.

Singh (1989) estimates that N, P and K released through decomposition in Indian deciduous forest is 95% (or more) of the amount returned through litterfall. Such releases represent nearly 43% of the N uptake required for net production, 46% of P and 37% of K. In another deciduous forest in India decomposition accounts for at least 11.6% of the N and 11% of the P of the annual vegetative uptake (Singh, 1975). Lugo & Murphy (1986) estimate that N and P released through decomposition represent only 2.3% (N) and 38.5% (P) of the nutrient uptake from the soil.

Nutrient release in soil

Although several authors consider that soil nutrient pools are large for dry and deciduous forests (R. P. Singh, 1975; Lugo & Murphy, 1986; K. P. Singh, 1989) such values are difficult to relate to actual availability (Vitousek & Sanford, 1986). Seasonal measurements in both forest types show that soil ammonium (NH_4) and nitrate (NO_3) (García-Méndez *et al.*, 1991), phosphate-P (PO_4) (Singh *et al.*, 1989), and mineral elements in general (Lambert, Arnason & Gale, 1980) increase during the dry season, presumably because of a decrease in plant nutrient demand and an increase in supply due to microbial death (see below). Few studies, however, have measured nutrient mineralization in soils of deciduous forest. Singh *et al.* (1989) determined *in situ* nitrification and García-Méndez *et al.* (1991) measured mineralization and nitrification with laboratory incubations. Both studies showed that the highest nitrification rate occurs during the rainy season and that nitrification is very low at the end of the dry season. At the Chamela site in México, net N mineralization peaks with the onset of rain (García-Méndez *et al.*, 1991).

Given the relatively high proportion of biomass in roots of dry and deciduous forests (see Table 13.2 in Martínez-Yrízar, Chapter 13) and their considerable fine root production (Singh & Singh, 1981; Kummerow *et al.*, 1990), fine root mortality and decomposition should represent an important pathway for nutrient cycling in these forests. Vogt *et al.* (1986) and Vitousek & Sanford (1986) show that more N is incorporated into the soil via fine root turnover than through litterfall for a number of forests. However, Singh (1989) estimated that for deciduous forest in India, fine-root turnover contributed approximately half of the return of N and P via litterfall. This result is indicative of the lack of nutrient limitation at that site and fits with the hypothesis that fine-root production may be reduced when nutrients are not limiting in an ecosystem (Gower, 1987).

Mycorrhizae may also be important for soil nutrient cycling in seasonally dry forests. Considerable work in Africa has shown that endomycorrhizal species tend to predominate in dry forests (Högberg, 1986; see Cuevas, Chapter 15). Most tropical trees are known to form vesicular-arbuscular (VA) mycorrhizal associations which aid them not only in exploring a larger soil volume and thereby obtaining labile P, but also in making low solubility forms of soil P available (Alexander, 1989). Furthermore, enhanced ammonium absorption by VA mycorrhizae may be important in situations where nitrification is reduced because competition for ammonium with nitrifiers may diminish potential N losses from the ecosystem (Alexander, 1989). Very little is known on the role of mycorrhizae in nutrient cycling of deciduous and dry forests of the neotropics.

The microbial hypothesis

Soil microbial biomass of deciduous forest has been recently proposed as the major nutrient source for initiation of plant growth and as a conservation sink during the dry season (Singh *et al.*, 1989; Raghubanshi *et al.*, 1990). Singh *et al.* (1989) found that the highest concentrations of microbial C, N and P occur during the dry season while the lowest microbial concentrations occur during the rainy season when plant growth is active and N mineralization is high. Raghubanshi *et al.* (1990) proposed that nutrients are released with the onset of rain, possibly for two reasons: plasmoptysis, due to a sudden increase in soil water potential; and expansion of microbivorous populations. Kundu (1990) maintains, on the contrary, that the most important nutrient pulse comes from litter decomposition.

Microbial control of nutrient release and availability is not a new concept. Chapin, Barsdate & Barel (1978) suggest that regulation of P availability and cycling in a wet meadow tundra ecosystem is mediated by crashes in microbial populations occurring during the growing season which subsequently release nutrients to the soil. They estimate that P released by a 90% decline in microbial biomass could satisfy the entire annual requirement of the vascular vegetation. Furthermore, steep microbial population declines would release plants from strong competition with soil fungi and bacteria for nutrients. The central idea of both hypotheses is that changes in microbial population size increase both nutrient release to the soil and plant nutrient uptake. In the case of tropical deciduous forest, Raghubanshi *et al.* (1990) suggest these changes are the result of increased soil moisture at the beginning of the rainy season. Recent evidence has shown a negative correlation between soil water and microbial C, N and P within an annual cycle (Srivastava, 1992).

We suggest that mineralization of microbial biomass represents an important pathway for nutrient cycling in deciduous forest **throughout** the growing season. This is particularly the case for nutrients such as P that may cycle independently of organic matter decomposition (Vitousek & Sanford, 1986) and that is limiting for plant growth in many lowland forests (Vitousek, 1984). Our hypothesis is based on the following observations. Air-drying of soil and subsequent microbial cell death increase available soil N (Sparling & Ross, 1988; Okano, 1990); soil-drying may also elicit active cellular responses which allow micro-organism survival of drought periods. Survival depends not only on the intensity of water stress but on the soil type and the drying speed (Stolp, 1988). Sudden re-wetting of dry soils releases more soil microbial biomass C and nutrients than air-drying because changes in soil water potential occur faster and this results in sudden plasmoptysis of sensitive microbial populations (Kieft, Soroker & Firestone, 1987). Frequent soil drying and wetting may thus result in 'release–immobilization' cycles that would significantly affect nutrient movement through the soil–microbe–plant system (Figure 14.1).

Rainfall in most deciduous forests is seasonal, providing a strong water pulse with the onset of rain and triggering a variety of functional processes. However, rainfall during the growing season is not continuous (Bullock, 1986) and this may produce variations in soil moisture (García-Méndez *et al.*, 1991). Therefore, the processes outlined above probably occur throughout the growing season. Furthermore, soil moisture fluctuation may occur at the onset of the **dry** season due to late isolated showers that result in concurrent increases in net N-mineralization (García-Méndez *et al.*, 1991). We suggest that the nutrients so released and those released as a consequence of soil drying after the rainy season can be taken up by plants and resistant microbial populations.

Ecophysiological evidence suggests that plants can utilize such nutrients before water shortage is excessive and before plants entirely abscise their leaves. In drying soil, osmotic adjustment may allow roots to move nutrients to the shoots (Bradford & Hsiao, 1982). Moreover, assimilates may be available for root growth because under moderate water stress photosynthetic rates are reduced less than leaf expansion (Schulze *et al.*, 1987). Nutrients taken up by plants at this time could be stored after

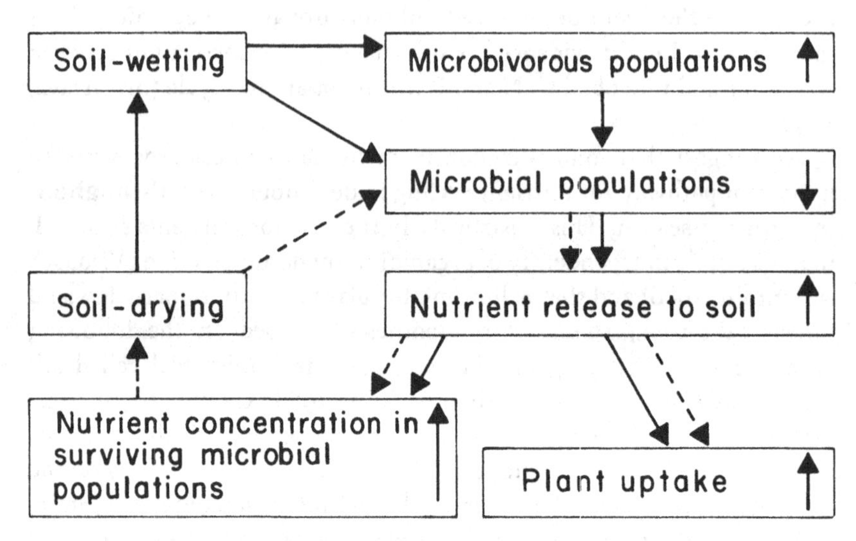

Figure 14.1. The microbial hypothesis: soil microbial biomass dynamics and nutrient release in response to soil wetting and dry cycles. The cycle following the solid-lined arrows occurs at the onset of the rainy season. The cycle through the broken-lined arrows takes place at the onset of the dry season. A combination of both may occur throughout the growing season. Arrows inside the boxes indicate that the populations or the processes considered increase (upward) or decrease (downward).

retranslocation from leaves and used for growth the following season.

This scenario provides a mechanistic explanation for some of the observations regarding microbial biomass in deciduous forest. The hypothesis explicitly defines plant physiological–soil nutrient relationships controlled by microbial processes. However, except for the studies on the effect of soil wetting and drying cycles on microbial biomass (see above), as yet there is no direct experimental evidence to confirm or reject this scenario for dry forest.

Summary

Nutrient cycling processes in tropical and subtropical deciduous and dry forests have not been studied as extensively as in tropical moist forest; most information has been collected in India and México.

As in other ecosystems, leaf N and P concentrations tend to be higher in deciduous than in evergreen and sclerophyllous plants. The highest values reported are 3.64% and 0.28% foliar N and P, respectively. Compared to moist forest in moderately fertile sites, leaf N tends to be lower and P similar. Litterfall N and P concentrations in the Mexican deciduous forest are among the highest for tropical forests. Root N and P concentrations appear to be in the same range as in tropical moist forest, but estimates are scarce. Nutrient cycling from root decomposition may be important because root biomass and turnover are high; in India this pathway contributes about half of the return of N and P.

Nutrient use-efficiency estimates suggest these forests are not nutrient limited. Phosphorus use-efficiency is the highest among the nutrients studied, implying a possible key role in these ecosystems. Assessment of the relative importance of nutrient uptake and immobilization indicates that plant uptake of P, K and Mg prevails at a Mexican site, while in Puerto Rico there is immobilization of both N and P. Nitrification rates are highest during the rainy season and very low at the end of the dry season.

We discuss the hypothesis of soil microbial biomass control on nutrient release and availability. We propose that this mechanism operates not only at the onset of the rainy season but also throughout and at the end of the growing season. However, experimental studies are needed of the relative importance of major pathways of nutrient cycling: litter decomposition, microbial biomass, fine root turnover, and plant translocation.

References

Adams, M. A., Attiwill, P. M. & Polglase, P. J. (1989). Availability of nitrogen and phosphorus in forest soils in northeastern Tasmania. *Biology and Fertility of Soils* **8**: 212–18.

Alexander, I. (1989). Mycorrhizas in tropical forests. In *Mineral Nutrients in Tropical Forest and Savanna Ecosystems*, ed. J. Proctor, pp. 169–88. Blackwell Scientific Publications, Oxford.

Anderson, J. M. & Swift, M. J. (1983). Decomposition in tropical forests. In *Tropical Rainforest: ecology and management*, ed. S. L. Sutton, T. C. Whitmore and A. C. Chadwick, pp. 287–309. Blackwell Scientific Publications, Oxford.

Bradford, K. J. & Hsiao, T. C. (1982). Physiological responses to moderate water stress. In *Encyclopedia of Plant Physiology, New Series, Vol. 12B, Physiological Plant Ecology II, Water Relations and Carbon Assimilation*, ed. O. L. Lange, P. S. Nobel, C. B. Osmond and H. Ziegler, pp. 263–324. Springer-Verlag, Berlin.

Bruijnzeel, L. A. (1991). Nutrient input-output budgets of tropical forest ecosystems: a review. *Journal of Tropical Ecology* **7**: 1–24.

Bullock, S. H. (1986). Climate of Chamela, Jalisco, and trends in the south coastal region of México. *Archives for Meteorology, Geophysics and Bioclimatology. Series B* **36**: 273–316.

Castellanos, J., Maass, M. & Kummerow, J. (1991). Root biomass of dry deciduous tropical forest in México. *Plant and Soil* **131**: 225–8.

Chapin, F. S. III. (1980). The mineral nutrition of wild plants. *Annual Review of Ecology and Systematics* **11**: 233–60.

Chapin, F. S. III, Barsdate, R. J. & Barel, D. (1978). Phosphorus cycling in Alaskan coastal tundra: a hypothesis for the regulation of nutrient cycling. *Oikos* **31**: 189–99.

Chapin, F. S. III, Vitousek, P. M. & Van Cleve, K. (1986). The nature of nutrient limitation in plant communities. *American Naturalist* **127**: 48- 58.

Chapin, F. S. III, Johnson, D. A. & McKendrick, J. D. (1980). Seasonal movement of nutrients in plants of differing growth form in an Alaskan tundra ecosystem: implications for herbivory. *Journal of Ecology* **68**: 189–209.

Chapin, F. S., III, Bloom, A. J., Field, C. B. & Waring, R. H. (1987). Plant responses to multiple environmental factors. *BioScience* **37**: 49–57.

Cole, C. V. & Sanford, R. L., Jr (1989). Biological aspects of the phosphorus cycle. In *Phosphorus Cycling in Terrestrial and Aquatic Ecosystems*, ed. H. Tiessen, pp. 10–16. University of Saskatchewan, Saskatoon.

García-Méndez, G., Maass, J. M., Matson, P. A. & Vitousek, P. M. (1991). Nitrogen transformations and nitrous oxide flux in a tropical deciduous forest in México. *Oecologia* **88**: 362–6.

Gower, S. T. (1987). Relations between mineral nutrient availability and fine root biomass in two Costa Rican tropical wet forests: a hypothesis. *Biotropica* **19**: 171–5.

Högberg, P. (1986). Soil nutrient availability, root symbioses and tree species composition in tropical Africa: a review. *Journal of Tropical Ecology* **2**: 359–72.

Janzen, D. H. (1988). Tropical dry forest: the most endangered major tropical ecosystem. In *Biodiversity*, ed. E. O. Wilson, pp. 130–7. National Academy Press, Washington.

Jaramillo-Villalobos, V. (1994). *Revegetación y Reforestación de las Areas Ganaderas en las Zonas Tropicales de México.* Secretaría de Agricultura y Recursos Hidráulicos, México.

Kauffman, J. B., Sanford, R. L., Jr & Sampaio, E. V. S. B. (1990). Biomass burning in Brazilian dry forest: carbon and nitrogen emissions. In *Abstracts of Papers. Chapman Conference on Global Biomass Burning: Atmospheric, Climatic, and Biospheric Implications*, Williamsburg.

Kieft, T. L., Soroker, E. & Firestone, M. K. (1987). Microbial biomass response to a rapid increase in water potential when dry soil is wetted. *Soil Biology and Biochemistry* **19**: 119–26.

Kummerow, J., Castellanos, J., Maass, M. & Larigauderie, A. (1990). Production of fine roots and the seasonality of their growth in a Mexican deciduous dry forest. *Vegetatio* **90**: 73–80.

Kundu, D. K. (1990). Tropical leaf-litter nutrients. *Nature* **344**: 203.

Lambert, J. D. H., Arnason, J. T. & Gale, J. L. (1980). Leaf-litter and changing nutrient levels in a seasonally dry tropical hardwood forest, Belize, C.A. *Plant and Soil* **55**: 429–33.

Lauenroth, W. K., Dodd, J. L. & Sims, P. L. (1978). The effects of water- and nitrogen-induced stresses on plant community structure in a semiarid grassland. *Oecologia* **36**: 211–22.

Lugo, A. E. & Murphy, P. G. (1986). Nutrient dynamics of a Puerto Rican subtropical dry forest. *Journal of Tropical Ecology* **2**: 55–72.

Malaisse, F. (1979). The miombo ecosystem. In *Tropical Forest Ecosystems: a state-of-knowledge report*, pp. 589–604. UNESCO/UNEP/FAO, Paris.

Medina, E. (1984). Nutrient balance and physiological processes at the leaf level. In *Physiological Ecology of Plants in the Wet Tropics*, ed. E. Medina, H. A. Mooney, and C. Vázquez-Yánes, pp. 139–54. Dr W. Junk, The Hague.

Misra, R. (1972). A comparative study of net primary productivity of dry deciduous forest and grassland of Varanasi, India. In *Tropical Ecology with an Emphasis on Organic Productivity*, ed. F. B. Golley and R. Misra, pp. 279–93. Institute of Ecology, University of Georgia, Athens.

Mooney, H. A. (1972). The carbon balance of plants. *Annual Review of Ecology and Systematics* **3**: 315–46.
Moore, A. W., Russell, J. S. & Coaldrake, J. E. (1967). Dry matter and nutrient content of a subtropical semiarid forest of *Acacia harpophylla* F. Muell. (Brigalow). *Australian Journal of Botany* **15**: 11–24.
Murphy, P. G. & Lugo, A. E. (1986). Ecology of tropical dry forest. *Annual Review of Ecology and Systematics* **17**: 67–88.
Okano, S. (1990). Availability of mineralized N from microbial biomass and organic matter after drying and heating of grassland soils. *Plant and Soil* **129**: 219–25.
Raghubanshi, A. S., Srivastava, S. C., Singh, R. S. & Singh, J. S. (1990). Nutrient release in leaf litter. *Nature* **346**: 227.
Rosswall, T. (1982). Microbiological regulation of the biogeochemical nitrogen cycle. *Plant and Soil* **67**: 15–34.
Rzedowski, J. (1978). *La vegetación de México*. Editorial LIMUSA, México.
Salcedo, I. H., Elliott, E. T. & Sampaio, E. V. S. B. (1991). Mechanisms controlling phosphorus retention in the litter mat of Atlantic coastal forests. In *Phosphorus Cycles in Terrestrial and Aquatic Ecosystems. Regional Workshop 3: South and Central America*, ed. H. Tiessen, D. López-Hernández and I. H. Salcedo, pp. 109–20. SCOPE, UNEP and Saskatchewan Institute of Pedology, Saskatoon.
Schulze, E.-D., Robichaux, R. H., Grace, J., Rundel, P. W. & Ehleringer, J. R. (1987). Plant water balance. *BioScience* **37**: 30–7.
Shaver, G. S., Fetcher, N. & Chapin, F. S. III. (1986). Growth and flowering in *Eriophorum vaginatum*: annual and latitudinal variation. *Ecology* **67**: 1524–35.
Singh, J. S., Raghubanshi, A. S., Singh, R. S. & Srivastava, S. C. (1989). Microbial biomass acts as a source of plant nutrients in dry tropical forest and savanna. *Nature* **338**: 499–500.
Singh, K. P. (1969). Nutrient concentration in leaf litter of ten important tree species of deciduous forests at Varanasi. *Tropical Ecology* **10**: 83–91.
Singh, K. P. (1989). Mineral nutrients in tropical dry deciduous forest and savanna ecosystems in India. In *Mineral Nutrients in Tropical Forest and Savanna Ecosystems*, ed. J. Proctor, pp. 153–68. Blackwell Scientific Publications, Oxford.
Singh, K. P. & Singh, R. P. (1981). Seasonal variation in biomass and energy of small roots in tropical dry deciduous forest, Varanasi, India. *Oikos* **37**: 88–92.
Singh, R. P. (1975). Biomass, nutrient and productivity structure of a stand of dry deciduous forest of Varanasi. *Tropical Ecology* **16**: 104–9.
Sparling, G. P. & Ross, D. J. (1988). Microbial contributions to the increased nitrogen mineralization after air-drying of soils. *Plant and Soil* **105**: 163–7.
Srivastava, S. C. (1992). Microbial C, N, and P in dry tropical soils: seasonal changes and influence of soil moisture. *Soil Biology and Biochemistry* **24**: 711–14.
Stolp, H. (1988). *Microbial Ecology: organisms, habitats, and activities*. Cambridge University Press, Cambridge.
Vitousek, P. M. (1984). Litterfall, nutrient cycling, and nutrient limitation in tropical forests. *Ecology* **65**: 285–98.
Vitousek, P. M. & Matson, P. A. (1984). Mechanisms of nitrogen retention in forest ecosystems: a field experiment. *Science* **225**: 51–2.
Vitousek, P. M. & Sanford, R. L., Jr (1986). Nutrient cycling in moist tropical forest. *Annual Review of Ecology and Systematics* **17**: 137–67.
Vogt, K. A., Grier, C. C. & Vogt, D. J. (1986). Production, turnover, and nutrient dynamics of above- and belowground detritus of world forests. *Advances in Ecological Research* **15**: 303–77.
Wood, T., Bormann, F. H. & Voigt, G. K. (1984). Phosphorus cycling in a northern hardwood forest: biological and chemical control. *Science* **223**: 391–3.

15

Biology of the belowground system of tropical dry forests

ELVIRA CUEVAS

Introduction

Tropical forests have received considerable attention recently, stimulated in part by the high rate at which they are being modified or completely destroyed. Emphasis has been given to their potential effect on the global carbon balance and on biogenic emissions (García-Méndez *et al.*, 1991; Matson & Vitousek, Chapter 16). Most studies have focused on tropical forests growing in wet or humid climates which account for 58% of tropical forests (Brown & Lugo, 1982), while seasonal forests growing in drier climates, which represent the remaining 42%, have been very little studied (Murphy & Lugo, 1986a). Tropical dry forests represented nearly 20% of the total biomass of forests in the world in the early 1970s (Persson, 1974). But due to deforestation, and extensive use of the area for intensive agriculture, pastures and shifting cultivation among other uses, the present area is considerably reduced.

Constraints on soil biological activity and their effect on ecosystem production, soil organic matter formation and nutrient cycling have been little studied in dry forests (Anderson & Flannagan, 1989). Most of the work on this topic has been carried out in India with considerably less information available from dry forests of the neotropics.

In this review I will discuss how water availability primarily constrains soil biological activity in tropical dry forests, in order to highlight its importance in the dynamics of organic matter and nutrients in the ecosystem.

Climatic characteristics of tropical dry forests

Strong rainfall seasonality is the overriding macro-determinant in dry forests. Seasonality and, most important, the duration and intensity of dry and wet periods, exert a strong influence on the biological activity of both the above- and belowground parts of the ecosystem. Temperature can also play an important role in soil biological activity. A characteristic of seasonal forests is their total or almost total leaf fall during the dry season, which produces a concomitant increase in surface soil temperatures. However, the extent to which these climatic variables affect the system is dependent on the interaction with other variables such as soil type, texture and structure, plant cover, depth of litter layer, and macrofaunal activity.

Soil respiration and its determinants

Soil respiration gives a good indication of biological activity because it is the sum of respiratory activity of microorganisms, mesofauna and roots (Medina & Zelwer, 1972; Singh & Gupta, 1977). There is a strong seasonality in soil respiration in areas with a seasonal pattern of rainfall. Soil CO_2 efflux varies with rainfall: more persistent and higher rates of biological activity are found in areas with higher rainfall and longer wet seasons (Medina & Zelwer, 1972; Singh & Ambasht, 1980; Singh, 1984; Rajvanshi & Gupta, 1986; Rout & Gupta, 1989; Holt, Hogden & Lamb, 1990). An example from areas differing in annual rainfall and extension of the rainy season shows that the forest with a longer rainy season has higher rates of soil respiration for more extended periods (Figure 15.1).

Soil respiration can be correlated with soil moisture, or soil water availability as expressed by Bailey's Moisture Index (Figure 15.2). Bailey's Index increases linearly with precipitation, but decreases exponentially with temperature (Bailey, 1979; Medina, Chapter 9). Months with Bailey's indices below 0.53 have strong limitation in soil water availability (Bailey, 1979). Soil moisture explains between 75 and 90% of the variability in soil CO_2 efflux (Singh & Ambasht, 1980; Singh, 1984; Singh, Srivastava & Singh, 1984; Rajvanshi & Gupta, 1986; Rout & Gupta, 1989; Holt, Hogden & Lamb, 1990). The very rapid flux of CO_2 after wetting of dry soils found by Matson & Vitousek (Chapter 16) also indicates how fast soil biological activity is stimulated by moisture availability.

Soil moisture is not only a function of rainfall and evaporation. Factors such as soil texture, structure, and organic matter content and quality are determinants of the water-holding capacity of soils (Russell, 1973). These factors in turn condition the rate and duration of soil biological activity. Dkhar & Mishra (1987) found that CO_2 evolution during the

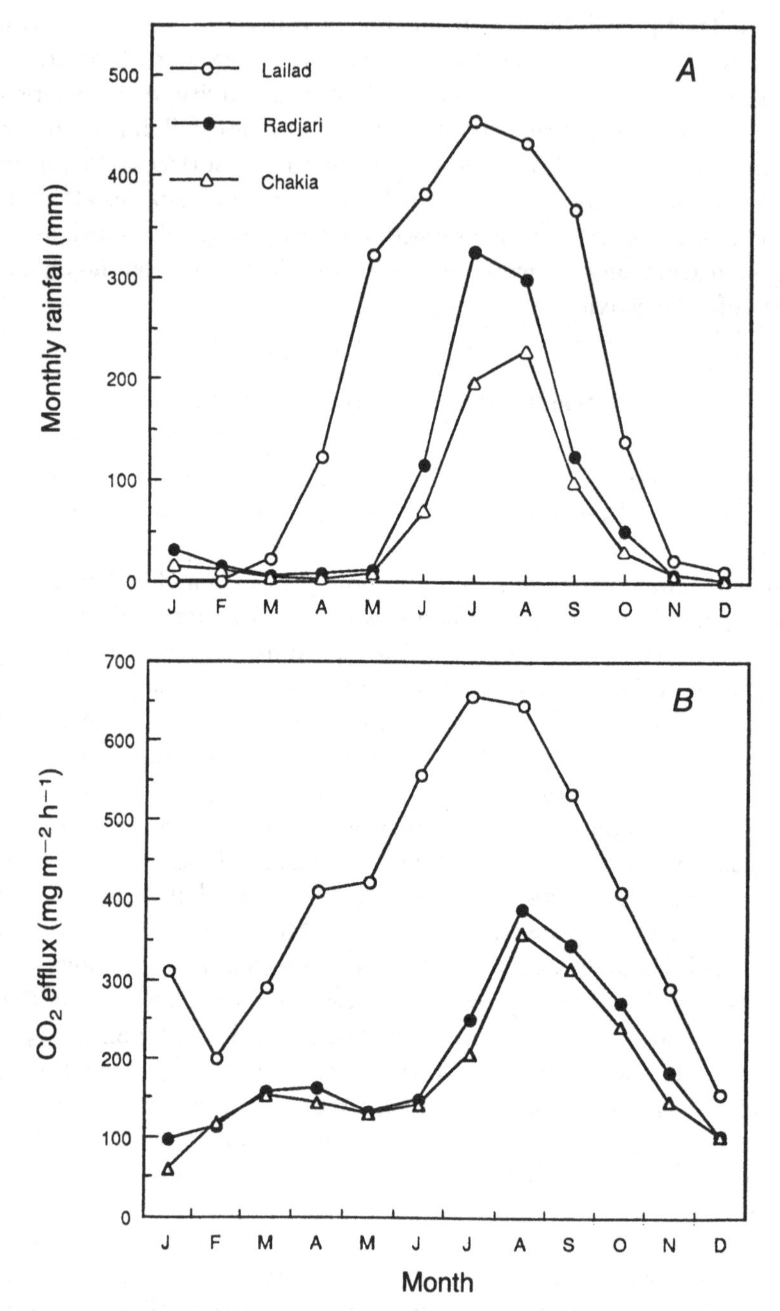

Figure 15.1. Annual variation of soil respiration with respect to rainfall from sites differing in annual precipitation. *A*. Monthly rainfall in mm. *B*. Monthly soil CO_2 efflux. Sources: Singh *et al*. (1984), Singh (1984).

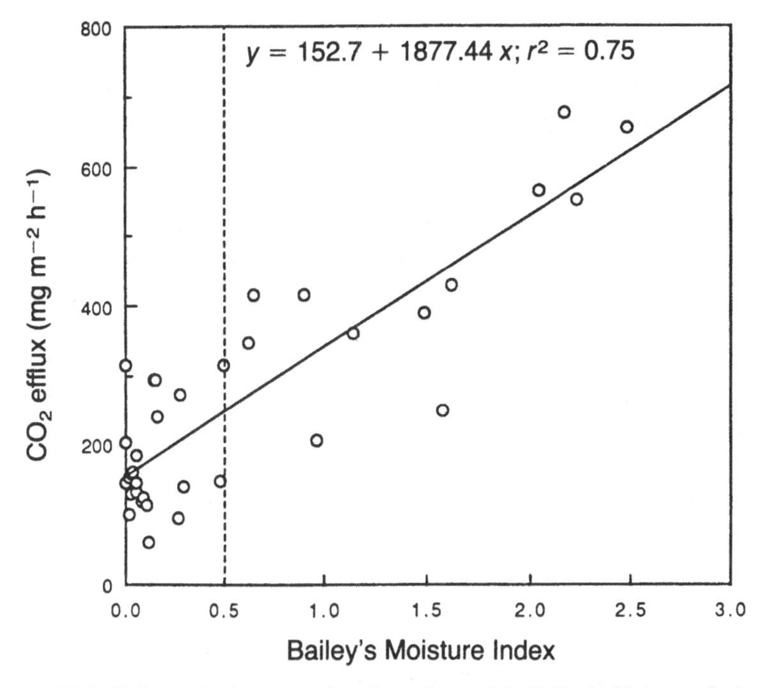

Figure 15.2. Soil respiration as a function of monthly Bailey's Moisture Index in seasonal tropical forests varying in total rainfall and extent of dry period. Sources: Singh *et al.* (1984), Singh & Shekhar (1986).

rainy season was positively correlated with organic C, N, microbial populations and moisture content of the soils. The two drier sites presented in Figure 15.1 had consistently lower organic C contents than the wetter site (Singh *et al.*, 1984; Singh, 1984), thus indicating substrate limitation on microbial activity.

Presence, depth and moisture-holding capacity of the litter layer are also important factors for soil biological activity. Rajvanshi & Gupta (1986) working in a deciduous forest dominated by *Dalbergia sissoo* showed that soil respiration could be predicted from a multiple regression based on soil water content, litter moisture and soil temperature [ln soil respiration = 1.5975 + 0.8856 (ln soil water %) − 0.0862 (ln litter moisture %) + 0.8255 (ln soil temperature °C) ($p < 0.001$, df = 42, r^2 = 0.92)]. Predicted respiration based on this equation was similar to measured values, indicating the importance of the litter layer. Singh & Ambasht (1980) also found for a *Tectona grandis* plantation at Varanasi,

India, that litter respiration and decomposition closely followed litter moisture content: moisture explained 99% of the variability in respiration. Rout & Gupta (1989), working in deciduous, scrub and pine forests at 1100 m elevation in India, also found that soil respiration was best correlated with litter humidity. Soil respiration varied according to forest type with higher respiration in the deciduous forests, followed by pine and scrub forests, indicating effects of substrate quality and edaphic condition. Thus the composition, quality and water-holding capacity of the overlying litter layer can influence the level of activity of microorganisms, which is expressed in the magnitude of soil respiration.

Resource quality and its interaction with soil fauna and microbial activity also has a bearing on the quality of soil organic matter formed and its rate of decay (Bernhard-Reversat, 1987). In dry forests, as in wet tropical areas, soil biological activity has been correlated with several characteristics of decomposing residues: C:N ratios (Ramakrishnan & Toky, 1981; Sandhu, Sinha & Ambasht, 1990) initial lignin and N concentrations (Laishram & Yadava, 1988; Rout & Gupta, 1989), and soluble and labile C compounds (Bernhard-Reversat, 1987). The proportion of deciduous to evergreen forms changes along a gradient of water availability (Medina, Chapter 9), with an increase of evergreen species at the lower extreme of water availability. With increasing proportion of evergreen components, there is also an increase in the degree of sclerophylly of the organic residues (as defined by low specific leaf area, $cm^2 g^{-1}$ dry weight, and low N and P concentrations: Sobrado & Medina, 1980). This can result in slower rates of decomposition due to reduced palatability for the soil fauna and reduced soluble and labile C compounds as energy sources for microbial activity. With the increase in evergreen components there is slower C mineralization and nutrient release, and more recalcitrant forms of soil organic matter predominate in the soil profile so rates of biological activity can be affected.

Mass and dynamics of roots in dry forests, as in wet forests, exert a strong influence on soil respiration and C balance (Medina & Zelwer, 1972; Singh & Gupta, 1977; Keller, Kaplan & Wofsy, 1986; Raich & Nadelhoffer, 1989; Cuevas, Brown & Lugo, 1991). The overall response can be affected by the composition of the community, such as the ratio of deciduous to evergreen life forms as indicated by Medina (Chapter 9). Soil respiration can be significantly correlated with fine root mass (Singh *et al.*, 1984); however, the significance of the correlation varies according to species (Upadhyaya, Siddiqui & Singh, 1981).

Root dynamics are also relevant in the seasonal variation of soil CO_2

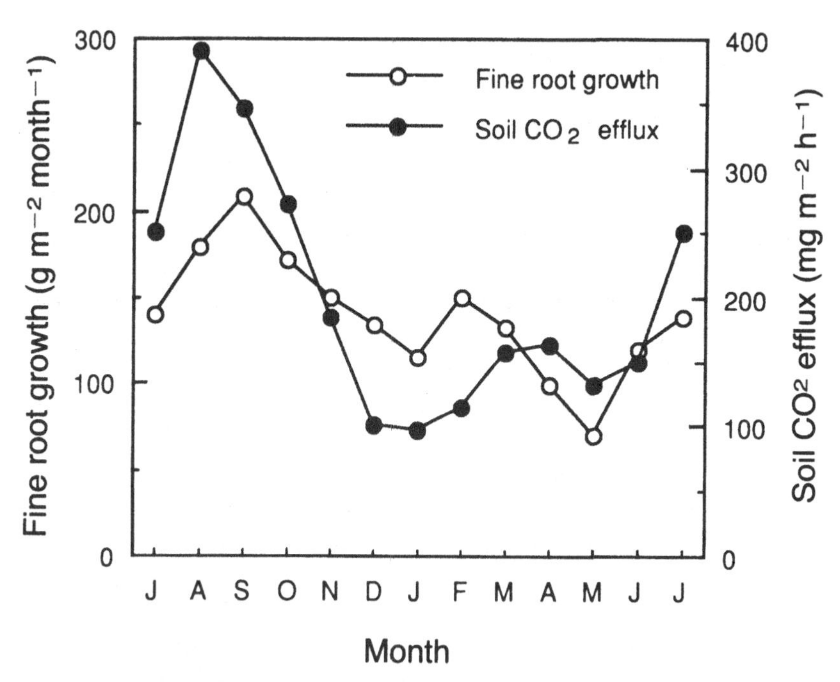

Figure 15.3. Relationship between fine root growth and soil CO_2 efflux in a 29 year old *Tectona grandis* plantation in India. Source: Singh *et al.* (1984).

efflux. Singh *et al.* (1984) and Singh & Shekhar (1986) measured maximum root respiration rates during the wet season and the lowest values during the dry season. The highest values of CO_2 efflux were related to the increased proportion of meristematic tissue present during the rainy season (Singh & Srivastava, 1985), a result of biosynthetic processes associated with growth (Marshall & Waring, 1985). An example of this relationship between fine root growth and soil respiration throughout the year for a 29 year old teak plantation in India is presented in Figure 15.3.

Fine root dynamics

Root plasticity and regulation of the shoot:root ratio according to growing conditions is well known (Russell, 1977). There is a functional balance between active leaf area and active root surface area. The total root surface area required for adequate water uptake is at least as great as the leaf surface area, but because roots have more surface area per unit weight than leaves, shoot:root ratios on a dry weight basis can be as high as 20 under optimal conditions (de Willigen & van Noordwijk, 1987).

Vertical growth and depth of rooting are not only age and species dependent (Sen & Tanwar, 1983; Castellanos, Maass & Kummerow, 1991) but also depend on transport properties of the soil. Texture modifies the root environment by differentiation of sorption capacity and moisture retention ability of the soil (Glinski & Lipiec, 1990). Optimal density of fine roots depends on soil adsorption characteristics. Sandy soils require less root density while more clayey soils require higher densities for better extraction (de Willigen & van Noordwijk, 1987), because in a coarser textured soil there is lower unsaturated hydraulic conductivity and lower storage capacity.

Fine roots represent a large and dynamic portion of the belowground biomass and nutrient capital in forest ecosystems and they account for a significant fraction of net primary production (Santantonio, Hermann & Overton, 1977; Vogt, Grier & Vogt, 1986; Cuevas *et al.*, 1991). Growth and maintenance of roots can exert a strong drain on C from the aboveground part. Thus, along a water availability gradient, aboveground biomass will be limited towards the extreme of water availability, resulting in trees of short stature, and higher C allocation belowground (Medina, 1983). Under a suboptimal supply of water and nutrients, shoot:root ratios equal to one on a dry weight basis can be obtained in seasonally dry forests, such as at Guánica, Puerto Rico and Chamela, México (Murphy & Lugo, 1986b; Castellanos *et al.*, 1991).

Biomass of fine roots in dry forests is towards the lower extreme compared to evergreen forests (Singh & Singh, 1981; Murphy & Lugo, 1986a; Srivastava, Singh & Upadhyay, 1986; Castellanos *et al.*, 1991; Cuevas *et al.*, 1991). Among the dry forests there are differences in C allocation below ground along a water gradient (Table 15.1). Small root biomass increases with decreasing water availability, indicating a strategy of sustaining suitable water absorption area.

The horizontal distribution of roots in dry forests is a function of species composition and age. Singh & Srivastava (1984) found that in teak trees less than six years old, higher biomass of fine roots was found at 50 cm distance from tree, and that lateral distribution increased with increasing age. In older trees higher fine root density was found again closer to the trees, probably as a result of decreased competition with roots of other species, greater nutrient availability through stemflow, and better moisture conditions. Kummerow *et al.* (1990) found similar but

Table 15.1. *Seasonal variation of small roots in tropical dry forests and plantations. Definition of small roots varies among authors*

Forest type	Annual rainfall (mm)	Root mass (g m^{-2} ± SE)		Source
		Wet season	Dry season	
Dry mixed deciduous, fenced	1264	932 ± 18	654 ± 82	1
Dry mixed deciduous, unfenced	1264	774 ± 103	532 ± 64	1
Teak plantation, 19 years old	814	1124 ± 94	689 ± 106	2
Teak plantation, 29 years old	838	1269 ± 130	869 ± 137	2
Dry mixed deciduous	707	1355	947	3

Sources: 1. Singh & Singh (1981). Small roots value includes roots of <1–25 mm diam. Roots <6 mm diam. account for 43–71% of total small roots biomass during the wet season. 2. Singh & Srivastava (1984). Small roots <1–8 mm diam. 3. Kummerow *et al.* (1990). Small roots <1–5 mm diam. Values are sum of live and dead fractions.

not significant tendencies in horizontal distribution of fine roots for two species in the deciduous forest at Chamela.

High surface temperatures and dryness can limit fine root growth in the superficial soil layer. Where surface temperatures increase considerably during the dry season, the majority of fine roots are found between 10 and 20 cm depth (Srivastava *et al.*, 1986). This temperature effect was not observed in a deciduous forest at Chamela, where more than 50% of all fine roots were found in the upper 10 cm (Kummerow *et al.*, 1990).

In dry forests, root densities and distribution within the soil system can depend on seasonal resource availability, e.g. water regime, N movement in the soil profile and P status of the soil. An exploring root system with a large surface area (which may include the mycorrhizal component) is important in water- or P-limited situations, whereas such properties are less important for mobile ions (such as nitrate: Persson, 1990). Van Noordwijk (1983) has proposed that optimal root density for water or nutrient uptake will vary according to the variable involved. Root density required for water uptake, even when free water is available, generally exceeds root density required for nutrient uptake (de Willigen & van Noordwijk, 1987), so water supply acts as a natural limit on reduction in root size.

Fine root growth in dry forests can be highly opportunistic and fitted to the hydrological regime. There are seasonal variations in root mass, with a rapid increase during the rainy season and a gradual decline during the dry season (Table 15.1). Stronger seasonal effects are observed in the finer root fractions, with ephemeral roots (<1 mm diameter) showing the biggest differences between dry and wet seasons (Figure 15.4) (Singh & Singh, 1981; Srivastava *et al.*, 1986; Kummerow *et al.*, 1990). Seasonal variations in soil moisture and temperature explain 67–84% of the seasonal variability in root biomass (Singh *et al.*, 1984).

In dry forests on stabilized sand dunes at Estación Biológica La Mancha in Veracruz, México, Kavanagh & Kellman (1992) found that water availability was the trigger for root growth. There was no endogenous cue for growth initiation and no effects were found with increased nutrient availability alone. Cumulative root length over the first three months of the wet season (15,716 ± 4149 m m^{-2}) was not different from annual totals. Root proliferation was concentrated in the first 24–36 days after rains started, with large proliferations always associated with soil layers remaining wet during longer periods. Percolation started around 45 days after the first rains in these sandy soils (Kellman & Roulet, 1990), so an extensive root system could be developed to make use of the water and nutrients present. The capacity of species to respond rapidly to increased soil water availability in both space (synlocalization) and time

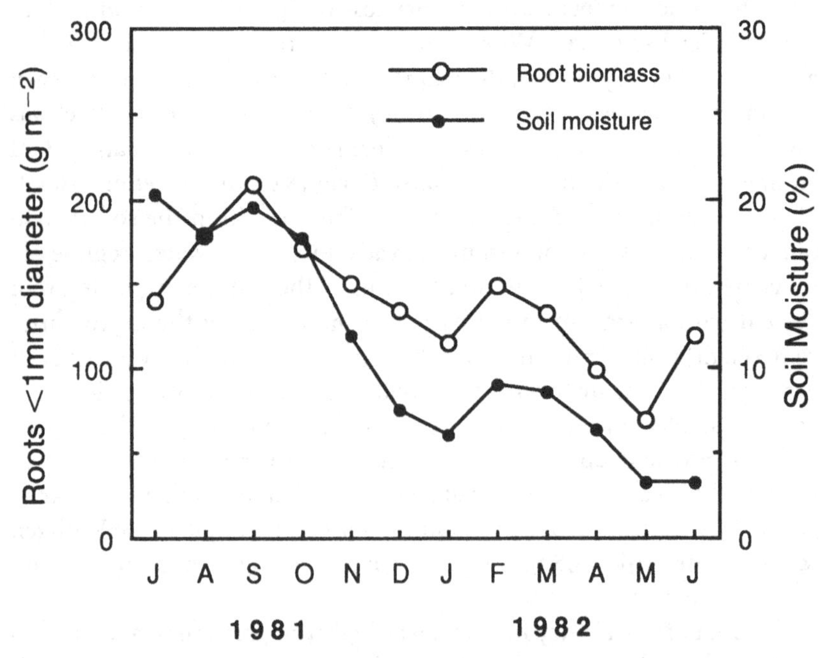

Figure 15.4. Relationship between fine root biomass and soil moisture throughout the year in a 29 year old *Tectona grandis* plantation in India. Source: Singh *et al.* (1984).

(synchronization) allows them to grow under extreme conditions of water and nutrient availability.

The presence of live root tips in the soil profile indicates the potential for fine root growth and activity, especially of very fine roots (<1 mm diameter: Singh & Srivastava, 1985; Kummerow *et al.*, 1990). Most root tips are present in the top 20 cm of soil (60–80%: Singh & Srivastava, 1985; Kummerow *et al.*, 1990). In a teak plantation, seasonal soil moisture changes explained at least 70% of the variability in root tip density (Singh & Srivastava, 1985). The highest number of root tips was measured in mid rainy season with a steady decline with increasing dryness. The largest variation between wet and dry seasons was found in the superficial soil layer, while minimum seasonal variations were measured at 10–20 cm depth. Large numbers of growing tips are present on recently developed roots at the beginning of the rainy season, so high root densities can be achieved in a short time which in turn allow for better water and nutrient uptake during the period of higher availability (Kavanagh & Kellman, 1992).

Non-structural carbohydrates constitute the main source of energy for meristematic growth for both the shoot (Gordon & Larson, 1970) and the root (Ford & Deans, 1977; Ericsson & Persson, 1980). Singh & Srivastava (1985) studied the changes in fine root biomass and total non-structural carbohydrates (TNC) as related to size, depth and seasonality in teak roots. They found that live fine root biomass and TNC were inversely related throughout the year. Although the seasonal changes were observed at all soil depths, the highest TNC and biggest changes were found in roots from 0 to 20 cm, where the majority of fine roots were present. Fine root TNC decreased from the annual maximum in May to the minimum in July. During the same period fine root biomass increased rapidly, reaching a peak in September at the end of the rainy season. Singh & Singh (1981) also found lower caloric values in fine roots of deciduous forests during the rainy season compared with the dry season. This indicates a major use of stored reserves for the initial buildup of a belowground absorption surface that will support the functioning of the photosynthetic surfaces by the middle of the rainy season. Lower TNC levels during the rainy season could be a result of higher respiration due to increased requirements for elongation and maintenance (Marshall & Waring, 1985). Higher TNC in roots of the upper soil layers can be a result of increased allocation in order to meet respiratory needs, and greater translocation from senescent roots to live roots, thus assuring faster root growth after soil wetting.

Seasonality of root growth may result in periods of efficient nutrient uptake when root mass is high and synthesis of new root material is occurring. Singh & Singh (1981) estimated a fine root production during the rainy season of about 40 kg ha^{-1} d^{-1}. Increasing root density, especially in the more superficial horizons where accumulated organic residues are being mineralized, results in a more efficient utilization of the soil resources, because the rate of ion movement to the root surface often limits the uptake rate (Bowen, 1973). Faster uptake of nutrients by plants due to increased root growth, as opposed to immobilization by microorganisms, could be a mechanism for efficient utilization of available nutrients during the rainy season.

Roots as a source of nutrients

Seasonal trends in fine root nutrient concentrations in dry forests show trends similar to TNC: higher concentrations of N, P and K at the end of the dry season and lower concentrations during the rainy season (Pandey, 1980; Singh *et al.*, 1984; Singh, 1989). Two possible explanations for these differences are a dilution effect where biomass increase exceeds nutrient uptake, or nutrient translocations from senescent to live roots. The consistently lower concentrations in dead roots compared to live roots (Singh *et al.*, 1984; Singh, 1989) points toward the importance of nutrient withdrawal from senescent roots as a nutrient conserving mechanism, identical to retranslocation from senescent leaves.

Root decomposition in dry forests is affected by size and chemical composition (substrate quality), depth location, micro-floral and faunal activity, and seasonal variations in soil moisture and temperature (Singh *et al.*, 1984; Singh & Shekhar, 1989a, b). The ratio of live to dead roots (L:D) is always higher during the wet season than during the dry season, but considerable variation in the range of values can be found depending on the species involved (Singh & Singh, 1981; Srivastava *et al.*, 1986; Kummerow *et al.*, 1990). For teak roots L:D ratios are 4.5 and 1 during the wet and dry season, respectively, while for roots from deciduous forests at Chamela, L:D ratios are 2.6 and 0.4, respectively, indicating an effect of resource quality on decomposition and turnover of fine roots.

Mineralization of above- and belowground residues occurs during the period of higher soil moisture (e.g. Madge, 1965; Hopkins, 1966; Singh, 1968; Bernhard, 1970; Egunjobi, 1974). With decreasing soil moisture there is a delay in decomposition with a concomitant increase in the amount of root necromass and organic residues present in the soil.

Residence time of roots also depends on diameter, and structural and chemical characteristics of the decomposing material. Larger diameter roots have longer residence times than fine, ephemeral roots due to differences in the amount of structural components such as cellulose and lignin.

Decomposing roots can also be an important source of nutrients to plants and soil microbes. Singh (1989) estimated an annual release of N, P and K by fine root decomposition of 300, 20 and 220 g m^{-2}, respectively, which were 41, 33 and 61% of the nutrients released by litter decomposition. Although litter decomposition is the main source of nutrients to the system, nutrients from fine roots still account for a considerable portion of the nutrients that are being cycled. Fine root production ranges from 50 to 180% of aboveground litter production in dry forests and plantations (Singh & Singh, 1981; Srivastava *et al.*, 1986; Kummerow *et al.*, 1990). In fact, in deciduous forest at Chamela annual fine root production (423 g m^{-2}) was similar to annual litter production (523 g m^{-2}; Kummerow *et al.*, 1990) indicating that within the complex of dry forests nutrient input from roots can be as important as litter inputs (Jaramillo & Sanford, Chapter 14). Although turnover of small roots (1–5 mm diameter) can be slower than litter turnover on an annual basis, fine roots (<1 mm) can decompose at rates greater than or similar to leaves during periods of high water availability (Srivastava *et al.*, 1986; Singh & Shekhar, 1989a; Sandhu *et al.*, 1990).

Nitrogen and P contribution via fine root turnover in temperate evergreen and tropical wet forests is higher than the amount contributed via litter decomposition (Vogt *et al.*, 1986; Sánchez *et al.*, 1989). There seems to be a similar situation in tropical dry forests. However, the relative importance of the fine root component as a source of nutrients across a range of sites differing not only in the intensity and duration of dry and wet periods but also on soil nutrient availability still needs to be assessed.

Microbial populations

There is a paucity of information on the composition and dynamics of microbial populations in tropical ecosystems as a whole, and especially in dry forests. The available information points to the determining role of water availability for microbial activity. Singh & Shukla (1977) measured highest CO_2 efflux from the soil of a deciduous forest in Varanasi during the months with highest precipitation. Highest respiration values coincided with peaks of mesofauna and mycoflora. Soil humidity was linearly and significantly correlated with mesofaunal and mycofloral numbers ($r = 0.80$ and 0.68, respectively). In derived systems such as maize fields, fungal and microbial populations also had a positive response to increasing water availability (Dkhar & Mishra, 1987). Microbial activity on decomposing substrates in the soil accounted for 43–89% of total respiration with highest values coinciding with periods of highest soil moisture (Singh & Shekhar, 1986).

Seasonal changes in precipitation and number of days without rain can also affect the microbial biomass in the soil and litter (O'Connell, 1990). Gómez (1992) found that microbial biomass in the surface soils of teak plantations in Venezuela was negatively affected by either too few or too many rainy days, which was reflected in soil moisture near either the wilting point or saturation. Soil CO_2 efflux followed the same trend as microbial biomass. Thus, microbial activity (biomass increase) will depend not only on seasonal changes in soil moisture but the duration of optimal conditions as well.

Güerere (1992) measured microbial biomass during the wet season along a gradient from savanna to deciduous forests at Calabozo, Venezuela. Microbial biomass increased with soil humidity throughout the gradient. Although the correlation was significant, soil humidity accounted for only 19% of the variability. Other factors such as amount of organic C, available P and exchangeable Ca were also significantly correlated with microbial biomass. Again, quality of the soil substrate is another determinant of microbial biomass and activity.

Microbial activity also may be related to plant phenology. Higher amounts of rhizodeposits, either as exudates or sloughed cells, are released during the period of maximum root growth (Billès & Bottner, 1981). The concomitant increase in microbial biomass and activity caused by increased high energy C sources results in an increased degradation of organic matter (Martens, 1990). Exudation by roots is strongly reduced by moisture stress (Martin, 1977), hence the effect of rhizodeposits on mineralization will be greater during the wet period.

Symbiotic relationships

It has already been established that nutrient availability and root symbioses interact and strongly affect the composition of plant communities (Janos, 1983; Högberg, 1986a, b; Alexander, 1989). The occurrence and potential roles played by mycorrhizas in tropical forests in general have been extensively reviewed by Janos (1983) and Alexander (1989). Högberg (1986a, b, 1989, 1992) reviewed the occurrence and function of N-fixing trees in forests, woodlands and savannas.

Högberg (1992) presents the following generalities for dry forests in Africa where both ecto- (ECM) and vesicular-arbuscular (VAM) mycorrhizal symbioses are present: (a) community composition is dominated by VAM species, (b) ECM species sometimes attain dominance, as in the miombo woodlands, but more often occur in low numbers, and (c) nodulated tree species are rare, but exceptions may occur. The number of N-fixing tree species varies across a moisture gradient, indicating limitation by water availability. Non-nodulated ECM and VAM species are highly competitive where the availability of P is low. In contrast to ectomycorrhizae from temperate zones, the fungal mantle or sheath formed around the root in ECM from wet and dry tropical forests and savannas may account for more than 50% of the root weight (Alexander & Högberg, 1986). This fungal sheath has a large water retention capacity and stores considerable amounts of both C compounds and nutrients (Harley & Smith, 1983). Nitrogen-fixing tree species form VAM, whereas many of the non-nodulated legumes, particularly the subtribe Amherstidae (Caesalpinoideae), form ECM (Alexander, 1989).

Why are there so few N-fixing legumes in tropical dry forests? Nitrogen fixation is advantageous for plants when soil N availability is low (Dommergues *et al.*, 1984), hence it could become a discriminating factor during the processes of natural succession and forest establishment. Once plants have reached maturity and soil N levels have increased sufficiently through biological activity, N fixation may be inhibited. However, N fixation might be important at the seedling and sapling stages.

Nitrogen fixation could also be limited by edaphic chemical conditions, such as P availability. This situation is more accentuated in dry forests, due to the water availability constraint on root growth. Högberg (1986a) assessed N fixation by ^{15}N natural abundance and acetylene reduction in trees of deciduous forests in Africa and correlated it with edaphic chemical conditions. Nitrogen-fixing species seemed to be limited by P, while non-fixing species were found to be limited by N. Table 15.2 presents the soil chemical characteristics and percentage of nodulated species in deciduous forests (Högberg, 1986a). There was a higher percentage of nodulated species as pH, exchangeable Ca, sum of bases and available P increased across the sites. The ratio of nodulated to non-nodulated species was negatively correlated with exchangeable aluminum.

Table 15.2. *Soil chemical characteristics and percentage of nodulated species in deciduous forest sites in Africa with similar sampling size*

Soil characteristics	Site III	Site IV	Site V
pH (water)	5.48	5.92	6.14
Exchangeable Al (meq/100 g)	9.97	2.37	2.19
Exchangeable Ca (meq/100 g)	2.6	2.08	3.80
Sum of bases	5.89	4.23	5.61
Nitrogen (%)	0.1	0.1	0.08
Phosphorus (mg kg^{-1})	1.2	2.0	4.8
Nodulated (%)	31	50	64

Source: Högberg (1986b).

Medina & Bilbao (1991), working in tropical savannas, also found that the percentage of legume species was correlated with percentage sand and exchangeable Ca in the soil solution, and negatively correlated with total N, cation exchange capacity and exchangeable Al. Phosphorus availability is regulated by the same parameters, so the relationships obtained could be considered as indirect evidence of the limiting role of P availability in N fixation.

The two situations presented above are representative of soils with high P fixation by Al and Fe and a low sum of exchangeable bases. But dry forests can be found on soils with neutral pH and high base content, especially Ca. It needs to be determined if nodulation is present on soils where low base status is not a constraint.

The balance between the supply of N and other elements, especially P, is crucial. Mycorrhizae have often been found to enhance uptake of P, but sometimes also organic N and other nutrients. In deciduous miombo woodlands in Africa, with low P and N availability, non-nodulating leguminous trees forming ECM are dominant. The ECM allow the trees to utilize additional organic forms of N, and thus increase N cycling (Högberg, 1990). Because of the large storage capacity of their mantles, ECM are thought to be advantageous when nutrient availability occurs in flushes, as in seasonal climates (Högberg, 1986b).

Salcedo, Elliott & Sampaio (1991), working in an Atlantic coastal forest in Recife, Brazil, showed with the use of ^{32}P additions, that P from the litter/fermentation layer is cycled back into the vegetation by VAM-mediated mechanisms. But there also is considerable movement down into the top mineral soil (20% of tracer recovered in the humus, 61% in the mineral soil) where P in the soil solution is controlled by microbial activity. Thus, in relatively richer soils mycorrhizae are important for cycling of organic P from the humus layer, but the bulk of P movement is via the soil solution and determined by microbial biomass and activity.

Mycorrhizae also are important for plant water uptake (Read, 1986). It has been demonstrated that severing of mycorrhizal hyphae rapidly decreases water supply to plants (Brownlee *et al.*, 1983; Hardie, 1986).

Soil fauna

Soil fauna composition and trophic structure vary with ecosystem type and rainfall patterns (Bouche, 1971; Lavelle, Maury & Serrano, 1981; Lavelle, 1984; Lavelle & Kohlmann, 1984). Little is known about how these variations determine the dynamics of C and nutrients in tropical deciduous forests along a gradient of water availability.

Soil organisms can control organic matter dynamics in some tropical environments (as reviewed by Anderson & Flannagan, 1989). Reddy & Venkataiah (1989) and Singh & Shekhar (1989a, b) have demonstrated the relative importance of microarthropod populations in residue decomposition in deciduous forests. As with root growth and aboveground production, increases in faunal populations and activity are synchronized with rainfall (Mishra & Dash, 1984; Badejo, 1990). For example, earthworm populations in deciduous forests respond directly to soil moisture in both time and space. Mishra & Dash (1984) found the highest density of earthworms at peak of soil moisture. Vertical distribution changed with season, being shallowest during the rainy season and deepest during the dry season, due to drying of the superficial soil layer and higher temperatures. Seventy-five per cent of the earthworms were found at 20–40 cm during the dry season. Faunal activity in dry forests is predominantly restricted to periods of increased water availability, and this synchrony is another factor in the dynamics of residues and soil organic matter.

Conclusions

Synchronization and synlocalization of biological processes with changes in water availability and nutrient release are determining factors in the maintenance of ecosystem productivity. Tropical dry forests are good examples of systems where a high degree of synchrony is achieved.

The interactive effects of resource quality and water availability on the dynamics of the soil system in dry forests is an important question. This is especially pertinent along water and productivity gradients where there are concomitant changes in the proportion of deciduous and evergreen species. The importance of residue quality on nutrient mineralization, soil organic matter production, and soil biological activity needs to be better recognized for tropical systems in general, and especially for dry forests.

The relative importance of symbionts and microbial biomass dynamics in the movement of nutrients in space and time is another open question. This is critical to understanding ecosystem dynamics because dry forests are found on soils with very different nutrient availabilities. One interesting aspect is the role of nitrogen fixation in the growth of early stages of leguminous trees that are non-fixing as adults.

Other important questions concern the role of fungal and bacterial populations in residue mineralization and immobilization under varying conditions of water and nutrient availability, and the effect of rhizodeposits under natural conditions in forest ecosystems. The severe restraints on root production and mineralization imposed by strong seasonality in water and C sources make dry forests excellent systems for such research.

Much remains to be learned on the belowground C and nutrient dynamics in dry forests, especially in the neotropics. This information is essential to understanding their ecological role on a regional and global basis, and also to developing viable management alternatives for these fast-disappearing systems.

Summary

Biological activity in the belowground system is strongly determined by water availability. There seems to be a tight coupling between available water and the activities of fine roots, microbial flora and micro-, meso- and macrofauna which results in an efficient utilization of available resources such as C and nutrients. Fine root growth in tropical dry forests seems to be highly opportunistic and fitted to the hydrological regime. Increased uptake of nutrients by plants, as opposed to immobilization by microorganisms, acts as a mechanism for efficient utilization of available nutrients during the rainy season. Gradients in seasonal water availability should indirectly affect soil biological activity via changes in decomposition rates as affected by changes in litter quality (from deciduous vs evergreen species) along the gradients. The relative importance of symbionts, such as nitrogen-fixing organisms and mycorrhizae, during different stages of growth and within seasonal cycles in tropical dry forests, needs to be further addressed.

References

Alexander, I. (1989). Mycorrhizas in tropical forests. In *Mineral Nutrients in Tropical Forests and Savanna Ecosystems*, ed. J. Proctor, pp. 169–88. Blackwell Scientific Publications, Oxford.

Alexander, I. & Högberg, P. (1986). Ectomycorrhizas of tropical angiospermous trees. *New Phytologist* **102**: 541–9.

Anderson, J. M. & Flannagan, P. W. (1989). Biological processes regulating organic matter dynamics in tropical soils. In *Dynamics of Soil Organic Matter in Tropical Ecosystems*, ed. D. C. Coleman, J. M. Oades and G. Uehara, pp. 97–123. University of Hawaii Press, Honolulu.

Badejo, M. A. (1990). Seasonal abundance of soil mites (Acarina) in two contrasting environments. *Biotropica* **22**: 382–90.

Bailey, H. P. (1979). Semi-arid climates: their definition and distribution. In *Agriculture in Semi-Arid Environments*, ed. A. E. Hall, G. H. Cannell and H. W. Lawton, pp. 73–97. Springer-Verlag, Berlin.

Bernhard, F. (1970). Étude de la litière et de sa contribution au cycle des éléments minéraux en forêt ombrophile de Côte-d'Ivoire. *Oecologia Plantarum* **5**: 247–66.

Bernhard-Reversat, F. (1987). Litter incorporation to soil organic matter in natural and planted tree stands in Senegal. *Pedobiologia* **30**: 401–17.

Billès, G. & Bottner, P. (1981). Effet des racines vivantes sur la decomposition d'une litière racinaire marquée au ^{14}C. *Plant and Soil* **62**: 193–208.

Bouche, M. B. (1971). Relations entre les estructures spatiales et fonctionnelles des écosystèmes illustrées par le rôle pédobiologique des vers de terre. In *La Vie dans les Sols, Aspects Nouveaux, Études Expérimentales*, ed. P. Pesson, pp. 189–209. Gauthiers-Villars, Paris.

Bowen, G. D. (1973). Mineral nutrition of ectomycorrhizae. In *Ectomycorrhizae: their ecology and physiology*, ed. G. C. Marks and T. T. Kozlowski, pp. 151–205. Academic Press, New York.

Brown, S. & Lugo, A. E. (1982). The storage and production of organic matter in tropical forests and their role in the global carbon cycle. *Biotropica* **14**: 161–87.

Brownlee, C., Duddridge, J. A., Malibari, A. & Read, D. J. (1983). The structure and function of mycelial systems of ectomycorrhizal roots with special reference to their role in forming interplant connections and providing pathways for assimilate and water transport. *Plant and Soil* **71**: 433–43.

Castellanos, J., Maass, M. & Kummerow, J. (1991). Root biomass of a tropical deciduous forest. *Plant and Soil* **131**: 225–8.

Cuevas, E., Brown, S. & Lugo, A. E. (1991). Above and belowground organic matter storage and production in a tropical pine plantation and a paired broadleaved secondary forest. *Plant and Soil* **135**: 257–68.

de Willigen, P. & van Noordwijk, M. (1987). Roots, plant production and nutrient use efficiency. Ph.D. thesis, Landbouwuniversiteit Wageningen, Wageningen.

Dkhar, M. S. & Mishra, R. R. (1987). Microbial population, fungal biomass and CO_2 evolution in maize (*Zea mays* L.) field soils. *Plant and Soil* **99**: 277–84.

Dommergues, Y. R., Diem, H. G, Gauthier, D. L., Dreyfus, B. L. & Cornet, P. (1984). Nitrogen-fixing trees in the tropics: potentialities and limitations. In *Advances in Nitrogen Fixation Research*, ed. C. Veeger and W. E. Newton, pp. 7–13. Martinus Nijhoff/Dr W. Junk, The Hague.

Egunjobi, J. K. (1974). Dry matter, nitrogen and mineral element distribution in an unburnt savanna during the year. *Oecologia Plantarum* **9**: 1–10.

Ericsson, A. & Persson, H. (1980). Seasonal changes in starch reserves and growth of fine roots of 20 year-old Scots pine. In *Structure and Function of Northern Coniferous Forests: an ecosystem study*, ed. H. Persson. Ecological Bulletin (Stockholm) **32**: 239–50.

Ford, E. D. & Deans, J. D. (1977). Growth of a Sitka spruce plantation: spatial distribution and seasonal fluctuations of lengths, weights, and carbohydrate concentrations of fine roots. *Plant and Soil* **47**: 463–85.

García-Méndez, G., Maass, J. M., Matson, P. A. & Vitousek, P. M. (1991). Nitrogen transformations and nitrous oxide flux in a tropical deciduous forest in México. *Oecologia* **88**: 362–6.

Glinski, J. & Lipiec, J. (1990). *Soil Physical Conditions and Plant Roots.* CRC Press, Inc., Boca Ratón.

Gómez, O. (1992). Dinámica de la respiración edáfica y masa microbiana en una cronosecuencia de teca, *Tectona grandis*, en la Reserva Forestal de Ticoporo, Edo. Barinas, Venezuela. Tesis de Licenciatura, Universidad Central de Venezuela, Caracas.

Gordon, J. C. & Larson, P. R. (1970). Redistribution of ^{14}C-labelled reserve food in young red pines during shoot elongation. *Forest Science* **16**: 14–20.

Güerere, I. (1992). Comparación de parámetros químicos, físicos y de la masa microbiana del suelo entre una sabana protegida del fuego y una sabana quemada anualmente. Tesis de Licenciatura, Universidad Central de Venezuela, Caracas.

Hardie, K. (1986). The role of extraradical hyphae in water uptake by vesicular-arbuscular mycorrhizal plants. In *Physiological and Genetical Aspects of Mycorrhizae*, ed. V. Gianinazzi-Pearson and S. Gianinazzi, pp. 651–5. INRA, Paris.

Harley, J. L. & Smith, S. E. (1983). *Mycorrhizal Symbiosis.* Academic Press, London.

Högberg, P. (1986a). Nitrogen-fixation and nutrient relations in savanna woodland trees (Tanzania). *Journal of Applied Ecology* **23**: 675–88.

Högberg, P. (1986b). Soil nutrient availability, root symbioses and tree species composition in tropical Africa: a review. *Journal of Tropical Ecology* **2**: 359–72.

Högberg, P. (1989). Root symbioses of trees in savannas. In *Mineral Nutrients in Tropical Forest and Savanna Ecosystems*, ed. J. Proctor, pp. 121–38. Blackwell Scientific Publications, Oxford.

Högberg, P. (1990). ^{15}N natural abundance as a possible marker of the ectomycorrhizal habit of trees on mixed African woodlands. *New Phytologist* **115**: 483–6.

Högberg, P. (1992). Root symbioses of trees in African dry tropical forests. *Journal of Vegetation Science* **3**: 393–400.

Holt, J. A., Hogden, M. J. & Lamb, D. (1990). Soil respiration in the seasonally dry tropics near Townsville, Queensland. *Australian Journal of Soil Research* **28**: 737–45.

Hopkins, B. (1966). Vegetation of the Olokemeji Forest Reserve, Nigeria. IV. The litter and soil with special reference to their seasonal changes. *Journal of Ecology* **54**: 687–703.

Janos, D. P. (1983). Tropical mycorrhizas, nutrient cycles and plant growth. In *Tropical Rain Forest: ecology and management*, ed. S. L. Sutton, T. C. Whitmore and A. C. Chadwick, pp. 327–45. Blackwell Scientific Publications, Oxford.

Kavanagh, T. & Kellman, M. (1992). Seasonal pattern of fine root proliferation in a tropical dry forest. *Biotropica* **24**: 157–65.

Keller, M., Kaplan, W. A. & Wofsy, S. C. (1986). Emissions of N_2O, CH_4 and CO_2 from tropical forest soils. *Journal of Geophysical Research* **91**: 11791–802.

Kellman, M. & Roulet, N. (1990). Nutrient flux and retention in a tropical sand dune succession. *Journal of Ecology* **78**: 664–76.

Kummerow, J., Castellanos, J., Maass, M. & Larigauderie, A. (1990). Production of fine roots and the seasonality of their growth in a Mexican deciduous dry forest. *Vegetatio* **90**: 73–80.
Laishram, I. D. & Yadava, P. S. (1988). Lignin and nitrogen in the decomposition of leaf litter in a subtropical forest ecosystem at Shiroy Hills in northeastern India. *Plant and Soil* **106**: 59–64.
Lavelle, P. (1984). The soil system in the humid tropics. *Biology International* **9**: 2–17.
Lavelle, P. & Kohlmann, B. (1984). Étude quantitative de la macrofaune du sol dan un fôret tropicale humide du Mexique (Bonampak, Chiapas). *Pedobiologia* **27**: 377–93.
Lavelle, P., Maury, M. A. & Serrano, V. (1981). Estudio cuantitativo de la fauna del suelo en la región de Laguna Verde, Veracruz. Epoca de lluvias. *Instituto de Ecología Publicaciones* **6**: 75–105.
Madge, D. S. (1965). Leaf fall and litter disappearance in a tropical forest. *Pedobiologia* **5**: 273–88.
Marshall, J. D. & Waring, R. H. (1985). Predicting fine root production and turnover by monitoring root starch and soil temperature. *Canadian Journal of Forest Research* **15**: 791–800.
Martens, R. (1990). Contribution of rhizodeposits to the maintenance and growth of soil microbial biomass. *Soil Biology and Biochemistry* **22**: 141–7.
Martin, J. K. (1977). Effect of soil moisture on the release of organic carbon from wheat roots. *Soil Biology and Biochemistry* **9**: 303–4.
Medina, E. (1983). Adaptation of tropical trees to moisture stress. In *Tropical Rain Forest Ecosystems: structure and function*, ed. F. B. Golley, pp. 225–37. Elsevier, Amsterdam.
Medina, E. & Bilbao, B. (1991). Significance of nutrient relations and symbiosis for the competitive interaction between grasses and legumes in tropical savannas. In *Modern Ecology: basic and applied principles*, ed. G. Esser and D. Overdieck, pp. 295–319. Elsevier, Amsterdam.
Medina, E. & Zelwer, M. (1972). Soil respiration in tropical plant communities. In *Tropical Ecology, with Emphasis on Organic Matter Production*, ed. P. Golley and F. B. Golley, pp. 245–69. University of Georgia Press, Athens.
Mishra, P. C. & Dash, M. C. (1984). Population dynamics and respiratory metabolism of earthworm population in a subtropical dry woodland of Western Orissa, India. *Tropical Ecology* **25**: 103–16.
Murphy, P. G. & Lugo, A. E. (1986a). Ecology of tropical dry forest. *Annual Review of Ecology and Systematics* **17**: 67–88.
Murphy, P. G. & Lugo, A. E. (1986b). Structure and biomass of a subtropical dry forest in Puerto Rico. *Biotropica* **18**: 89–96.
O'Connell, A. M. (1990). Microbial decomposition (respiration) of litter in Eucalypt forests of south-western Australia: an empirical model based on laboratory incubations. *Soil Biology and Biochemistry* **22**: 153–60.
Pandey, O. N. (1980). Cycling of nitrogen, phosphorus and potassium in the soil vegetation system of tropical dry deciduous forests of Chandraprabha region, Varanasi. Ph.D. thesis, Banaras Hindu University, Varanasi.
Persson, H. (1990). Methods of studying root dynamics in relation to nutrient cycling. In *Nutrient Cycling in Terrestrial Ecosystems: field methods, application and interpretation*, ed. A. F. Harrison, P. Ineson and O. W. Heal, pp. 198–217. Elsevier Science Publishers, Essex.
Persson, R. (1974). World Forest Resources: review of the world's forest resources in the early 1970's. Rapporter Och Uppsatser, Institutionen for Skogstaxering, No. 17.
Raich, J. W. & Nadelhoffer, K. J. (1989). Belowground carbon allocation in forest ecosystems: global trends. *Ecology* **70**: 1346–54.

Rajvanshi, R. & Gupta, S. R. (1986). Soil respiration and carbon balance in a tropical *Dalbergia sissoo* forest ecosystem. *Flora* **178**: 251–60.

Ramakrishnan, P. S. & Toky, O. P. (1981). Soil nutrient status of hill-agroecosystems and recovery pattern after slash and burn agriculture (jhum) in North-Eastern India. *Plant and Soil* **60**: 41–64.

Read, D. J. (1986). Non-nutritional effects of mycorrhizal infection. In *Physiological and Genetical Aspects of Mycorrhizae*, ed. V. Gianinazzi-Pearson and S. Gianinazzi, pp. 169–76. INRA, Paris.

Reddy, M. V. & Venkataiah, B. (1989). Influence of microarthropod abundance and climatic factors on weight loss and mineral nutrient contents of *Eucalyptus* leaf litter during decomposition. *Biology and Fertility of Soils* **8**: 319–24.

Rout, S. K. & Gupta, S. R. (1989). Soil respiration in relation to abiotic factors, forest floor litter, root biomass and litter quality in forest ecosystems of Siwaliks in northern India. *Oecologia Plantarum* **10**: 229–44.

Russell, E. W. (1973). *Soil Conditions and Plant Growth*. 10th edition. Longman, London.

Russell, R. S. (1977). *Plant Root Systems: their function and interaction with the soil*. McGraw-Hill, London.

Salcedo, I. H., Elliott, E. T. & Sampaio, E. V. S. B. (1991). Mechanisms controlling phosphorus retention in the litter mat of Atlantic coastal forests. In *Phosphorus Cycles in Terrestrial and Aquatic Ecosystems. Regional Workshop 3: South and Central America*, ed. H. Tiessen, D. López-Hernández and I. H. Salcedo, pp. 109–20. SCOPE, UNEP and Saskatchewan Institute of Pedology, Saskatoon.

Sánchez, P. A., Palm, C. A., Szott, L. T., Cuevas, E. & Lal, R. (1989). Organic residue management in tropical agroecosystems. In *Dynamics of Soil Organic Matter in Tropical Ecosystems*, ed. D. C. Coleman, J. M. Oades and G. Uehara, pp. 125–52. University of Hawaii Press, Honolulu.

Sandhu, J., Sinha, M. & Ambasht, R. S. (1990). Nitrogen release from decomposing litter of *Leucaena leucocephala* in the dry tropics. *Soil Biology and Biochemistry* **22**: 859–63.

Santantonio, D., Hermann, R. K. & Overton, W. S. (1977). Root biomass studies in forest ecosystems. *Pedobiologia* **17**: 1–31.

Sen, D. N. & Tanwar, G. S. (1983). Arid environment and root behaviour. In *Root Ecology and its Practical Applications*, ed. W. Böhm, L. Kutschera and E. Lichtenegger, pp. 185–206. Bundesanstalt Gumpenstein, Irdning.

Singh, A. K. & Ambasht, R. S. (1980). Production and decomposition rate of litter in a teak (*Tectona grandis*) plantation at Varanasi (India). *Revue d'Ecologie et de Biologie du Sol* **17**: 13–22.

Singh, J. (1984). Effects of temperature, rainfall and soil moisture on soil respiration and litter decomposition in a sub-tropical humid forest ecosystem. *Acta Botanica Indica* **12**: 167–73.

Singh, J. S. & Gupta, S. R. (1977). Plant decomposition and soil respiration in terrestrial ecosystems. *Botanical Review* **43**: 449–528.

Singh, K. P. (1968). Litter production and nutrient turnover in deciduous forests of Varanasi. In *Proceedings of the Symposium Recent Advances in Tropical Ecology*, ed. R. Misra and B. Gopal, pp. 655–65. International Society of Tropical Ecology, Varanasi, India.

Singh, K. P. (1989). Mineral nutrients in tropical dry deciduous forest and savanna ecosystems in India. In *Nutrient Cycling in Tropical Forests and Savanna Ecosystems*, ed. J. Proctor, pp. 153–68. Blackwell Scientific Publications, Oxford.

Singh, K. P. & Shekhar, C. (1986). Seasonal pattern of total soil respiration, its fractionation and soil carbon balance in a wheat–maize rotation cropland at Varanasi. *Pedobiologia* **29**: 305–18.

Singh, K. P. & Shekhar, C. (1989a). Weight loss in relation to environmental factors during the decomposition of maize and wheat roots in a seasonally-dry tropical region. *Soil Biology and Biochemistry* **21**: 73–80.

Singh, K. P. & Shekhar, C. (1989b). Concentration and release patterns of nutrients (N, P and K) during decomposition of maize and wheat roots in a seasonally-dry tropical region. *Soil Biology and Biochemistry* **21**: 81–5.

Singh, K. P. & Singh, R. P. (1981). Seasonal variation in biomass and energy of small roots in tropical dry deciduous forests, India. *Oikos* **37**: 88–92.

Singh, K. P. & Srivastava, S. K. (1984). Spatial distribution of fine roots in young trees (*Tectona grandis*) of varying girth sizes. *Pedobiologia* **27**: 161–70.

Singh, K. P. & Srivastava, S. K. (1985). Seasonal variations in the spatial distribution of root tips in teak (*Tectona grandis* Linn. f.) plantations in the Varanasi Forest Division, India. *Plant and Soil* **84**: 93–104.

Singh, K. P., Srivastava, S. K. & Singh, R. K. (1984). Analysis of seasonal dynamics and nutrient relations of tree roots in tropical deciduous forests: final technical report. Centre for Advanced Study in Botany, Banaras Hindu University, Varanasi, India.

Singh, U. R. & Shukla, A. N. (1977). Soil respiration in relation to mesofaunal and mycofloral populations during rapid course of decomposition on the floor of a tropical dry deciduous forest. *Revue d'Ecologie et de Biologie du Sol* **14**: 363–70.

Sobrado, E. & Medina, E. (1980). General morphology, anatomical structure and nutrient content of sclerophyllous leaves of the 'Bana' vegetation of Amazonas. *Oecologia* **45**: 341–5.

Srivastava, S. K., Singh, K. P. & Upadhyay, R. S. (1986). Fine root growth dynamics in teak (*Tectona grandis* Linn. f.). *Canadian Journal of Forest Research* **16**: 1360–4.

Upadhyaya, S. D., Siddiqui, S. A. & Singh, V. P. (1981). Seasonal variation in soil respiration of certain tropical grassland communities. *Tropical Ecology* **22**: 157–61.

van Noordwijk, M. (1983). Functional interpretation of root densities in the field for nutrient and water uptake. In *Root Ecology and its Practical Applications*, ed. W. Böhm, L. Kutschera and E. Lichtenegger, pp. 207- 26. Bundesanstalt Gumpenstein, Irdning.

Vogt, K. A., Grier, C. C. & Vogt, D. J. (1986). Production, turnover, and nutrient dynamics of above- and belowground detritus of world forests. *Advances in Ecological Research* **15**: 303–77.

16

Nitrogen trace gas emissions in a tropical dry forest ecosystem

PAMELA A. MATSON & PETER M. VITOUSEK

Introduction

The production and emission of trace gases from tropical forests are of interest for calculating global budgets, for determining local ecosystem mass balances, and for the insight they provide into ecosystem processes. Globally, tropical dry forests are widespread and functionally different from other biomes, and hence need to be characterized directly. Moreover, the extent and intensity of human modification of dry forests (in comparison to other tropical biomes: Murphy & Lugo, 1986) raises the possibility that land use change in this biome could be globally significant in the concentration or distribution of one or more gases.

At the ecosystem level, loss to the atmosphere can be a major pathway of transfer of nitrogen and sulfur from terrestrial ecosystems (Likens *et al.*, 1977; Bowden, 1986), and hence a possible determinant of the long-term fertility and potential productivity of terrestrial ecosystems. Since gaseous emissions are not normally included in watershed-level measurements of nutrient inputs and outputs, they can provide a useful complement to watershed studies. Finally, on the process level the trace gases emitted from terrestrial ecosystems represent either end products or by-products of fundamental metabolic processes that occur within terrestrial ecosystems. Their magnitude, timing and regulation provide insight into the operation of those processes even where emissions are not quantitatively significant in either global or local element budgets (Matson, Vitousek & Schimel, 1989).

None of these reasons is specific to tropical dry forests; all apply to any biome with a large areal extent and significant human modification. However, the dry forest biome is particularly interesting for three reasons. (1) It occupies an intermediate position on the great moisture gradient that underlies variation among tropical ecosystems (Medina,

Chapter 9). (2) The extent of human alteration is so substantial that little-modified reference ecosystems are difficult to locate in many areas (Gentry, Chapter 7). (3) There is a lack of information on biosphere–atmosphere exchange (Vitousek *et al.*, 1989), and indeed most aspects of ecosystem function in the biome (Murphy & Lugo, 1986; Janzen, 1988).

This chapter will emphasize the emissions of the N-containing trace gases nitrous oxide (N_2O) and nitric oxide (NO) from tropical dry forest. Both are produced by similar processes, being by-products or intermediate products of nitrification and denitrification (Firestone & Davidson, 1989), and end products of abiotic decomposition of nitrite and of biomass burning (particularly for NO) (Lobert *et al.*, 1990). However, the two gases differ dramatically in their fate and consequences in the atmosphere. Nitrous oxide is stable in the troposphere, with an atmospheric lifetime of approximately 160 years. Its concentrations are increasing at 0.2–0.3% y^{-1} globally (Houghton, Jenkins & Emphraums, 1990; Prinn *et al.*, 1990), and it is radiatively active (a 'greenhouse gas'), contributing a small but significant amount to the ongoing enhancement of the greenhouse effect (Lashof & Ahuja, 1990). In the stratosphere, breakdown products of N_2O react with and destroy stratospheric ozone (O_3). In contrast, NO is highly reactive in the troposphere. Its lifetime is short (hours to days), but it acts as a regulator of atmospheric photochemistry. When NO concentrations are low, the oxidation of atmospheric hydrocarbons consumes tropospheric O_3; where NO concentrations are high, tropospheric O_3 is produced (Crutzen & Andreae, 1985). Nitric oxide and particularly its reaction products (NO_x) are taken up by vegetation (Johansson, 1989; Bakwin *et al.*, 1990), while N_2O is not. Despite these substantial differences, and additional ones in techniques of measurement, the similarity in the processes producing these gases makes it worthwhile to evaluate them together.

Nitrogen trace gases in the tropics

Interest in nitrogen trace gases in tropical dry forest grows out of suggestions that tropical forest soil is a substantial source of N_2O, probably the greatest single source globally, and a likely source of the ongoing global increase in N_2O. Several lines of evidence support these suggestions as they apply to tropical moist/wet forests. Measured fluxes of N_2O are substantially higher in such forests (on average) than in any other biome (Keller *et al.*, 1983; Keller, Kaplan & Wofsy, 1986; Matson & Vitousek, 1987; Goreau & de Mello, 1988). These high fluxes are consistent with observations of rapid N transformations in soil and circulation through vegetation within intact tropical moist/wet forests in comparison with other biomes (Vitousek & Sanford, 1986; Vitousek & Matson, 1988). Indeed, N appears to function as an excess nutrient in the majority of lowland moist/wet forests. Moreover, estimates of flux that are based on the latitudinal distribution of N_2O concentrations in the troposphere (Cicerone, 1989; Prinn *et al.*, 1990) demonstrate that tropical latitudes are the major source for N_2O globally, and that they are at least partially responsible for the ongoing increase in N_2O concentrations.

Early flux estimates (Keller *et al.*, 1983, 1986) from tropical forests were summarized in a global source budget by McElroy & Wofsy (1986). They estimated that the 2.5×10^7 km^2 of tropical forest emit approximately 7.5 teragrams (Tg; 1 Tg = 10^6 metric tons or 10^{12} g) of N as N_2O annually, or more than 70% of background fluxes. (Background fluxes are defined as those that are unaltered or little altered by human activity.) Matson & Vitousek (1990) used additional flux estimates (bringing the total to 32 sites in all) and a stratification of tropical moist/wet forest soils based upon soil fertility to calculate that the 1.5×10^7 km^2 of intact moist/wet forest emit 2.4 Tg of N_2O-N annually. This lower estimate in part reflects the incorporation of results from measurements in low-fertility sites; in part, it also reflects a decision to exclude dry forests entirely, as at that time no measurements had been reported from such forests. Matson & Vitousek (1990) also estimated that conversion of moist forest to pasture could account for approximately 0.7 Tg of additional N_2O emission (Luizão *et al.*, 1989); the possible contribution from tropical agriculture was not determined.

Sanhueza *et al.* (1990) recently estimated the total emission of N_2O-N from tropical savanna and grassland ecosystems at 1.4 Tg y^{-1}, based on a set of measurements in Venezuela. This leaves dry forests as a substantial, indeed glaring, gap in our understanding of N_2O emissions from natural tropical ecosystems. This gap is crucial for three reasons. (1) Dry forests cover a large area. (2) The limited available information on nutrient cycling suggests that dry forests cycle relatively large amounts of N at relatively high concentrations (Raman, 1975; Gessel *et al.*, 1980; Esteban, 1986; Lugo & Murphy, 1986; Patiño, 1990), like moist/wet forests, and unlike savannas. (3) The extent and intensity of anthropogenic land use change in dry forest is such that if it causes increased N_2O emission, land use change could be a significant cause of the global increase in N_2O

Our understanding of global sources and sinks for NO is still less certain than that for N_2O, but the limited available evidence suggests that tropical soils are disproportionately important as background sources for NO, just as they are for N_2O (Johansson, Rodhe & Sanhueza, 1988; Johansson & Sanhueza, 1988; Kaplan *et al.*, 1988; Bakwin *et al.*, 1990; Sanhueza *et al.*, 1990). Moreover, NO emission is often maximized at lower soil moisture contents than is N_2O, in part because nitrification is a more efficient source of NO, and in part because emissions of reactive NO are more dependent on rates of diffusivity through soil. Additionally, much of the NO emitted under an active plant canopy is absorbed by foliage before reaching the free troposphere, but that absorption is less important in dry forest during the leaf-free period. Consequently, while the global context in which NO flux measurements are placed is less clear than that for N_2O, there is reason to think that emissions of NO in dry forest are at least as important as those for N_2O.

N_2O and NO emissions from a tropical dry forest

Our research approach was first to determine spatial and temporal patterns in emissions of N-containing trace gases. This was done by surveying fluxes in a number of sites in intact dry forest and human-altered areas in and near the Estación de Biología Chamela, Jalisco, México (19° 30′ N, 105° 03′ W), and by following fluxes through the seasons in a smaller number of sites. Second, soil moisture and N parameters were measured in conjunction with gas flux measurements so that the pattern of correlations and putative causes of N gas emissions could be identified. Finally, the mechanisms controlling flux were determined through experimental studies under laboratory and field conditions. Due to equipment requirements, field measurements were made in more sites and at more times during the year for N_2O than NO.

Results of the initial survey of N_2O fluxes in a number of sites during the wet season were summarized by Vitousek *et al.* (1989). The eight forest sites (six within the experimental watersheds) and four upland pastures on comparable soils had similar fluxes, with most of the values falling between 0.5 and 1.0 ng cm^{-2} h^{-1} (5 and 10 μg m^{-2} h^{-1}). These values are lower than those in the majority of moist/wet forests (Matson & Vitousek, 1990), but they are substantially greater than those from most temperate forests. Two maize fields sampled in their first wet season following clearing and burning had higher fluxes (2.5 ng cm^{-2} h^{-1}) while the one intensively managed lowland pasture had a mean flux of nearly

10 ng cm^{-2} h^{-1} (comparable to the Amazonian pastures reported by Luizão *et al.*, 1989). Measurements of emissions over an annual cycle in two forest and one upland pasture sites yielded lower fluxes (generally <0.25 ng cm^{-2} h^{-1}) during the dry season and variable, moisture dependent fluxes during the wet season (Figure 16.1) (García-Méndez *et al.*, 1991). NO flux across the soil–air interface was also relatively low during the dry season and higher during the wet season (Table 16.1) (Davidson *et al.*, 1991). Where both N_2O and NO were measured at the same time, site means were positively correlated (Figure 16.2; $r = 0.638$, $p < 0.05$) as were fluxes in individual chambers ($r = 0.481$, $p < 0.001$).

Comparisons of N gas emissions during the dry versus wet seasons ignored the possible effect of wetting events on gas fluxes. Wetting soil after an extended dry period is known to cause a substantial pulse of nitrogen gas emissions (cf. Slemr & Seiler, 1984; Hao *et al.*, 1988), and conditions at Chamela (a long completely dry period preceding the onset of rain: Bullock, 1986) are particularly conducive to this effect. We evaluated this pulse through experiments in which soils were wetted late in the dry season, and gas emissions were measured at intervals after

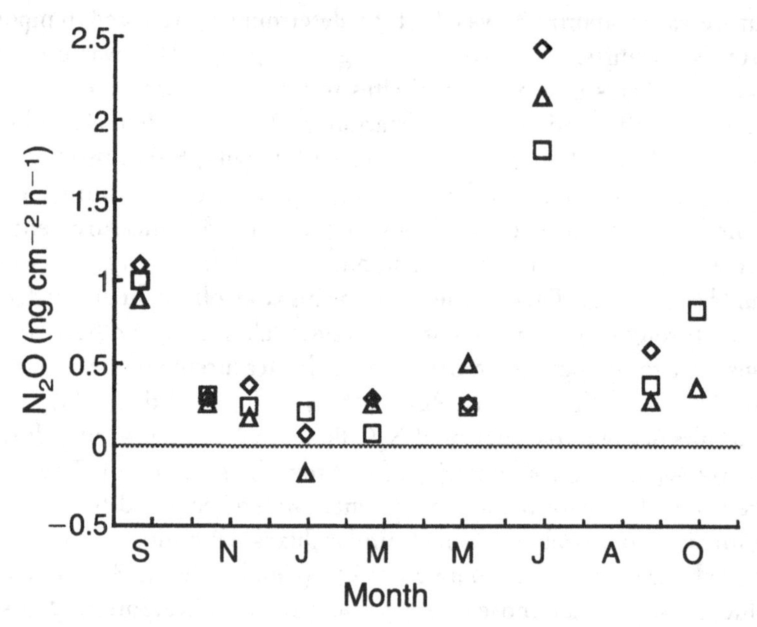

Figure 16.1. Annual variation in N_2O-N flux from three seasonally dry sites near Chamela, México. The wet season was from late June until November. □, intact drought-deciduous forest on WS 1; ◇, intact drought-deciduous forest on WS 4; △, 12 year old pasture. Data from García-Méndez *et al.* (1991). WS, watershed.

Table 16.1. *Average N_2O-N and NO-N fluxes (in ng cm^{-2} h^{-1} ± SE; the number of chamber measurements is given in parentheses) for Chamela sites in May 1990 (dry season) and October 1990 (wet season). The Verdín and WS 1 forest sites are both intact, upland, relatively level sites*

Site	Dry season		Wet season	
	N_2O-N	NO-N	N_2O-N	NO-N
Forest				
Verdín	0.17 ± 0.08 (8)	0.14 ± 0.03 (14)	0.73 ± 0.18 (8)	0.73 ± 0.10 (28)
WS 1	0.24 ± 0.05 (8)	0.52 ± 0.22 (7)	0.82 ± 0.05 (8)	1.27 ± 0.11 (36)
Pasture				
Upland	0.50 ± 0.13 (7)	0.09 ± 0.02 (4)	0.34 ± 0.06 (7)	0.72 ± 0.07 (8)
Lowland	0.54 ± 0.09 (8)	3.57 ± 0.51 (6)	1.05 ± 0.06 (8)	9.21 ± 1.28 (9)
Crop				
Maize	0.85 ± 0.17 (8)	0.98 ± 0.29 (7)		
Sorghum			0.39 ± 0.10 (14)	0.47 ± 0.26 (16)

wetting. The standard water addition that we used was equivalent to a 20 mm rain, although some work was done with greater and lesser additions of water.

Results of these experiments at Chamela are summarized in Figure 16.3 (from Davidson *et al.*, 1993). Wetting causes a very rapid and substantial pulse of CO_2; NO and N_2O emissions occurred slightly later (2–6 h after wetting). The enhanced emissions of both N gases were substantial but short-lived. NO responded to wetting to a much greater extent than N_2O, yielding a much larger NO:N_2O ratio than under unmodified conditions. The high flux of NO upon wetting is particularly interesting because under natural conditions it would occur when the forest is virtually leafless, so most of the NO emitted from soil should reach the free troposphere above the canopy. Wetting experiments carried out during a relatively dry time during the wet season and repeated wetting of plots late in the dry season yielded no substantial increase in flux (Davidson *et al.*, 1993); it appears that very high fluxes occur only after the first substantial rain at the beginning of the wet season.

Ammonium and nitrate concentrations in surface soil were relatively low at the end of the wet season; they accumulated gradually through the dry season (García-Méndez *et al.*, 1991). Soil nitrite had also accumulated to relatively high concentrations late in the dry season in these sites (Davidson *et al.*, 1993). In contrast, potential net N mineralization and net nitrification (measured under laboratory conditions) were greater in the wet than the dry season (García-Méndez *et al.*, 1991). The experimental wetting of dry soil with a simulated 20 mm rain caused rapid changes in soil N pools and transformations; within 24 hours, soil ammonium concentrations decreased and nitrate concentrations increased significantly. Moreover, ^{15}N isotope dilution measurements of gross mineralization demonstrated that rates of N transformations were orders of magnitude greater than the net changes in N pools immediately after wetting (Davidson *et al.*, 1993). Overall, N trace gas fluxes were better correlated with soil moisture and its dynamics than they were with parameters related to soil N; it appears that soil moisture dynamics are the major control over soil N pools, N transformations, and N trace gas emissions in this highly seasonal system.

Experiments on the microbial mechanisms controlling N trace gas

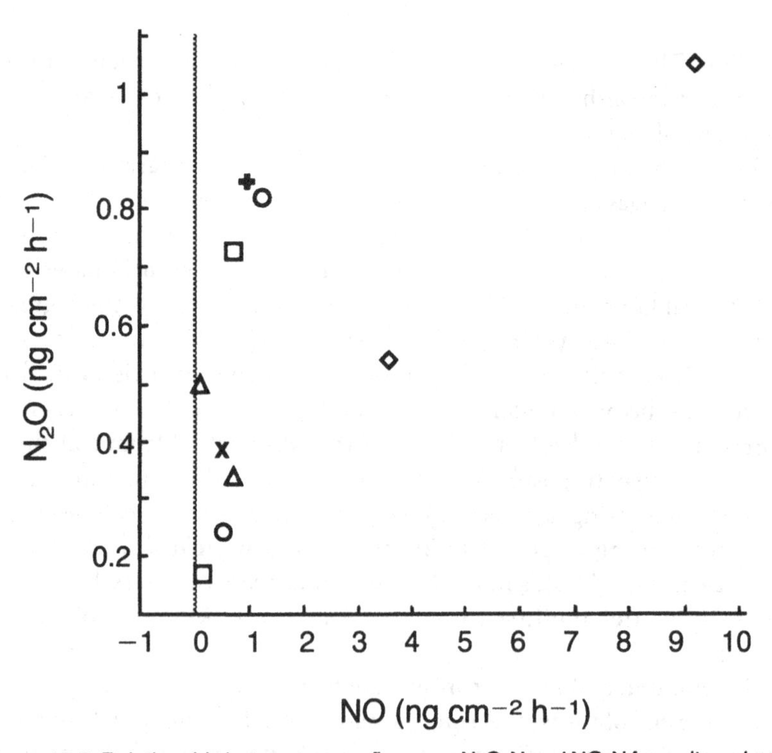

Figure 16.2. Relationship between mean fluxes on N_2O-N and NO-N from sites where both were measured at the same time. ○, WS 1 forest; □, intact forest on Verdín trail; △, upland pasture; ◇, floodplain pasture; +, maize field; ×, sorghum field.

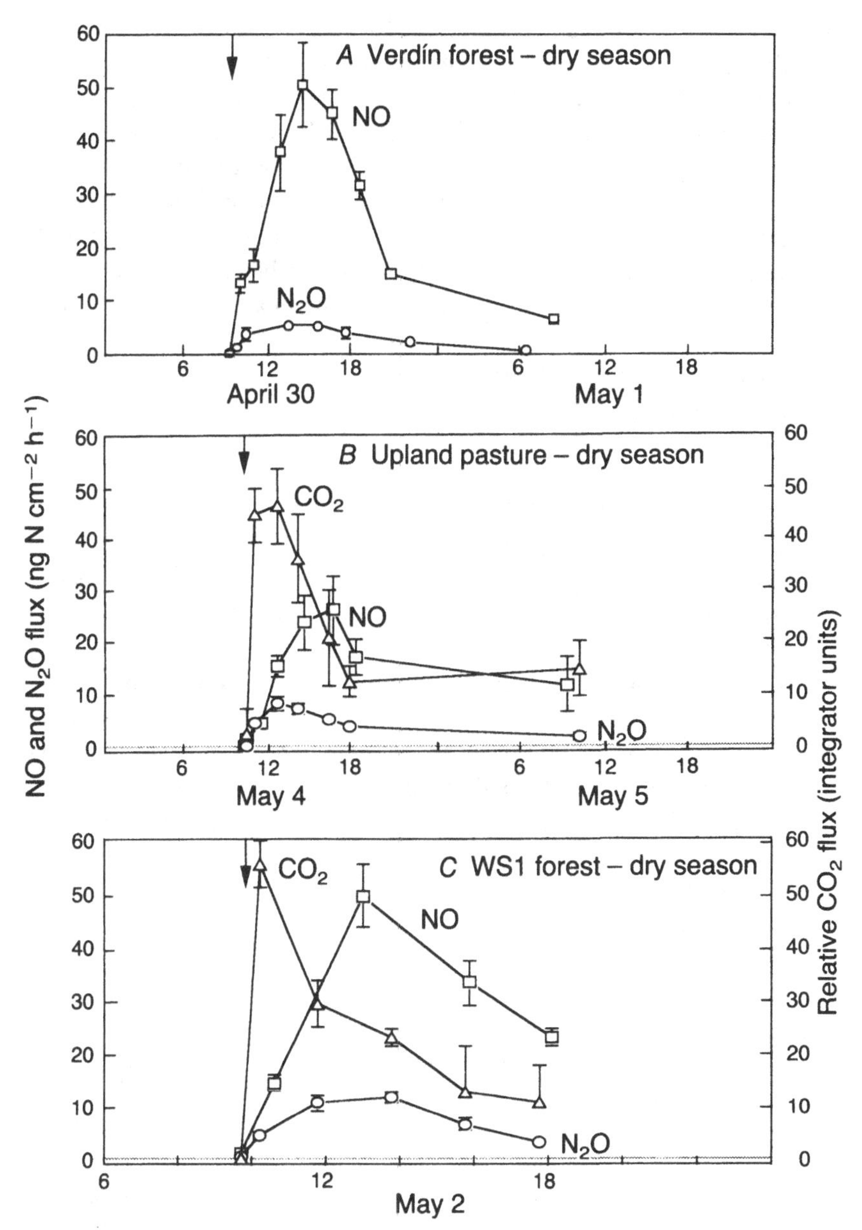

Figure 16.3.Fluxes of CO_2 (△), NO (□) and N_2O (○) following simulated wetting events in the dry season in *A*, an intact forest, *B*, upland pasture and *C*, WS 1 forest. Arrows indicate when 20 mm water was added. The CO_2 peaks observed during gas chromatographic analyses were not calibrated, so only relative fluxes in arbitrary integrator units can be given. Each point is a mean and each bar is the standard error of four replicate measurements. Modified from Davidson *et al*. (1993).

emissions demonstrated that nitrifying bacteria are responsible for most NO flux in both unmodified and wetted soils. Nitrifying bacteria are also the source of N_2O produced in recently wetted soils. For both gases, low concentrations of acetylene (which inhibit nitrifying but not denitrifying bacteria) eliminated gas emissions. Similar results were obtained for NO during the wet season, but the results for N_2O during the wet season were less clear. N_2O emission was only partly inhibited by either low or high concentrations of acetylene (high concentrations block the reduction of N_2O to N_2 during denitrification), suggesting that some of the N_2O emitted during the wet season is produced by denitrifying bacteria, and that N_2O is the major end product of denitrification under these conditions (Davidson *et al.*, 1993).

Annual budgets and global significance

Annual emissions of nitrogen as N_2O and NO were estimated by multiplying the average flux (by site) during the wet and the dry season by the length of each season. We also included the amounts emitted in the first rains at the beginning of the wet season, which were calculated using the emissions that occurred following wetting experiments late in the dry season. These estimates are summarized in Table 16.2; they are based on several measurements per season for N_2O and one sampling period per season for NO (Davidson *et al.*, 1991; García-Méndez *et al.*, 1991). The enhanced fluxes following wetting of dry soils are a significant component of the annual budget for NO but not N_2O, and the absence of a plant canopy at this time means that a widespread rain at the end of the dry season could represent a significant perturbation to regional tropospheric chemistry.

We calculate overall N emissions from soil (in both gases combined) at 1.0–1.6 kg ha^{-1} y^{-1}, although much of the NO emitted during the wet season may be taken up again in the forest canopy (Bakwin *et al.*, 1990). This gaseous flux compares with precipitation inputs of approximately 2 kg N ha^{-1} y^{-1} at Chamela, and N losses in streamwater of 0.24 kg N ha^{-1} y^{-1} (J. M. Maass, personal communication). Nitrogen-containing gases clearly represent a relatively significant pathway of N loss from forest watersheds, and one that would be worth examining following forest cutting or other disturbances.

These results can also be used to make a tentative first step towards assessing the global significance of tropical dry forests in N trace gas emissions. If dry forests generally have emissions similar to those of

Table 16.2. *Calculated seasonal and annual flux of N_2O-N and NO-N (in mg m^{-2}) from the repeatedly sampled sites at Chamela*[a]

Site	Wet season	Dry season	Wet-up	Total
N_2O-N				
Forest				
WS 1	40	10	0.9	51
WS 4	56	12	0.9	68
Pasture				
Upland	37	10	0.8	47
NO-N				
Forest				
WS 1	61	20	11	92
Verdín	35	5	11	51
Pasture				
Upland	64	3	6	73

[a] WS 1, WS 4, and Verdín are all intact, upland, level sites that were sampled as replicates. Experimental wet-up was done only in WS 1.
Sources: N_2O data from García-Méndez *et al.* (1991); NO data from Davidson *et al.* (1991).

Chamela, and if there are approximately 1×10^7 km^2 of such systems globally, then dry forests emit 0.5–0.7 Tg of N_2O-N and 0.5–0.9 Tg of NO-N annually. Soils at Chamela are relatively dry and infertile compared with most tropical dry forests, and N_2O emissions (at least) are positively correlated with soil fertility in moist/wet forests (Matson & Vitousek, 1987). We believe direct extrapolation using results from Chamela should therefore be taken as a lower bound for dry forests. If dry forests generally have emissions similar to moist forests on fertile soils during the wet season (and emissions close to zero in the dry season), then they could produce as much as 1.5 Tg of N_2O-N annually. This probably represents an upper bound. These emissions compare with estimates of 2.4 Tg of N_2O-N from moist/wet forests (Matson & Vitousek, 1990) and 1.4 Tg of N_2O-N from savannas (Sanhueza *et al.*, 1990); because of the lack of data, similar comparisons are not yet practical for NO-N.

Could land use change of the sort that is widespread in the Chamela region be a regionally or globally important source of change in N trace gas budgets? The most common land use practice in the area involves clearing and burning forest, planting maize on the cleared site for one year (occasionally two), and then converting the site to pasture for several to many years (de Ita-Martínez, 1983). Our measurements show that maize fields have elevated N_2O emissions, but both the enhancement (*c.* 3-fold) and the area involved at any one time are relatively small. Pastures are a longer-term land use and so occupy a much greater area, but the upland pastures we sampled have N gas fluxes no greater than forests. We conclude that regional fluxes of N_2O and NO are little altered by land use change in the Chamela area (except for biomass burning discussed below). The one improved pasture we measured (fertilized, and in a fertile lowland site) did have very high N_2O and NO fluxes; such pastures (and the land to support them) are uncommon near Chamela, but to the extent that they are more common elsewhere they could represent a significant source of alteration to regional or even global N gas fluxes.

Other processes and other gases

A process that is likely to be important in changing gas fluxes, but that was not included in our study at Chamela, is biomass burning. In the Chamela region, forest cutting occurs in the early or middle part of the dry season, followed by burning of residue late in the dry season. Once pastures are established on cleared sites, they are burned annually late in the dry season (de Ita-Martínez, 1983). The intact forests themselves do not burn, even with large pasture or land-clearing fires on their margin. Biomass combustion produces a variety of trace gases, including large quantities of NO and relatively small amounts of N_2O (Lobert *et al.*, 1990; Crutzen & Andreae, 1990). The NO produced during biomass burning has been demonstrated to alter tropospheric photochemistry over large areas of the tropics during the dry season (Fishman, Minnis & Reichle, 1986; Jacob & Wofsy, 1988). Plumes of NO_x, carbon monoxide (CO), and O_3-enriched air reach for hundreds of kilometers from their source areas, and the O_3 in particular is likely to have a significant negative effect on natural systems and agricultural productivity where its concentration is elevated (Reich & Amundson, 1985).

Biomass burning also produces a large number of other gases and aerosols, most notably particulate carbon (soot) (which absorbs incoming solar radiation), sulfur dioxide, methane (a greenhouse gas), and of course carbon dioxide (Crutzen & Andreae, 1990). The relative amounts of gases produced depend on the amount of biomass consumed and the nature of combustion: vigorous flaming produces a different suite of gases than smouldering combustion (Crutzen & Andreae, 1990). The magnitude of biomass burning emissions in the dry forest biome is not known well, but the atmospheric consequences of forest and pasture burning discussed above already have been identified clearly (Jacob & Wofsy, 1988). Most likely, human-set fires are the most significant agent of alteration of biosphere–atmosphere interactions in tropical dry forest areas globally.

Other than emissions during burning, very little is known about the emission or uptake of gases (other than NO and N_2O) in plants or soils of dry forest. Ammonia, the other major N-containing gas, would be worth examining if appropriate methods were available. The soils at Chamela are circumneutral in pH (in contrast to the acidic soils of many moist/wet forests), and soil ammonium concentrations are relatively high, particularly late in the dry season; both of these factors favor ammonia volatilization. Indeed, any gas for which rapid increases in moisture following a long dry season could alter emissions (most C, N and S gases) would be interesting to examine in dry forests (Cuevas, Chapter 15). Such systems offer the opportunity of examining moisture itself as a control of gas emissions (as distinct from temperature or temperature–moisture interactions) in a relatively straightforward way.

Ecosystem conservation and global change

One important impediment to evaluating the global significance of the tropical dry forest ecosystem is that there is relatively little of it left in an undisturbed state. As Janzen (1988) pointed out, only 2% remains of the 5.5×10^5 km^2 of dry forest that once reached from México to Colombia along the Pacific coast; dry forests elsewhere in the neotropics are being degraded by clearing and unrestricted grazing (Gentry, Chapter 7; Sampaio, Chapter 3). Similarly, much of the dry forest zone of West Africa is now derived savanna.

The dearth of intact dry forests makes it difficult to know how deforestation in the seasonal tropics has contributed to global change, or how reforestation could reverse change. We can measure current biosphere–atmosphere interactions in disturbed sites, but such measurements have relatively little value in assessing change unless they can be compared to good reference ecosystems. A few areas are in reasonably good condition, and they must be protected uncompromisingly, but even they usually have survived intact for special reasons (such as infertile or extremely shallow soils) that make them unrepresentative of dry forests in general. Forests on the most fertile soils, those that might respond most to human alteration, may already be gone.

This problem is not unique to tropical dry forests. The tall grass prairie ecosystem of central North America is another system for which most of the original area has been lost. However, the areal extent, magnitude of change, and degree of human dependence on dry forest make it a particularly important case. We need to work for the preservation of remaining areas of the dry forest biome not just for its endemic species, or evolutionary marvels, or potentially useful pharmaceutical compounds, or overall biological diversity. We should also be working to preserve it because we need it in order to understand how Earth functions as a system, and how that system is being altered globally by human activity.

Summary

Tropical dry forests are widespread and have been subject to extensive alteration by human activities, yet the importance of these ecosystems in the global budget of atmospheric trace gases is little understood. We have examined the emission of nitrogen-containing trace gases and the soil processes that control them in natural and disturbed dry forests at Chamela, México. Our data suggest that nitrous oxide and nitric oxide fluxes are strongly related to soil moisture, with highest fluxes occurring during the first rains at the end of the dry season. We estimate that tropical dry forests produce between 0.7 and 1.5 Tg of nitrous oxide nitrogen annually. Conversions from forest to maize and unfertilized pasture, as are typical of the Chamela area, do not significantly alter gas fluxes. However, to the extent that fertilization of pastures is important in other areas of the tropics, it may represent a significant source of both nitrous oxide and nitric oxide.

References

Bakwin, P. S., Wofsy, S. C., Fan, S.-M., Keller, M., Trumbore, S. E. & daCosta, J. M. (1990). Emission of nitric oxide (NO) from tropical forest soils and exchange of NO between the forest canopy and atmospheric boundary layers. *Journal of Geophysical Research* **95**: 16755–64.

Bowden, W. B. (1986). Gaseous nitrogen emmissions (*sic*) from undisturbed terrestrial ecosystems: an assessment of their impact on local and global budgets. *Biogeochemistry* **2**: 249–79.

Bullock, S. H. (1986). Climate of Chamela, Jalisco, and trends in the south coastal region of México. *Archives for Meteorology, Geophysics, and Bioclimatology, Series B* **36**: 297–316.

Cicerone, R. J. (1989). Analysis of sources and sinks of atmospheric nitrous oxide (N_2O). *Journal of Geophysical Research* **94**: 18265–72.
Crutzen, P. J. & Andreae, M. O. (1985). Atmospheric chemistry. In *Global Change*, ed. T. F. Malone and J. G. Roedereu, pp. 75–113. Cambridge University Press, Cambridge.
Crutzen, P. J. & Andreae, M. O. (1990). Biomass burning in the tropics: impact on atmospheric chemistry and biogeochemical cycles. *Science* **250**: 1669–78.
Davidson, E. A., Matson, P. A., Vitousek, P. M., Riley, R., Dunkin, K., García-Méndez, G. & Maass, J. M. (1993). Process regulation of soil emissions of NO and N_2O in a seasonally dry tropical forest. *Ecology* **71**: 130–9.
Davidson, E. A., Vitousek, P. M., Matson, P. A., Riley, R., García-Méndez, G. & Maass, J. M. (1991). Soil emissions of nitric oxide in a seasonally dry tropical forest of México. *Journal of Geophysical Research* **96**: 15439–45.
de Ita-Martínez, C. (1983). Patrones de producción agrícola en un ecosistema tropical estacional en la costa de Jalisco. Tesis de Licenciatura, Facultad de Ciencias, Universidad Nacional Autónoma de México, México.
Esteban, R. (1986). Contenido mineral de la hojarasca en una selva baja caducifolia en la costa de Jalisco, México. Tesis de Licenciatura, Facultad de Ciencias, Universidad Nacional Autónoma de México, México.
Firestone, M. K. & Davidson, E. A. (1989). Microbiological basis of NO and N_2O production and consumption in soil. In *Exchange of Trace Gases Between Terrestrial Ecosystem and the Atmosphere*, ed. M. O. Andreae and D. S. Schimel, pp. 7–21. John Wiley & Sons, Chichester.
Fishman, J., Minnis, P. & Reichle, H. G., Jr (1986). Use of satellite data to study tropospheric ozone in the tropics. *Journal of Geophysical Research* **91**: 14451–65.
García-Méndez, G., Maass, J. M., Matson, P. A. & Vitousek, P. M. (1991). Nitrogen transformations and nitrous oxide flux in a tropical deciduous forest in México. *Oecologia* **88**: 362–6.
Gessel, S. P., Cole, D. W., Johnson, D. W. & Turner, J. (1980). The nutrient cycles of two Costa Rican forests. In *Progress in Ecology*, ed. V. P. Agarwal and V. K. Sharma, pp. 23–4. Today and Tomorrow's Printers and Publishers, New Delhi.
Goreau, T. J. & de Mello, W. Z. (1988). Tropical deforestation: some effects on atmospheric chemistry. *Ambio* **17**: 275–81.
Hao, W. M., Scharffe, D., Crutzen, P. J. & Sanhueza, E. (1988). Production of N_2O, CH_4, and CO_2 from soils in the tropical savanna during the dry season. *Journal of Atmospheric Chemistry* **7**: 93–105.
Houghton, J. T., Jenkins, G. J. & Emphraums, J. J. (ed.) (1990). *Climate Change: the IPCC scientific assessment*. Cambridge University Press, Cambridge.
Jacob, D. J. & Wofsy, S. C. (1988). Photochemistry of biogenic emissions over the Amazon forest. *Journal of Geophysical Research* **93**: 1477–86.
Janzen, D. H. (1988). Tropical dry forests: the most endangered major tropical ecosystem. In *Biodiversity*, ed. E. O. Wilson, pp. 130–7. National Academy Press, Washington.
Johansson, C. (1989). Fluxes of NO_x above soil and vegetation. In *Exchange of Trace Gases Between Terrestrial Ecosystems and the Atmosphere*, ed. M. O. Andreae and D. S. Schimel, pp. 229–46. John Wiley & Sons, New York.
Johansson, C., Rodhe, H. & Sanhueza, E. (1988). Emission of NO in a tropical savanna and a cloud forest during the dry season. *Journal of Geophysical Research* **93**: 7180–92.
Johansson, C. & Sanhueza, E. (1988). Emission of NO from savanna soils during rainy season. *Journal of Geophysical Research* **93**: 14193–8.
Kaplan, W. A., Wofsy, S. C., Keller, M. & daCosta, J. M. (1988). Emission of NO and deposition of O_3 in a tropical forest system. *Journal of Geophysical Research* **93**: 1389–95.

Keller, M. K., Goreau, T. J., Wofsy, S. C., Kaplan, W. A. & McElroy, M. B. (1983). Production of nitrous oxide and consumption of methane by forest soils. *Geophysical Research Letters* **10**: 1156–9.
Keller, M., Kaplan, W. A. & Wofsy, S. C. (1986). Emissions of N_2O, CH_4, and CO_2 from tropical soils. *Journal of Geophysical Research* **91**: 11791–802.
Lashof, D. A. & Ahuja, D. R. (1990). Relative contributions of greenhouse gas emissions to global warming. *Nature* **344**: 529–31.
Likens, G. E., Bormann, F. H., Pierce, R. S., Eaton, J. S. & Johnson, N. M. (1977). *Biogeochemistry of a Forested Ecosystem*. Springer-Verlag, New York.
Lobert, J. M., Scharffe, D. H., Hao, W. M. & Crutzen, P. J. (1990). Importance of biomass burning in the atmosphere budget of nitrogen-containing trace gases. *Nature* **346**: 552–4.
Lugo, A. E. & Murphy, P. G. (1986). Nutrient dynamics of a Puerto Rican subtropical dry forest. *Journal of Tropical Ecology* **2**: 55–72.
Luizão, F., Matson, P. A., Livingston, G., Luizão, R. & Vitousek, P. M. (1989). Nitrous oxide flux following tropical land clearing. *Global Biogeochemical Cycles* **3**: 281–5.
McElroy, M. B. & Wofsy, S. C. (1986). Tropical forests: interactions with the atmosphere In *Tropical Rain Forests and the World Atmosphere*, ed. G. T. Prance, pp. 33–60. Westview Press, Boulder.
Matson, P. A. & Vitousek, P. M. (1987). Cross-system comparisons of soil nitrogen transformation and nitrous flux in tropical forest ecosystems. *Global Biogeochemical Cycles* **1**: 163–70.
Matson, P. A. & Vitousek, P. M. (1990). Ecosystem approach to a global nitrous oxide budget. *Bioscience* **40**: 667–72.
Matson, P. A., Vitousek, P. M. & Schimel, D. S. (1989). Regional extrapolation of trace gas flux based on soils and ecosystems. In *Exchange of Trace Gases Between Terrestrial Ecosystems and the Atmosphere*, ed. M. O. Andreae and D. S. Schimel, pp. 97–108. John Wiley & Sons, New York.
Murphy, P. G. & Lugo, A. E. (1986). Ecology of tropical dry forest. *Annual Review of Ecology and Systematics* **17**: 67–88.
Patiño, M. C. (1990). Variación espacial y temporal de la capa de hojarasca (mantillo) en una selva baja caducifolia en Chamela, Jalisco, México. Tesis de Licenciatura, Facultad de Ciencias, Universidad Nacional Autónoma de México, México.
Prinn, R., Cunnold, D., Rasmussen, R., Simmonds, P., Alyea, F., Crawford, A., Fraser, P. & Rosen, R. (1990). Atmospheric emissions and trends of nitrous oxide deduced from 10 years of ALE-GAGE data. *Journal of Geophysical Research* **95**: 18369–85.
Raman, S. S. (1975). Primary production and nutrient cycling in tropical deciduous forest ecosystems. *Tropical Ecology* **16**: 140–6.
Reich, P. B. & Amundson, R. G. (1985). Ambient levels of ozone reduce net photosynthesis in tree and crop species. *Science* **230**: 566–70.
Sanhueza, E., Hao, W. H., Scharffe, D., Donoso, L. & Crutzen, P. J. (1990). N_2O and NO emissions from soils of the northern part of the Guayana Shield, Venezuela. *Journal of Geophysical Research* **95**: 22481–8.
Slemr, R. & Seiler, W. (1984). Field measurements of NO and NO_2 emissions from fertilized and unfertilized soils. *Journal of Atmospheric Chemistry* **2**: 1–24.
Vitousek, P. M. & Matson, P. A. (1988). Nitrogen transformations in tropical forest soils. *Soil Biology and Biogeochemistry* **20**: 361–7.
Vitousek, P. M. & Sanford, R. L., Jr (1986). Nutrient cycling in moist tropical forests. *Annual Review of Ecology and Systematics* **17**: 137–67.
Vitousek, P. M., Matson, P. A., Volkmann, C., Maass, J. M. & García, G. (1989). Nitrous oxide flux from dry tropical forests. *Global Biogeochemical Cycles* **3**: 375–82.

17

Conversion of tropical dry forest to pasture and agriculture

J. MANUEL MAASS

Introduction

Conversion of tropical dry forest to agriculture and pasture is occurring at alarming rates, and entails nearly total destruction of forest structure and composition, and disruption of ecosystem functions. In this chapter the driving forces of this process are briefly discussed, the general pattern of the transformation is described, and the consequences of the conversion for ecosystem functioning are analysed in some detail. Particular attention is given to specific environmental conditions and management practices that determine or alter the extent of the perturbation.

Causes and patterns of forest conversion

Driving forces

Janzen (1986) considers dry forest as the 'most endangered major tropical ecosystem' because the distribution of these forests has been reduced to a small fraction of the original area. Less than 0.1% of the original dry forest has conservation status in Pacific Mesoamerica. The status of dry forest is just as critical, or worse, for regions of Australia, Southeast Asia, Africa, and major parts of South America (Janzen, 1986, 1988). Actual rates and the extent of dry forest conversion are analysed by Murphy & Lugo (Chapter 2). They have pointed out that we will never know the true original or potential extent of dry forest because many savannas and scrub or thorn woodlands are thought to be derived from disturbed dry forest (Murphy & Lugo, 1986; see Menaut, Lepage & Abbadie, Chapter 4).

There are ecological, social, political and economic factors involved in forest clearing for agricultural purposes. Often subsistence farmers are blamed for the massive deforestation in the tropics, but other groups are also involved in this process. According to Thapa & Weber (1990), those responsible for deforestation can be broadly categorized into three main groups. (1) Shifting cultivators are responsible for 45% of the elimination of closed tropical forests, according to the FAO; a large proportion of these occupy marginal hilly areas. (2) Spontaneous settlers are groups or individuals with political and/or economic power which are the most successful in obtaining large tracts of land. (3) The governments of countries in the tropics settle farmers in large projects, with the aims of population redistribution, land resources utilization, poverty eradication, national security, and resettlement of reservoir-area evacuees.

In many societies, the size of a livestock herd is a mark of social prestige and economic well-being, and this promotes cattle ranching and overstocking (Hudson, 1980). Also, government policies to increase food production encourage over-exploitation of natural resources (Collins, 1986; Carabias, 1990).

In Africa, India and South America, firewood extraction and charcoal production have been an important cause of forest degradation. However, the recent explosive pasture expansion is considered to be the main cause of forest destruction in tropical areas, particularly in subhumid areas of tropical America. According to Toledo (1992), in Central America dry forest has been more extensively affected (Honduras, El Salvador and Nicaragua) than forest in humid areas (see Table 17.1).

In México also, expansion of cattle ranching has been explosive during the last 50 years, causing serious deforestation of tropical forests. Toledo *et al.* (1989) gave three main reasons to explain this. (1) Cattle ranching is a legally and socially accepted way to accumulate extensive land areas, since maximum quotas of land ownership are determined by the amount of land required to support a specific number of head (500). Larger cattle ranches exist in subhumid regions because more surface is required for each head of cattle than in the more humid areas. (2) Cattle raising is mainly an extensive system being highly profitable because of its low labor and production costs. (3) In the last 50 years, rural areas have received large influxes of national and international credits for agriculture and cattle raising. Most of these loans were promoted by the large demand for beef by the USA. Similar factors applied to cattle expansion in the rest of tropical America (V. M. Toledo, personal communication).

Table 17.1. *Area (10^2 km^2) of tropical forest converted to pasture in three major geographical regions of Latin America*

Area	Date	Dry tropics[a]	Humid tropics[b]	Total
México	1980–1	293	382	675
Central America		687	430	1117
Guatemala	1978	78	119	197
Honduras	1978	214	23	237
El Salvador	1978	63	6	69
Nicaragua	1978–9	245	43	288
Costa Rica	1978	50	126	176
Panamá	1982	37	113	150
South America		?	1495	>1495
Brazilian Amazonia	1980		1075	
Peruvian Amazonia	1979		76	
Colombian Amazonia	1980		300	
Total		980	2307	>3287

[a] Areas with <2000 mm annual precipitation and/or > 3 dry months.
[b] Areas with >2000 mm annual precipitation and/or ≤ 3 dry months.
Source: Toledo (1992).

Patterns of forest conversion

The pattern of dry forest conversion depends largely on local conditions such as topography, soils, length of the dry season, the existence of local traditional knowledge, economic possibilities of the farmers, the type of land tenure system, and local economic and population pressures. The magnitude of conversion also varies from area to area, varying from shifting cultivation systems to large land settlements created to develop new 'agriculture emporia'. The former is characterized by small pieces of forest land that are cleared manually and cropped for a few years followed by long fallow periods. The latter is characterized by areas of several square kilometres that are cleared with bulldozers, establishing continuous cultivation systems.

The combination of these factors produces a variety of conversion patterns, which are difficult to describe in detail (e.g. Frissel, 1978; Lambert & Arnason, 1982; de Ita-Martínez, 1983; Goedert, 1983; Adedeji, 1984; Murphy & Lugo, 1986; Hardesty, 1988; Nyerges, 1989; Singh & Singh, 1989; Toledo *et al.*, 1989; Stromgaard, 1991). However, some general aspects and trends can be detected among these conversions. (1) Most of the time the original forest vegetation is completely removed. (2) Fire is extensively used as a primary management tool. (3) Maize, beans, sorghum and rice are the most common crops, particularly in the neotropics. (4) Areas with the best production potential for agriculture are cleared first, whereas poor and sloping lands are cropped by subsistence farmers with much less technical support and efficiency (Posner, 1982). (5) Cropping:fallow ratios tend to increase toward a continuous cropping system. (6) Crop varieties with short growth periods are selected because rain-fed agriculture predominates. (7) There is a strong tendency towards the establishment of extensive pastures. (8) African grasses and the Indian zebu cattle are introduced.

Loss of biotic resources

The biodiversity of dry forest is documented and discussed in previous chapters (Gentry, Chapter 7; Ceballos, Chapter 8). At present, forest conversion is the most important cause of reduction in biodiversity in the tropics (Lugo, 1986; Toledo, 1988). Conversion of dry forest for agriculture and pasture causes a great reduction in species diversity because the entire forest is removed; the reduction is 90–95% (Janzen, 1986). According to Murphy & Lugo (1986), 1–3 ha of dry forest commonly contain between 35 and 90 tree species, whereas pasture land occasionally has one to five tree species that are left standing for shading or future use for construction (de Ita-Martínez, 1983). Species reintroduction is not an easy task, and the restoration of an ecosystem is even harder (Janzen, 1988).

Although moist/wet forests have a greater species diversity than the dry forests, the latter have a larger number of endemic species (Toledo, 1988). Furley & Ratter (1988) noted that the Brazilian cerrado has approximately 800 species of trees and large shrubs, the vast majority being endemic species. This situation applies to animals as well as to plants: for México, Flores-Villela & Gerez (1988) report a higher number of endemic vertebrates in dry than in wet forest (229 vs 197 species, respectively; see Ceballos, Chapter 8). Therefore, when conversion to agriculture or pasture occurs in a dry forest area, the risk of losing species is high, even in small transformed areas. In such situations, local extinction of a species may indeed be the final extinction of that species.

Tropical forests have been recognized as rich reservoirs of valuable resources such as timber, food, medicine and industrial goods including tannins, waxes and dyes (Toledo, 1992; Sampaio, Chapter 3; Bye, Chapter 18). As Cobley & Steele (1976) have noted, the earliest farming communities and the earliest civilizations developed in the seasonally dry tropics and subtropics. There is a very good possibility that several of today's major tropical crops and food animals originated from dry forest ecosystems. Janzen (1986) suggests that dry forests (rather than wet forests) could provide plantation timber species that will grow throughout the year in wet forests yet withstand the droughts produced by agricultural clearing. This is the case with *Cordia alliodora* and *Tectona grandis*, already widespread plantation trees.

The list of potentially useful species in dry forest is enormous (Bye, Chapter 18); unfortunately many of these resources are lost when the forest is converted to agriculture and pasture. Sometimes these natural products are used when a few tree species are harvested before burning the land, or are left standing for future use. At the present rate of forest destruction, many of these species will disappear before we even know of their existence or properties.

Changes in the physical environment and soil

Albedo

The shortwave reflectivity of a surface is known as albedo and is expressed as a percentage of total incoming shortwave radiation. The albedo value sets the limit to the surface energy balance, and the interrelated water balance; therefore it controls the thermal and moisture climate of the surface and the adjacent air and soil layers (Oke, 1978). Changes in albedo occur naturally in the dry forest. At Chamela, state of Jalisco, México, average albedo during the period with leaves is 18%, and increases to 25% during the leafless season (Barradas, 1991). Changes in plant cover as a result of forest conversion can lead to important changes in albedo. A bare soil can reach albedo values of 40% when dry, and as low as 5% when saturated (Oke, 1978; Nicholson, 1988). Climatic changes resulting from albedo modifications range from minor micro-climatic alterations of local effect (e.g. soil temperature variations), up to regional climatic disruptions. The latter may be the case with the Sahel drought. It has been suggested that an increase in albedo in the Sahel region, as a result of overgrazing, could have caused a shift in the Intertropical Convergence Zone to the south, decreasing the local rainfall (Charney, Stone & Quirk, 1975). Although this idea is still in debate, Nicholson (1988) points out that numerical models consistently predict that changes of surface characteristics would alter the atmospheric flow patterns and moisture fields and, in turn, would modify climate and weather (Bruijnzeel, 1990). For further discussion on forest conversion effects on plant–atmosphere interactions see Matson & Vitousek (Chapter 16).

Hydrological processes

Water is the primary limiting factor in dry forest. Potential evapotranspiration is higher than precipitation, making evapotranspiration the major route of water movement in the ecosystem. Generally, overland runoff and base flow (stream flow during rainless periods) contribute little to the water budget, and become important only during sporadic events of high rainfall (Cervantes, Maass & Domínguez, 1988). Forest conversion to agriculture and pasture seriously disrupts the water cycle, changing the relative importance of the major fluxes of water in the ecosystem.

Soil evaporation usually increases after clear-cutting as a result of higher soil temperatures. However, both transpiration and canopy interception are greatly reduced, making the whole evapotranspirative flux of water to the atmosphere much lower than under unperturbed conditions. This results in higher stream flows after removal of the natural forest cover (Bruijnzeel, 1990).

Reduction in vegetative cover, as a result of land clearing and burning, also diminishes protection of the soil from the impact of rain drops and promotes soil crusting and a significant reduction of soil infiltration capacity (Greenland & Lal, 1979). Compared with runoff under forest, runoff from pasture lands and maize fields averages about six and 28 times higher, respectively (Maass, 1985).

Maass, Jordan & Sarukhán (1988) examined the changes in soil cover and runoff on small experimental plots under different crop systems (maize and two different species of introduced grasses). Soil vegetative cover on all treatments was negligible (less than 10%) after clear-cutting and burning of the original dry forest. Soil vegetative cover partially recovered 30 days after seeding, when it reached about 50%. In the first year of the study, the average water runoff on treatment plots (maize and grass) was 33 times higher than for forest control plots (13.3 and 0.4% of annual precipitation, respectively).

Management can significantly alter runoff in transformed ecosystems. Mulching on croplands can dramatically reduce runoff by improving soil cover, particularly at the beginning of the rainy season when the crop is still young. Maass *et al.* (1988) reported that runoff on maize plots was reduced 78% by placing a thin layer of forest leaf litter as protective mulch.

Overstocking increases runoff due to soil compaction and soil cover reduction. In the seasonally dry tropics, overstocking is common because the dry season reduces pasture availability.

Increase in runoff not only reduces water availability to plants, but also decreases the recharge of groundwater reservoirs, increases flood risk, and promotes soil erosion.

Soil erosion

Soil erosion is one of the most damaging processes during ecosystem transformation. Rates of soil formation are so low that once the topsoil is removed by erosion, opportunities for forest restoration are highly reduced (Lal, 1990).

It is well established that soil erosion rates depend on several related factors: rainfall erosivity, soil erodibility, soil cover, topography (length and degree of slope) and management practices (Morgan, 1979; Hudson, 1981). It is also well documented that soil erosion in tropical areas is much greater than in temperate zones, as a result of higher and more intense rainfall (Greenland & Lal, 1979; El-Swaify & Dangler, 1982; Maass & García-Oliva, 1990).

Most soil erosion studies in the tropics refer to humid areas and there are only a few examples from the seasonally dry tropics (Goedert *et al.*, 1982, cited by Goedert, 1983; Martínez-Menez & Fernández, 1984; Maass *et al.*, 1988). Vegetative cover is usually lower in the seasonally dry tropics than in wet areas because the former are generally less productive (Martínez-Yrízar, Chapter 13) and have been more intensively perturbed. Also, lower rainfall and the extended period of annual drought greatly reduce soil cover in these areas. As a result, erosion rates are higher in these areas. In a nationwide survey of sediment load of the major rivers in México, Martínez-Menez & Fernández (1984) measured higher loads in tropical dry watersheds than in moist ones (>5 and 4–5 Mg ha^{-1} y^{-1}, respectively).

Maass *et al.* (1988) made a detailed analysis of the effect on soil erosion rates of dry forest conversion to agriculture and pasture at Chamela. Using small plots (20–40 m^2) they measured soil losses as high as 59 Mg ha^{-1} on a maize plot for a single storm event. The maximum loss was 130 Mg ha^{-1} y^{-1}, also for a maize plot. In contrast, soil loss from forest

Table 17.2. *Soil erosion losses ($g\ m^{-2}\ y^{-1}$) as a result of deciduous forest conversion at Chamela, Jalisco, México. Mean ± SE. n = 3 except n = 6 for grass*

Year	Grass[a]	Maize	Maize with mulch[b]	Forest
1983	7638 ± 1212	10203 ± 1542	513 ± 88	<50
1984	2273 ± 545	3915 ± 731	596 ± 61	<50

[a] Average of treatments with *Panicum maximum* Jacq. and *Cenchrus ciliaris* L.
[b] Forest leaf litter was used as protective mulch.
Source: Maass *et al.* (1988).

control plots never reached more than 0.50 Mg ha^{-1} y^{-1} (Table 17.2). These accelerated soil erosion rates were high in comparison with other studies. This is not surprising, because most erosion data come from studies carried out on sites with gentler slopes (<15%) than in this study (41% slopes).

The major risk of erosion occurs during and immediately after crop seeding, particularly when fire is used to prepare the land for planting. At this time an almost bare soil receives the first rains of the season, suffering severe soil erosion. Management practices such as mulching, cover crops, contour farming, terracing, minimum tillage, fire suppression, etc. can substantially alter the rates of erosion following forest conversion. Maass *et al.* (1988) found that it was possible to reduce soil erosion by more than 90% and increase crop productivity by almost 30%, by placing a thin forest leaf litter mulch on maize plots. Similar results were obtained by Goedert *et al.* (1982, cited by Goedert, 1983) in the cerrado region of Brazil, where soil erosion was reduced from 130 to less than 2 Mg ha^{-1} y^{-1} by covering the soil with a soybean crop established with a no-tillage system.

Grasses provide a much better soil cover than most row crops. Average soil erosion in maize fields is 17.5 times higher than in pasture lands (1.4 and 24.6 Mg ha^{-1} y^{-1}, respectively; Maass, 1985). However, in overstocked conditions the situation can be reversed, because soil compaction and soil cover reduction can substantially increase runoff and erosion. Unfortunately, because good land is scarce, overstocking associated with farming and cattle-ranching on steep slopes is becoming a common practice. This is particularly true in México, Central America and the Caribbean, where hilly topography is abundant.

Soil nutrient status

Table 17.3 summarizes soil nutrient status changes after burning and cropping dry forest. Data are scarce and highly variable, therefore only a limited comparison can be made. With the exception of soil organic C, all nutrient elements increased in the upper soil horizon as a result of burning the original forest vegetation. On average, exchangeable Ca and Mg had the highest increments, followed by K and available P. Total N and total P had only a small increase. One study in particular (Budelman & Zander, 1990) reported enormous increases in cation availability after burning (13 times more Ca, 6 times more Mg and 5 times more K than the original vegetation). The investigators attributed this to the fact that, unlike N and P, most cations in tropical ecosystems on acid soils are found in the biomass rather than in the soil itself. However, it is important to mention that this particular study was in a semideciduous forest, which had much higher biomass than the others noted in Table 17.3.

The changes in soil nutrient status immediately after a fire are caused by the combined effect of two processes: changes of soil material from heating, and the addition of mineral elements from the burned vegetation (Kauffman, Till & Shea, 1992). In relation to changes due to soil heating, Andriesse & Koopmans (1984) found, by controlled laboratory burning of selected tropical soils, that the most significant changes appear to take place within the temperature range of 150–250 °C. At this temperature range, most organic materials present in the soil were combusted; available P increased as the organic P was mineralized; pH increased almost one unit; exchangeable Na and K increased, whereas Ca and Mg decreased; and N decreased as a result of volatilization.

Soils are generally poor heat conductors, restricting the high soil temperatures generated during fires to the first few centimeters of the soil profile. This was shown by Kauffman *et al.* (1993) who studied the effects of slash fire on the nutrient dynamics of caatinga in northeastern Brazil. They found that the top one centimeter of the soil reached peak temperatures of 316 °C, high enough to produce significant changes in the soil. Heat fluxes below 5 cm were negligible. Although concentration of soil nutrients were highest at the surface (top 2 cm) only a few differences were found in belowground nutrient concentration or mass following fire. As will be discussed next, most of the immediate soil chemistry changes were attributed to ash input from the burned vegetation. Not all of the nutrients released by burning the vegetation are available for crop

Table 17.3. *Changes in soil properties as a result of dry forest conversion*

Parameter	Units	Notes	Forest (a)	Right after burning (b)	<5 y cropping (c)	>5 y cropping (d)	Ratios		
							b:a	c:a	d:a
Bulk density	g cm^{-3}	2	1.23			1.32			1.1
	g cm^{-3}	5[a]	0.80			1.30			1.6
Average									1.3
Water-holding capacity		2	42.2			29.1			0.7
Organic matter	%	1	3.11	3.32	1.69		1.1	0.5	
	%	3	2.89	3.25	2.32		1.1	0.8	
	%	6	2.04		2.58	2.07		1.3	1.0
Average							1.1	0.9	1.0
Organic C	%	2	2.32			0.87			0.4[b]
	%	4	2.07	1.54	1.40		0.7	0.7	
	g m^{-2}	2	2854			1148			0.4
	g m^{-2}	5[a]	4500			4500			1.0
	g m^{-2}	7	1801	1877			1.0		
Average							0.9	0.7	0.7
Total N	ppm	1	1100	1800	1200		1.6	1.1	
	ppm	2	2140			790			0.4
	ppm	4	1500	2500	700		1.7	0.5	
	ppm	6	901		1067	1223		1.2	1.4
	g m^{-2}	2	263			104			0.4
	g m^{-2}	5[a]	680			480			0.7
	g m^{-2}	7	158	165			1.0		
Average							1.4	0.9	0.7
Total P	ppm	1	137	202	163		1.5	1.2	
	ppm	6	240		234	159		1.0	0.7
	g m^{-2}	7	22.3	25.0			1.1		
Average							1.3	1.1	0.7

Available P	ppm	1	15.0	24.0	13.0		1.6	0.9	
	ppm	3	22.5	39.9	33.6		1.8	1.5	
	ppm	4	12.0	43.5	19.0		3.6	1.6	
	ppm	6	9.4		19.8	4.4		2.1	0.5
Average							2.3	1.5	0.5
Exchangeable K	ppm	1	35.2	179.9	35.2		5.1	1.0	
	ppm	3	162.8	251.0	175.9		1.5	1.1	
	ppm	4	97.8	215.0	58.6		2.2	0.6	
	ppm	6	105.2		401.2	188.7		3.8	1.8
Average							3.0	1.6	1.8
Exchangeable Ca	ppm	1	191	2510	1509		13.1	7.9	
	ppm	3	1744	3144	2319		1.8	1.3	
	ppm	4	280	818	700		2.9	2.5	
Average							6.0	3.9	
Exchangeable Mg	ppm	1	63.2	388.8	298.9		6.2	4.7	
	ppm	3	434.6	657.7	357.9		1.5	0.8	
	ppm	4	203.1	260.2	85.1		1.3	0.4	
	ppm	6	259.2		257.5	178.4		1.0	0.8
Average							3.0	1.7	0.8
Exchangeable Na	ppm	3	21.5	40.6			1.9		
	ppm	4	4.6	4.6	5.8		1.0	1.3	
	ppm	6	78.0		125.7	100.5		1.6	1.4
Average							1.4	1.4	1.4
Cation exchange capacity	mmol kg^{-1}	1	59.1	73.1	71.4		1.2	1.2	
	meq/100 g	4	8.2	11.7			1.4		
Average							1.3	1.2	
Base saturation	%	1	15	100	68		6.7	4.5	
Soil biomass C									
Winter	μg g^{-1}	2	609			218			0.4
Summer	μg g^{-1}	2	711			289			0.4
Wet season	μg g^{-1}	2	373			180			0.5
Average									0.4

Table 17.3. (*cont.*)

Parameter	Units	Notes	Forest (a)	Right after burning (b)	<5 y cropping (c)	>5 y cropping (d)	Ratios b:a	c:a	d:a
pH[c]									
in water		1	4.3	7.3	6.9				
in water		2	6.3			6.6			
in KCl		1	3.9	6.5	6.1				
in KCl		3	6.2	7.4	6.7				
in $CaCl_2$		4	5.2	5.2	4.8				

[a] Data obtained from a graph.
[b] Data not included in the parameter average.
[c] Soil pH was measured in water, KCl, or $CaCl_2$ solution.

Sources and notes: 1. Budelman & Zander (1990). Top 20 cm. (a) Secondary semideciduous forest, (b) after burning before cultivation, (c) after 9 months of cultivation. Tai region, Ivory Coast.
2. Srivastava & Singh (1989). Top 10 cm. (a) dry forest (sal), (d) 15 y old (never manured) crop field (maize and *Hordeum vulgare*). Uttar Pradesh, India.
3. J. M. Maass (personal observations). Top 10 cm. (a) Deciduous forest, before burning, (b) the day after burning, (c) 2 y after cropping (maize and grass). Chamela, México.
4. Stromgaard (1991). Top 10 cm. (a) Miombo woodland, undisturbed before burning, (b) the day after burning, (c) 3.5 y after burning. Kasama, Zambia.
5. Brown & Lugo (1990). Top 25 cm. (a) Subtropical dry forest, (d) 60 y old cropland. St John, US Virgin Islands.
6. García-Oliva (1992). Top 12 cm. (a) Deciduous forest, (c) after pasture conversion, (d) 11 y after pasture conversion. Chamela, México.
7. Kauffman *et al.* (1993). Top 10 cm. (a) Dry forest, the day before burning, (b) the day following fire. Serra Talhada, Brazil.

uptake. Volatilization of nutrients takes place during fire, ashes are lost by wind and soil erosion, and leaching of excess cations is promoted when soil adsorption complexes are saturated with exchangeable bases.

Not all nutrients are equally prone to volatilization. Nitrogen, organic P, and S start to volatilize at low temperatures (<200, 360 and 444 °C, respectively), whereas K, inorganic P and Na require about 800 °C, and Ca and Mg higher than 1100 °C (Wright & Bailey, 1982; Raison, Khanna & Woods, 1985). In the caatinga, Kauffman *et al.* (1993) reported that as much as 96% of the pre-fire aboveground N and C pool and 56% of the pre-fire aboveground P pool was lost during combustion. That is, only 4% of the N and C, and 46% of the P released by burning the vegetation were potentially incorporated to the soil. The actual incorporation was lower because they observed that 57% of the ash disappeared within 17 days after burning, primarily as a result of wind. They also found that volatilization of N, P and C increases as fire severity increases. This trend was clearly reflected in the ash Ca:N, Ca:P and Ca:C ratios from fires of different intensities. Ash Ca:P ratios of 21.7, 47.5 and 48.9 were obtained on sites where fire consumed 78, 88 and 95% of the aboveground biomass, respectively. The authors pointed out that volatilization losses of P in their study site were as high as 21 kg ha^{-1}, two orders of magnitude higher than the natural rates of P replacement in that system, suggesting that it could take well over a century to replenish the quantity of P lost by combustion alone.

Leaching losses following forest conversion have been monitored by Adedeji (1984) in a semideciduous forest in Nigeria. In this study, lysimeters installed below the root zone (15 cm) collected percolated water during three years at biweekly intervals. During the first year after the clear-cut and burn, peak nutrient concentrations in lysimeters were 9–10 times higher for the cultivated areas than for the unperturbed forest. At the end of the study (2 years after conversion), concentrations two times higher were still observed.

Nutrient losses as a result of runoff and soil erosion in dry forest conversions have also been quantified by Kang & Lal (1979), and by Maass *et al.* (1988). In the first case, the authors evaluated the effect of different types of land clearing and soil management treatments on the quality of runoff water (pH, conductivity and nutrient concentration). They used small catchments located in southwest Nigeria, which had 15 year successional dry forest at the time of clearing. A summary of their results is given in Table 17.4. Manual clearing coupled with no-tillage produced the smallest runoff and nutrient losses, whereas the mechanical

Table 17.4. *Runoff and nutrient losses from small watersheds in southwest Nigeria cleared by different methods. CT, conventional tillage; NT, no tillage*

Clearing method	Runoff (mm)	Nutrient loss (g m^{-2})					
		NO_3-N	NH_4-N	PO_4-P	K^+	Ca^{++}	Mg^+
Manual–CT	2.399	0.023	0.010	0.001	0.237	0.156	0.040
Manual–NT	0.531	0.004	0.003	0.001	0.096	0.087	0.009
Tree pusher–CT	9.462	0.165	0.042	0.007	1.000	0.794	0.130
Tree pusher–NT	6.674	0.051	0.056	0.008	0.671	0.891	0.136
Blade bulldozer	3.124	0.066	0.037	0.001	0.366	0.194	0.053
Traditional	1.275	0.014	0.010	0.001	0.112	0.212	0.046

Source: Kang & Lal (1979).

Table 17.5. *Soil erosion nutrient losses (g m^{-2} y^{-1}) as a result of deciduous forest conversion at Chamela, México*

Nutrient	Grass[a]	Maize	Maize with mulch[b]	Forest
Total N	13.92	12.68	1.17	0.01
Total P	1.82	1.77	0.15	0.01
Exchangeable K^+	2.11	2.15	0.17	0.01
Exchangeable Ca^{++}	23.46	27.52	2.39	0.03
Exchangeable Mg^{++}	4.86	4.71	0.38	0.01

[a] Average treatments with *Panicum maximum* Jacq. and *Cenchrus ciliaris* L.
[b] Forest leaf litter was used as protective mulch.
Source: Maass *et al.* (1988).

clearing coupled with conventional tillage produced 17.8 times higher runoff and substantially higher nutrient losses.

Smaller scale, plot experiments at Chamela (Maass *et al.*, 1988) showed severe nutrient losses during forest conversion as a result of soil erosion. Table 17.5 shows a summary of rates of nutrient losses in the eroded soil particles. As was explained earlier, these losses occur when natural dry forest is converted to maize and pasture fields. Nutrient losses as dissolved elements in the runoff water were one to two orders of magnitude lower than those in the eroded particles. Thus, in maize and grass plots, 96% of the nutrients lost were adsorbed on the eroded soil particles, whereas less than 5% of the nutrients lost were dissolved in the runoff water. The tremendous nutrient losses in maize and grass plots contrast strikingly with the almost negligible losses under forest cover. Using forest leaf litter as protective mulch on maize plots was also efficient at controlling nutrient losses from the system (see Table 17.5). But even with a 90% reduction of soil erosion by mulching, nutrient loss is still high compared with forest.

Finally, nutrient removal in harvested products, and denitrification also contribute to diminished fertility of the system. Losses by these means can be substantial. For a moist tropical site, Sánchez (1982) cited results that a total of 1916 kg ha^{-1} of N were removed by crops during a period of 8 years (an average of 249 kg ha^{-1} y^{-1}), in a rice–maize–soybean rotation in Yurimaguas, Perú. Losses in dry forest sites may be lower because of their lower crop productivity. Husz (1978) analysed several tropical and subtropical small farming systems for South America. He calculated that an average of 34.1, 6.8 and 68 kg ha^{-1} y^{-1} of N, P and K, respectively, were lost from the cropping system with primary products harvested.

Vitousek *et al.* (1989) reported nitrous oxide (N_2O) fluxes from deciduous forest and from several maize and pasture lands at Chamela. Fluxes from four of the five pastures sampled were similar to those from intact forests (average of 0.88 and 0.91 ng cm^{-2} h^{-1}, respectively). However, the fifth pasture sample (an intensively managed floodplain site) had very high N_2O fluxes (9.95 ng cm^{-2} h^{-1}). The maize fields sampled had relatively high N_2O fluxes, with a mean of 2.64 ng cm^{-2} h^{-1} (Matson & Vitousek, Chapter 16).

The net effect of nutrient input (as ash) and output (by erosion, volatilization and leaching) on soil fertility status is positive immediately after clearing and first burning, and negative thereafter. Total P, total N and organic matter return to initial conditions within the first 3.5 years of cropping (Figure 17.1). It takes longer for exchangeable Ca, K and Mg, as well as for available P to return to the pre-burn levels. Only with detailed long-term studies would it be possible to establish how many years are actually required, and certainly this will vary with the type of crop, topography, management and soil characteristics. However, preliminary data for the Chamela region show that 11 years after conversion, soil available P and exchangeable Mg are less than half and 20% lower, respectively, than their original forest values (García-Oliva, 1992).

Soil organic carbon

The low thermal conductivity of soils restricts the high temperatures generated at the soil surface during fire. Only when biomass is piled up, does the temperature at a depth of 5 cm reach values as high as 150–250 °C

(Stromgaard, 1991). As a result, only small changes in soil organic matter content occur immediately after burning.

Data on changes in soil C during cultivation are not very consistent. Whereas some authors measure high reductions (Srivastava & Singh, 1989), others report no significant changes (Lugo, Sánchez & Brown, 1986; Brown & Lugo, 1990; García-Oliva, 1992). In order to compare data, soil C content changes during cultivation should be analysed in the context of life zone, agricultural practice, and length of time the soil has been cultivated (Lugo *et al.*, 1986). Also, differences in texture must be considered because high soil clay content can influence soil C losses by inhibiting decomposition and soil erosion.

Srivastava & Singh (1989) studied the effect of cultivation on microbial C and N in deciduous tropical forest soils of Uttar Pradesh, India. They observed a reduction of 57% in soil organic C in the 0–15 cm layer after

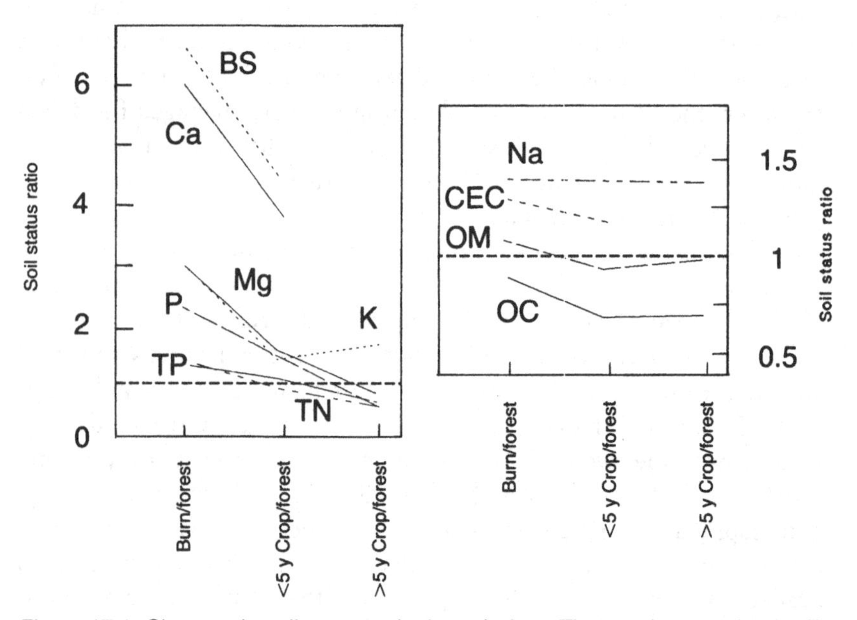

Figure 17.1. Changes in soil status ratio through time. Three ratios are plotted: (i) Burn/forest (soil status right after burning the forest/soil status under unperturbed conditions); (ii) <5 years crop/forest (soil status during the first 5 years of cropping/soil status under undisturbed conditions); (iii) >5 years crop/forest (soil status after >5 years of cropping/soil status under undisturbed conditions). Symbols: BS, base saturation; Ca, exchangeable calcium; Mg, exchangeable magnesium; K, exchangeable potassium; P, extractable phosphorus; TP, total phosphorus; TN, total nitrogen; Na, exchangeable sodium; CEC, cation exchange capacity; OM, organic matter; OC, organic carbon. Data obtained from the average values in Table 17.3.

15 years of cultivation. They attribute the depletion to changes in the magnitude of biological and physical processes in the soil during cultivation, associated with a reduced input of C into the soil.

It is well known that tropical soils, subject to relatively higher temperatures throughout the year, lose more organic C following cultivation than temperate soils. Lugo *et al.* (1986), using results of a soil survey in Puerto Rico (1940s, 1960s and 1980s), found that sites in the wet and moist life zones lost more organic C under cultivation than those in the dry life zones. They also found that pastures accumulated more soil C than cultivated sites, suggesting that only cultivated soils are consistent sources of carbon dioxide, whereas pastures may be sinks. Similarly, at Chamela, no significant changes in soil organic matter were observed on a chronosequence of three, five, seven and 11 years since conversion to pasture (García-Oliva, 1992).

Soil compaction

Bulk density increases during dry forest conversion to agriculture have been reported by Srivastava & Singh (1989) and Brown & Lugo (1990). In the latter study, the conversion of forest to agriculture had no effect on soil bulk density in the wet life zone sites, whereas in the dry and moist zones agricultural sites had significantly higher bulk densities than the control forests. Soil structure dissimilarities between wet and dry sites, as well as mechanization and trampling by cattle, may be responsible for these differences.

Soil compaction reduces the volume and continuity of the larger pores, diminishing the movement of water and air through the soil profile. Also, root elongation is reduced because they are unable to decrease in diameter to enter pores narrower than their root caps. Thus, if roots are to grow through compacted soil they must displace soil particles to widen the pores by exerting a pressure greater than the soil's mechanical strength (Hillel, 1980).

Changes in productivity

Forest conversion has a direct effect on ecosystem productivity. As a general principle, gross productivity of transformed ecosystems does not exceed that which can be found in the original ecosystem (Odum, 1971). However, net primary productivity (NPP) can be substantially higher in the transformed ecosystem, mostly through energy subsidies that reduce

Table 17.6. *Biomass and net primary productivity (NPP) of deciduous forest and grasslands of Varanasi, India*

Community	Biomass ($g\ m^{-2}\ y^{-1}$)	NPP ($g\ m^{-2}\ y^{-1}$)
Tropical deciduous forest	23967	1550
Dry savanna		
(*Prosopis* species, grass)	2680	1450
Grasslands		
Protected		
Heteropogon contortus	2928	2880
Dichanthium annulatum	2970	1838
Grazed		
Medium	1739	853
Heavy	956	742
Protected and irrigated		
Savanna region		
Sehima species	–	4900
Heteropogon contortus	–	3400
Old alluvium		
Dichanthium annulatum	2970	1838
Dichanthium desmodium	1789	2852

Compiled by: Misra (1972).

both autotrophic and heterotrophic consumption. Data on productivity changes as a result of conversion of dry forest are scanty (Murphy & Lugo, 1986). Table 17.6 presents a comparative analysis of NPP of the deciduous forest and of the grassland derived from forest conversion at Varanasi, India (Misra, 1972). The table shows that NPP in grasslands can be as much as 119 and 216% higher than the nearby forest (for the *Heteropogon contortus* and *Sehima* species grasslands, respectively). However, both cases consist of sites irrigated and protected from grazing. When grazing is not controlled, NPP in the grassland is lower than in the original forest. Data from grasslands with medium and heavily grazed conditions show, respectively, 45 and 52% lower NPP than the forest. Site conditions as well as species composition also influence NPP changes after conversion. For example, sites on old alluvium soils support less productive grasslands than those in the young savanna region, both being irrigated and protected areas (Table 17.6).

Crop productivity during the first few years after clear-cutting and burning is usually good as a result of increasing nutrient, light and water availability. However, productivity declines with time of use as a result of soil degradation. Long-term monitoring of crop and grass productivity after forest conversion is essential but has not been well documented for dry forest. However, evidence of productivity decline can be detected observing land use patterns after forest conversion. In the Chamela region, for example, maize is grown for two years at the most; grasses are cultivated thereafter because soil conditions can no longer support maize development (de Ita-Martínez, 1983). Under subsistence conditions, pasture production lasts only for 10 years and thereafter is abandoned allowing forest succession to proceed. Under better site conditions (flat lands with irrigation and fertilization possibilities), pasture can last for many years, depending on management.

Comparison of conversions of dry and humid tropical forests

Climatic as well as structural and functional differences between deciduous and humid tropical forests make them differ in their vulnerability and response to conversion. The most important differences are noted below. Some have already been discussed and will only be briefly mentioned for completeness.

1. Climate in subhumid tropical areas is more favorable for humans and cattle than the hot and wet weather of the humid tropics (Murphy & Lugo, 1986). As a result, higher population density and, therefore, higher pressure for conversion is found in dry forest regions.
2. A severe dry season makes burning easier and more efficient during conversion and cultivation than in the wet tropics (Janzen, 1986).
3. During fires, most aboveground biomass burns whereas root biomass is protected from combustion. Because the root:shoot biomass ratio is higher in dry than wet forests (R. L. Sanford, personal communication), dry forest may be less vulnerable to nutrient losses and erosion during clear cutting and burning since roots decompose *in situ*.
4. Drought and fire greatly reduce soil cover in dry forest areas converted to agriculture. Low vegetative cover at the beginning of the wet season promotes soil erosion on these open agricultural areas (Maass *et al.*, 1988).
5. The annual drought causes shortages of forage and water during the dry period. This promotes overgrazing and, therefore, low soil cover that also increases the risk of soil erosion.
6. A strong dry season restrains biological activity for several months, lowering pest problems.

7 In wet forests, high rainfall and continuous cultivation accelerate the leaching losses of nutrients as well as nutrient depletion due to harvesting (Jordan, 1985). A shorter growing season in dry forest areas reduces cropping possibilities, but also reduces the risk of soil degradation as a result of continuous farming and leaching.
8 Soils in the humid tropics are more susceptible to soil compaction because wet soil has less resistance to the process than dry soils, provided the wet soil is not at or near saturation.
9 The biomass of dry forest is less than moist/wet forests (see Martínez-Yrízar, Chapter 13), which makes them easier to clear-cut and burn.
10 Dry forest has a higher coppicing capacity than wet forest (Nyerges, 1989). Coppicing keeps roots alive for longer periods, reducing nutrient leaching (promoted by root decomposition), and increasing nutrient conservation through root uptake. Living roots are also important in reducing the risk of soil erosion by providing a retention matrix for soil particles. Coppicing appears to be particularly advantageous in dry habitats where the difficulty of seedling establishment may be great (Nyerges, 1989). Low stem diameter, combined with high coppicing capacity, makes dry forest trees good sources of firewood and forage (Hardesty, 1988).
11 Tropical moist/wet forests have higher species diversity than dry forests, but the latter have larger numbers of endemic species (Gentry, Chapter 7; Ceballos, Chapter 8). Taking this into consideration, it seems that the risk of losing species during land transformation is higher in seasonally dry habitats, even on small perturbed areas.
12 Lower diversity as well as lower biomass of dry in contrast to moist/wet forests, makes dry forest a lesser source of potential natural resources.

All these differences emphasize the importance of developing appropriate management schemes specific for tropical dry forest.

Summary

Clearing of tropical dry forest is occurring at alarming rates. Pasture expansion has been considered one of the main causes of this forest destruction. Patterns of conversion vary greatly depending on local conditions such as topography, soils, length of dry season, the existence of traditional knowledge, economic possibilities of the farmers, land tenure system, and local economic and population pressures.

A general pattern of conversion involves the complete removal of the original forest vegetation by cutting and fire. Conversion starts with sites of highest potential, leaving poor sloping lands to subsistence farmers with lower technical support and efficiency. Short growth period varieties of maize, beans, sorghum and rice are commonly cultivated on a rain-fed agriculture system. There is a strong tendency towards the establishment of extensive pasture systems, where African grasses and the Indian zebu cattle are introduced.

This conversion pattern entails nearly total destruction of forest structure and composition, and disrupts functioning of the ecosystem. Among the most important ecological consequences of conversion are: (1) reduction in species diversity; (2) reduction in soil vegetative cover, which promotes soil crusting and significant reduction of soil infiltration capacity; (3) disruption of the water cycle, increasing runoff, soil erosion and floods risks, and reducing water availability to plants, as well as the recharge of groundwater reservoirs; (4) changes in soil nutrient status, due to the combined effect of the addition of mineral elements from the burned vegetation, and changes of soil material from heating; (5) losses of nutrients from the ecosystem as a result of mineral element volatilization during fires, ashes blown away by the wind, soil erosion, and leaching of excess cations which is higher with over-saturated adsorption complexes. The extent of these changes depends on management strategies.

Management and site conditions influence changes in net primary productivity after conversion. Crop productivity during the first few years after clear-cutting and burning is usually good as a result of increasing nutrient, light and water availability. However, productivity declines with time of use as a result of soil degradation.

Climatic, as well as structural and functional differences between dry and humid tropical forests make them differ in their vulnerability and response to conversion. These differences stress the point that proper management strategies also vary between ecosystems.

Sustainable land use systems are urgently needed for dry forest. These systems must be based on detailed knowledge of forest structure and function, and response to perturbation. Further studies are needed for a better understanding of the ecology of this ecosystem. Particularly important are long term studies that include rare events or extreme years which may be of great significance in controlling the structure and function of the tropical dry forest.

Acknowledgements

The author is deeply grateful to S. Bullock, H. Mooney, R. Sanford and J. Whitbeck for reviewing the manuscript. Technical support from Rocío Esteban is also acknowledged. This is a contribution from the Ecosystem Group of the Centro de Ecología, UNAM, which has received financial support from the Consejo Nacional de Ciencia y Tecnología and the Dirección General de Asuntos del Personal Académico, UNAM.

References

Adedeji, F. O. (1984). Nutrient cycles and successional changes following shifting cultivation practice in moist semi-deciduous forest in Nigeria. *Forest Ecology and Management* **9**: 87–99.

Andriesse, J. P. & Koopmans, T. T. (1984). A monitoring study on nutrient cycles in soils used for shifting cultivation under various climatic conditions in tropical Asia. I. The influence of simulated burning on form and availability of plant nutrients. *Agriculture, Ecosystems and Environment* **12**: 1–16.

Barradas, V. (1991). Radiation regime in a tropical dry deciduous forest in western México. *Theoretical and Applied Climatology* **44**: 57–64.

Brown, S., & Lugo, A. E. (1990). Effects of forest clearing and succession on the carbon and nitrogen content of soils in Puerto Rico and US Virgin Islands. *Plant and Soil* **124**: 53–64.

Bruijnzeel, L. A. (1990). *Hydrology of Moist Tropical Forest and Effects of Conversion: a state of knowledge review.* UNESCO International Hydrological Programme, Free University, Amsterdam.

Budelman, A., & Zander, P. M. (1990). Land-use by immigrant Baoulé farmers in the Tai region, South-west Côte d'Ivoire (Ivory Coast). *Agroforestry Systems* **11**: 101–23.

Carabias, J. (1990). Las políticas de producción agrícola, la cuestión alimentaria y el medio ambiente. In *Medio Ambiente y Desarrollo en México*, Vol. I, ed. E. Leff, pp. 329–62. Universidad Nacional Autónoma de México, and Editorial Porrúa, México.

Cervantes, L., Maass, J. M. & Domínguez, R. (1988). Relación lluvia-escurrimiento en un sistema pequeño de cuencas de selva baja caducifolia. *Ingeniería Hidráulica en México. Segunda época*, Vol. III(1): 30–42.

Charney, J., Stone, P. H. & Quirk, W. J. (1975). Drought in the Sahara: a biogeophysical feedback mechanism. *Science* **187**: 434–5.

Cobley, L. S., & Steele, W. M. (1976). *An Introduction to the Botany of Tropical Crops.* Longman, New York.

Collins, J. L. (1986). Smallholder settlement of Tropical South America: The social causes of ecological destruction. *Human Organization* **45**: 1–10.

de Ita-Martínez, C. (1983). Patrones de producción agrícola en un ecosistema tropical estacional en la costa de Jalisco. Tesis de Licenciatura, Facultad de Ciencias, Universidad Nacional Autónoma de México, México.

El-Swaify, S. A. & Dangler, E. W. (1982). Rainfall erosion in the tropics: a state-of-the-art review. In *Soil Erosion and Conservation in the Tropics*, ed. W. Kussow, S. A. El-Swaify and J. Mannering, pp. 1–25. American Society of Agronomy, and Soil Science Society of America, Madison.

Flores-Villela, O. & Gerez, P. (1988). *Conservación en México: Síntesis sobre vertebrados terrestres, vegetación y uso del suelo.* INIREB, and Conservación Internacional, México.
Frissel, M. J. (ed.). (1978). *Cycling of Mineral Nutrients in Agricultural Ecosystems.* Elsevier, Amsterdam.
Furley, P. A. & Ratter, J. A. (1988). Soil resources and plant communities of the central Brazilian cerrado and their development. *Journal of Biogeography* **15**: 97–108.
García-Oliva, F. (1992). Erosión y pérdida de nutrientes del suelo en una cronosecuencia de uso agropecuario en un ecosistema tropical estacional mexicano. Tesis Doctoral, Centro de Ecología, Universidad Nacional Autónoma de México, México.
Goedert, W. J. (1983). Management of the cerrado soils of Brazil: a review. *Journal of Soil Science* **34**: 405–28.
Greenland, D. J. & Lal, R. (ed.) (1979). *Soil Conservation and Management in the Humid Tropics.* John Wiley and Sons, Chichester.
Hardesty, L. H. (1988). Multiple-use management in the Brazilian caatinga. *Journal of Forestry* **86**(8): 35–7.
Hillel, D. (1980). *Fundamentals of Soil Physics.* Academic Press, New York.
Hudson, N. W. (1980). Social, political and economic aspects of soil conservation. In: *Soil Conservation: problems and prospects*, ed. R. P. C. Morgan, pp. 45–54. John Wiley and Sons, Chichester.
Hudson, N. W. (1981). *Soil Conservation.* Cornell University Press, Ithaca.
Husz, G. S. (1978). Nutrient cycling data of agro-ecosystems in South America. In *Cycling of Mineral Nutrients in Agricultural Ecosystems*, ed. M. J. Frissel, pp. 244–76. Elsevier, Amsterdam.
Janzen, D. H. (1986). Tropical dry forests: the most endangered major tropical ecosystem. In *Biodiversity*, ed. E. O. Wilson, pp. 130–7. National Academy Press, Washington.
Janzen, D. H. (1988). Management of habitat fragments in a tropical dry forest: growth. *Annals of the Missouri Botanical Garden* **75**: 105–16.
Jordan, C. F. (1985). *Nutrient Cycling in Tropical Forest Ecosystems.* John Wiley and Sons, Chichester.
Kang, B. T. & Lal, R. (1979). Nutrient losses in water runoff from agricultural catchments. In *Tropical Agricultural Hydrology, Watershed Management and Land Use*, ed. R. Lal and E. W. Russell, pp. 1953–62. John Wiley and Sons, Chichester.
Kauffman, J. B., Sanford, R. L., Jr, Cummings, D. L., Salcedo, I. H. & Sampaio, E. V. S. B. (1993). Biomass and nutrient dynamics associated with slash fires in neotropical dry forests. *Ecology* **74**: 140–51.
Kauffman, J. B., Till, K. M. & Shea, R. W. (1992). Biogeochemistry of deforestation and biomass burning. In *The Science of Global Change: the impact of human activities on the environment*, ed. D. A. Dunnette and R. J. O'Brien, pp. 426–56. American Chemical Society, Washington.
Lal, R. (1990). *Soil Erosion in the Tropics: principles and management.* McGraw-Hill, New York.
Lambert, J. D. H. & Arnason, J. T. (1982). Nitrogen distribution in hybrid and local corn varieties and its possible relationship to a declining soil nitrogen pool under shifting agriculture at Indian Church, Belize. *Plant and Soil* **67**: 119–27.
Lugo, A. E. (1986). Estimating reductions in the diversity of tropical forest species. In *Biodiversity*, ed. E. O. Wilson, pp. 58–70. National Academy Press, Washington.
Lugo, A. E., Sánchez, M. J. & Brown, S. (1986). Land use and organic carbon content of some subtropical soils. *Plant and Soil* **96**: 185–96.
Maass, J. M. (1985). Soil erosion and nutrient losses in seasonal tropical agroecosystems under various management techniques. Ph.D. thesis, University of Georgia, Athens.

Maass, J. M., & García-Oliva, F. (1990). La conservación de suelos en zonas tropicales: el caso de México. *Ciencia y Desarrollo* **15**(90): 21–36.

Maass, J. M., Jordan, C. F. & Sarukhán, J. (1988). Soil erosion and nutrient losses in seasonal tropical agroecosystems under various management techniques. *Journal of Applied Ecology* **25**: 595–607.

Martínez-Menez, M. R. & Fernández, V. J. (1984). Evaluación de la degradación específica a nivel nacional a partir de cuencas hidrológicas. *Memorias del XVII Congreso Nacional de la Ciencia del Suelo*, p. 73. Sociedad Mexicana de la Ciencia del Suelo, México.

Misra, R. (1972). A comparative study of net primary productivity of dry deciduous forest and grassland of Varanasi, India. In *Tropical Ecology, with Emphasis on Organic Production*, ed. P. Golley and F. B. Golley, pp. 279–93. International Society of Tropical Ecology, New Delhi.

Morgan, R. P. C. (1979). *Soil Erosion*. Longman, London.

Murphy, P. J. & Lugo, A. E. (1986). Ecology of tropical dry forest. *Annual Review of Ecology and Systematics* **17**: 67–88.

Nicholson, S. E. (1988). Land surface atmosphere interaction: physical processes and surface changes and their impact. *Progress in Physical Geography* **12**: 36–65.

Nyerges, A. E. (1989). Coppice swidden fallows in tropical deciduous forest: biological, technological, and socio-cultural determinants of secondary forest successions. *Human Ecology* **17**: 379–400.

Odum, E. P. (1971). *Fundamentals of Ecology*, 3rd edition. Saunders, Philadelphia.

Oke, T. R. (1978). *Boundary Layer Climates*. Methuen, London.

Posner, J. A. (1982). Cropping systems and soil conservation in the hill areas of tropical America. *Turrialba* **32**: 287–99.

Raison, R. J., Khanna, P. K. & Woods, P. V. (1985). Mechanisms of element transfer to the atmosphere during vegetation fires. *Canadian Journal of Forest Resources* **15**: 132–40.

Sánchez, P. A. (1982). Nitrogen in shifting cultivation systems of Latin America. *Plant and Soil* **67**: 91–103.

Singh, V. P. & Singh, J. S. (1989). Man and forests: a case-study from the dry tropics of India. *Environmental Conservation* **16**: 129–36.

Srivastava, S. C. & Singh, J. S. (1989). Effect of cultivation on microbial carbon and nitrogen in dry tropical forest soil. *Biology and Fertility of Soils* **8**: 343–8.

Stromgaard, P. (1991). Soil nutrient accumulation under traditional African agriculture in the miombo woodland of Zambia. *Tropical Agriculture (Trinidad)* **68**: 74–80.

Thapa, G. B. & Weber, K. E. (1990). Actors and factors of deforestation in 'Tropical Asia'. *Environmental Conservation* **17**: 19–27.

Toledo, V. M. (1988). La diversidad biológica de México. *Ciencia y Desarrollo* **8**: 7–16.

Toledo, V. M. (1992). Bio-economic cost. In *Development or Destruction? The Conversion of Tropical Forest to Pasture in Latin America*, ed. T. Downing, S. Hecht and H. Pearson, pp. 63–71. Westview Press, New York.

Toledo, V. M., Carabias, J., Toledo, C. & González-Pacheco, C. (1989). *La Producción Rural en México: alternativas ecológicas*. Fundación Universo Veintiuno, México.

Vitousek, P., Matson, P., Volkmann, C., Maass, J. M. & García, G. (1989). Nitrous oxide flux from dry tropical forest. *Global Biogeochemical Cycles* **3**: 375–82.

Wright, A. H. & Bailey, A. W. (1982). *Fire Ecology, United States and Southern Canada*. John Wiley & Sons, New York.

18

Ethnobotany of the Mexican tropical dry forests

ROBERT BYE

Introduction

Human life is sustained and enriched by the surrounding vegetation. The exploitation of the local vegetation is one aspect of human relationship with plants. The interactions between humans and plants may result in beneficial as well as deleterious effects.

Ethnobotany may be defined as the study of the biological bases of plant–human interactions and relationships at different levels of organization (e.g. ecosystem, community, individual), over geographic and social space, and along the evolutionary time scale. On one hand, the lifestyle and form of subsistence of inhabitants can be strongly influenced by the exploitation of local plants as well as by introduced elements (e.g. cultigens) and processes (e.g. livestocking). The diversity and availability of vegetal resources depends upon the local demand to make a living as well as the external demands for raw material. Intensification of certain extractive or ecologically non-compatible processes often destroys over time valuable and promissory resources. For the purpose of this chapter, I consider only the spontaneous plants which are exploited for local consumption rather than the cultivated exotics or the systems that modify the habitat for ecologically non-compatible production.

The tropical dry forest (tropical deciduous forest, 'selva baja caducifolia' or 'bosque tropical caducifolio' and associated vegetation types) extends along the length of México, principally parallel to both the Gulf of México and the Pacific Ocean coasts (see Murphy & Lugo, Chapter 2). It ranges from 0 to 1900 m in elevation and covers 157,800 km^2 (or 8% of the Mexican territory) (Rzedowski, 1978). Flores *et al.* (1971), on the other hand, estimate that the Mexican tropical dry forest ('selva baja caducifolia') occupies 315,213 km^2 (or 16% of the country). This figure can be expanded to 20.8% if the associated vegetation types of 'selva baja subperennifolia' (with 61,543 km^2) and of 'selva mediana caducifolia' (with 33,533 km^2) are added. Based upon the analysis of the vegetation and soil use maps of the National Atlas (Secretaría de Programación y Presupuesto, 1980), Toledo *et al.* (1989) estimate that México has 24.47 $\times 10^6$ ha (14.6% of the territory) of tropical dry forest ('selva baja caducifolia') of which 72% was reportedly undisturbed in the 1970s. The seasonal fluctuation of moisture is dramatic and the median annual temperature is high (Rzedowski, 1978). The dry season is without precipitation for 5–8 consecutive months during which most of the trees are leafless. During the wet season, between 300 and 1200 mm of rain falls in 4–7 months. The median annual temperature is between 20 and 29 °C. The characteristic trees include *Alvaradoa, Amphipterygium, Bursera, Ceiba, Cordia, Cyrtocarpa, Haematoxylon, Ipomoea, Lysiloma, Piscidia* and *Pithecellobium* (Flores *et al.*, 1971; Rzedowski, 1978).

Status of ethnobotanical studies

Ethnobotanical information can be handled at two levels. General patterns of land use relative to economic activities and cultural background of the human population are useful in generating the basic trends of plant–human relationships. Specific studies by community, cultural group or botanical taxon provide greater profundity in understanding plant–human interactions.

During the last two decades, considerable federal support has been extended to inventory México's natural and human resources. Few studies have evaluated the details. Unfortunately, the ecological analysis of rural production in México based upon governmental documents (Toledo *et al.*, 1989) does not readily distinguish the dry forest which appears to be included under one of the three categories 'zona tropical húmeda', 'zona tropical subhúmeda', or 'zona multizonal'. As a consequence, the relative importance of human population, agriculture, livestock and forestry practices on a national level cannot be reliably determined.

Until recently, few ethnobotanical studies in México focused on areas with dry forests. In order to bring these data to light and to promote systematic research on useful plants in the dry forests, the Instituto de Biología of the Universidad Nacional Autónoma de México organized the first meeting of ethnobotanists in 1989 (Reunión Nacional sobre Investigaciones Etnobotánicas de Selva Baja Caducifolia de México, or

RNIESBCM: Cervantes & Bye, 1989a). Current research was summarized and common goals were established in order to make available information on useful plants of this vegetation, to publicize the importance of this flora and to promote its conservation. The second meeting, which was co-sponsored by the Instituto Nacional de Investigaciones Forestales y Agropecuarias (INIFAP) and the Universidad Estatal de Morelos, evaluated methodologies that are appropriate to ethnobotanical studies in this ecological unit. The following is a summary of some of the studies under way on useful plants.

Many of the inhabitants of the Mexican dry tropics are mestizos who have immigrated within the last 100 years. Hence, there is little indigenous knowledge of the flora based upon generations of cultural experience and coevolution. The interaction has been based upon extractive practices and concepts of habitat modification imported from other areas of México and which are not compatible with dry forest ecosystems.

Case studies of commercial products

Although dry forest occupies about 8–20% of México, it is not an important source of commercial timber. Only two species of the 49 commercially important timber trees originate from this forest type (F. Solares Arenas, personal communication, and based upon data from the Cámara Nacional de la Industria Forestal). These trees are 'guayacán' (*Conzattia multiflora*) and 'cuachalalate' (*Amphipterygum adstringens*).

Although commercial products derived from dry forest are not registered, recent studies suggest that many products enter the regional, national and international markets. Our continuing ethnobotanical survey of markets in México and adjacent United States of America (Bye & Linares, 1983) have catalogued several medicinal and edible plants (Table 18.1). A selection of dry forest plants which have received attention in recent years is briefly summarized below. They represent examples of support stakes for the expanding vegetable agriculture ('vara' and 'estacón' from *Croton*), fruits of 'cirián' (*Crescentia*) that are transformed into 'maraca' rattles, palm leaves used in the manufacture of brooms, and two popular medicinal plants, 'cancerina' and 'cuachalalate'. (Note: monetary values are reported here in the new currency of 'nuevos pesos ': 1 NP = 10^3 pre-1993 pesos.)

The rapid expansion of commercial irrigated agriculture in northwestern México since 1950 has generated an increased demand for support material derived from the adjacent dry forests (Vega &

Table 18.1. *Products of the dry forest in Mexican and international markets (R, regional; N, national; I, international)*

Scientific and common species names	Market	Parts utilized and notes
Edible plants		
Byrsonima crassifolia – nanche	R N	Fruit
Cnidoscolus chayamansa – chaya	R N	Leaf
Crotalaria species – chipil	R N	Leaf
Erythrina americana – colorín	R N	Flowers
Escontria chiotilla – chotillia	R	Fruit
Leucaena species – guaje	R N	Fruit (seed), toasted seed, sprouts, flower buds, galls
Pithecellobium dulce – guamúchil	R N	Fruit (aril)
Pleurotus ostreatus – oreja de cazahuate	R N	Fungus cultivated on *Ipomoea murucoides*
Porophyllum ruderale macrocephalum – papaloquelite	R N	Plantlet and branch (leaf)
Psidium sartorianum – arrayán	R N	Fruit
Capsicum annuum glabrisculum – chiltepín	R N I	Fruit
Sideroxylon capiri – tempesquite	R	Fruit
Stenocereus species – pitaya	R N	Fruit
Medicinal plants		
Aristolochia brevipes – guaco	R N	Root
Amphipterygium adstringens – cuachalalate	R N I	Bark
Basacantha echinocarpa – grangel	R N I	Fruit
Cordia boissieri – nacahua	R N I	Flower
Eysenhardtia polystachya – palo dulce	R N I	Wood
Guazuma ulmifolia – guasuma	R N I	Fruit
Haematoxylon brasilleto – palo brasil	R N I	Wood
Hemiangium excelsum – cancerina	R N I	Root bark
Hintonia latiflora – copalquín, quina amarilla	R N I	Bark; 'copangel'
Serjania triquetra – tres costillas	R N I	Stem
Tecoma stans – tornadora	R N I	Flower, leaf
Zornia species – hierba de la víbora	R N	Whole plant
Specialty and craft plants		
Bursera species – copal	R N	Resin
Bursera aloexyon – olinaloe	R N I	Resin, wood
Ceiba acuminata – pochote	R N	Bark
Crescentia alata – cirián	R N I	Fruit: bowls, maracas
Ficus and substitutes – amate	R N I	Bark

Hernández, 1989). Two types of support stakes are required: the estacón which measures 2 m long with a diameter of 2.5–6.0 cm and the vara of the same length but with a diameter of 1.2–2.4 cm. Vara blanca (*Croton* species, usually *C. reflexifolius* HBK., *C. alamosanus* Rose, *C. flavescens*

Greenm., *C. morifolius* Willd., *C. adspersus* Benth. and *C. fragilis* HBK.) has become the dominant support stake for cultivated tomato (*Lycopersicon esculentum* Mill.), eggplant (*Solanum melongena* L.), chile (*Capsicum annuum* L.), and cucumber (*Cucumis sativus* L.) because of its availability nearby and its durability (usually 3–5 years). These cultigens cover 37,963 ha in Sinaloa and need on the average 2587 estacones and 9292 varas ha^{-1}. In the 1989–90 growing season in Sinaloa, the commercial farmers required 98,226,595 estacones and 271,102,000 varas which resulted in a cost of 130,444,000 NP (or US $53,112,377) including transportation and handling. Collectors of these support stakes charged 0.50 NP ($0.20) and 0.30 NP ($0.12) for each estacón and vara, respectively. Until recently up to 30% of the natural populations were exploited.

With the increased demand, not only has this small tree been over-exploited but other genera with less durability and less capacity for regeneration are being cut. The problems caused by the reduction in number of stems (of vara blanca and associated trees), coupled with the reduction of stem sprout regeneration, are compounded because these woody plants usually grow on easily eroded slopes. Hence, not only is there a reduction in the number of individuals of the species in the area but also there is a general degradation of the whole environment. Researchers at the Escuela Superior de Agricultura de la Universidad de Sinaloa are currently studying the situation and proposing rational exploitation programs.

The densely-leaved spreading branches make *Crescentia alata* HBK. ('cuatecomate' or 'cirián') one of the characteristic trees of the dry forest. In the state of Morelos in central México (at the northern limit of the Río Balsas drainage), the density varies from 1 to 8 trees ha^{-1}. Approximately 8140 trees cover an area of 7375 ha in the whole state. The fruits are collected for sale in the local and national markets (1 NP or US $0.41 per fruit) as medicine for kidney and respiratory ailments. Specifically shaped spherical and ellipsoidal fruits selected to produce rattles or maracas can cost between 0.60 and 0.90 NP per fruit ($0.24–0.37). Animal saddles made from the wood cost between 60 and 300 NP ($24-$122) each.

In Morelos, families of 18 craftsmen and 50 fruit collectors depend economically upon the natural production of these fruits. These rattles generate about 10,080 NP per week (US $4104 per week or $213,408 per year) and create a demand of 25,200 fruits per year. On an annual average, a tree bears 116 fruits. Thus, the Morelos collectors of maraca fruits need 10,440 trees, 2300 more than occur in the state. Because of this increased demand, they now import them from the adjacent state of Guerrero. Obviously, this is an underestimate because it does not take into account fruit loss, those collected for medicine and those fruits that produce shells for 'jícaras' or cups and containers. The situation of multiple use and resource competition (e.g. living tree to produce economic fruits annually vs timber tree for one-time exploitation for wooden saddles) coupled with increased demand for marketable products has stimulated the local branch of the INIFAP to inventory, monitor and promote the rational use of this tree that once was considered a disturbance in land cleared for cattle grazing (Solares, 1989).

In the Pacific coastal state of Colima, the 'palmilla' (*Cryosophila nana* (HBK.) Blume), a short stature palm of the dry forest, is highly prized for the production of brooms that are called 'escoba verde' (Vergara, 1990). About 158 craftsmen in 66 production units depend economically on the production of about 262,640 brooms a year. In 1988 a palm leaf broom sold for 1.025 NP (US $0.45) in the local market and daily individual production on the average generated 2.5 minimum salaries for each worker. The regeneration of harvestable leaves varies between two or three months during the rainy season and four or more months during the dry season. The future of this industry is threatened due to over-collection of leaves (sometimes harming the regenerative processes) and the destruction of the natural populations as land is converted into farmland, as livestock grazing on tender shoots expands and as unnatural fire damages the palm during the sensitive stage of growth (in its resistant stage, fire tends to favor the palm by reducing competition). The increased demand for edible palm hearts ('palmitos') promotes the one-time harvest of the palm plants, especially considering that each palmito brings in about 2 or 3 NP ($0.88-1.33) or 29–43% of the minimum daily salary.

The dry forest produces many medicinal plants which enter the national and international market. The flattened, stringy pieces of cancerina (*Hemiangium excelsum* (HBK.) A.C. Smith) are gaining popularity (Barreto, Linares & Bye, 1989). The root bark of this vine is highly esteemed in medicinal mixtures for healing skin wounds and gastric ulcers. Because the harvesting technique destroys the plant and due to the increased demand for the roots, the woody vine has become extinct in many localities in the states of Morelos, Guerrero and Puebla. Recent studies show the root bark to be one of the most effective natural insecticides in the treatment of bean and corn weevils (*Acanthoscelides obtectus, Prostephanus truncatus, Zabrotes subfasciatus, Sitophilus zeamais*: Lagunes, 1989) and to have one of the highest contents of gutta rubber (11% on a dry weight basis: Palacios *et al.*, 1989). In 1987, 1 kg of dry root bark sold for 20 NP ($14.37). The price increased at each stage of the commercial process. The gatherers earned 3 NP per kg ($2.15). The individuals who concentrate the processed root bark obtain 8 NP per dry kg ($5.75) in the wholesale market. Intermediate merchants sell it to resalers and consumers for 15–20 NP per kg ($10.78–14.37). The gatherer, who is responsible for the immediate destruction of the resource in order to glean extra income, gains only 15% of the retail value of the plant. With such little absolute and relative return for the material, he is forced to remove all the plants possible to make his time worthwhile and to prevent others from depriving him of his find. Because the leaves remain green during the dry season, when unoccupied subsistence farmers dig out the roots to supplement their income, the plants are easily located by the conspicuous greenery in the canopies of the dry forest.

Another popular vegetal remedy from the dry forest is derived from the tree 'cuachalalate' (*Amphipterygium adstringens* (Schlecht.) Schiede) (Soberanes, 1989). The bark is prized as the major ingredient in medicinal preparations for 12 ailments and is credited with anti-inflammatory properties. In a survey of 25 retailers in three local market towns in Morelos, over 456 kg were sold per month. The national and international consumption is much higher but exact figures are not available. The removal of the bark, if done incorrectly, is fatal to the tree. The future production of this wild resource in the tropical forest is threatened. The natural populations are thin. Based upon a sample of 0.005% of the 103,375 ha of dry forest of Morelos, the average density is 5 trees ha^{-1}. Habitat destruction by clear-cutting and livestock grazing was found on 55% of the sites sampled in the state of Morelos. The regeneration of the population (based upon the presence of stems less than 5 cm diameter) was documented in only half of the sites. INIFAP of Morelos, the Universidad de Morelos and the Universidad de Chapingo are conducting studies on bark harvesting and seed germination in order to promote the rational use of this tree.

Regional studies of indigenous uses

In addition to commercial exploitation of plants for use outside the dry forest, this vegetation provides many useful products for consumption by

Table 18.2. *Numbers of kinds of uses and of taxa used for plants of the dry forests of México. SUU, species utilization unit; Taxa, species with families in parentheses*

State	Medicine	Food[a]	Wood[b]	Other	Total	Culture	Geography
Chihuahua							
SUU	29%	37%	9%	25%	471	Tarahumara	Barranca–Pacific slope
Taxa	45%	58%	14%	38%	305 (65)		
Tamaulipas							
SUU	43%	20%	22%	15%	268	Mestizo	Atlantic slope
Taxa	70%	34%	35%	43%	164 (59)		
Jalisco							
SUU	42%	19%	31%	8%	198	Mestizo	Coast
Taxa	54%	25%	40%	10%	153 (64)		
Guerrero							
SUU	52%	14%	18%	16%	151	Mestizo	Interior
Taxa	64%	17%	23%	20%	121 (44)		
Morelos							
SUU	51%	15%	20%	14%	307	Mestizo	Interior
Taxa	61%	18%	23%	17%	259		

[a] Food uses include food, beverages and condiments.
[b] Wood uses include construction, implements and fuel.
Sources: Chihuahua: Bye (1989); Jalisco: Cervantes & Bye (1989b); Guerrero: Gispert *et al.* (1989); Tamaulipas: González-Romo *et al.* (1989); Morelos: Maldonada (1989).

the inhabitants. At present, the information gathered at various Mexican sites is incomplete and inconsistent. In the future, participants in the RNIESBCM plan to use comparable methods to collect and analyse data on local utilization and management of dry forest resources. Preliminary results derived from existing data sets suggest some interesting trends. Non-cultivated plants from five regions were evaluated as to taxa (number of botanical families, number of species), species utilization unit for medicinal, food (including beverages and condiments), wood (as raw material for construction, implements and fuel) and other uses (based on two state-wide surveys (Tamaulipas: González-Romo *et al.*, 1989; Morelos: Maldonada, 1989) and three regional studies (Chihuahua: Bye, 1989; Jalisco: Cervantes & Bye, 1989b; and Guerrero: Gispert, Gómez & Rodríguez, 1989)). The forest in each area is different with respect to geographic features (Chihuahua with barrancas or deep canyons near headwaters of the rivers on the Pacific slope; Morelos and Guerrero, in the upper and middle Río Balsas basin; Jalisco, along the Pacific coast; and Tamaulipas, in northeastern México with drainages emptying into the Gulf of México). The Chihuahua region is inhabited by Tarahumara Indians; the other regions have mestizo populations.

Despite the inconsistencies in the original data gathering and a superficial analysis, the heterogeneity of exploitation of the non-cultivated flora of the dry forest is evident (Table 18.2). The region with inhabitants which retain their pre-hispanic culture exploits more plants as well as having the greatest number of species utilization units. The Tarahumara use 305 taxa while the mestizo regions report between 121 and 259. The percentage of species for the use categories is different. The Tarahumara eat 58% of the useful flora while the mestizos consume only between 17 and 34%. This pattern is also reflected in the relative proportions of species utilization units where the mestizos employ about 40 to 50% of the flora for medicinal purposes while the Tarahumara use less than 30%. This is not to say that the indigenous culture has fewer plant species, because the Tarahumara have more remedial species than the other sites (except Morelos, which has a number of introduced plants not present in Chihuahua).

The pattern of greater relative importance of medicinal plants among mestizo populations living in the dry forest, and of more edible resources known to indigenous peoples may have a cultural as well as a biological basis. Because the mestizo people have lived in the dry forests of Tamaulipas, Jalisco, Guerrero and Morelos for only a few generations, they are probably less familiar with the edible plants. They brought with them an agricultural system which is concentrated near settlement centers and which sustains them for most of the year. Their diet is supplemented by certain plants managed and gathered nearby. On the other hand, the Tarahumara Indians do not live in villages. This non-centralized settlement pattern as well as the scattered cultivated fields on the barranca slopes require constant movement. Seasonally available resources can be obtained *en route* and consumed along the way or shared at home. The information and experience of hundreds of generations of Tarahumara ancestors who lived off the land out of necessity and/or for diversification of the seasonal diet have been transmitted to contemporary inhabitants of the dry forest. In addition to the native species, exotic plants augment the exploitable flora. These non-native floristic elements in particular are associated with mestizos who unconsciously or willingly move them as they invade and transform a new portion of the dry forest.

Plants that are of direct benefit to the local people often have more than one use. A detailed analysis is not possible due to the generalized treatment of the data. Nonetheless, each plant species has on average has between 1.2 and 1.6 uses.

Comparison of dry and wet tropical forests

In order to measure the relative importance of the dry forest as a source of plant resources for humans, it is instructive to compare the local flora and its serviceable elements. The floristic fabric provides the backdrop upon which humans extract primary material for their sustenance and well-being. The recent elaboration of floristic checklists for the biological stations Chamela (in dry forest along the Pacific coast of Jalisco: Lott, 1985) and Los Tuxtlas (in wet forest along the Gulf of México coast of Veracruz: Ibarra & Sinaca, 1987) permits the comparison of the basic flora. The sites are similar in latitude, elevation, distance to the coast and temperature, but the mean annual precipitation is six times greater at Los Tuxtlas than at Chamela (Ibarra & Sinaca, 1987; Bullock, 1988). Although the dry forest has fewer vascular plant families, genera and species than the wet forest, the differences are not dramatic. Chamela has 10 fewer families (107), 52 fewer genera (445) and 32 fewer species (780). The taxon ratios are comparable with more than four genera per family, more than 1.5 species per genus and about seven species per family. The two areas share 175 of the 767 total non-cultivated genera present. At the specific level, 65 of a total of 1526 species are found in both forests.

One form of comparing the plant–human interaction in the dry forest with that in the wet forest is to focus on the shared taxa and their utilization. Of the genera exploited near Chamela, 50 are reported at both stations. In order to avoid taxonomic confusion, only the shared species are considered at this time. Of the 153 species employed in the Chamela region (Cervantes & Bye, 1989b), 19 are found at Los Tuxtlas (Table 18.3). Only five species are generally used or named in a similar manner.

Summary

Ethnobotanical research in the Mexican tropical dry forest is in its initial stage of basic investigation. Until recently, this vegetation had not been the focus of systematic inventories even though it occupies up to 20% of the national territory. Partial studies indicate that many woody species enter into the regional, national and international markets. Local inhabitants not only derive a living from these commercial resources but also depend upon them as a part of their lifestyle. It appears that people with a long cultural tradition in the dry forest have a closer tie to the forest, especially in terms of the greater number of species used and a greater contribution of edible plants to their diet. Based upon floristic studies at two biological stations, it appears that the taxonomic diversity of the dry and wet forests is about the same. Even though the lack of reliable ethnobotanical data does not permit a comparison between the dry and wet forests, a pattern emerges. Only 8% of the species are shared by both floras. Of those common taxa, only 7% (or 0.6% of the total flora) have the same general ethnobotanical characteristics.

Acknowledgements

The following people provided specific data and encouragement for this paper: Victor M. Barreto, José C. Boyas Delgado, Luis Cervantes Servín, Nelly Diego, Montserrat Gispert Cruells, Armando Gómez Campos, Claudia González-Romo, Edelmira Linares Mazari, Luis Hernández, Belinda Maldonada Almanza, Miguel Angel Martínez Alfaro, T. P. Ramamoorthy, Juan Manuel Rodríguez Chávez, Noe E. Soberanes Cespedes, Fortunato Solares Arenas, Rito Vega Avina, Martha Vergara Santana. I also thank other members of the Group of Etnobotánica de Selva Baja Caducifolia de México: J. M. Aguilar P., M. Anaya C., B. Benz, J. R. Bonilla B., M. A. Cervantes S., N. Cervantes, D. Deniz L., A. Flores M., C. Ita M., G. I. Manzanero M., M. A. Martínez A., S.

Table 18.3. *Useful plants of the dry forest of Chamela, Jalisco, compared to the same species in the wet forest of Los Tuxtlas, Veracruz*

	Tropical dry forest		Tropical wet forest	
Taxa	Common name	Use	Common name	Use
Asclepiadaceae				
Asclepias curassavica	*calderona*	Medicine	*guaco*	Medicine
Boraginaceae				
Cordia alliodora	*botoncillo, hormiguillo*	Construction	*súchil, laurel rayado*	Construction, flooring, furniture, boats, tools, fuel
Heliotropium indicum	*hierba del alacrán*	Medicine		
Cucurbitaceae				
Melothria pendula	*pepinillo del monte*	Food		
Momordica charantia	*alvellana*	Food	*Cundeamor, cola, rabo de faisán*	Medicinal
Euphorbiaceae				
Euphorbia heterophylla	*hierba de la golondrina*	Medicine		
Leguminosae				
Gliricidia sepium	*cacahuanance*	Essence	*cocuite*	Ritual medicine
Flacourtiaceae				
Casearia corymbosa	*cuatalata, chamizo, mataperro*	Medicine		
Meliaceae				
Trichilia havanensis	*cedro blanco*	Construction		

Moraceae				
Brosimum alicastrum	*mojote, capomo*	Medicine, food, fodder	*ojite, ojoche*	Medicine, food
Ficus pertusa	*camichín*	Medicine, food		
Nyctaginaceae				
Pisonia aculeata	*garabato, garabato prieto*	Medicine		
Phytolaccaceae				
Petiveria alliacea	*hierba del zorillo*	Medicine	*hierba del zorillo*	Medicine
Polygonaceae				
Coccoloba barbadensis	*cahuil, Juan Pérez, jubero*	Medicine		
Schizaeaceae				
Lygodium venustum	*hierba de la víbora*	Medicine		
Solanaceae				
Capsicum annuum glabrisculum	*chile de monte*	Condiment	*chile piquín*	Ritual medicine
Ulmaceae				
Celtis iguanaea	*granjen*	Food, beverage		
Vitaceae				
Cissus sicyoides	*parrilla*	Medicine		
Verbenaceae				
Lantana camara	*hierba de siete colores*	Medicine		

Sources: Jalisco: Cervantes & Bye (1989b); Veracruz: Amo (1979); Echenique *et al.* (1972); Gispert *et al.* (1977); Gómez (1980), Lazos & Alvarez-Búylla (1983); Williams (1980).

Medellín M., M. Mendoza, R. Monroy, A. Ortiz S., F. J. Palma C., J. A. Pérez de la Rosa, L. Robles Hernández, F. J. Santana M., G. F. Tavera A., L. Trejo H. and C. Vázquez.

References

Amo R., S. del (1979). *Plantas Medicinales del Estado de Veracruz*. Instituto Nacional de Investigaciones sobre Recursos Bióticos, Xalapa.

Barreto, V. M., Linares, E. & Bye, R. (1989). Comercialización de la cancerina (*Hippocratea excelsa* HBK.) en México. In *Programa y Resúmenes, Reunión Etnobotánica Ecológica Regional de Selvas Bajas Caducifolias (Bosque Tropical Caducifolio) y Vegetación Asociada en México*, ed. L. Cervantes and R. Bye, pp. 27–9. Instituto de Biología, Universidad Nacional Autónoma de México, México.

Bullock, S.H. (1988). Rasgos del ambiente físico y biológico de Chamela, Jalisco, México. *Folia Entomológica Mexicana* 77: 5–17.

Bye, R. (1989). Plantas útiles de la bosque tropical caducifolia de Chihuahua, México. In *Programa y Resúmenes, Reunión Etnobotánica Ecológica Regional de Selvas Bajas Caducifolias (Bosque Tropical Caducifolio) y Vegetación Asociada en México*, ed. L. Cervantes and R. Bye, pp. 19–20. Instituto de Biología, Universidad Nacional Autónoma de México, México.

Bye, R. & Linares, E. (1983). The role of plants found in the Mexican markets and their importance in ethnobotanical studies. *Journal of Ethnobiology* **3**: 1–13.

Cervantes, L. & Bye, R. (ed.) (1989a). *Programa y Resúmenes, Reunión Etnobotánica Ecológica Regional de Selvas Bajas Caducifolias (Bosque Tropical Caducifolio) y Vegetación Asociada en México*. Instituto de Biología, Universidad Nacional Autónoma de México, México.

Cervantes, L. & Bye, R. (1989b). Plantas útiles de la costa central de Jalisco. In *Programa y Resúmenes, Reunión Etnobotánica Ecológica Regional de Selvas Bajas Caducifolias (Bosque Tropical Caducifolio) y Vegetación Asociada en México*, ed. L. Cervantes and R. Bye, pp. 17–18. Instituto de Biología, Universidad Nacional Autónoma de México, México.

Echenique Manrique, R., Calderón y Sánchez, C., Carrasco Gómez, J. A., Escudero Díaz, M., Ferrara Guerrero, M. J., Riemann Ramírez, P., Saitz Ceballos, S. & Sánchez Mora, M. C. (1972). El aprovechamiento de los recursos forestales en la región de Los Tuxtlas, Veracruz: reconocimiento preliminar. In *Problemas Biológicos de la Región de Los Tuxtlas, Veracruz*, ed. V. M. Toledo, A. Lot Helgueras, C. Juárez López, J. J. Martínez, and J. Zamacona, pp. 105–32. Facultad de Ciencias, Universidad Nacional Autónoma de México, México.

Flores Mata, G., Jiménez López, J., Madrigal Sánchez, X., Moncayo Ruiz, F., & Takaki Takaki, F. (1971). *Tipos de Vegetación de la República Mexicana*. Secretaría de Recursos Hidráulicos, México.

Gispert Cruells, M., Diego Pérez, N., Jiménez Ramírez, J., García Barrios, L. & Gómez Campos, A. (1977). Etnobotánica de las plantas medicinales empleadas en Balzapote, Veracruz. In *Estudios sobre Etnobotánica y Antropología Médica II*, ed. C. Viesca Treviño, pp. 83–96. Instituto Mexicano para el Estudio de las Plantas Medicinales and Centro de Estudios Económicos y Sociales del Tercer Mundo, México.

Gispert Cruells, M., Gómez Campos, A. & Rodríguez Chávez, J. M. (1989). Manejo tradicional de los recursos vegetales en el Cañon del Zopilote: el caso de Xochipala, Guerrero. In *Programa y Resúmenes, Reunión Etnobotánica Ecológica Regional de Selvas Bajas Caducifolias (Bosque Tropical Caducifolio) y Vegetación Asociada en México*, ed. L. Cervantes and R. Bye, pp. 15–16. Instituto de Biología, Universidad Nacional Autónoma de México, México.

Gómez Campos, A. (1980). Estudio etnobotánico de las plantas medicinales de la selva alta perennifolia en Balzapote, Veracruz. Tesis de Licenciatura, Facultad de Ciencias, Universidad Nacional Autónoma de México, México.

González-Romo, C. E., Medellín-Morales, S., Trejo Hernández, L. & Tavera, G. (1989). El manejo tradicional de la selva baja caducifolia en Tamaulipas. In *Programa y Resúmenes, Reunión Etnobotánica Ecológica Regional de Selvas Bajas Caducifolias (Bosque Tropical Caducifolio) y Vegetación Asociada en México*, ed. L. Cervantes and R. Bye, p. 21. Instituto de Biología, Universidad Nacional Autónoma de México, México.

Ibarra Manríquez, G., & Sinaca Colín, S. (1987). *Listados Florísticos de México. VII. Estación de Biología Tropical Los Tuxtlas, Veracruz.* Instituto de Biología, Universidad Nacional Autónoma de México, México.

Lagunes Tejeda, A. (1989). Búsqueda de tecnología apropiada para el combate de plagas del maíz almacenado en condiciones rústicas. Report to CONACyT, project PVT/AI/NAL/85/3149, from the Colegio de Postgraduados, Chapingo.

Lazos Chavero, E. & Alvarez-Búylla Roces, M. E. (1983). Estudio Etnobotánico en Balzapote, Veracruz: Los Solares. Tesis de Licenciatura, Facultad de Ciencias, Universidad Nacional Autónoma de México, México.

Lott, E. J. (1985). *Listados Florísticos de México. VI. La Estación de Biología Chamela, Jalisco.* Instituto de Biología, Universidad Nacional Autónoma de México, México.

Maldonada Almanza, B. (1989). Etnobotánica de la selva baja caducifolia en Morelos. In *Programa y Resúmenes, Reunión Etnobotánica Ecológica Regional de Selvas Bajas Caducifolias (Bosque Tropical Caducifolio) y Vegetación Asociada en México*, ed. L. Cervantes and R. Bye, p. 13. Instituto de Biología, Universidad Nacional Autónoma de México, México.

Palacios, J., Mata, R., López, R., Linares, E. & Bye, R. (1989). *Hippocratea excelsa* (Hippocrataceae), a new source of *trans* polyisoprene. *Economic Botany* **43**: 508–9.

Rzedowski, J. (1978). *Vegetación de México.* Editorial LIMUSA, México.

Secretaría de Programación y Presupuesto (1980). *Atlas Nacional del Medio Físico.* Secretaría de Programación y Presupuesto, México.

Soberanes Cespedes, N. E. (1989). Contribución al conocimiento de la ecología y etnobotánica del cuachalalate (*Amphipterygium adstringens* (Schlecht.) Schiede) en el estado de Morelos. In *Programa y Resúmenes, Reunión Etnobotánica Ecológica Regional de Selvas Bajas Caducifolias (Bosque Tropical Caducifolio) y Vegetación Asociada en México*, ed. L. Cervantes and R. Bye, p. 23. Instituto de Biología, Universidad Nacional Autónoma de México, México.

Solares Arenas, F. (1989). Aspectos etnobotánicos, ecológicos y tecnológicos de *Crescentia alata* (cuatecomate, cirián) en el estado de Morelos. In *Programa y Resúmenes, Reunión Etnobotánica Ecológica Regional de Selvas Bajas Caducifolias (Bosque Tropical Caducifolio) y Vegetación Asociada en México*, ed. L. Cervantes and R. Bye, pp. 24–5. Instituto de Biología, Universidad Nacional Autónoma de México, México.

Toledo, V. M., Carabias, J., Toledo, C. & González-Pacheco, C. (1989). *La Producción Rural en México: alternativas ecológicas.* Fundación Universo Veintiuno, México.

Vega A., R. & Hernández, L. (1989). Vara blanca (*Croton* spp.) usada como tutor de hortalizas en Sinaloa. In *Programa y Resúmenes, Reunión Etnobotánica Ecológica Regional de Selvas Bajas Caducifolias (Bosque Tropical Caducifolio) y Vegetación Asociada en México*, ed. L. Cervantes and R. Bye, p. 43. Instituto de Biología, Universidad Nacional Autónoma de México, México.

Vergara Santana, M. I. (1990). La Palmilla (*Cryosophila nana* (HBK.) Blume) en el ploceso de producción de la 'escoba verde' en el estado de Colima. *Avances de Investigación* No. 11. Universidad de Colima, Colima.
Williams Linera, M. G. (1980). Estudio etnobotánico de agunas plantas rituales utilizadas por un curandero de Santiago Tuxtla, Veracruz. *Cuadernos Tuxtecos* 3, Instituto Nacional de Antropología e Historia, México.

Index

Bold type refers to figures; *italic* type refers to tables.